Legacies OF THE Turf

Legacies OF THE Turf

A Century of Great
Thoroughbred Breeders (Vol. 2)

Edward L. Bowen

EP
ECLIPSE
PRESS

Lexington, Kentucky

Library of Congress Control Number: 2004104476

ISBN 1-58150-117-X

Printed in the United States
First Edition: November 2004

Distributed to the trade by
National Book Network
4501 Forbes Boulevard, Suite 200, Lanham, MD 20706
1.800.462.6420

a division of
Blood-Horse Publications
PUBLISHERS SINCE 1916
ECLIPSE
PRESS

CONTENTS

(Continued on next page)

CONTENTS
(CONTINUED)

Introduction

This volume is a companion and sequel to *Legacies of the Turf* (Vol. I), published by Eclipse Press in 2003. It completes a review of great and influential Thoroughbred breeders in North America during the twentieth century. In general, the list of breeders was divided chronologically for the two books, although some breeders, such as Paul Mellon, Ogden Phipps, Alfred Vanderbilt, and Calumet Farm, had such lengthy careers that placing them in the "first half" or "last half" of the twentieth century was more a coin-flip than an assessment.

As in the first volume, the intent was to concentrate on the domestic scene. Breeders who mostly produced horses for racing and/or the market in America were most strongly considered for inclusion. This became less of a clear line as the century progressed. By the final three decades a long-standing pattern in the bloodstock world had reversed. Whereas for many years American breeders had looked to England for the best bloodstock, the success of Sir Ivor in the 1968 English Derby was a catalyst to a trend of European horsemen looking to the American sales for top prospects. The American horses did not fail. Canada's Northern Dancer sired three English Derby winners, as did his son Nijinsky II. American-breds Mill Reef and Roberto were English Derby winners, as well, and the success of American-breds abroad became common. (The fact that one horseman, Vincent O'Brien, trained Sir Ivor, Nijinsky II, and Roberto gives that scholarly Irishman a singular status in the history of world-wide bloodstock trade. Were the U.S. government more attuned to matters of the Turf, O'Brien might be awarded a medal by those bureaucrats concerned with balance of import/export activity.)

The Sir Ivor syndrome caused emphasis on North Americans breeding for the North American scene to become less distinct as we worked our way through the latter part of the century. E.P. Taylor's great Windfields Farm operations in Canada and Maryland, for example, bred a great many horses for racing and for sales from Ontario to Kentucky. So many of the Windfields-breds were eagerly scooped up by international buyers, however, that Taylor virtually became, de facto, a greater force in Europe than at home. Nonetheless, he was not "penalized" for this wide influence and is included in this volume.

The criteria for inclusion in these volumes were not strict, for quality comes in differing flow charts. Nevertheless, what we looked for in making selections was generally narrowed to the following: reigning as leading breeder in earnings in North America on at least one occasion; breeding of domestic champions and classic winners particularly, and of a large number of what today are known as grade I winners. (In individual chapters foreign champions are also noted, but American champions were more important in driving the selection process.) Lastly, the breeding of horses that had influence on succeeding generations was a major consideration in choices.

In some cases the opportunity for geographic diversity had an influence, and, admittedly, some of the subjects no doubt had an edge in the sheer fascination of their careers.

Some breeders, of course, were stronger in some of these measures than others. Captain Harry Guggenheim, for example, bred relatively few stakes winners, but they included the following: Never Bend, an American champion and lasting influence; Red God, who as the sire of Blushing Groom authored considerable influence of his own; and Riverman, himself a distinguished sire. Captain Harry also was the leading owner once, with a stable of homebreds, and he made the cut. Likewise, C.T. Chenery's record was not high in numbers, but we saw no logic in excluding the breeder of Secretariat, Riva Ridge, Sir Gaylord, Cicada, First Landing, etc., the man who plucked Hildene and Imperatrice from sale rings and masterfully directed their genetic destinies.

As was true in the first volume, the careers of breeders included herewith represent a wide range of tales within the broad tapestry of American life. There are those such as Guggenheim, William S. Farish, Paul Mellon, Bunker Hunt, and Seth Hancock whose family backgrounds provided a boost of one sort or another, but who understood that inheritance does not equal entitlement to success. Work, knowledge, and investment — sometimes daring investment — are the tools of the Thoroughbred breeder, large and small.

Some subjects were self-made entrepreneurs who turned to the Thoroughbred, perhaps as a sideline, and then became swept up in the challenges and glories of the Turf. These include the likes Taylor of Windfields and Chenery of Meadow Stud, Maxwell Gluck of Elmendorf, John Mabee of Golden Eagle, Allen Paulson of Brookside, W.L. McKnight of Tartan Farms, John W. Galbreath of Darby Dan, Louis Wolfson of Harbor View, W.T. Young of Overbrook, and centenarian Fred W. Hooper.

Other subjects in this volume fit no pattern, and so are categories unto themselves. Ogden Phipps was influential as a young man in encouraging his mother to become involved in racing, and then he soon set up his own stable and eventually lived to see two more generations share his sporting passions. Rex Ellsworth was a cattle rancher who turned to horses. Leslie Combs II set his cap at being a promoter, and used persuasive people skills to buttress a high-energy commercial breeding/selling operation. Robert Kleberg Jr. of King Ranch is another who fits no pattern, unless you happen to know several geneticist cowboys who developed their own new breed of cattle, profoundly influenced the Quarter Horse, and raised a Texas-bred Triple Crown Thoroughbred!

Omissions

Through the last two decades of the twentieth century, the prominence of the Maktoum family of Dubai and the success of Saudi Prince Khalid Abdullah's Juddmonte Farm were among the most important and engaging developments on the international Thoroughbred scene. Their stories blur the distinction between what is an "American" story and what is not. With establishment of handsome breeding farms in Kentucky, these sportsmen were certainly American breeders in one sense, but for some time they tended to use those farms as a base for international operations. More recently, that distinction, too, has been blurred, as both began to concentrate on the American racing scene as well as the international circuits. Surely, in the first four years of the twenty-first century, Abdullah particularly has been amazingly successful as a breeder raising horses to run in his own colors, with major results on these shores. As of the cutoff date of 2000 for selection, however, it was decided that putting these operations into context as forces on the American racing/breeding scene was not practical, although they are certainly poised to be important chapters of the early decades of the new century.

Some of the most famous and visible of American horsemen are not found in this volume, due to our purposefully narrow focus. They include revered sportsmen and women who have made indelible marks in various aspects of the Turf — from racing, to shaping the sport, to syndication — and who bred a fair number of good horses, too. Yet, in the strict sense being addressed here, i.e., specific achievements as a Thoroughbred breeder (owner of the mare when she foals), they were not selected. No doubt they sleep better in the understanding of their own accomplishments than does the undersigned in his tortured decision to exclude them!

Edward L. Bowen
Versailles, Kentucky, 2004

CHAPTER 1

Christopher T. Chenery and Meadow Stud

American racing fans came to revere the name of the Chenery family's Meadow Stud in Virginia when Riva Ridge won two-thirds of the Triple Crown in 1972 and then Secretariat sealed the deal in 1973. Secretariat rose to such fame and acclaim that at the end of the century he was ranked, along with Man o' War and Citation, as one of the three best of the hundred years just completed.

Long before he bred classic winners in consecutive crops at the twilight of his life, Meadow Stud founder Christopher T. Chenery had made two of the most productive broodmare purchases in history with the bargain-basement Hildene and the upscale Imperatrice. The first gave him two champions and the dam of a third, and the other mare gave him the dam of Secretariat.

Chenery's family represented the form of American success story that achieved, lost, and regained. One of his ancestors, Charles Dabney Morris, built a home he called The Meadow on his Virginia land, near the North Anna River. The home was built in 1810, but after the Civil War the family lost the property during hard times. Born near Richmond in 1886, Christopher T. Chenery launched into a vigorous career, achieving enough success to regain the family homestead by the time he was fifty, in 1936.

He started working at age sixteen as an assistant surveyor for the Virginia Railway, and he put away sufficient savings to continue his education. He attended Randolph-Macon and then Washington and Lee University, from which he graduated Phi Beta Kappa in 1909 with a degree in engineering. He worked on engineering projects in Virginia before heading to the Northwest and on to Alaska for additional engineering work. World War I took him back to Virginia, where as a major he commanded training facilities at Camp Humphries.

Continuing his business career after the war, he formed the Federal Water Service Corporation. By 1936, the year he purchased The Meadow and a core of surrounding property, he became board chairman of the Southern National Gas Company. He also became board chairman of additional public utilities companies, including Apex Gas, New York Water Service Corporation, Rochester and Lake Ontario Water Service Corporation, Scranton-Spring Book Water Service Company, South Bay Consolidated Water Company, Southern Production Company, and Western New York Water Company. *Who's Who* gave him the tag line "corporate executive." His son and two daughters were encouraged to use the more precise description "public-utilities executive."

For many years prior to obtaining his ancestral estate, Chenery had various connections to horses of different kinds, and in different pursuits. As early as 1911 he organized a drag hunt (similar to a fox hunt, but the hounds follow an artificial scent or "drag" laid prior to the hunt) in Portland, Oregon, and three years later he helped move a pack train to what became Anchorage, Alaska. He also raised and rode polo ponies, and one of his preferred equine maneuvers was preserved in a family photo in which he sits astride a horse as it swims down a river.

The same year Chenery purchased The Meadow, he also bought a Thoroughbred stallion. This was not a grand entrance into the world of breeding racehorses but had redolence of distant glories. The beast was Whiskaway, who had won the rich ($50,000) Kentucky Special in 1922, when he was regarded as the best three-year-old colt in America. Whiskaway, a son of the vaunted Whisk Broom II, raced first for Harry Payne Whitney and was sold for $125,000 to continue racing for C.W. Clark. By 1936, however, he was a seventeen-year-old stallion who had sired only four stakes winners. Chenery bought Whiskaway for one-thousandth of his earlier price, paying $125. That same year the stallion's foals included Dolly Whisk, who would win the 1938 Debutante Stakes and later produce the high-class runner Palestinian. Whiskaway wound up as the sire of eight stakes winners, of which Chenery bred two, Cherrydale and Hornbeam. Both were out of the *Bright Knight mare Annie R. Cherrydale was foaled in 1939 and Hornbeam in 1940, and both became moderate but hardy campaigners.

If his first modest purchase of a stallion had produced a few dividends, only three years later he purchased one of the two most significant animals of his career, the great mare Hildene. In 1939 Chenery sent Norman Tallman to the dispersal of the late Edward F. Simms' horses at Xalapa Farm in Kentucky. Of the twenty-eight newly turned yearlings in the January 19 sale, thirteen were by 1926 Kentucky Derby winner Bubbling Over, and Tallman bought three of them among the six horses he got for Chenery. The others included the sire himself, Bubbling Over, who had gotten another Derby winner in Burgoo King but by and large was not a major sire. One of the Bubbling Over yearlings, to be named Hildene, was out of Fancy Racket, a daughter of the important sire *Wrack. Fancy Racket, who had failed to earn a part of the purse in each of her four starts, had foaled a very modest stakes winner at the time Hildene was bought for Chenery.

This mélange of pedigree ingredients in the form of Hildene cost $750, for which price Chenery put himself on the map as a major breeder. As a race filly, Hildene was only a small improvement on her dam. She failed to win in any of her eight starts but snuck into the earnings category enough to take home a hundred dollars. Chenery bred her first to Flares, William Woodward Sr.'s Ascot Gold Cup winner, and she foaled Sunset Bay, who outdid her first two dams by winning a race. Then came Mangohick, Hildene's 1944 gelding whose sire Sun Beau in his day was the all-time leading money earner but who had not become much of a sire. Mangohick campaigned for seven years and won twenty-three of ninety-seven races, including a pair of stakes. He earned $115,115, the most of any of the first three stakes winners Chenery bred.

The breakthrough for The Meadow and Hildene came in 1946, when Chenery availed himself of a new young stallion at the Hancock family's Ellerslie Farm outside Charlottesville. He did not have to ship the mare out of state to get to this Cup horse for $250. "Cheap" and "standing nearby" are keys to many a breeding decision, but they are not generally looked upon as the best formula for classic success. In Hildene's 1947 colt, Hill Prince, however, *Princequillo was siring his first American champion. His first English star came in the same crop, as Woodward had agreed to

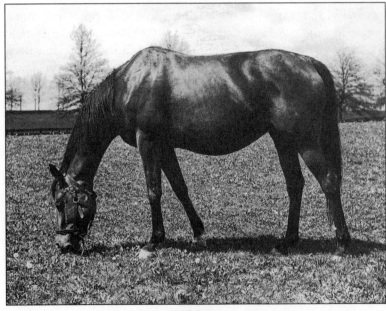

Hildene

THE BLOOD-HORSE

send a mare to *Princequillo and got Prince Simon.

Hill Prince won the World's Playground and Cowdin stakes and Babylon Handicap at two and gained a share of championship honors in the major polls of the time. At three in 1950, he won the Wood Memorial, and for the first time Chenery had a prominent Kentucky Derby contender. The *Princequillo colt finished second to Middleground, but he defeated that colt in the Withers and again in the Preakness. He had not won the Derby, but he gave Chenery his first classic victory. Middleground prevailed, however, in the Belmont Stakes, but Hill Prince took another turn at beating him in the Jerome and also added the American Derby before defeating the older champion *Noor in the Jockey Club Gold Cup. Hill Prince was voted Horse of the Year as well as champion three-year-old for 1950, and Hildene, the erstwhile $750 yearling, was named Broodmare of the Year. Hill Prince had a brief campaign at four and shared older male championship honors with Citation. Hill Prince won seventeen of thirty races and earned $422,140.

Hill Prince was conceived the year that the Hancock family sold Ellerslie. *Princequillo was moved to the family's Kentucky spread, Claiborne, but even without the sale of the Virginia farm he no doubt would have earned a stall at the more prestigious Bluegrass address. Upon retirement, Hill Prince joined his sire at stud at Claiborne and got twenty-three stakes winners (7 percent), of which Chenery bred three: Pepperwood, Imperial Hill, and Salt Lake. Hill Prince's greatest contribution was in siring major fillies and producers Bayou and Levee. Of Chenery's three, Pepperwood was the best, winning the 1966 Gardenia Stakes at nearly 12-1. With a purse of $75,000 added, the one and one-sixteenth mile Gardenia was the richest race for two-year-old fillies at the time. With placings in three other stakes that season, Pepperwood was one of several fillies ranked two pounds below champion and top-weight filly Regal Gleam on the Experimental Free Handicap.

Tourmaline, dam of Pepperwood, illustrated Chenery's interest in, and willingness to utilize, bloodlines that were well shy of the high fashion to which Hill Prince and Hildene attained. Chenery not only reached out to such bloodlines but also gained reward from doing so. Tourmaline, winner of one race, was by the imported German stallion

Christopher T. Chenery

*Nordlicht, as were the Chenery-bred stakes winners Travertine and Cartagena. Conversely, Travertine's dam, Chelita, was by the leading sire *Sir Gallahad III out of the prominent La Chica, by Sweep. La Chica foaled 1936 juvenile colt champion El Chico and became the second dam of the great champion Native Dancer and of the blue hen producer Grey Flight.

Anthemion, dam of Tourmaline and thus second dam of Pepperwood, was another of Chenery's astute early purchases. Two years after paying $750 for Hildene, Chenery bought Anthemion from the Claiborne Farm consignment at Saratoga for slightly more, one thousand dollars. Two years later she won the Gazelle Stakes, becoming the first stakes winner to race in Chenery's name. Bred to the leading sire *Blenheim II in 1946, Anthemion produced Bryan G., a hard-knocking homebred who won the 1951 Pimlico Special, two runnings of the Aqueduct Handicap, and two other stakes and earned $165,625. Bryan G. had fourteen wins in sixty-two starts. He stood for years at The Meadow and did not do much in a statistical sense, but Chenery got plenty out of him. Of the six stakes winners (3 percent) sired by Bryan G., Chenery bred four, one of which was one of his greatest treasures, the three-time champion filly Cicada.

Chenery again illustrated his faith in his own convictions by standing a horse he had named for

Hill Prince

Doswell, Virginia, the closest municipality to The Meadow. Although he was doing very well with a $750 filly here, a $250 stud fee there, Chenery also was willing to reach deeper. Doswell was a yearling by Bull Lea—Highclere, by Jack High, for whom Chenery paid $20,000. The horse won only one race in twelve starts, but Chenery believed in him enough to stand him. The stallion sired fifty-one foals, of which two, both homebreds for Chenery, won stakes, including the filly Willamette. From Doswell's first crop Willamette was out of the *Blenheim II mare Sister Union, who had cost Chenery $4,800 in 1948. Willamette gave Chenery a major prize when she defeated champion Bayou in the 1957 Coaching Club American Oaks.

In the meantime Hildene had been turning out foals with sires ranging from the steeplechase specialist *Hunters Moon IV to the nondescript *Flushing II and Bossuet, to the upscale *Princequillo and Triple Crown winner Count Fleet. Her success in those years came only with *Princequillo, as two more *Princequillo—Hildene foals followed Hill Prince as stakes winners. One was Prince Hill, who was gelded despite his pedigree and won the 1955 American Bred Stakes and Longfellow Handicap while earning $98,300. The next was named in honor of his birthright, Third Brother. He was not Hill

Prince but was better than Prince Hill. Third Brother, a foal of 1953, won four nice stakes, placed in a number of top-class races, and earned $310,787. (He later sired the champion Roman Brother.)

In 1957, the year the somewhat off-bred Willamette won the CCA Oaks, Chenery's yearling crop included a Hildene colt sired by the decidedly upscale *Turn-to. The latter had won the Garden State Stakes and Flamingo for Captain Harry Guggenheim, with whom Chenery had a far greater business connection than one of just sending mares to the other's stallion.

By 1953 New York racing had been allowed to languish to the extent that Ashley Trimble Cole, chairman of the New York Racing Commission, made it clear to the sport's leaders that if they did not do something to correct the slide, he would. Ogden Phipps, as vice chairman of The Jockey Club, appointed Chenery, Guggenheim, and John W. Hanes as a committee to develop and recommend a solution.

It is tough to grasp that a man who was breeder of record of a Secretariat might have done something even more significant for racing. Nevertheless, in developing the difficult plan that became the New York Racing Association, Chenery and his associates more or less saved the status of the New York Turf. They prescribed a reduction in number of tracks, the consolidation of Belmont Park, Aqueduct, and Saratoga under one organization (NYRA), whose directors would not benefit financially, and the replacing of the old Aqueduct with a huge, modern plant. Although off-track betting and other political hurdles would be placed in front of those who ran NYRA in the succeeding years, the committee could be said to have set New York back on course in keeping with its proud history.

On the racetrack the son of Guggenheim's *Turn-to and Chenery's Hildene made his debut in 1958, the year before the new Aqueduct opened. Named First Landing, he was close to perfect as a two-year-old.

Under management of Chenery's long-time trainer, Casey Hayes, First Landing ripped off seven victories in his first seven starts. The prestige rider of the East at the time, Eddie Arcaro, had hopped aboard from the start, and he guided the handsome colt to scores in such major summer stakes as the Great American, Saratoga Special, and Hopeful. After being upset by Intentionally in the Futurity, First Landing righted himself to win the Champagne and Garden State stakes over *Tomy Lee and set an earnings record for a two-year-old at $396,460. He was the overwhelming choice as champion juvenile colt, and he topped the Experimental Free Handicap at a rarified 128 pounds.

First Landing had various setbacks in the winter and spring of 1959 but was second in the Wood Memorial and then won the Derby Trial. Chenery had had a 5-2 shot in Hill Prince on Derby Day nine years earlier (second choice to Your Host), and now that horse's half brother, First Landing, was the post-time favorite. Victory again eluded Chenery, for First Landing finished third behind his old rival *Tomy Lee and Sword Dancer.

First Landing never regained leadership of his crop but came back at four as an admirable handicap horse, when his wins included the Santa Anita Maturity and Monmouth Handicap, and went to stud with nineteen wins in thirty-seven starts and earnings of $779,577. He was not a great sire but got twenty-seven stakes winners (6 percent). Four of his stakes winners were bred by Meadow Stud Inc.,

which Chenery took as his corporate breeding operation name in the late 1950s, when he also began racing in the name of Meadow Stable rather than in his own name. The best of First Landing's horses was Riva Ridge, a two-time champion who ended the family's Kentucky Derby frustration. (First Landing had been foaled from Hildene when the mare was eighteen and was her next-to-last foal. The final one, Goodspeed, by Tom Fool, won three races.)

The year after First Landing's retirement, however, the Hildene saga generated a new chapter when her granddaughter, the lithe little filly Cicada, began winning in February of 1961, her two-year-old year. She didn't stop winning until she had become the first filly in history to be American champion at two, three, and four. Although by then Hildene was a queen of the stud book, the breeding of Cicada in a sense was another illustration of Chenery's rambling up and down the ranks of stallions and finding pearls in various price ranges. Cicada's dam was one of Hildene's least distinguished runners, Satsuma, winner of one of eight races. Satsuma was a foal of 1949 and was by Bossuet, who had been a nice racehorse, renowned for arriving in line with Brownie and Wait a Bit at the finish of the 1944 Carter Handicap — still the only three-way dead heat in a major stakes in American racing history. He was a young sire in 1948, so no one could have known that he would sire only three stakes winners in his career. Of course, in choosing him as a mate for Hildene in 1948, Chenery could not know just what he had wrought earlier by

			Pharos (Phalaris—Scapa Flow)
		Nearco, 1935	Nogara (Havresac II—Catnip)
	*Royal Charger, 1942	Sun Princess, 1937	Solario (Gainsborough—Sun Worship)
			Mumtaz Begum (*Blenheim II—Mumtaz Mahal)
*Turn-to, 1951		Admiral Drake, 1931	Craig an Eran (Sunstar—Maid of the Mist)
	*Source Sucree, 1940		Plucky Liege (Spearmint—Concertina)
		Lavendula, 1930	Pharos (Phalaris—Scapa Flow)
			Sweet Lavender (Swynford—Marchetta)
FIRST LANDING		*North Star III, 1914	Sunstar (Sundridge—Doris)
	Bubbling Over, 1923		Angelic (St. Angelo—Fota)
		Beaming Beauty, 1917	Sweep (Ben Brush—Pink Domino)
Hildene, 1938			Bellisario (Hippodrome—Biturica)
		*Wrack, 1909	Robert Le Diable (Ayrshire—Rose Bay)
	Fancy Racket, 1925		Samphire (Isinglass—Chelandry)
		Ultimate Fancy, 1918	Ultimus (Commando—Running Stream)
			Idle Fancy (Ben Brush—*Fair Vision)

crossing the mare with *Princequillo.

Satsuma was bred to good-class Chenery home-bred Bryan G. in 1958 and foaled Cicada the next year. On the other end of the fashion scale, the Chenery-bred foals that same year included another son of *Turn-to, who by then was established as not only a brilliant racehorse but also a major young sire. This was Sir Gaylord, whose granddam, Imperatrice, was to rank with Hildene as the most famous of Chenery's numerous successful acquisitions.

Like Cicada, Sir Gaylord was out of one of the lesser foals from a blue hen mare. His dam, Somethingroyal, ran once and did not place. Somethingroyal was a foal of 1952, a *Princequillo filly born after her dam, Imperatrice, had already produced three stakes winners and had three more to go. Imperatrice was a 1938 foal by Caruso—Cinquepace, by Brown Bud and was bred in New Jersey by William H. LaBoyteaux. She won the Test Stakes, Fall Highweight Handicap, and two other stakes, and her second foal, Scattered (by Whirlaway), won the CCA Oaks for King Ranch in 1948. The year before, LaBoyteaux had died and his horses were put up for auction. Chenery, the sometime bargain hunter, was wearing his high-roller hat, and he paid $30,000 for Imperatrice.

In addition to Scattered, Imperatrice had two other foals, both by Piping Rock, that would win stakes: Imperium and Squared Away. Once Chenery owned Imperatrice, however, it would be eight years before she foaled another stakes winner, Yemen, by Bryan G. Chenery also bred from her the stakes-winning Hill Prince filly Imperial Hill and the stakes-winning Bold Ruler filly Speedwell.

Somethingroyal, however, was by all odds the most important of Imperatrice's foals. Sir Gaylord, Somethingroyal's 1959 colt, was by First Landing's sire *Turn-to. When the homebred grandson and granddaughter of blue hens were two, Cicada won eleven of sixteen races, including the Gardenia (by ten lengths), Frizette, Matron, Spinaway, and four other stakes, and was the overwhelming choice as champion two-year-old filly. Sir Gaylord prevailed for some months as leader pro-tem of the juvenile colt division, with victories in the Sapling and three other stakes. Sir Gaylord faded somewhat late in the year, but the following winter defeated the

previously unbeaten Ridan in the Bahamas and added the Everglades Stakes. Despite some setbacks he was the Kentucky Derby favorite as the race neared. Meanwhile, in a dazzling sequence for the Chenery blue-and-white-block silks, Cicada was winning again and was barely beaten in a scintillating duel with Ridan in the Florida Derby.

While Hill Prince technically had not been the favorite for the Kentucky Derby, emotionally Chenery must have felt that in Sir Gaylord he arrived at Derby Week with the favorite for the third time. Would it be a charm?

Well, no. Sir Gaylord was lame when he returned to his barn on the Friday before the race. X-rays revealed a hairline fracture of a sesamoid. He was retired with ten wins from eighteen starts and earnings of $237,404. Had trainer Hayes known the day before that he had no Derby colt, he would likely have opted for a Derby filly. However, Cicada had already had her feed drawn, for she was scheduled to run in the Kentucky Oaks that afternoon. She knew the routine enough to recognize race day, so she was kept in the Oaks, which she duly won. Cicada also won the Acorn, Mother Goose, and Beldame, and repeated as champion. At four she added a third

Cicada and trainer Casey Hayes

HAL BORGSOHN

14

championship, and she was retired with twenty-three wins in forty-two starts and earnings of $783,674. Naturally, she was bred to Sir Gaylord and foaled the stakes winner Cicada's Pride, but thereafter she was beset by reproductive problems. Eventually, Cicada had six foals, but none other of racing distinction.

Sir Gaylord stood at Claiborne for most of his career. However, he also stood in France under lease, returned to stand again at Claiborne, and then was sold back again

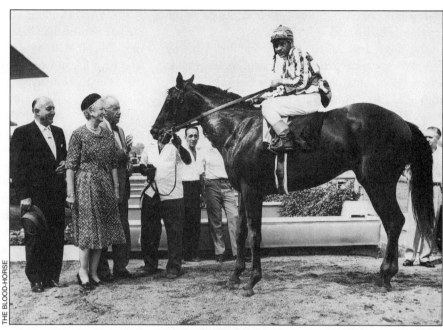

Sir Gaylord

to France, where he died in 1981. He sired sixty stakes winners (13 percent), the bellwether among them being Sir Ivor. As winner of the 1968 Epsom Derby, Sir Ivor proved the stamina potential of the *Turn-to line and launched a sequence of American-bred success in the Derby and other major European races that profoundly affected the direction of international commerce in the Thoroughbred world. The Americans, long importers, became exporters of yearlings highly sought by Europeans.

In addition to Cicada's Pride, Chenery bred one other stakes winner by Sir Gaylord. This was the filly Gay Matelda, whose dam, Hasty Matelda, was another bargain purchase. A daughter of Hasty Road, Hasty Matelda was purchased from Mrs. George Greenhalgh at the 1962 Saratoga yearling sale for eight thousand dollars. Hasty Matelda won the Matron Stakes for her only win. Her Sir Gaylord filly Gay Matelda won the rich Gardenia Stakes in 1967, as well as the 1968 Alabama Stakes, among her five stakes wins before retiring with earnings of $409,945. Gay Matelda was Chenery's third winner of the Gardenia, following Pepperwood and Cicada.

In 1965, the same year Sir Gaylord's daughter Gay Matelda was foaled, Sir Gaylord's half sister Syrian Sea was also in the Chenery crop. Syrian Sea was by Bold Ruler out of Somethingroyal, a pedigree combination that would have overwhelming importance

within the decade. Syrian Sea won the Selima, Astarita, and Colleen Stakes at two and earned $178,245 during a two-year racing career. Gay Matelda and Syrian Sea were ranked second and third below champion Queen of the Stage among fillies on the 1967 Experimental Free Handicap. (Somethingroyal also foaled a good one for Chenery when bred to the home star, First Landing. That colt, First Family, won the 1966 Gulfstream Park Handicap and three other stakes, finished third in the 1965 Belmont, and earned $188,040.)

Chenery had important success in breeding to the Phippses' champion Bold Ruler, who stood at Claiborne. In addition to Syrian Sea, he had bred another major juvenile filly by Bold Ruler in Bold Experience. Syrian Sea was out of a non-winning daughter of Imperatrice, and Bold Experience was out of a non-winning daughter of Chenery's other great producer, Hildene. First Flush, by *Flushing II, failed to win in three starts and earned nothing, somewhat in the tradition of Hildene and her dam, Fancy Racket. First Flush's 1962 Bold Ruler filly, Bold Experience, won the Sorority Stakes, then a key summer test for juvenile fillies. Bold Experience and her full brother, Virginia Delegate, were among the five stakes winners by Bold Ruler bred by Chenery. Bold Experience produced the Meadow homebred Round Table colt Upper Case, who won the Florida

Derby and Wood Memorial as Riva Ridge's over-shadowed stablemate in 1972.

The revered homebred champion First Landing, one of three Chenery horses that raised high Kentucky Derby hopes, sired Riva Ridge, the horse who broke through for the family as a Derby winner. Foaled in 1969, Riva Ridge was one of the many Chenery horses born at Claiborne and thus officially Kentucky-breds. Riva Ridge was a lanky, narrow colt, whose dam, Iberia, was yet another important figure Chenery had purchased. Iberia, by the two-time leading sire *Heliopolis, was acquired as a yearling from the 1955 Saratoga consignment of Larry MacPhail for $15,000. She was out of War East, an Easton mare whose dam, Warrior Lass, by Man o' War, had foaled 1939 Metropolitan Handicap winner Knickerbocker. Moreover, Warrior Lass was also the dam of Marching Home, who produced 1944 Belmont Stakes winner Bounding Home. Iberia won three of eleven races and foaled three stakes winners for Meadow Stud in Potomac (by First Landing); $277,958-earner Hydrologist (by *Tatan); and Riva Ridge.

By the time Riva Ridge got to the races, Chenery was a widower confined to a nursing home. Tears were said to have rolled down his cheeks when a nurse told him of Riva Ridge's victory in the 1972 Kentucky Derby. One of Chenery's daughters, Helen (Penny) Chenery, was operating the family farm and stable. (She was married at that time to John Tweedy, so that the legions soon to hail her did so as Penny Tweedy.)

The First Landing colt Riva Ridge, trained by Lucien Laurin, had been the 1971 champion two-year-old colt, with a campaign not unlike that of his sire. Riva Ridge won seven of nine starts, including the Flash, Futurity, Pimlico-Laurel Futurity, Champagne Stakes, and Garden State Stakes. The following year he won the Hibiscus and Blue Grass stakes before heading into the Kentucky Derby as the fourth favorite, or near favorite, for the Chenerys in the last twenty-three runnings. Riva Ridge dominated, winning by three and a quarter lengths. Roses long denied were draped over his neck.

After floundering in the mud in the Preakness, Riva Ridge rebounded to win the Belmont Stakes. By the end of the year, he had waned a bit and Key to the Mint came on to earn honors as champion three-year-old colt. This was 1972, when Riva Ridge's brash stablemate Secretariat soared into the nation's sporting consciousness and grasped Horse of the Year honors from his elders. At four in 1973, Riva Ridge was overshadowed by Secretariat's Triple Crown spectacle, but he set a world record for a mile and three-sixteenths and won the Brooklyn, Stuyvesant, and Massachusetts handicaps to earn the championship for older males. Riva Ridge's runner-up effort in the first Marlboro Cup, when Secretariat set the world record for nine furlongs, was one of his nobler moments. Riva Ridge won seventeen of thirty races and earned $1,111,497. He was syndicated for $5.12 million and entered stud at Claiborne Farm along with Secretariat. Riva Ridge got twenty-nine stakes winners (8 percent).

We turn full attention now to the singular Secretariat. It was a sad irony that Chenery's failing health made him unable to delight in the marvel he had bred. Chenery had bred stakes winner Speedwell from the first crop of Bold Ruler. (Foaled from Imperatrice, Speedwell later became the second dam of the major winner and sire Cure the Blues.) He had gone back to that sensational stallion frequently, and the best of his Bold Rulers had been Syrian

Riva Ridge (left front) and Secretariat (left rear)

Sea, from Sir Gaylord's dam, Somethingroyal. In 1968 the status of Bold Ruler was such that the Phipps family, who owned him, asked Claiborne owner A.B. (Bull) Hancock Jr. to develop a plan whereby exceptional outside mares would be bred to the horse for two years, with the mares' owners and the Phippses to take one foal each. There would be no stud fee paid in such cases, but the Phippses would thus have access to selected outside bloodlines that otherwise would be unavailable or illogically expensive.

Having foaled a high-class Bold Ruler filly in Syrian Sea in earlier years, Somethingroyal was accepted into this program. This cross involved what was widely hailed as a nick, i.e., the *Nasrullah line on *Princequillo mares. This pattern was virtually inevitable given that both *Nasrullah and *Princequillo stood at Claiborne. Amusingly, at the time of Secretariat's Triple Crown, Bold Ruler's percentage of stakes winners when bred to *Princequillo mares was slightly under his percentage when bred to all other mares! Still, a nick is a nick if a believer in nicks says it is, and when a pattern produces a Triple Crown winner, who's to argue? Somethingroyal's first Bold Ruler foal under the new arrangement vis-à-vis Bold Ruler was a filly that went to Ogden Phipps by coin flip. She was named The Bride. She failed to win but foaled two stakes winners. The second foal in the renewed assignations between Bold Ruler and Somethingroyal was a bristling chestnut colt foaled at Meadow Stud who became the pride of Virginia before being adopted by one and all as the pride of America.

Delivered to the stable of Lucien Laurin, Secretariat soon was teamed with Riva Ridge's jockey, Ron Turcotte. So, in the midst of their adventure with Riva Ridge as a not-quite Triple Crown winner in 1972, Penny Tweedy, Laurin, and Turcotte stepped onto a newer and more dazzling roller coaster that ascended far more often than it dropped.

Secretariat had a devastating turn of foot and

The Chenery family, Lucien Laurin (second left), and Ron Turcotte (front row center)

loved to run on the turns, giving ground by going outside but racing past his rivals with vigorous finality. At two he lost only in his maiden voyage and by a disqualification in the Champagne Stakes. He romped home in the Sanford Stakes, Hopeful Stakes, Belmont Futurity, Laurel Futurity, and Garden State Stakes. He went First Landing one pound better on the Experimental, being assigned 129 pounds.

Eclipse Awards had come into being the previous year, and in 1972 there was a legion of them awaiting the Chenery combine: champion two-year-old colt and Horse of the Year went to Secretariat, champion older male to Riva Ridge, best trainer to Laurin, and owner-breeder to Meadow Stable.

On January 3, 1973, Christopher Chenery passed away. The realities of dealing with his estate intruded into the already bittersweet feelings the family had dealt with in having these sensational successes at a time the father could not exult in the apex of his years as a Thoroughbred breeder.

To the families of the many Thoroughbred breeders and owners who had enjoyed a long association — give and take — with Bull Hancock of Claiborne Farm, looking to Hancock for guidance would be a natural. However, Bull Hancock had died after a short illness only months before Chenery's death. His twenty-three-year-old son,

Seth, was thrust into a leadership position at the time-honored farm. As set out by Tim Capps in Eclipse Press' *Secretariat*, in the Thoroughbred Legends series, there was another obvious source of advice in the Thoroughbred market. Humphrey Finney and his son John of Fasig-Tipton were called in. The elder Finney, along with Bull Hancock, had been asked by the William Woodward family estate to put values on the Belair Stud broodmare band and, above all, the single champion Nashua, in 1955. Finney and Hancock had arrived at very similar evaluations. Over many years the Finneys and Fasig-Tipton had built and earned a reputation for these and other matters in the Thoroughbred market. They were consulted by Penny Tweedy, the one among Chenery's children who had the most passion for her father's pastime of breeding and racing horses.

It was decided that syndicating Secretariat at that time was wisest. Looking back, it might seem a cautious decision. After all, imagine the world beating its way to his doorstep after he won the Belmont by thirty-one lengths to complete the Triple Crown! At the time, however, twenty-five years had passed since Citation had won the Triple Crown, and no horse had done so in the interim. Moreover, there was a prevalent doubt about Bold Ruler's offspring being well suited for one and a quarter miles, the Kentucky Derby distance. Although a number of his offspring had won over that route, several of his better sons had failed in the Derby.

Hancock, in a major life move of his own, stepped up to the plate and syndicated Secretariat on the basis of $190,000 a share (thirty-two shares), or a total valuation of $6.08 million. This was within the parameters of possibility set out by the Finneys, but, nonetheless, was a bold, record price.

So, the deal was struck. Secretariat was under the

Champions Bred

Bred by C.T. Chenery's Meadow Stud

Hill Prince
1949 champion two-year-old colt
1950 champion three-year-old colt
1950 Horse of the Year
1951 champion handicap male

First Landing
1958 champion two-year-old colt

Cicada
1961 champion two-year-old filly
1962 champion three-year-old filly
1963 champion handicap female

Riva Ridge
1971 champion two-year-old colt
1973 champion handicap male

Secretariat
1972 champion two-year-old colt
1972 Horse of the Year
1973 champion three-year-old colt
1973 champion turf male
1973 Horse of the Year

Bred by Mrs. Helen (Penny) Chenery

Saratoga Dew
1992 champion three-year-old filly

management of the Meadow Stud heirs through his three-year-old season. He won early, suffered a stunning setback in the Wood Memorial, and then erupted into a Triple Crown for the ages: the Derby in record time, the Preakness (probably in record time, although a teletimer malfunction denied that officially), and then the Belmont by thirty-one lengths in world-record time on the dirt of 2:24 for one and a half miles. *Time*, *Newsweek*, and *Sports Illustrated* had featured him on their covers, and he had not let them down.

Secretariat lost twice afterward but won four additional races in spectacular style: a special race at Arlington Park, the first Marlboro Cup, and then the Man o' War Stakes and Canadian International on grass. He went to Claiborne wrapped in the glory of a second Horse of the Year campaign, accompanied, of course, by the three-year-old championship, too. He had won sixteen of twenty-one races and earned $1,316,808. Meadow Stable won a second Eclipse award for owner-breeder, and Somethingroyal was a rather obvious choice as Broodmare of the Year.

Secretariat's early stallion career was marked by a brief concern over his fertility. Then he settled into his career as a celebrity at Claiborne, albeit one cut short by laminitis. He died at nineteen in 1989. Secretariat was a good sire, but his record was not as good as knowledgeable horsemen might have hoped or his idolaters might have taken for granted. He sired fifty-six stakes winners (8 percent), of which Lady's Secret was 1986 Horse of the Year and Risen Star won the 1988 Preakness and Belmont and was champion three-year-old colt.

Three years after his death Secretariat became the leading broodmare sire, his daughters' offspring including that year's Horse of the Year, A.P. Indy. The weight of genetic excellence in his mates is a key element in a broodmare sire's career. In that avenue of the Turf, Secretariat was both a catalyst and a beneficiary. He is the broodmare sire of 161 (as of mid-2004) stakes winners, with a record far from complete. One of his daughters, Weekend Surprise, was named Broodmare of the Year in 1992, she the dam of classic winners A.P. Indy and Summer Squall.

If one assigns to Chenery status as breeder of all those foals born during his lifetime — irrespective of whether he were still in a decision-making situation — he and Meadow Stud were the breeders of forty-three stakes winners. This is not a number that ranks with the career figures of most individuals in this volume and its predecessor, but in terms of quality, Chenery's record was difficult to overstate. One can put it this way: Of his forty-three stakes winners, five (12 percent) were champions, of which two were Horse of the Year. Three of them won a total of six Triple Crown races. Four are in the Hall of Fame (Hill Prince, Cicada, Riva Ridge, Secretariat). Three of Chenery's mares were Broodmares of the Year (Hildene, Iberia, Somethingroyal). Not quite fitting into any of these categories was Sir Gaylord, a sire of international consequence.

Often quoted is Chenery's response when, as he grew older, his family suggested he should retire and enjoy his horses full time: "How can I make a full-time occupation out of something I have always done with one hand?" With a bit more devotion of time and energy, imagine what he could have done!

Epilogue

From 1973 through 1978, there were ten additional stakes winners bred in the name of Meadow Stud or the estate of Christopher T. Chenery. At the Keeneland November sales of 1978 and 1979, the Chenery estate sold a total of eighteen broodmares through Jonabell Farm, as agent, for a total of $5,941,000, averaging $540,000. Syrian Sea, stakes-winning sister to Secretariat, topped these sales at $1.6 million. This broke the previous record of the time — $1 million for Queen Sucree — by a margin reminiscent of a Secretariat race.

Penny Chenery has graciously assented to three decades of the role of "Secretariat's mother," fielding it, loving it, convincing you she has never tired of it. Although the tides of life meant she was no longer managing the handiwork of her dear father, per se, she has created her own testimony to his life as a horseman and an inspiration, breeding a half-dozen stakes winners in her own name. In 1992 one of these, a three-year-old filly, added the lovely sporting name of Saratoga Dew to the list of champions the family had bred.

Secretariat

Stakes Winners Bred by Christopher T. Chenery and Family

Cherrydale, ch.f. 1939, Whiskaway—Annie R., by *Bright Knight
Hornbeam, ch.g. 1940, Whiskaway—Annie R., by *Bright Knight
Mangohick, b.g. 1944, Sun Beau—Hildene, by Bubbling Over
Bryan G., ch.c. 1947, *Blenheim II—Anthemion, by Pompey
Hill Prince, b.c. 1947, *Princequillo—Hildene, by Bubbling Over
Tripoli, b.g. 1948, Bossuet—Meadow's First, by Sun Beau
Guayana, b.f. 1951, *Hunters Moon IV—Tringle, by Stimulus
Permian, br.c. 1951, Bossuet—Consolida, by Blue Larkspur
Prince Hill, b.g. 1951, *Princequillo—Hildene, by Bubbling Over
Travertine, ro.g. 1951, *Nordlicht—Chelita, by *Sir Gallahad III
Scansion, b.g. 1953, Heliodorus—Chelita, by *Sir Gallahad III
Third Brother, b.c. 1953, *Princequillo—Hildene, by Bubbling Over
North End, b.c. 1954, Doswell—Virginian Sea, by *Blenheim II
Willamette, b.f. 1954, Doswell—Sister Union, by *Blenheim II
Cartagena (stp), ch.g. 1955, *Nordlicht—Pomparray, by Pompey
Yemen, ch.c. 1955, Bryan G.—Imperatrice, by Caruso
First Landing, b.c. 1956, *Turn-to—Hildene, by Bubbling Over
Imperial Hill, b.f. 1956, Hill Prince—Imperatrice, by Caruso
*Palladio (stp), ro.g. 1956, Tabriz—*Giralda, by El Greco (Bred in France)
Salt Lake, b.f. 1957, Hill Prince—Yarmouth, by *Watling Street
Cicada, b.f. 1959, Bryan G.—Satsuma, by Bossuet
Sir Gaylord, b.c. 1959, *Turn-to—Somethingroyal, by *Princequillo
Belfort, b.c. 1960, Third Brother—French Cuff, by Count Fleet
Speedwell, b.f. 1960, Bold Ruler—Imperatrice, by Caruso
Suruga Bay (stp), b.c. 1961, Third Brother—Good Morrow, by *Nasrullah
Bold Experience, ch.f. 1962, Bold Ruler—First Flush, by *Flushing II
First Family, ch.c. 1962, First Landing—Somethingroyal, by *Princequillo
Lake Chelan, b.f. 1964, Bryan G.—Queens Moon, by *Hunters Moon IV
Pepperwood, b.f. 1964, Hill Prince—Tourmaline, by *Nordlicht
Copper Canyon, ch.f. 1965, Bryan G.—First Flush, by *Flushing II
Gay Matelda, dk.b/br.f. 1965, Sir Gaylord—Hasty Matelda, by Hasty Road
Potomac, ch.c. 1965, First Landing—Iberia, by *Heliopolis
Syrian Sea, b.f. 1965, Bold Ruler—Somethingroyal, by *Princequillo
Cicada's Pride, b.c. 1966, Sir Gaylord—Cicada, by Bryan G.
Hydrologist, ch.c. 1966, *Tatan—Iberia, by *Heliopolis
Pleasant Harbour, ro.g. 1966, First Landing—Manotick, by Double Jay
Queen's Double, b.f. 1966, Double Jay—Queens Moon, by *Hunters Moon IV
Virginia Delegate, b.g. 1966, Bold Ruler—First Flush, by *Flushing II
Riva Ridge, b.c. 1969, First Landing—Iberia, by *Heliopolis
Upper Case, b.c. 1969, Round Table—Bold Experience, by Bold Ruler
Secretariat, ch.c. 1970, Bold Ruler—Somethingroyal, by *Princequillo
Lefty, ch.c. 1972, Prince John—Kushka, by First Landing

Somethingroyal

Slip Screen, b.f. 1972, Silent Screen—Orissa, by First Landing
Romantic Lead, b.c. 1973, Silent Screen—Golden Spike, by Sir Gaylord

Estate of Christopher T. Chenery

Perils of Pauline, ch.f. 1974, Silent Screen—Hasty Landing, by First Landing
Spirit Level, b.c. 1974, Quadrangle—Due Dilly, by Sir Gaylord
Bemis Heights, b.f. 1975, *Herbager—Orissa, by First Landing
Ring of Light, b.g. 1975, In Reality—Due Dilly, by Sir Gaylord
Shelter Half, b.c. 1975, Tentam—Gay Matelda, by Sir Gaylord
Alada, b.f. 1976, Riva Ridge—Syrian Sea, by Bold Ruler
Ms. Ross, b.f. 1976, Hoist the Flag—Bold Experience, by Bold Ruler
Dr. Blum, ch.c. 1977, Dr. Fager—Due Dilly, by Sir Gaylord
Who's to Answer, b.f. 1978, Secretariat—Orissa, by First Landing

Helen Chenery

Appalachian Spring, b.f. 1982, Honest Pleasure—Frontier Nurse, by Dr. Fager
Go for It Zen, dk.b.c. 1986, Zen—Sheltered Harbor, by First Landing
Gold Oak, b.g. 1986, Gold Exchanged—Singing Oak, by Tarleton Oak
Burning Oak, b.g. 1987, Believe the Queen—Singing Oak, by Tarleton Oak
Erchless, ch.c. 1988, Entropy—Princess Ebony, by Black Mountain
Saratoga Dew, b.f. 1989, Cormorant—Super Luna, by In Reality

Source: *The Blood-Horse* Archives

CHAPTER

2

Tartan Farms

Central Florida emerged in the latter half of the twentieth century as one of the world's most important breeding centers. Marylander Joseph M. O'Farrell led the charge after Ocala's live oak trees, lush pastures, gentle gradients, limestone sub-strata, and climate won him over.

James Bright, Carl Rose, William Leach, and Dan Chappell had pioneered Florida Thoroughbred breeding before him, but it was O'Farrell's drive and promotional skills, along with horse savvy and exceptional timing, that pushed Florida breeding from a pleasant departure for a sleepy agricultural environment into a superb and enduring industry.

Jack Dudley's and Bonnie Heath's Florida-bred Needles won the Kentucky Derby in 1956, around the time O'Farrell discovered Ocala. Then, as founding general manager of Ocala Stud, O'Farrell bred champions My Dear Girl, Roman Brother, and Office Queen, but more pivotal to the history of the state, he brought Rough'n Tumble from Maryland to stand at Ocala Stud. With the Rough'n Tumbles leading the way, O'Farrell helped turn the annual auction of state-bred two-year-olds at Hialeah from a regional affair to a nationally important market.

Ocala Stud was owned by various investors and hit rocky financial waters at times, and it was no easy thing for O'Farrell's son, Michael, to right the ship and maintain it as a thriving operation into the present century. Along the way the success of Ocala Stud helped lure to the Ocala area a wide array of proven horsemen and new farm owners. The names of Jack

Dreyfus, Fred Hooper, Liz Whitney-Tippett, P.A.B. Widener III, Grant Dorland, Butch Savin, Joseph Brunetti, Bruce Campbell, and Charles F. Kieser were in the next wave to establish farms. Many of them stayed and thrived, and they were joined by additional waves of farm purchasers.

Among the best was Tartan Farms Inc., established by one of America's industrial giants and destined to breed one of America's Thoroughbred giants.

William L. McKnight thought he had gotten his fill of horses in the struggle of early 1900s farm life in White, South Dakota. He looked beyond his hometown for something larger, and he got as far as Duluth, Minnesota. With money saved from a threshing job, McKnight attended a business school in Duluth. He left school when, upon a third application, a struggling local outfit called Minnesota Mining and Manufacturing hired him as an assistant bookkeeper. He earned $11.55 a week.

McKnight's acumen helped push the company, to become known as 3M, up the ladder of business success and himself up the ladder within the firm. By 1909 he had become its cost accountant, and within two years the company's health had improved to the point that McKnight became manager of the Chicago sales office.

The promotions continued: general manager in 1914, president in 1929, and, twenty years later, chairman of the board. When one of the employees fiddled with an idea for improved tape in a handy roll dispenser, McKnight was adroit at marketing it.

He looked to his own family heritage for a name and dubbed it Scotch Tape. Like the product, the name stuck, becoming almost synonymous with such tape, regardless of brand.

As 3M prospered and diversified, McKnight's personal wealth reached the point by 1957 that *Fortune* magazine listed him in the $100 million to $200 million level. By the latter 1960s he was up a couple of notches, to the $300 million to $500 million echelon.

When McKnight turned seventy, in 1957, a group of employees hit on the idea of giving their boss a racehorse. By then, he had reacquainted himself with the horse in a manner that bespoke leisure rather than labor and enjoyed attending the races as a two-dollar bettor. Trainer John Sceusa was given the assignment of finding a horse, and as the story has been told over the years, had $6,500 to spend. He spotted a filly named Aspidistra and for about that amount purchased for McKnight the dam of future champions Dr. Fager and Ta Wee and the foundation of ongoing successful generations.

The next year McKnight made a larger commitment on his own, purchasing a pair of imported runners, *Meeting and *Munch, both of whom won stakes for him. (Hialeah publicist Everett Clay used to say he had encouraged McKnight to buy his first horses on the basis that an owner's badge would enable him to get into the racetrack free.)

Aspidistra had a number of positives. She had been bred by Robert J. Kleberg Jr. under the successful banner of King Ranch and was by the goodish King Ranch stallion Better Self, by champion Bimelech. The dam of Aspidistra, Tilly Rose, was not from an old King Ranch family, but her good form at Keeneland had impressed the ranch's Kentucky division manger, J. Howard Rouse. At Rouse's urging Kleberg bought the filly from Warren Douglas for $25,000, a very high price for the early 1950s. Tilly Rose won the Prioress Stakes and Colonial Handicap and earned $45,017, but her broodmare career consisted solely of foaling Aspidistra before her death. In order to be available for a birthday present costing $6,500, Aspidistra, of course, had been no exceptional runner. McKnight, in fact, risked her for a $6,500 claiming price a couple of times in her three races for his stable — he chose the name Tartan Stable —- and then she was retired with two wins from fourteen starts and earnings of $5,115.

The next step into the vortex of heavy investment

LEO FRUTKOFF

(from left) Dr. Fager's first jockey, David Hidalgo; owner William McKnight; and trainer John Nerud

was property in Ocala, where McKnight established Tartan Farms Inc. He needed outside expertise, and, after closely checking on the reputation of John Nerud with racing secretaries around the country, he hired him. Nerud, who had become famous as trainer of the Belmont Stakes winner and luckless Kentucky Derby runner-up *Gallant Man, was not only to train but also to oversee the acquisition of bloodstock. To clear the land and develop the farm, McKnight and Nerud hired a big, energetic Irishman, John Hartigan. "Anytime you worked for McKnight," Hartigan said later, "it was a question of being told what the policy was, what they wanted, and you had to produce."

Nerud and Hartigan, both strong-minded men, probably had their edgy moments. (Hartigan once showed us through a house Nerud had built for his own use on the farm and scoffed that one of the bedrooms was large enough to be "a paddock.") At any rate, Tartan's success eventually allowed Hartigan to launch out on his own, at Cashel Stud. He was one of the key figures in Florida breeding until his and his wife's deaths in a private airplane crash.

In 1960 Nerud purchased for Tartan additional acreage from the neighboring Bonnie Heath Farm (where Needles stood) and later parcels raised the total to 930 acres by the end of the 1960s. A total of sixty people worked on the farm. Clay had come aboard the 3M team in a public relations capacity, and a PR representative was sometimes sent to the farm to accompany Hartigan in showing the media the operation. (Tartan and Clay got considerable mileage from photos showing three young contemporaries at the farm in homebreds Dr. Fager and Minnesota Mac and the boarder In Reality.)

McKnight also became drawn into racing from a track ownership perspective. He wound up owning Calder Race Course, where a 3M product was the foundation for the racing surface for many years. He and son-in-law James Binger again showed acumen in whom they selected for management, bringing aboard Kenny Noe Jr. to run Calder.

While Aspidistra was a gift horse that became a great genetic resource, Tartan early on acquired another outstanding producer by more conventional means with the purchase of Cequillo. Although she had failed to win in eight starts, Cequillo was a prized prospect for any breeder believing in a strong female family. She was by the leading sire *Princequillo, and her family went back in stakes-producing generations to the great French producer Plucky Liege, dam of *Sir Gallahad III, *Bull Dog, Bois Roussel, and Admiral Drake. The sequence of dams from Cequillo back was, Cequillo's dam, Boldness (by *Mahmoud), who was out of Acorn Stakes winner Hostility (by Man o' War), who was out of *Marguerite de Valois (*Teddy—Plucky Liege), who was a full sister to *Sir Gallahad III and *Bull Dog. For the rest of Tartan's existence, and beyond, the succeeding generations of this family would continue their influence on the breed.

Cequillo was purchased in 1954 along with four other three-year-olds in a package deal from Eugene Mori and Jimmy Jones. She produced four stakes winners for Tartan: Hot Dust (by Jet Action) won the 1965 Hialeah Turf Cup and three other stakes and earned $263,642; Grand Splendor (by Correlation) won a division of the 1965 Pageant Handicap and earned $44,684; Tequillo (by Intentionally) won the 1966 Choice Stakes and two other stakes and earned $133,615; and Ruffled Feathers (by Rough'n Tumble) won the 1967 Man o' War Stakes and one other stakes and earned $228,904.

The sires of several of those took part in the growing importance of Florida. Correlation had been brought into the state by Grant Dorland and, although he was not a major success, had been a Florida Derby winner and a beaten favorite during the 1954 Triple Crown and thus represented a whiff of prestige for the state at its fledgling stage of development. (In a similar vein, Alsab, a champion in the 1940s, stood late in his career at Bonnie Heath Farm, but by then his reputation at stud had faded.) Intentionally was one of Nerud's key acquisitions.

Bred by Harry Isaacs' Brookfield Farm, Intentionally was the champion sprinter of 1959 as a three-year-old. A son of the War Relic horse Intent—My Recipe, by Discovery, Intentionally owned a highly important victory in the Futurity over champion First Landing when they were two, and he had stretched out enough to win that year's mile and a sixteenth Pimlico Futurity. His sprint title came largely off a three-length win in the Withers and a ten-length score in the Jerome, both at a mile. As we shall see, prowess at a mile was to be an important element again in Nerud's mind.

Intentionally had continued racing successfully at four and five, and then Nerud acquired him as a Tartan sire prospect. In a thirty-share syndicate arrangement based on a $750,000 valuation, Tartan

HAWTHORNE RACE COURSE

Dr. Fager

bought eighteen, Nerud got one, and the remaining shares went to Isaacs, Ocala Stud, Louis Wolfson, Josephine Abercrombie, and Rigan McKinney. In the winter of his six-year-old season, Intentionally, with Nerud then the trainer, defeated Florida-bred champion Carry Back in both the seven-furlong Palm Beach Handicap and mile and an eighth Seminole Handicap at Hialeah. The most important handicap at Hialeah, the Widener, beckoned, but it was at a mile and a quarter, so Nerud convinced the syndicate to go ahead and retire the horse. Intentionally had won eighteen of thirty-four starts and earned $652,258.

As a stallion prospect, Intentionally had a great deal to offer: early maturity, plenty of speed, and yet a pedigree with stamina elements. He was a significant success, getting twenty stakes winners (11 percent), including Tartan's champion sprinter Ta Wee. Intentionally's best horse, In Reality, was overshaded by two great ones in Dr. Fager and Damascus, but since Dr. Fager was a Tartan homebred, McKnight and Nerud were hardly disappointed.

Like Cequillo, the original gift horse Aspidistra produced four stakes winners by a variety of sires. Illustrating her ascension, her first mate was the obscure Argentine import *Esmero, whereas late in her career she was sent to the Phipps family's great Buckpasser at Claiborne Farm in Kentucky. Aspidistra's stakes winners were A. Deck (by First Cabin), winner of the Ponce de Leon Stakes and earner of $126,185; Chinatowner (by Needles), win-

ner of the Canadian Turf Handicap and earner of $83,305; Dr. Fager (by Rough'n Tumble), Horse of the Year for 1968 and a Hall of Fame member; Ta Wee (by Intentionally), two-time sprint champion and a Hall of Fame member.

Dr. Fager was named for Dr. Charles Fager, the neurosurgeon who operated on Nerud after an accident suffered on the backstretch. Dr. Fager took Rough'n Tumble's reputation from being a remarkable sire to the status of having fathered a truly great horse. Rough'n Tumble had come to Florida via a circuitous route. Purchased from his breeder, Dr. Charles Hagyard, by Mr. and Mrs. Harold Genter (like McKnight and 3M, based in Minnesota), he had won the Santa Anita Derby in 1951. Hagyard agreed to stand the horse at his Kentucky farm, but apparently with no real conviction to do so. Joe O'Farrell and his brother Tom needed a stallion for their Windy Hill Farm in Maryland. Cromwell Bloodstock Agency connected them to the Genters, and Rough 'n Tumble headed east. After two seasons in Maryland, he headed south, to Ocala Stud. Before he sired Dr. Fager, he had sired champion My Dear Girl, Kentucky Oaks winner Wedlock, and the solid handicap horse Conestoga and had become the mainstay of the Florida-bred juvenile sales.

Dr. Fager came out blazing at two, when he won his first three races by a minimum of seven lengths each, including his first stakes, the World's Playground, by twelve. He hooked up with the Genters' Tartan-raised In Reality (by Intentionally

Rough'n Tumble, 1948	Free For All, 1942	Questionnaire, 1927	Sting (Spur—Gnat)
			Miss Puzzle (Disguise—Ruby Nethersole)
		Panay, 1934	*Chicle (**Spearmint**—Lady Hamburg II)
			Panasette (**Whisk Broom II**—Panasine)
	Roused, 1943	*Bull Dog, 1927	*Teddy (Ajax—Rondeau)
			Plucky Liege (**Spearmint**—Concertina)
		Rude Awakening, 1936	Upset (**Whisk Broom II**—Pankhurst)
			Cushion (Nonpareil—Hassock)
DR. FAGER	Better Self, 1945	Bimelech, 1937	Black Toney (Peter Pan—Belgravia)
			*La Troienne (*Teddy—Helene de Troie)
		Bee Mac, 1941	War Admiral (Man o' War—Brushup)
			Baba Kenny (Black Servant—Betty Beall)
Aspidistra, 1954	Tilly Rose, 1948	Bull Brier, 1938	*Bull Dog (*Teddy—Plucky Liege)
			Rose Eternal (Eternal—Rose O Roses)
		Tilly Kate, 1935	Draymont (Wildair—Oreen)
			Teak (Tea Caddy—Fricassee)

and from Rough'n Tumble's first champion My Dear Girl) in the Cowdin. Dr. Fager beat the other horse by nearly a length. In the Champagne a championship was available, but Dr. Fager was rank early and was beaten by Successor, who had been third in the Cowdin. Successor went on to further success and won the title of champion two-year-old colt.

The following year, 1967, Dr. Fager began a rivalry with a late-developing sort in Edith W. Bancroft's Damascus. Mrs. Bancroft was a daughter of William Woodward Sr. of Belair Stud and represented a half-century of family prominence in breeding and racing; Dr. Fager represented another sportsman's operation launched only a decade before.

Dr. Fager won the Gotham over Damascus, but Nerud demurred on asking the headstrong, run-them-all-to-death sort try a mile and a quarter. So, instead of heading for the Kentucky Derby, Dr. Fager demolished his fields in the one-mile Withers and the mile and an eighth Jersey Derby. In the Jersey Derby, however, his number was taken down in favor of In Reality's for some mischief born of his rage to run on the first turn.

Then followed a ten-length score in the Arlington Classic, again going a mile, and a four and a quarter-length victory in the Rockingham Special. Finally, on September 25, Nerud asked his three-year-old to race a mile and a quarter, and Dr. Fager defeated In Reality by slightly more than a length in the New Hampshire Sweepstakes Classic.

This sent him into one of racing's apparent dream match-ups, and he and Preakness-Belmont winner Damascus faced the older champion Buckpasser in the Woodward Stakes at a mile and a quarter. Damascus and Buckpasser each had a pacesetter, known by then colloquially as a "rabbit." If Nerud's sporting nature were affronted by this, he could scarcely utter any complaint; ten years earlier, he had virtually revived the concept insofar as American racing is concerned when he sent Bold Nero to run Bold Ruler off his feet in the Belmont Stakes, abetting *Gallant Man's powerful rally to victory.

When two rabbits whipsaw a giant, one might presume he has wandered into a fairy tale. In the Woodward, though, with the pacemakers' jockeys whooping and hustling, Dr. Fager's blood and speed hit boil right out of the gate and the result was grim. He set himself up for eventual exhaustion. Damascus rushed by to win by ten, while Buckpasser got past

Dr. Fager to finish second by a half-length. Advantage (and Horse of the Year) Damascus.

Dr. Fager finished the year winning the mile and a quarter Hawthorne Gold Cup and seven-furlong Vosburgh.

At four Dr. Fager prevailed. True weight carriers have been rarities in the past four decades. Dr. Fager and Damascus were among them. Dr. Fager started the year with a sprint victory in the seven-furlong Roseben Handicap under 130 pounds. A number of Eastern horses have come a cropper when shipped to California, but Nerud had been successful in that sort of cross-country adventure with *Gallant Man, and he took Dr. Fager west for the mile and a sixteenth Californian. The Tartan colt defeated the champion filly Gamely while carrying 130 and giving her fourteen pounds of actual weight.

Back east Dr. Fager traded hours of greatness with his old foe Damascus. In the Suburban Handicap at Aqueduct, Dr. Fager took up 132 pounds, getting one from Damascus, and blazed him in track-record-equaling time of 1:59 3/5. Between the Suburban and Brooklyn handicaps, Damascus lost the Haskell, so the weights were 135 on Dr. Fager and 130 on the defending Horse of the Year. Damascus' trainer, Frank Whiteley, reverted to the rabbit option for the Brooklyn, putting Hedevar in the race. Run, rabbit, run was not novel anymore, but it worked. After Hedevar goaded Dr. Fager, Damascus came along to win by two and a half lengths, breaking the mark that had been tied in the Suburban by getting the mile and a quarter in 1:59 1/5.

The pair did not meet again. They stood at two and two, and Dr. Fager's fans undoubtedly took on the mantra that when there were no pacemakers, their guy won; when he lost, there had been rabbits afoot.

Dr. Fager had more challenges, but nothing else brought him down. In the Whitney he won by eight under 132 pounds. In Chicago he blazed a mile in a world-record 1:32 1/5 in winning the Washington Park Handicap under 134 pounds. This record has been broken on grass and equaled on dirt, but as of early 2004, it still stood after thirty-six years.

Nerud was game to ask Dr. Fager new questions. Raced on grass for the first time, the colt took up 134 pounds again for the mile and three-sixteenths United Nations Handicap and got home by a neck over the high-class Advocator (112). For Dr. Fager's farewell Nerud gave the horse, himself, and fans the treat of putting his idol in a context where speed could be pure. In the seven-furlong Vosburgh, Dr. Fager carried 139 pounds and led throughout winning by six lengths in 1:20 1/5 to set an Aqueduct track record.

He was all conquering on the track and in the ballot box. Dr. Fager swept year-end honors as Horse of the Year, older male champion, sprint champion, and turf champion, a collation without equal before or since. McKnight, who was eighty at the time, was elected to The Jockey Club that year.

Dr. Fager had won eighteen of twenty-two races and earned $1,002,642. At the end of the century, a panel put together by *The Blood-Horse* voted him sixth best to Damascus' sixteenth. Among those placed in the top twenty, there was no other twosome who had met as often as four times.

Dr. Fager was a great horse, but not without chinks as a stallion prospect. His female family at that time looked good, but not exceptional; his dam had been a claimer, and Rough'n Tumble was a sire who had

outbred himself. Also, while Florida had emerged as a ratified site for raising major horses, standing there did not equal standing in Kentucky in a commercial/opportunity sense. Taking all this into account, Dr. Fager must be counted a very good success, and he might have risen higher but for dying as a young stallion, at twelve, of a twisted intestine. He sired thirty-five stakes winners (14 percent), including the champion two-year-old filly Dearly Precious, champion sprinter Dr. Patches, and Canadian Horse of the Year L'Alezane. The *Daily Racing Form* regarded Dr. Fager as the leading North American sire of 1977 (although *The Blood-Horse*'s more inclusive international statistics pointed to Northern Dancer as the leader that year). In terms of lingering influence within the ongoing Tartan history, Dr. Fager would be counted a patriarch, for his bloodlines had much more to give.

No mare could be asked to foal a Dr. Fager and then something better, but Aspidistra did have another amazing foal. This was the Intentionally filly

Ta Wee (center) with her dam Aspidistra (left) and half brother Dr. Fager (right)

Ta Wee, who was trained by Scotty Schulhofer since Nerud had given up the day-to-day training while retaining management responsibilities for Tartan's operation. Ta Wee, at three in 1969, beat older males in the Vosburgh to give Aspidistra's foals three consecutive wins in that autumn sprint. At four she won five of seven races, ending with two victories that bespoke her gameness and Tartan's sporting willingness to accept any challenge. These came in the six-furlong Fall Highweight under 140 pounds and the six-furlong Interborough Handicap under 142. She was voted champion sprinter both years and was retired to Tartan with fifteen wins from twenty-one starts and earnings of $284,941.

Ta Wee produced four stakes winners: Great Above (by Minnesota Mac), winner of the Toboggan Handicap and one other stakes and earner of $331,377; Tweak (by Secretariat), winner of the Fair Lawn Stakes and earner of $148,597; Tax Holiday (by What a Pleasure), winner of two stakes and earner of $254,938, and Entropy (by What a Pleasure), winner of the Paumonok Handicap and two other stakes and earner of $293,999.

Her first, Great Above, was of particular interest. He was sired by Minnesota Mac, who was a Tartan foal of 1964 and was named for the founder of the feast, Mr. McKnight. Minnesota Mac was by Rough'n Tumble and out of *Cow Girl II, who in turn was a full sister to *Munch; one of the early imports bought by McKnight, Munch had won the Atlantic City Handicap and Cleopatra Handicap in 1958.

*Cow Girl II and *Munch were Irish-breds by Mustang—Ate, by Phideas. Before foaling Minnesota Mac, *Cow Girl II foaled Western Warrior, who was by the sprinter Decathlon but won the mile and three-sixteenths United Nations Handicap and four other stakes.

Although he was overshadowed by Dr. Fager — who wouldn't be? — Minnesota Mac added to Tartan's and Rough'n Tumble's legacy. He won the 1967 Chicagoan Handicap among four wins in eleven starts to earn $63,275. He then sired 1978 Turf champion MacDiarmida, as well as the stakes winner and producer Katonka and Great Above. The all-Tartan-bred Great Above sired Holy Bull, the Horse of the Year of 1994, and became the broodmare sire of two-time sprint champion Housebuster (who thus mimicked the record of great-granddam Ta Wee).

W. L. McKnight died at ninety during the winter of 1978 in LaGorce Island, Florida, where he maintained a home. His daughter, Virginia Binger, had been given control of the farm in 1974. Son-in-law Jim Binger became the most visible in management of Tartan, while Nerud continued as president of the firm. Mrs. Binger, who died in 2003, was involved but tended to stay out of the limelight.

In the year of the founder's death, Tartan raced another homebred champion, Dr. Patches, who was named the co-champion Eclipse Award winner as sprinter along with J. O. Tobin. Dr. Patches represented an amalgam of Tartan history from its early days. He was a son of Aspidistra's great son Dr.

MINNESOTA MAC				
Rough'n Tumble, 1948	Free For All, 1942	Questionnaire, 1927	Sting (Spur—Gnat)	
			Miss Puzzle (Disguise—Ruby Nethersole)	
		Panay, 1934	*Chicle (**Spearmint**—Lady Hamburg)	
			Panasette (**Whisk Broom II**—Panasine)	
	Roused, 1943	*Bull Dog, 1927	*Teddy (Ajax—Rondeau)	
			Plucky Liege (**Spearmint**—Concertina)	
		Rude Awakening, 1936	Upset (**Whisk Broom II**—Pankhurst)	
			Cushion (Nonpareil—Hassock)	
*Cow Girl II, 1949	Mustang, 1941	Mieuxce, 1933	Massine (Consols—Mauri)	
			L'Olivete (Opott—Jonicole)	
		Buzz Fuzz, 1933	The Recorder (Captain Cuttle—Lady Juror)	
			Lady Buzzer (Honey Bee—Lady Derelict)	
	Ate, 1941	Phideas, 1934	Pharos (Phalaris—Scapa Flow)	
			Imagery (Gainsborough—Sun Worship)	
		Messe, 1933	Buchan (Sunstar—Hamoaze)	
			*Messaline (Caligula—Monisima)	

Fager and out of a daughter of Cequillo. In addition to her four stakes winners, Cequillo's fourteen winners from twenty foals also included Expectancy, a winner by Intentionally. Expectancy foaled Dr. Patches. The gelding Dr. Patches had started out in Nerud's stable, had been sent west to up-and-coming trainer D. Wayne Lukas, and then was sent back east. In the summer of 1978, he put together five consecutive excellent races that earned him champion status. At Saratoga he won a six-furlong overnight race in 1:08 3/5 in his last start for Lukas. Back with Nerud, he won the mile and an eighth Paterson Handicap at Meadowlands, defeating Seattle Slew. This marked only the second defeat in the career of Triple Crown winner Slew, who was coming back after a long layoff at that time. Dr. Patches dropped back to seven furlongs and dropped the Celanese Cup by a nose to Buckfinder, then at the same distance defeated the defending sprint champion, What a Summer, in the Tartan-friendly Vosburgh. He followed with a four-length victory at a mile and a quarter in the Meadowlands Cup, completing an unusual distance pattern. With Seattle Slew having rebounded to win the Woodward, Dr. Patches had little chance at the over-all older male title, but voters found him a spot in the sprint category.

Dr. Patches raced on into his eight-year-old season, being trained by Nerud's son, Jan. He won no further stakes and was retired with seventeen wins in forty-seven starts and earnings of $737,612. His dam, Expectancy, foaled additional stakes winners Imminence and Who's for Dinner.

In 1980, two years after McKnight's death, Tartan accomplished something that Dr. Fager might have been capable of doing had Nerud chosen to try, i.e., victory in an American classic. The vehicle of this milestone was Codex, winner of the Preakness.

Codex went back to a good Calumet Farm family. Tartan had bought into this family when it acquired Heliolight, daughter of 1950s champion distaffer Real Delight. Heliolight, bred by Calumet, was one of the lesser foals of Real Delight, but her Tartan-bred daughter Minnetonka, though unraced, became an important producer. After acquisition by Tartan, Heliolight foaled the Rough'n Tumble stakes winner Lonesome River, who won the Sport Page Handicap and fourteen other races.

To the cover of Minnesota Mac, Minnetonka foaled stakes winner Katonka, and to the cover of Raise a Native she foaled stakes winner Barrera. Roundup Rose, dam of Codex, was another Minnesota Mac foal from Minnetonka. She won at first asking, at Calder, as a two-year-old in 1973 but bowed a tendon

Codex

in the process and was retired. Codex, the second foal from Roundup Rose, was by Arts and Letters, a *Ribot colt who was Horse of the Year for Paul Mellon in 1969. Codex was among the drafts of Tartan horses sent to trainer Lukas, a former Quarter Horse trainer who was making a name in Thoroughbreds and whose hard work, organizational skills, and burgeoning success impressed Nerud.

The colt blossomed under Lukas in the winter of 1980 in California, where he won the Santa Anita Derby and Hollywood Derby over Rumbo, who was destined to be second in the Kentucky Derby. Codex had not been nominated for the Kentucky Derby, but he showed up for the Preakness and defeated Derby-winning filly Genuine Risk. With the controversial Angel Cordero aboard, and with a move in the stretch turn that some saw as fouling the filly, Codex consigned his Preakness to controversy, and a commission hearing, but the victory stood. It was the first classic win for Lukas, who within twenty years had won a total of thirteen Triple Crown events to tie Sunny Jim Fitzsimmons for most victories in the series.

Codex won six of fifteen races and earned $534,576. He sired only ten stakes winners (9 percent), but the high-class campaigner Lost Code was among them.

Another significant Tartan-bred stakes winner of the early 1980s was Maudlin, a son of Foolish Pleasure—Zonta, by Dr. Fager, who won the Sport

Dr. Patches

Page, Bold Ruler, and Forego handicaps among eleven wins in thirty-seven starts and earned $423,789. Maudlin sired the full siblings Beautiful Pleasure, John Oxley's 1999 Breeders' Cup Distaff winner and Eclipse Award-winning older filly, and Mecke, winner of the Arlington Million.

One of the ironies of racing is that today's rival can frequently become tomorrow's accomplice. A case in point was Ogygian, one of the most brilliant of Tartan homebreds. A foal of 1983, Ogygian was by Dr. Fager's rival Damascus. His dam, Gonfalon (by Francis S.), was out of Grand Splendor, who will be recalled as one of the stakes winners from early foundation mare Cequillo. The precocious colt won the Futurity at two but did not make the Triple Crown. During his three-year-old season, however, Ogygian won the Riva Ridge, Dwyer, and Jerome before his first defeat, in the Pegasus. Ogygian won seven of ten races and earned $455,520. Jan Nerud was the trainer.

In addition to his role with Tartan Farms, John Nerud was allowed to breed and race on his own. He bred two colts that became outstanding stallions, each on the basis of access to major female families belonging to clients. From the Joseph M. Roebling family of Portage, Nerud bred Cozzene (*Caro—Ride the Trails, by Prince John) whom son Jan sent out to win the Breeders' Cup Mile and Eclipse Award for turf males in 1985. Cozzene's sixty-two stakes winners (8 percent) include fellow Breeders' Cup winners Alphabet Soup and Tikkanen, and he led the North American sire list in 1996.

Unbridled

The other key stallion bred and raced by Nerud was Fappiano, who benefited from and contributed to his Tartan legacy. The pedigree of Fappiano takes us back again to Cequillo and then combines the horse Nerud always has idolized, Dr. Fager. Cequillo's foals included a 1962 Correlation filly, Grand Splendor, who, in turn, became the dam of a 1970 Dr. Fager filly, Killaloe. Killaloe produced five stakes winners, among whom was Fappiano. The latter was sired by Mr. Prospector and, as a foal of 1977, was bred with the knowledge of that stallion's brilliance but not necessary the realization of his potential to get distance horses (Conquistador Cielo, Forty Niner, etc.)

Nerud's homebred Fappiano was named for longtime *New York Times* racing writer Joe Nichols, who chose to be known that way rather than by the Giuseppe Carmine Fappiano on his birth certificate. Nerud liked the rhythm of Fappiano.

He also liked the horse. Fappiano won ten of seventeen races and earned $370,213. He won four stakes, of which the signature victory was his win in the one-mile Metropolitan Handicap. Just as Nerud liked the speed of Intentionally, he felt the mile distance of great significance. After all, Dr. Fager had set the world record at that route, and going further back in his career, the stayer *Gallant Man had defeated the speedy Bold Ruler for Nerud in the Met

Mile. Nerud knew that since mile races are often contested with similar strategy as sprints in this country, the distance can indicate stamina as well as speed. The Met Mile, in fact, has taken on a reputation as somewhat of an indicator of potential sire power, and its winners of the last two decades or so include Conquistador Cielo, Gulch, In Excess, Holy Bull, Dixie Brass, You and I, Honour and Glory, and Langfuhr. Fappiano also won the Discovery at a mile and an eighth but was unplaced in his final start, in the Marlboro Cup over an additional furlong.

The contribution Tartan stock made personally to Nerud was returned when Fappiano went to stud, although the farm did not reap full benefit. Through the 1980s Tartan continued to turn out a succession of stakes horses, but by 1987 Binger thought it was time to cease.

"John Nerud is seventy-four, and I'm not far behind," he said. "I have no children who have shown any interest in assuming this business." Also, he said, in an increasingly difficult economic situation the farm's recent profitability had rested on the sale of a high-price horse "every year or so. Without such a sale, the operation would show a loss. That was a consideration."

An announcement was made in June that Tartan would disperse most of its horses. The yearlings were

Maudlin

sold at Fasig-Tipton Kentucky in September, fifty-five head grossing $5,634,500 to average $102,445. The mares and weanlings were sold at Fasig-Tipton Kentucky that autumn. Nerud also dispersed his own horses at that autumn sale. The Tartan-Nerud combination included 194 horses that grossed $25,634,000 to average $132,134.

The yearlings included a $300,000 Fappiano colt from the winning Dr. Fager mare Demure. This colt, who was to be named Quiet American, was inbred closely, 3x2, to Dr. Fager. He was yet another descendant of Cequillo, his third dam. Quiet American scored his most important win at the anointed distance of a mile, in the grade I NYRA Mile, for the Maktoum stables and was retired to one of the Maktoum's Kentucky farms, Gainsborough. Quiet American is the sire of the 1998 Kentucky Derby and Preakness winner, Real Quiet, and of 1997 older filly champion Hidden Lake, thus continuing the impact of various Tartan elements.

The key individual coming out of the Tartan autumn consignment represented another layer of the interwoven connections of Tartan and the Genter family of Minnesota. Genter had passed away, but his wife continued their family breeding and racing stable under the name Frances Genter Stable. For $70,000, Mrs. Genter purchased one of Tartan's weanling colts at the dispersal. He was a son of Fappiano and was out of Gana Facil, yet another Tartan-bred who traced directly through the whole of the farm's history. Gana Facil's third dam was none other than Aspidistra. Gana Facil, by French classic horse *Le Fabuleux, was from the winner Charedi, who was sired by the Tartan-raised Genter homebred In Reality. The next dam was Magic, an unraced filly who resulted from the crossing of the once low-profile Aspidistra to the champion Buckpasser.

Mrs. Genter named the Fappiano colt Unbridled. Trained by Carl Nafzger, Unbridled prompted one of

Champions Bred

Dr. Fager
1967 champion sprinter
1968 champion grass horse
1968 champion handicap male
1968 Horse of the Year
1968 champion sprinter

Ta Wee
1969 champion sprinter
1970 champion sprinter

Dr. Patches
1978 champion sprinter

Unbridled
1990 champion three-year-old colt

the touching moments in racing when the trainer was televised giving an impromptu call to the aged Mrs. Genter as the colt swept to victory in the Kentucky Derby of 1990. Late that year Unbridled scored in the Breeders' Cup Classic to secure the three-year-old championship. He thus joined Dr. Fager, Ta Wee, and Dr. Patches as the fourth champion bred by Tartan, and followed Codex as its second classic winner. Unbridled had a career record of eight wins in twenty-four starts and earnings of $4,489,475.

Ironically, three years after its dispersal, Tartan Farms Inc., was North America's leading breeder in earnings, with $6,930,043 in 1990. Results since the dispersal raised the total number of stakes winners bred by Tartan to 103.

Genter Stable also purchased Unbridled's dam, Gana Facil, and her next Fappiano colt, Cahill Road, won the 1991 Wood Memorial for the stable although injuring himself in the running.

Fappiano, who stood at Lane's End Farm in Kentucky, died young but was a significant sire who got Cryptoclearance, Rubiano, Tasso, and Defensive Play, in addition to Unbridled, Quiet American, and Cahill Road. Unbridled, who stood at Gainesway Farm and then at Claiborne Farm, also died young but in ten crops left a sterling line-up of important winners. The Tartan-bred imparted an admixture of all those enduring bloodlines to Kentucky Derby winner Grindstone, Preakness winner Red Bullet, Belmont Stakes winner Empire Maker, three-year-old filly champion Banshee Breeze, Breeders' Cup Juvenile winners Unbridled's Song and (two-year-old champion) Anees, Breeders' Cup Juvenile Fillies winner and champion Halfbridled, Travers winner Unshaded, plus Broken Vow, Niigon, and Manistique. Already in the next generation, Grindstone has sired a classic winner in 2004 Belmont Stakes hero Birdstone.

And it all started with a birthday present.

Stakes Winners Bred by Tartan Farms

A. Deck, b.g. 1961, First Cabin—Aspidistra, by Better Self

Chinatowner, b.c. 1962, Needles—Aspidistra, by Better Self

Grand Splendor, b.f. 1962, Correlation—Cequillo, by *Princequillo

Base Leg (stp), b.c. 1963, Jet Pilot—Hannah Dustin, by *Princequillo

Tequillo, b.c. 1963, Intentionally—Cequillo, by *Princequillo

Dr. Fager, b.c. 1964, Rough'n Tumble—Aspidistra, by Better Self

Minnesota Mac, b.c. 1964, Rough'n Tumble—Cow Girl II, by Mustang

Ruffled Feathers, b.c. 1964, Rough'n Tumble—Cequillo, by
   *Princequillo

Mountain Man, b.g. 1965, *Gallant Man—Molecomb Peak, by Nearco

Black Pipe, dk.b/br.g. 1966, Intentionally—Dream On, by Fair Ruler

Lonesome River, dk.b/br.c. 1966, Rough'n Tumble—Heliolight, by
   Helioscope

Shelter Bay, dk.b/br.g. 1966, Intentionally—De Hostess, by Your Host

Shut Eye, dk.b/br.g. 1966, Intentionally—Languid, by *Tudor Minstrel

Ta Wee, dk.b/br.f. 1966, Intentionally—Aspidistra, by Better Self

Arachne, ch.f. 1967, Intentionally—Molecomb Peak, by Nearco

Play Pretty, dk.b/br.f. 1968, Needles—Wild Cherry, by Nashua

Willmar, dk.b/br.g. 1968, Boldnesian—Pillow Fight, by Combat

Worldling, b.f. 1969, Never Bend—Sofisticada, by Timor

Actuality, dk.b/br.g. 1970, In Reality—Admiral's Dancer, by War Admiral

Lonetree, b.g. 1970, Restless Wind—Arrangement, by Intentionally

Tastybit, b.f. 1970, *Noholme II—Munch, by Mustang

Tin Goose, dk.b/br.g. 1970, Intentionally—Translucent, by *Royal
   Charger

Heloise, ch.f. 1971, Dr. Fager—Fleet II, by Immortality

With Aplomb, b.g. 1971, Dr. Fager—She's Very Ultra, by Olympia

Finery, dk.b/br.f. 1972, *Prince Taj—Bit of Style, by Intentionally

Great Above, dk.b/br.c. 1972, Minnesota Mac—Ta Wee, by
   Intentionally

Hunka Papa, b.c. 1972, Dr. Fager—Quilting, by *Princequillo

Imminence, gr.f. 1972, Native Charger—Expectancy, by Intentionally

Land Girl, dk.b/br.f. 1972, Sir Ivor—Arrangement, by Intentionally

Arachnoid, b.c. 1973, Dr. Fager—Arachne, by Intentionally

Dr. Patches, ch.g. 1974, Dr. Fager—Expectancy, by Intentionally

Magnificence, b.f. 1974, Graustark—Magic, by Buckpasser

Kam Tam Kan, dk.b/br.g. 1975, Tentam—Kamakura II, by Faristan

Quip, dk.b/br.c 1975, Hoist the Flag—Play Pretty, by Needles

Great Neck, dk.b/br.c. 1976, Tentam—Turtle Cove, by Dr. Fager

R. F. Dee, dk.b/br.c. 1976, Buckpasser—Mail Box, by Francis S.

Tax Holiday, b.f. 1976, What a Pleasure—Ta Wee, by Intentionally

Tweak, dk.b/br.f. 1976, Secretariat—Ta Wee, by Intentionally

Virilify, dk.b/br.c. 1976, Minnesota Mac-Quiet Char, by Nearctic

Codex, ch.c. 1977, Arts and Letters—Roundup Rose, by Minnesota Mac

Extra Shot, b.c. 1977, My Dad George—Heroine, by Vitriolic

Idyll, dk.b/br.c. 1977, Riva Ridge—Arrangement, by Intentionally

Open Gate, b.f. 1977, Dr. Fager—On Duty, by *Sea-Bird II

Acaroid, dk.b/br.c. 1978, Big Spruce—Arachne, by Intentionally

Maudlin, dk.b.c. 1978, Foolish Pleasure—Zonta, by Dr. Fager

Parcel, dk.b.g. 1978, *Noholme II—Mail Box, by Francis S.

Raisin Thunder, dk.b/br.c. 1978, Crozier—Thundering, by Native
   Charger

Cintula, b.g. 1979, Ramsinga—Thill, by Iron Ruler

Guyana, dk.b/br.c. 1979, Cutlass—Cult, by Dr. Fager

Muttering, ro.c. 1979, Drone—Malvine, by Gamin

Sepulveda, dk.b/br.c. 1979, First Landing—Carnet de Bal, by Dr. Fager

Who's for Dinner, dk.b/br.c. 1979, Native Charger—Expectancy,
   by Intentionally

Entropy, ch.c. 1980, What a Pleasure—Ta Wee, by Intentionally

Eskimo, b.c. 1980, Northern Dancer—Dr. Mary Lou, by Dr. Fager

Great Hunter, dk.b.g. 1980, Chieftain—Heliogram, by Native Charger

Hare Brain, b.f. 1980, Naskra—Milina, by Nijinsky II

Liturgism, gr.f. 1980, Native Charger—Cult, by Dr. Fager

Mezzo, ch.c. 1980, Ramsinga—Juliet's Aria, by Hasty Road

Agacerie, b.f. 1981, Exclusive Native—Quiet Charm, by Nearctic

Id Am Fac, ch.f. 1981, Nodouble—For the Occasion, by Dr. Fager

Lament, ch.f. 1981, *Noholme II—Montagues Lament, by T. V. Lark

Machalstva, b.g. 1981, Stage Door Johnny—When and If, by Dr. Fager

Rilial, b.g. 1981, Sir Ivor—Bite the Dust, by Dr. Fager

Coup de Fusil, b.f. 1982, Codex—Cult, by Dr. Fager

Equalize, gr.c. 1982, Northern Jove—Zonta, by Dr. Fager

Erstwhile, dk.b.f. 1982, Arts and Letters—Eyes, by Iron Ruler

Important Business, b.c. 1982, Codex—Suite of Dreams, by Tentam

Jeblar, b.c. 1982, Alydar—City Girl, by Lucky Debonair

Nyama, b.f. 1982, Pretense—Chorea, by Northern Dancer

Uene, b.c. 1982, Hold Your Peace—Mythographer, by Secretariat

Banbury Cross, dk.b.g. 1983, J. P. Tobin—Wanga, by Dr. Fager

Final Reunion, b.f. 1983, In Reality—Milina, by Nijinsky II

Kex, b.g. 1983, In Reality—Hemlock, by Big Spruce

Ogygian, b.c. 1983, Damascus—Gonfalon, by Francis S.

Sciustre, dk.b/br.c. 1983, Blushing Groom (Fr)—Tundra Queen, by Le
   Fabuleux (Fr) (SW in Italy)

Wakan Tanka, ch.f. 1983, Nodouble—Heofon, by In Reality

Ataentsic, b.f. 1984, Hold Your Peace—Mythographer, by Secretariat

Billie Osage, ch.c. 1984, Superbity—Windy and Mild, by Best Turn

Feudal, dk.b.g. 1984, Transworld—Suite of Dreams, by Tentam

Kapalua, ro.g. 1984, Caro (Ire)—Demure, by Dr. Fager

Pentelicus, b.c. 1984, Fappiano—Charedi, by In Reality

Portage, dk.b.g. 1984, Riverman—Open Gate, by Dr. Fager

Schism, b.c. 1984, Green Dancer—Susurrant, by What a Pleasure

Yaguda, b.f. 1984, Green Dancer—Bite the Dust, by Dr. Fager

Phantast, b.g. 1985, Superbity—Suite of Dreams, by Tentam

Primal, b.g. 1985, Maudlin—Perl, by Graustark

Zie World, b.c. 1985, Transworld—Ziebarth, by Proudest Roman

Chenin Blanc, gr.c. 1986, Lypheor (GB)—Mouthfull, by Caro (Ire)

Coolawin, dk.b/br.f. 1986, Nodouble—Magaro, by Caro (Ire)

Mr. Carvel, ch.g. 1986, Hold Your Peace—Cavalle, by Dart Board

Purl One, ch.f. 1986, Give Me Strength—Perl, by Graustark

Quiet American, b.c. 1986, Fappiano—Demure, by Dr. Fager

Rodeo Star (stp), ch.g. 1986, Nodouble—Roundup Rose, by
   Minnesota Mac (SW in England)

Surging, b.f. 1986, Fappiano—Lady Ice, by Vice Regent

Urbanity, b.f. 1986, Superbity—Urbacity, by Fappiano

Wind Shear, b.f. 1986, Fappiano—Ellie Milove, by Dr. Fager

Cove Dancer, dk.b/br.g. 1987, Sovereign Dancer—Turtle Cove, by Dr.
   Fager

Heart of Joy, dk.b/br.f. 1987, Lypheor (GB)—Mythographer, by
   Secretariat (SW in England/U.S.)

Izatys, b.c. 1987, Riverman—Artificial, by Dr. Fager

Jane Scott, b.f. 1987, Copelan—Arrange the Silver, by State Dinner

Miner's Dream, b.c. 1987, Fappiano—Lady Ice, by Vice Regent

Unbridled, b.c. 1987, Fappiano—Gana Facil, by Le Fabuleux (Fr)

Derivative, dk.b.g. 1991, Fappiano—Coup de Fusil, by Codex

Source: *The Blood-Horse* Archives

Robert J. Kleberg Jr.

King Ranch is the central locus of one of the legendary sagas of the American West. General Robert E. Lee advised Captain Richard King to "buy land and never sell," and land forever since has had virtually a chromosomal influence on King's descendants. One of Captain King's grandsons, Robert Justus Kleberg Jr., was the central figure in running King Ranch through the middle decades of the twentieth century.

King Ranch, the prototypical Texas ranching spread centered around Kingsville, Texas, was always, in Kleberg's eyes, primarily in the cattle business. Oil, however, drove a good portion of the expansion, which eventually saw King Ranch roaming over 11 million acres — said to be the largest chunk of the earth under the legitimate control of a single family — with vast parcels in South America, Australia, and Africa. While the knee-jerk reaction to such figures might summon fears of the slavering maw of American expansion run amok, in the case of King Ranch the wealth and determination brought to bear on expansion often turned theretofore useless land into a workable, arable condition; the feeding of a company's bottom line has seldom been so consistent with the feeding of mankind.

Through it all Kleberg remained a rancher, and therefore an animal breeder at heart. He was a rawboned, strapping sort of fellow. Photographed from certain angles, the toothiness of his big smile distracted. More striking, though, was Kleberg's rugged handsomeness; this resonated from the wrappings of

a business suit in the winner's circle as honestly as from under a battered cowboy hat that bespoke many an hour on dusty roundups of King Ranch cattle.

Kleberg was born in the waning years of the nineteenth century, on March 29, 1896, in Corpus Christi, Texas. He attended the University of Wisconsin for a couple of years, where he studied genetics "and not much of anything else. I wasn't interested in anything else. I took what courses they had, but I did a lot of the studying on my own, just reading, which I have done ever since — still do," as he recalled only a year before his death.

Kleberg left school to return to King Ranch and help his father, who was in failing health. He became president of King Ranch when an older brother died in 1932. He applied his book learning to his instincts and experience. Texas was good land for cattle, he was sure, but he was not sure the family's British beef breeds had the best genetic components for the environment. He worked out a superb percentage for crossing the British shorthorn with the Brahman from India, and by 1940 he could register his new breed under the name Santa Gertrudis. This breed, invested with a deep, viscerally earthy, reddish hue and powerful, efficient muscling, flourished in Texas as intended. The Santa Gertrudis also surprised its designer with its adaptability around the world, as King Ranch spread its vision from one continent to the next.

Kleberg was not around in time actually to invent the Quarter Horse, but his crossing patterns shaped the modern version of that versatile breed, which was originally meant to be — and remains — efficient in the workaday athleticism needed on a cattle ranch.

We do not shrink from the smug assertion that a fellow who loves, understands, and appreciates horses of any sort is unlikely to go a lifetime without falling for the Thoroughbred if given a proper introduction. This addition to his horse knowledge and appreciation began for Kleberg in 1934. As he recalled for us in an interview for *The Blood-Horse* in 1973, he was doing a bit of Texas horse trading with a neighboring spread in the 1930s. Kleberg went over with an eye on a certain Quarter Horse mare,

Robert J. Kleberg Jr.

one with a bit of age — fifteen or sixteen — but "the best one that had been produced up to that time." Adhering to the prescribed do-si-do of the horse trader, Kleberg feigned interest in a few Thoroughbreds, which the neighbor also raised.

"I knew I could not have bought her if I had just trotted out to look at the one mare," he explained, to one not tutored in such exchanges. "It would have been hopeless."

The trick was on the trickster. While Kleberg was at his neighbor's ranch, the horseman's eye of the King Ranch cattleman was struck not by the Quarter Horse but by a Thoroughbred. A horse of Whitney breeding, the horse was merely living up to his heritage in taking the eye of Kleberg.

"I saw a very fine Thoroughbred horse that was a little different from any I ever had seen," he recalled nearly forty years later. "The thing that interested me in him was that he was big horse, really he was bigger than I liked, but he had the same muscling that the Quarter Horse has. He was a horse called Chicaro."

Kleberg bought the son of *Chicle—Wendy, by Peter Pan, from J.W. Dial, and sixteen years later stood in the winner's circle after the Kentucky Derby with one of the horse's grandsons. The fact that Middleground was Kleberg's second Derby winner speaks to the rapidity with which the Quarter Horse

virtuoso had fine-tuned his curiosity about the Thoroughbred.

In 1973 Kleberg succinctly summarized a collation of knowledge developed through years of study and application: "To produce any particular quality, and, finally, to produce a whole bunch of characteristics that go together in any of these animals, the first thing you have to have is find animals who are dominant in the particular thing that you want to produce. You cannot produce anything from animals that are more or less recessive. You should start with very top individuals of any breed of animals if you want to get anywhere. Once you have, you do not need so many animals if they live long enough.

"Say you get a good male — I don't care if it is a bull or a Quarter Horse or a Thoroughbred — you get that top animal and you get a fair number of descendants. Then you can work through the sons and daughters of those animals for generations and you keep trying to make little changes that you think would be advantageous. You get a big pool with that kind of genes, and you get the thing in your own hands to work with. It is surprising how consistent they are after you get those characteristics concentrated in a certain number of males."

The man who had dealt with creating a consistency in a herd animal that was intended to produce a body type for the sake of body type (carcass weight, muscling quality, that sort of thing), was applying his knowledge to a breed in which physical type was only a clue to a different goal, i.e., racing speed, soundness, and desire.

In starting with a colt bred by Harry Payne Whitney, Kleberg was in a manner of speaking starting with quality. On the other hand, Chicaro had not been a distinguished racehorse, nor was he a good sire in the usual measure of racing quality of offspring. Chicaro sired 102 registered Thoroughbreds, of which only two won stakes. The second of these was bred by Kleberg, i.e., a 1937 colt named Maecaro (Chicaro—Cherry Rose, by General Roberts) who won the Miles Standish Stakes at two in 1939.

When he spotted Chicaro, Kleberg did more than buy a fascinating individual: "I went to Kentucky to see how a horse like that was produced. I hadn't seen one before. And I found out that he was from one of the best families in Kentucky. In looking around, I ran into a very good band of mares, belonging to Morton Schwartz."

Schwartz, who had a successful operation, had decided to sell virtually all his stock and retire from breeding, while retaining a few colts. In 1935 he made twenty horses available for auction at Saratoga, and the newcomer Kleberg sidestepped through the catalog with the efficiency of a champion cutting Quarter Horse separating a cow from a herd. He purchased the Man o' War mare Sunset Gun and her foal for $8,600, and the mare became the second dam of champions Stymie and High Gun; a Clock Tower—Gun Play, by Man o' War filly, for $4,100, and she became his first champion, Dawn Play; the mare Science (Star Master—Triangle, by *Omar Khayyam) and her foal, and she became the dam of Santa Anita Derby winner Ciencia.

"And, so, I was in the racehorse business," he recalled. Was he ever. (The King Ranch silks were brown and white and incorporated the King Ranch brand of the Running W.)

The evolution of the Thoroughbred generates importance in different bloodlines from time to time, and the importation of bloodlines from Europe to America was a key component to the history of the breed during the four decades Kleberg was active. Nonetheless, he never merely reached out for something different, willy nilly, to the exclusion of a belief in inbreeding.

"Since St. Simon, there has been a tendency in most countries, including this one, continually to want to outcross, bringing in fresh blood," he said in that 1973 interview. "But, as far as I can see, it is just accidental when it works — it could not be anything else. In more recent years, though, there has been a lot of close breeding, and I think that is what has improved the breed."

At the time of his entry into the game, Kleberg recalled, "I would say — and a lot of people will differ with it — the Commando and Domino blood appeared to be the best family in the world. It certainly was the best in this country, and there were not many of them abroad so there was no way to tell over there. But I got interested in it, and then I went back and bought Bold Venture (*St. Germans—Possible, by Ultimus)."

In selling Bold Venture to King Ranch, owner Schwartz was more or less completing his contribution to the establishment of Kleberg's new enterprise. Bold Venture had been one of the horses Schwartz had retained when he sold most of his horses in 1935. The next year, Bold Venture won the Kentucky Derby and Preakness in Schwartz' colors before bowing a tendon. Kleberg bought him for a reported $40,000 in 1939. (That was the same year Kleberg was elected to The Jockey Club.)

Kleberg went about putting his theories into motion, including the concept of "making little changes that you think would be advantageous." In the case of Bold Venture, this involved enhancing soundness. Kleberg recalled later that Bold Venture for some years was not as successful as had been hoped. "From my breeding experience, I had the idea — and it proved correct — that the principal reason for his being a failure was that he tended to get soft-boned horses," Kleberg recalled. "The unsoundness came through Ultimus, which was an inbred Domino horse." (Ultimus, broodmare sire of Bold Venture, was by Domino's son Commando, and Ultimus' dam, Running Stream, was by Domino.)

Rather than discard Bold Venture and look for another stallion prospect, Kleberg employed a compensating strategy: "I commenced to look around for a line of Domino blood that was hard and sound and in which that characteristic was dominant. And, as a matter of fact, in that breed — if you pick for it — soundness is more dominant than the softness. So, I went to Equipoise. That's how I bred Assault."

Assault, a Kleberg masterpiece as a Triple Crown winner, was foaled in 1943. By that time Bold Venture had been in the stud a half-dozen years, was generally disappointing, as Kleberg said, and had been moved from Kentucky to Texas. Assault's dam, Igual, was by the vaunted Whitney-bred champion Equipoise and was out of a mare by *Chicle (sire of Chicaro). Equipoise was known for gameness that overcame unsoundness, but Kleberg was not deterred by the latter. Equipoise brought his element of Domino blood to the pedigree through his sire, Pennant, who was by Peter Pan, son of the Domino stallion Commando. Thus, Assault had three crosses of Domino.

Assault's dam, Igual, was out of Incandescent, who in turn was out of Masda, a speedy but underachieving full sister to the great Man o' War (Fair Play—Mahubah, by *Rock Sand). Igual was so sickly as a foal that she was about to be destroyed, but King Ranch veterinarian J.K. Northway discovered an abscess under a stifle, and she improved after treatment. She never raced but was retained for breeding.

Igual's 1943 Bold Venture foal carried on the bad luck as a youngster, stepping on something sharp enough to puncture the frog of his right forefoot. The wound healed, but Assault picked up a habit of favoring it in the slower paces, a tendency that famously stayed with him during his racing days.

Max Hirsch, the veteran who had trained Bold Venture, was hired by Kleberg, and in 1946 he brought the promising but previously unconvincing Assault along to win the Kentucky Derby by eight lengths. The Preakness found the King Ranch colt opening a lead prematurely and then barely holding off Lord Boswell to win by a neck. As Assault attempted to become racing's seventh Triple Crown winner, he was not even given the courtesy of favoritism for the Belmont Stakes. Nevertheless, the steel quality was there, and Assault and jockey Warren Merhtens won comfortably by three lengths.

Having won racing's most glorious series of three races, Assault proceeded to win the Dwyer on June 15 and then lose his next six races. Had something halted his career at that point, he would have lived under the contradictory damnation of being "the worst Triple Crown winner in history." He bounced back, however, to win the Pimlico Special and Westchester Handicap and was voted Horse of the Year. By the time this winning streak reached seven the next year, he had diverted any criticism with such feats as carrying 133 pounds and beating Stymie in the Brooklyn Handicap and carrying 135 and defeating Stymie and Gallorette in the Butler Handicap. Assault also won the Suburban Handicap under 130 pounds and swapped short-lived reigns with Stymie as the all-time leading earner. At the end of the year, however, he was beaten by Armed in a match race for which the King Ranch champion was widely recognized as not at his best. Armed was logically voted older male champion and Horse of the Year, but Assault had ratified his ranking as a great horse.

Assault was injured early in his five-year-old season and retired, but he proved sterile and was returned to training. At six he won a second Brooklyn Handicap, and he added one more victory at seven before his final retirement with eighteen wins in forty-two starts and earnings of $675,470.

The fact that Assault would be scrimmaging with Stymie was a nettlesome irony to Kleberg and Hirsch. Technically, Stymie was bred by Hirsch, but only because of an inadvertent lateness in transfer of papers. The horse was essentially King Ranch-bred, and he embodied the Kleberg approach although he wound up as a rival. When Kleberg was honored as Testimonial Dinner Guest of the Thoroughbred Club of America in 1946, he summarized his strategy in a manner similar to his recollection many years later: "We have accumulated what we believe to be a good band of Thoroughbred broodmares, and are hopeful that by using the blood of Domino, principally through his son Commando, we have been able to strengthen and perpetuate the Commando line, which we believe to be one of the greatest bloodlines in the world. We had this in mind when we bred Stymie and again in the case of Assault. Assault had two crosses of Commando and one of Domino, sire of Commando; he greatly resembles Equipoise (broodmare sire). Stymie has three crosses of Commando and one of Domino (perhaps genetic hairsplitting since Domino was the sire of Commando). He also has two crosses of Broomstick and two of Man o' War. He is, therefore, line bred not only to Commando, but to our two other great American families."

In *American Race Horses of 1945*, Joe H. Palmer had addressed this term as applied to cattle: "The term (line breeding) has somewhat varying definitions, but in a general way it means that the breeding stock is established by taking bulls from lines which show the most desirable characteristics and breeding them to the best cows available from these same lines. This amounts to an attempt to concentrate and intensify these desirable characteristics by bringing them in again and again for generation after generation. The eventual result is a pedigree which shows an increasing tendency to become the same on the male and female side. This is a variety of inbreeding, of course, differing from ordinary inbreeding in that the intensification of one or more lines is more desired than the close-up repetition of the name of some prepotent individual."

The concept was difficult for Palmer to accept. He added, "Robert J. Kleberg, more interested than others of his family in the development of Thoroughbred horses, couldn't see why this shouldn't work with horses as well as cattle, and even after it had been explained to him he still couldn't see it." Palmer had been tutored by another drummer in pedigree philosophy, but did the journalistic fair-comment equivalent of throwing up one's hands, to ask rhetorically, "If Stymie was not the produce of a pedigree manipulation, where did he come from?"

Stymie's pedigree was like an exhibition of the lesser paintings by a great artist. His sire, Equestrian, a foal of 1936, was a lightly raced Whitney-bred who won two races and was acquired by King Ranch. He was easily seen as an underachiever, for in addition to being by the great runner Equipoise, he was from the Man o' War mare Frilette, a Beldame Handicap winner who also foaled major winners Jabot and Cravat. Stop Watch, dam of Stymie, was by On Watch, whose very existence as a son of the fertility challenged champion Colin constituted a case of beating the odds. Doubling up on the caprices of fate, Stop Watch was actually inbred to Colin, who was also the sire of her third dam.

The second dam of Stymie, coincidentally, was the early Kleberg purchase Sunset Gun, although she had foaled Stop Watch prior to the purchase. Sunset Gun was by Man o' War, meaning that Stymie was inbred to that great horse 3x3, a factor that leads the romantic in one to think it accounted for his high-headed running style.

Stymie raced briefly for King Ranch before being claimed by Hirsch Jacobs, who guided his career to championship status and retired him as the world-leading money earner with $918,485. That Stymie did not show officially on the breeding record of Kleberg was cause for regret; that he showed officially on the breeding record of Hirsch, but not his training record, gave the trainer ample reason simply not to want to talk about the damned horse.

Turf writer Palmer was still put off by Kleberg's approach when he had occasion a few years later to address the career of Middleground. As will be recalled, Middleground was from a daughter of Kleberg's original Thoroughbred purchase, Chicaro. Middleground was also by Assault's sire Bold Venture, who thus took on the distinction of being the only Kentucky Derby winner to sire two Kentucky Derby winners. More than a half-century later, this unique ground still belongs only to Bold Venture.

Middleground had seven crosses to Domino, when the pedigree is extended to eight crosses. "The mating that produced Middleground was effected through what one might call a footnote to the principal theory [described above]," commented Palmer in *American Race Horses of 1949*. To produce Middleground, *St. Germans' son Bold Venture was bred to the moderate winning King Ranch mare Verguenza, a daughter of Chicaro. Wendy, the dam of Chicaro, was out of Remembrance; Remembrance was the second dam of Twenty Grand, the spectacular Derby-Belmont winner of 1931. What was off-putting to Palmer was Kleberg's lack of distinction between the bottom females of a pedigree and the interlocking other individuals within the chart: "It will be noted that Mr. Kleberg paid no attention to the tail-female line which is ordinarily accepted as determining families. He merely got a mare that carried, through her sire in this case, the blood of Remembrance. Here again one may have some doubts about the theory, but there is no arguing with the result, for the result is Middleground."

The nomenclature of the Turf has come down in a way that the word "family" generally refers not to stallions' pedigrees, but specifically to the "bottom line," i.e., the dam, and her dam, and the next dam, etc., and the produce of same, or, in the common jargon employed by Palmer, "tail-female." As quoted above, Kleberg used the word "family" in the more inclusive sense that one might apply to humans, i.e., commenting that "the Commando and Domino blood appeared to be the best family in the world" —

whereas most would refer to the relationship not as a Thoroughbred "family" but as a "sire line" or the "tail-male" line. Kleberg apparently did not speak of it that way and did not think of a pedigree that way.

The Texas-bred Middleground (Bold Venture—Verguenza, by Chicaro) won three races at Saratoga in 1949, climaxed by a powerful triumph in the Hopeful Stakes. A suspicious ankle prompted Hirsch to stand Middleground in ice for two hours before the Hopeful. Prior to the race, Hirsch announced that it would be the colt's final start at two. Hill Prince and Oil Capitol shared year-end championship honors, but Middleground topped the Experimental Free Handicap.

The following year, Middleground and Hill Prince banged heads in the classics. Middleground won the Kentucky Derby in 2:01 3/5, only one-fifth of a second over the Derby record held at the time by Whirlaway. He thus became the second Derby winner for King Ranch. Bill Boland rode, following Ira Hanford, Bold Venture's jockey, as the second

Middleground

apprentice to win the Derby. Boland was a Texan, as was Hirsch, so it was a Lone Star State conclave with Texan Kleberg in the winner's circle.

Hill Prince then defeated Middleground in both the Withers Stakes and Preakness Stakes in the busy campaigning not atypical of the era, but the King Ranch colt defeated the other in the Belmont, again becoming the second winner for Kleberg of a highly coveted prize. Middleground won six of his fifteen races and earned $237,725.

Middleground was a useful sire for Kleberg, although he did not have strong fertility numbers. He got only seven stakes winners (5 percent), all bred by King Ranch. Three of them were Here and There, Resaca, and Ground Control, winners of major filly races for King Ranch, and another was Disperse, who was third in the Belmont Stakes. Resaca won the Coaching Club American Oaks; Here and There, the Alabama; and Ground Control, the Acorn.

In 1946 the death of Colonel E.R. Bradley had created a unique opportunity in bloodstock investments. Bradley's brother was left in charge of certain decisions and did not want to continue the breeding/racing operation famed for producing four Kentucky Derby winners. Kleberg learned of the opportunity to purchase the stock en masse, or at least in large blocks, rather than wait for the prized mares and their offspring to appear one by one at auction. At Saratoga, Kleberg approached Ogden Phipps with the idea of buying a draft of Bradley horses as partners, and Phipps suggested that his host of the moment, John Hay Whitney of Greentree Stud, be invited to join. So, three major breeders bought and divvied up the core of the Bradley bloodstock, whose chief distinctions included the great mare *La Troienne and an abundance of her issue.

Kleberg also dealt for a section of Bradley's Idle Hour Farm and thus established a Kentucky division of King Ranch. Kentucky's location relative to the major stallions of the industry prompted a pattern of only about 40 percent of the King Ranch Thoroughbreds being bred at the original ranch. Nevertheless, Kleberg continued to favor Texas as an environment: "I really feel that with a colt raised in Texas all the way through, that you have a better chance of raising a class horse than you do in Kentucky. For some reason or the other, I think horses raised in those big pastures and in that hot and dry climate are stronger. I don't know if they are sounder, but they are certainly stronger."

On the bloodstock front the King Ranch haul among the Bradley horses included a pair of three-year-old filly champions, one current the other future. Both were descendants of *La Troienne. Bridal Flower (*Challenger II—Big Hurry, by Black Toney) won the Beldame, Newcastle, Roamer, and Gazelle in 1946 and was the champion under the colors of Bradley. But Why Not (Blue Larkspur—Be Like Mom, by *Sickle) raced for King Ranch. She defeated males in the Arlington Classic and older females in the Beldame and Arlington Matron, as well as winning the Alabama, Pimlico Oaks, and Acorn as champion three-year-old (and, in the manner of the day, "handicap female") in 1947. Bridal Flower was out of *La Troienne's daughter Big Hurry; But Why Not was from Be Like Mom, she in turn from the great mare's champion daughter Black Helen. But Why Not was the second three-year-old filly champion to race for King Ranch after a Kleberg purchase. The first one, as noted above, was Dawn Play, purchased by King Ranch from the Schwartz sale in 1935 for $4,100. A daughter of Clock Tower—Gun Play, by Man o' War, Dawn Play gave Kleberg an early Coaching Club American Oaks victory and also won the Acorn Stakes and defeated males in the American Derby.

Included in the King Ranch portion of the Bradley stock was the stallion Blue Larkspur, although the horse was moved to Greentree Stud. The champion was twenty at the time and died in 1947, but King Ranch got the Experimental Free Handicap No. 2 winner Sonic, who also was disqualified after winning the Blue Grass Stakes. Sonic was from King Ranch's CCA Oaks runner-up Split Second, by Sortie.

Also in the Bradley draft acquired by King Ranch was Be Like Mom, dam of champion But Why Not. Be Like Mom was by *Sickle and from *La Troienne's champion daughter Black Helen, by Black Toney. For Kleberg, Be Like Mom in 1947 foaled the Blue Larkspur filly Renew, who won the Firenze and Top Flight handicaps. Renew, in turn, foaled the dam of the high-class King Ranch homebred Buffle, and the success of this family for other breeders included the champion Princess Rooney.

Yet another of the Bradley-cum-King Ranch mares was Bee Mac, who was by War Admiral and out of the Black Servant mare Baba Kenny. Though not King Ranch breeding, it went back to Commando

CARL SCHULTZ

Dawn Play

and became a major resource. Bee Mac had defeated males in the 1943 Hopeful Stakes and also won the Spinaway. For King Ranch she produced the stakes winners Mac Bea, Riverina, and Boyar. Her yearling at the time, Better Self, also went to King Ranch and won ten stakes from two through five, including the Saratoga, Carter, and Gallant Fox handicaps and the East View and Saratoga Special. Better Self won sixteen of fifty starts and earned $383,925. He was a useful sire, getting fourteen stakes winners (8 percent). The three of these for King Ranch included Selima Stakes and Vineland Handicap winner Tamarona, but Better Self's most important daughter was the Phipps family's producer Lady Be Good.

Bee Mac foaled a half brother to Better Self named Beau Max, who was by Bull Lea. While Beau Max did not win any stakes, he slightly edged Better Self numerically as a sire, getting fifteen stakes winners (5 percent). Nine of these were bred by King Ranch, ranking Beau Max second to Bold Venture with ten as the leading sire of Kleberg-bred stakes winners. Beau Max' horses for King Ranch included the $339,587-earner How Now and the $276,355-earner Golden Notes.

(As noted above, Middleground managed to sire seven King Ranch-bred stakes winners, and next with six was a horse named Depth Charge. Depth Charge was bred by Mrs. John D. Hertz and was out of Quickly, by Haste, and thus a half brother to her year-old future Triple Crown winner Count Fleet. Since Depth Charge was a son of Bold Venture, he was understandably appealing for Kleberg to acquire as a King Ranch stallion.)

As the filly counterpart to the Belmont, especially in the minds of New York-based stables, the CCA Oaks was a favored target. Kleberg was owner or breeder, or both, of five CCA Oaks winners, and Split Second was runner-up in the event. Originally, the Singleton Cup was the designated trophy for a CCA Oaks winner, but it was stipulated that the first owner to win three runnings would retire this trophy. Walter M. Jeffords Sr. got there first. Afterward the Coaching Club Cup was the trophy for the event.

Six years after Dawn Play first won the race for him, Kleberg won with one of his early homebred stakes winners, Too Timely. This was a filly by Alfred Vanderbilt's great handicap horse Discovery—On Hand, by Stymie's broodmare sire On Watch. A half sister to the top racehorse Market Wise, Too Timely defeated the previous year's juvenile filly champion Askmenow in the 1943 CCA Oaks.

A year younger than Dawn Play was another out-standing filly, Ciencia, who did not win the Oaks but would later be the second dam of a CCA Oaks winner. Technically, Ciencia was the first stakes winner bred by King Ranch, for her dam, Science, was carrying her when she was purchased from the Schwartz stock in 1935. Ciencia (*Cohort—Science, by Star Master) defeated colts in the Santa Anita Derby early in 1936. She later foaled Curandero, winner of the $100,000 Washington Park Handicap and two other stakes in the early 1950s. Ciencia would be the second dam of the King Ranch-bred CCA Oaks winner Miss Cavandish.

Five years after Too Timely's CCA Oaks, King Ranch won another edition, with Scattered. A daughter of the Triple Crown winner Whirlaway, Scattered was the first stakes winner from Imperatrice, who would acquire lasting fame as dam of six stakes winners and, more importantly, as second dam of Secretariat and Sir Gaylord. In Imperatrice's less-exalted days, Scattered was purchased by Kleberg for $23,000 at the 1946 Saratoga yearling sale. Scattered foaled Disperse, the Hempstead Handicap winner who placed in the Belmont for Kleberg, and she also foaled Here and There, who died in a stable fire hours after winning the historic Alabama at Saratoga in 1957.

The next CCA Oaks winner for King Ranch was Resaca, a mare so memorable that near the end of his life Kleberg ranked her second, slightly below Dawn Play, among the best fillies he had raced. (This is no small tribute, for he had bred and raced the stellar champion Gallant Bloom.) Resaca had a typical King Ranch pedigree in that she was the offspring of homebreds, not all of them particularly fashionable — but pleasing enough for Kleberg to retain them as breeding stock. She was by the Derby-Belmont winner Middleground and out of Retama, whose dam, Otra, was by the well-liked Equipoise. Retama was by the King Ranch stallion Brazado, who was another by the Colin stallion On Watch.

Brazado was a full brother to the

Dotted Line

BERT MORGAN

dam of Too Timely, being from Kippy, a Broomstick mare Kleberg had acquired. Another full sibling was Sortie, winner of the Brooklyn Handicap and sire of CCA Oaks runner-up Split Second. Brazado combined the revered cross of Domino and Broomstick. Brazado did not get to the stakes winner category but won four of five and was used at stud. Brazado's best son, the major winner Curandero, brought in another pedigree element as he was inbred to the early twentieth century leading sire *Star Shoot. Brazado sired Curandero and four other stakes winners (5 percent).

Resaca got good for Hirsch in the summer of her three-year-old year. She defeated the defending champion of the crop, Quill, in the CCA Oaks and then beat her again as well as defeating one of the co-champions of the same strong crop, Silver Spoon, in the Delaware Oaks. Having won the Santa Anita Derby with the filly Ciencia and the American Derby with Dawn Play some two decades earlier, Kleberg was not intimidated by running a female against males. Resaca was next set on the tough task of facing the budding champion Sword Dancer in the Travers Stakes at a mile and a quarter. Showing signs of unsoundness for the first time, she finished last.

In addition to being the broodmare sire of Resaca, Brazado sired the dam of another of the more distinguished among King Ranch's distaffers. Dotted Line was by *Princequillo and out of the Brazado mare Inscribe. Dotted Line scored her first major win in the Delaware Oaks of 1956. Three years later she defeated the high-class *Amerigo and other males on turf in a division of the first running of the Man o' War Stakes, at a mile and a half. In between she won three other stakes and crafted a career total of eleven wins in sixty-seven starts and earnings of $324,159.

The year Resaca was near the top of the three-year-old fillies, King Ranch seemingly had a classic and championship contender among the colts as well, in the form of Black Hills. A son of the Claiborne Farm stallion *Princequillo, Black Hills came from an Argentine family. Argentina was significant to Kleberg, for Santa Gertrudis flourished there. In the Thoroughbred sphere, Congreve was Argentina's epochal stallion, and Black Hills' dam, *Blackie II, was an Argentine classic winner by that singular sire. Black Hills developed slowly under Max Hirsch's deft hand, but when the colt won the Peter Pan

Handicap, he earned his way into the Belmont Stakes as a prime contender. Black Hills seemed to be in contention on the far turn, but then he suddenly fell. He struggled to his feet with a badly broken leg, although whether the injury had been caused by the fall or was the cause of it could not be determined. Lake Erie, who fell over him, got up and followed the field. Black Hills had to be euthanized. The promise of youth had been cruelly abrogated.

Kleberg set great store by the Belmont Stakes but was not beguiled by the distance specialist for American racing. "My preference is to have the horse that can run the honest one and a quarter miles," he said, "and then hope he can do anything else. I think Assault and Middleground would run one and a quarter miles or one and a half miles — or two miles, I guess."

Black Hills had seemed a prime candidate to follow Assault and Middleground as a third homebred winner of the Belmont. In the interim Kleberg had won it with a purchased colt named High Gun. That colt had a close connection to King Ranch breeding, although he was not King Ranch-bred. It will be recalled that the Man o' War mare Sunset Gun had been acquired by Kleberg during his raid on Morton Schwartz' sale in 1935. Coincidentally, the same mare was the second dam of the champion Stymie. If Stymie represented a star of this family lost to King Ranch, High Gun was a champion retrieved. Sunset Gun's Brazado filly Rocket Gun was unsuccessful as a race filly, and even though she had the Kleberg imprint as a descendant of both Colin and Man o' War, Kleberg dealt her off in 1947. Five years later, trainer Hirsch — breeder of record of Stymie — was impressed by Rocket Gun's Keeneland sale yearling, who had been bred by brothers Kellar M. and W. Paul Little in partnership with Cary Boshamer. Hirsch bought the colt for $10,200 on behalf of King Ranch. The loss of Stymie was about to be put right.

Although Kleberg concentrated on inbreeding to old established sources, he was not resistant to emerging horses or indifferent to their potential for his breeding program. In England, Lord Derby's Hyperion had been an acclaimed and versatile classic winner in the 1930s and was making a major impact in the stud. High Gun was by one of the important Hyperion stallions imported to this country, i.e., *Heliopolis. Being from the daughter of a Man o' War mare, High Gun represented a pattern that Kleberg came to feel was pivotal.

"I won't say Hyperion ever was a failure, because he wasn't," he recalled, "but for a time very few of his horses had run one and a quarter miles in England. They just did not want to. When they brought Hyperion blood to this country — *Heliopolis was the first one that I remember — that blood was mixed with the Man o' War blood, and it was just magic. They commenced to run, and that was what really made Hyperion. Strange thing."

The career of High Gun made the loss of Black Hills five years later all the more haunting. They had seemed to be progressing in a similar pattern. High Gun had placed in some major stakes but had won none until the Peter Pan Handicap on June 5, just a week before the Belmont Stakes. In the classic race he drove relentlessly to run down C.V. Whitney's classy little Fisherman to win by a neck. He followed Assault and Middleground as the third winner of the great target for Kleberg.

High Gun defeated Fisherman more authoritatively that autumn in the two-mile Jockey Club Gold Cup, and in the interim had won the Dwyer, Sysonby and Manhattan. He was the champion three-year-old colt of 1954, and he helped put King Ranch atop the national owner standings for the only time, with $837,615. High Gun earned $314,550 that year.

Another major King Ranch contributor was Rejected, with $276,800. Rejected was by the outside stallion Revoked. The latter had been purchased by Max Hirsch as a yearling in 1944 but was returned to consignor Eslie Asbury because Hirsch concluded he was a roarer. Revoked proved a major stakes winner that got away, but King Ranch embraced him as a sire. The archly named Revoked colt Rejected was from By Line, a *Blenheim II filly purchased as a yearling at Saratoga for $3,300 in 1941. By Line was out of a half sister to Belair Stud's 1936 Horse of the Year Granville. Rejected, inbred 3x3 to *Sir Gallahad III, won the Santa Anita Handicap in 1954 among a career total of seven important stakes. He won eleven of forty-seven races and earned a career total of $549,500. He was a moderate stallion, with seven stakes winners (3 percent).

The year after High Gun and Rejected boosted King Ranch to the top in owner earnings, High Gun flirted with recognition, not just as a champion but as a great horse. He won the 1955 Metropolitan Handicap under 130 pounds and the Brooklyn under 132. He was narrowly beaten in the Suburban and Monmouth handicaps, under 133 and 135, giving weight both times to winner Helioscope.

In the fall of the year, High Gun ran down the younger Horse of the Year Nashua to win the Sysonby again. Early in that race, he fell so far back that Kleberg was moved to excuse the guests in his box from even watching his horse since there was an interesting contest up front, but jockey Boland brought High Gun up to win by a head from Jet Action, with Nashua third. High Gun was voted the champion older horse of 1955 and was retired with eleven wins in twenty-four starts and earnings of $486,025. He proved virtually sterile, perhaps an even bigger blow than the sterility of Assault given the exceptional fashionability of High Gun's bloodlines; his sire, *Heliopolis, was the leading sire in 1954 for the second time.

Another example of Kleberg's enthusiasm for bringing in European blood by the 1950s was La Fuerza, whose sire, Never Say Die, was an Epsom Derby and St. Leger winner by *Nasrullah. Never Say Die was foaled in America but stood in England. La Fuerza's dam, *Solar System II, was an import by Hyperion. La Fuerza won the 1959 Selima Stakes, then one of the most important autumn races for juvenile fillies.

Kleberg's respect for Hyperion had led the King ranch owner to go "to see him in England when he was very old, and, of course, he had lost a lot of his appearance, but a lot of it remained. He was almost a typical Arab horse. Stymie was the same, a beautiful thing — high head carriage. I persuaded Lord Derby (owner of Hyperion) to let me send a couple of mares to Hyperion even at that age, one the first year and another the next. And I sent Fair Play/Man o' War-bred mares. Unfortunately, the first, Too Timely (by Discovery, by Display, by Fair Play) — she won the Coaching Club American Oaks — had a colt that was killed by lightning. It was a very nice colt everybody said, but I never saw it.

"Then I sent a Beau Max mare to Hyperion (Beau Max was from a daughter of Man o' War's son War Admiral). The mare was Timed, which was a half sister to Stymie, and she produced Zenith. He raced well, trained well, right up to the Derby, and then he bowed (1960). I believe he could have won it...He never had a pimple on him before he bowed. He was just a running machine. Then, of all the luck you can have, when I put him to stud he immediately got an infection. I took him to Texas and it was I don't know how many years before I got him really cleared up. …And, of course, he did produce that one top horse, Buffle. We had bad luck with him too, you know. He got a brain tumor and died."

Buffle, son of Zenith, was from the Bold Venture mare Refurbish, whose dam, Renew, was from the Bradley mare Be Like Mom. Buffle came along in the spring of 1966 to finish second to Amberoid in the Belmont. Then, Buffle became the first three-year-old to win the time-honored Suburban Handicap since Crusader in 1926. Carrying 110 pounds, Buffle defeated a field that included future champion Bold Bidder (123) and existing champion Bold Lad (135).

Buffle later defeated Belmont winner Amberoid in the New Hampshire Sweepstakes, which had a purse of $250,000, far above the $100,000-$150,000 of most of the other top races of the day. He could not handle Buckpasser in their meetings, but then neither could anyone else. Soon afterward, Buffle collapsed with partial paralysis, lost the ability to swallow, and was placed under care of veterinarians and brain surgeons. He died within a few days, and the autopsy revealed brain lesions, the origins of which could not be explained. He had won five of twenty-two races and earned $363,091.

Also during the middle 1960s, Kleberg bred another CCA Oaks winner. Miss Cavandish was sired by *Cavan, the Irish son of Mossborough (by Nearco) who burst onto the scene by defeating the injured

Tim Tam to end the Calumet Farm colt's Triple Crown bid in the Belmont in 1958. Miss Cavandish was out of New Weapon, who was by Bold Venture and thus in 1964 linked the stable's history to the 1936 Derby winner who had sired the 1946 and 1951 Derby winners. New Weapon was a non-winner, a 1942 foal from early Santa Anita Derby winner Ciencia and was nineteen when she foaled Miss Cavandish. Kleberg was a believer in the horse as a work of nature, but even he had his limits. Miss Cavandish toed in so badly that she was consigned to the Keeneland September sale. The rank and file of buyers and agents ran from the sight, but Harry Nichols bought her for $1,500. By the end of her two-year-old season, Miss Cavandish was a bargain who confounded the experts and had been narrowly beaten in the rich Gardenia Stakes. Under Roger Laurin's training, she came on at three in 1964 to defeat champion Castle Forbes in the CCA Oaks and also add the Delaware Oaks, Monmouth Oaks, and Alabama. That line-up would win the three-year-old filly championship in most years, but Miss Cavandish was outvoted by the brilliance of Tosmah. (Tosmah was by Tim Tam, so that old 1958 Belmont meeting had its ongoing overtones.)

Just as Kleberg had stepped up to breed to the European-bred Belmont winner *Cavan, he also patronized the Irish-bred 1957 Belmont winner, *Gallant Man. *Gallant Man defeated Bold Ruler in the final leg of the Triple Crown and also had the versatility to beat him later at a mile in the Metropolitan

Middleground, 1947	Bold Venture, 1933	*St. Germans, 1921	Swynford (John o'Gaunt—Canterbury Pilgrim)
			Hamoaze (Torpoint—Maid of the Mist)
		Possible, 1920	Ultimus (**Commando**—Running Stream)
			Lida Flush (*Royal Flush III—Lida H.)
	Verguenza, 1940	Chicaro, 1923	*Chicle (Spearmint—Lady Hamburg)
			Wendy (**Peter Pan**—Remembrance)
		Blushing Sister, 1932	Bubbling Over (*North Star III—Beaming Beauty)
			Lace (Bunting—Stickling)
RESACA	Brazado, 1936	On Watch, 1917	Colin (**Commando**—*Pastorella)
			Rubia Granda (*Greenan—The Great Ruby)
		Kippy, 1920	**Broomstick** (Ben Brush—*Elf)
Retama, 1946			Seamstress (*Star Shoot—Busy Maid)
	Otra, 1936	Equipoise, 1928	Pennant (**Peter Pan**—*Royal Rose)
			Swinging (**Broomstick**—*Balancoire II)
		Tenez, 1928	Friar Rock (*Rock Sand—*Fairy Gold)
			Glenrose (*Polymelian—*Kiss Again)

43

Handicap, shorten to sprints successfully, and stretch out to win the Sunset Handicap at a mile and five-eighths. In 1965 King Ranch sent Multiflora to the Spendthrift Farm stallion *Gallant Man, and the result was an athletic filly named Gallant Bloom. She represented the other side from Miss Cavandish in terms of fillies that won what generally would be a championship menu but were denied the title; in 1969, Gallant Bloom spotted Shuvee the New York Filly Triple Crown and the Alabama and still was so dominant late in the year as to wrest championship honors away.

Gallant Bloom's dam, Multiflora, was by the above-mentioned Beau Max (son of Bee Mac) and was out of Flower Bed. Flower Bed was by *Beau Pere and from the foundation matron *Boudoir II, by Epsom Derby winner *Mahmoud. Kleberg purchased Flower Bed privately from Isabel Dodge Sloane's Brookmeade Stable. Mrs. Sloane already had bred the major winner Flower Bowl and the additional stakes winner Floral Park from Flower Bed. For King Ranch, Flower Bed foaled the useful stakes winner Brambles, but it was her non-winning Beau Max filly Multiflora that paid the larger dividend.

Gallant Bloom was one of the last good horses trained by Max Hirsch, who passed away when she was three. Hirsch's son, Buddy Hirsch, who for some time had trained divisions for King Ranch, succeeded his father. At two in 1968, Gallant Bloom won the

Gallant Bloom

Miss Cavandish

Matron Stakes by nine lengths and the Gardenia by a length and a quarter. The *Daily Racing Form* voted her the juvenile filly division title, although Process Shot got the nod in the Thoroughbred Racing Associations ballot. At three Gallant Bloom won all her eight races (although the Delaware Oaks win got a boost via disqualification of Pit Bunny). She did not take on top-level competition until the Monmouth Oaks, in which she defeated Kentucky Oaks winner Hail to Patsy by a dozen lengths. After the Delaware Oaks (in which Shuvee was fourth), Gallant Bloom defeated Pit Bunny and Shuvee in the Gazelle, dashed the older champion Gamely by seven in the Matchmaker, and was such a favorite that her handy Spinster Stakes triumph came as a betless exhibition. The King Ranch filly was champion on both polls of the time.

The following winter at Santa Anita, Gallant Bloom won the Santa Maria and the Santa Margarita (129 pounds) before suffering two setbacks. From late in her two-year-old season through her first two races at four, Gallant Bloom won twelve consecutive races. She had a career mark of sixteen wins in twenty-two starts and earnings of $535,739. Despite our respect for one Robert J. Kleberg, we have trouble assimilating his thought that Dawn Play and Resaca were the better of Gallant Bloom!

Just one year before Gallant Bloom, the King Ranch foal crop had included Heartland, a filly by

ROBERT J. KLEBERG JR.

Bold Ruler. By then the Phipps family's Bold Ruler was highly visible in his early climb to distinction as eight-time leading sire. Heartland was another 1960s horse out of a Bold Venture mare, in this case Equal Venture. Over a decade Equal Venture foaled three stakes winners that had class and sentiment and irony connected to their careers. That Equal Venture was a full sister to Triple Crown winner Assault (Bold Venture—Igual, by Equipoise) added the poignancy.

Equal Venture's first stakes winner was Saidam, by Never Say Die. Raced in the name of Kleberg's daughter and son-in-law, Dr. and Mrs. J.D. Alexander, Saidam won the Quaker City Handicap at four in 1963 and the more important Grey Lag Handicap the following year. In 1965 Equal Venture foaled the Bold Ruler filly Heartland, who won the Test Stakes at three and added the Bed o' Roses and Distaff at four for King Ranch. She crossed the wire first in the 1968 Alabama Stakes but was disqualified for interference. Finally, in 1969, at the age of sixteen, Equal Venture foaled Prove Out, a son of Graustark who was by *Ribot, a scion of the St. Simon line and a European champion of particular interest to Kleberg. Prove Out was late in living up to his potential. He had been sold to Hobeau Farm and its masterful trainer, Allen Jerkens, before the King Ranch runner defeated the great Secretariat in the 1973 Woodward Stakes. Prove Out also won the Jockey Club Gold Cup and Grey Lag Handicap, and he surfaced internationally some years later as the broodmare sire of the international champion filly Miesque.

In 1974, the year of his death, the last foal crop officially bred by Robert J. Kleberg included the Hydrangea and Las Cienegas Handicap winner filly Pressing Date, by the modern stallion Never Bend. This filly was out of *Monade, who a dozen years before had represented another of Kleberg's decisions to acquire a major European. The filly *Monade (Klairon—Mormyre, by Atys) won the English Oaks and France's Prix Vermeille and fin-

ished second in the Prix de l'Arc de Triomphe in 1962. Purchased by King Ranch, she was to prove a key to the legacy in bloodstock Kleberg's family took over upon his death.

A bizarre pattern of bad luck had trailed along with the years of success of King Ranch's Thoroughbred "diversion" — as Kleberg himself called it even years later. It began early, when that first champion, Dawn Play, was struck by lightning soon after winning the American Derby; she did not produce a foal for five years. As the seasons rolled on, the misfortune saw Assault and High Gun prove sterile; Middleground prove a diminished breeder; Black Hills and Here and There were killed young; Resaca never produced a live foal; and Buffle died of a brain tumor. This woeful litany does not even take into account the more common, and less forlorn, aspect of horses sold early, i.e., Stymie and Prove Out being lost by claim or sale before proving their inherent worth.

Nevertheless, it was the repetitive success in major races from East to West that characterized Kleberg's forty years as a Thoroughbred owner and breeder. He died at seventy-eight on October 13, 1974, at St. Luke's Hospital in Houston. He had bred eighty-six stakes winners.

The personal success of Kleberg as a Thoroughbred breeder was but one of many elements of the man's career. His vision in guiding the far-flung business affairs of King Ranch earned entry into the American National Business Hall of Fame. Nor was his vigor devoted purely to personal success, for he was honored by the Philosophical Society of Texas, which had been formed in 1837 and whose motto was: "Texas has her captains; let her have her wise men." Kleberg clearly was seen as both. The Robert J. Kleberg and (wife) Helen C. Kleberg Foundation has been important to research on behalf of human health. In 2000, for example, this organization donated $4.4 million, ranking in the top thirty among foundations funding grants for medical research.

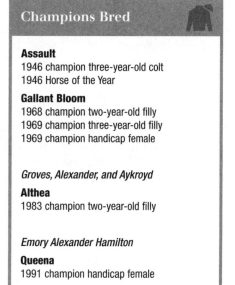

Champions Bred

Assault
1946 champion three-year-old colt
1946 Horse of the Year

Gallant Bloom
1968 champion two-year-old filly
1969 champion three-year-old filly
1969 champion handicap female

Groves, Alexander, and Aykroyd
Althea
1983 champion two-year-old filly

Emory Alexander Hamilton
Queena
1991 champion handicap female

In the more narrow fields of Thoroughbred/agriculture, Kleberg was an active trustee from the beginnings of the New York Racing Association, was chairman of the Thoroughbred Horse Association of Texas, and worked for some years toward the re-establishment of legalized racing in the state. Kleberg was a member of the Grayson Foundation, which funds equine research, and after his death, the Kleberg Foundation made a record contribution of $2 million to the foundation's current configuration, the Grayson-Jockey Club Research Foundation.

Epilogue

Kleberg's daughter Helen, seen in old Belmont Park photos as a child and then as a young lady with Dawn Play and Assault, grew into adulthood as an avid and knowledgeable horsewoman. Well into the 1990s, she had a horse-world identity ranging from breeding and owning Thoroughbred stakes winners to participating actively in championship cutting horse competition, and as Helen Groves she remains active in various branches of the equine industry early in the twenty-first century.

Moreover, Mrs. Grove's daughters have been highly successful along with her in following Kleberg's example. One of the daughters, Helen Alexander, managed King Ranch along the lines her grandfather had for some years and then segued into a market breeding posture. She was stunningly successful,

King Ranch leading Keeneland July consignors in average price four times during the 1990s and also sending top-level yearlings to the Fasig-Tipton Kentucky sales.

Helen Alexander later established her own farm, Middlebrook, in Lexington, and her clients include her sisters as well as her mother. All of the daughters — Helen, Emory, Dorothy (Didi), Henrietta, and Caroline — have been successful in racing/breeding.

Combining her years of management in the King Ranch name with various partnerships with her mother and sisters, Helen Alexander has bred forty-six stakes winners. Her purchase in partnership of the Never Bend mare Courtly Dee has produced a line of fillies and mares highlighted by the champion Althea. Courtly Dee eventually earned Broodmare of the Year honors for 1983. Grade I winners from these families include Arch and Aldiza, in addition to Althea.

Directing one of the most historic of old King Ranch families, Helen Alexander bred two stakes winners from Equal Change, an Arts and Letters filly from Fairness (by *Cavan), who was a daughter of Assault's sister Equal Venture. Equal Change, who finished second to Ruffian in the 1975 CCA Oaks, foaled stakes winners Make Change and Spur Wing.

One of the fillies bred in the King Ranch name on Helen Alexander's watch was Too Chic, by Blushing Groom and from Remedia, a daughter of the European classic winner *Monade, whom her grandfather had bought. Too Chic was a major winner in the name of one of Helen's sisters, Emory (Hamilton), and then proved a broodmare that would have been highly satisfactory to the grandfather. Too Chic's daughters include Emory Hamilton's Queena, a Mr. Prospector filly who was the champion older filly of 1991 and the dam of grade I winner Brahms. Some seventy years after Chicaro caught the eye of a King Ranch cowboy, the saga is still unfolding.

*Monade (far left) and her descendants: Too Chic, Chic Shirine, and Remedia and her foal by Blushing Groom

ANNE M. EBERHARDT

Stakes Winners Bred by Robert J. Kleberg Jr. and Family

Ciencia, br.f. 1936, *Cohort—Science, by Star Master
Maecaro, b.c. 1937, Chicaro—Cherry Rose, by General Roberts
Fuego, b.g. 1939, Bim Bam—Incandescent, by *Chicle
Too Timely, ch.f. 1940, Discovery—On Hand, by On Watch
Assault, ch.c. 1943, Bold Venture—Igual, by Equipoise
Flash Burn, b.c. 1943, Brazado—Incandescent, by *Chicle
Incline, ch.g. 1943, Bold Venture—Holua, by *Phalaros
Woven Web, ch.f. 1943, Bold Venture—Bruja, by Livery (SW in Mexico)
Mild Retort, ch.c. 1944, Bold Venture—Plain Talk, by Chicaro (SW in Mexico)
Pleasure Fund, b.c. 1944, Bold Venture—Morena, by *Chicle (SW in Mexico)
Flying Missel, ch.c. 1945, Equestrian—Momentum, by Boojum
Curandero, br.c. 1946, Brazado—Ciencia, by *Cohort
Re-Torta, ch.f. 1946, Bold Venture—Plain Talk, by Chicaro (SW in Mexico)
Middleground, ch.c. 1947, Bold Venture—Verguenza, by Chicaro
Renew, br.f. 1947, Blue Larkspur—Be Like Mom, by *Sickle
Syracuse Lad (stp.), b.c. 1947, Brazado—Perigee, by *Sir Gallahad III
Weatherdeck (stp.), ch.g. 1947, Contradiction—Jacquemein, by Johnstown
Encantadora, ch.f. 1948, Depth Charge—Bruja, by Livery
Reticule, ch.ro.f. 1948, Depth Charge—Occlusion, by Equestrian
Sonic, blk.c. 1948, Blue Larkspur—Split Second, by Sortie
Tuzado, ch.c. 1948, Brazado—Tatula, by Eight Thirty
Marcador, ch.c. 1949, Bold Venture—Maravilla, by *Sun Briar
Market Level, ch.c. 1949, Market Wise—Eye to Eye, by *Teddy
Baloma, ch.ro.f. 1950, Depth Charge—Woven Web, by Bold Venture
Beylerbey, br.g. 1950, War Admiral—Bridal Flower, by *Challenger II
Haunted, blk.f. 1950, Depth Charge—Bruja, by Livery
Mac Bea, b.f. 1950, Bimelech—Bee Mac, by War Admiral
Postillion, ch.c. 1950, Bold Venture—Igual, by Equipoise
Rejected, br.c. 1950, Revoked—By Line, by *Blenheim II
On Your Own, ch.f. 1951, Bold Venture—Igual, by Equipoise

Althea

Riverina, b.f. 1951, *Princequillo—Bee Mac, by War Admiral
Damocles, ch.g. 1952, Destino—Tense Moment, by Equestrian
Free Stride, b.g. 1952, Depth Charge—Inscoelda, by Insco
Dotted Line, ch.f. 1953, *Princequillo—Inscribe, by Brazado
How Now, dk.br.g. 1953, Beau Max—But Why Not, by Blue Larkspur
Foot Race, b.g. 1954, On the Mark—Daphne, by *Heliopolis
Golden Notes, b.g. 1954, Beau Max—Melodic, by Blue Larkspur
Hark, b.c. 1954, Beau Max—Alondria, by War Admiral
Here and There, ch.f. 1954, Middleground—Scattered, by Whirlaway
Anaqua, br.f. 1955, Beau Max—Lantana Queen, by Depth Charge
Chistosa, ch.f. 1955, Middleground—Friponne, by Shut Out
Macbern, b.g. 1955, *Bernborough—Mac Bea, by Bimelech
Margaretta, ch.f. 1955, Beau Max—La Marga, by Depth Charge
Sandusky, ch.g. 1955, Depth Charge—Retama, by Brazado
Well Away, b.f. 1955, Bimelech—Close, by Contest
Black Hills, b.c. 1956, *Princequillo—Blackie II, by Congreve
Call the Witness, b.c. 1956, Better Self—Your Witness, by *Beau Pere
Grand Wizard, ch.g. 1956, Poised—Baloma, by Depth Charge
Resaca, b.f. 1956, Middleground—Retama, by Brazado
Ring of Kerry, dk.b.g. 1956, Destino—Include, by Equestrian
Disperse, ch.c. 1957, Middleground—Scattered, by Whirlaway
El Humor, br.c. 1957, Destino—Mellow Mood, by Better Self
Espanto, ch.c. 1957, Beau Max—Baloma, by Depth Charge
La Fuerza, ch.f. 1957, Never Say Die—Solar System II, by Hyperion
Bal Musette, ch.c. 1958, Middleground—Friponne, by Shut Out
Beau Admiral, b.g. 1958, Beau Max—Caesar's Wife, by Better Self
Full Regalia, b.c. 1959, Middleground—Royal Bride, by *Princequillo
Pistol W., b.g. 1959, Destino—Script, by Bimelech
Solazo, b.c. 1959, Beau Max—Solar System II, by Hyperion
Tamarona, dk.b/br.f. 1959, Better Self—Retama, by Brazado
Brambles, b.c. 1960, Beau Max—Flower Bed, by *Beau Pere
Inclusive, dk.b/br.g. 1960, Curandero—Include, by Equestrian
Well Ordered, b.c. 1960, Better Self—On Order, by Discovery
Big Felly, b.c. 1961, Navigator—Otrora, by Equestrian
Boyar, b.c. 1961, *Princequillo—Bee Mac, by War Admiral
Miss Cavandish, ch.f. 1961, Cavan—New Weapon, by Bold Venture
Moss Vale, b.g. 1961, *Gallant Man—Riverina, by *Princequillo
Ground Control, b.f. 1962, Middleground—Black Panic, by Better Self
Buffle, ch.c. 1963, Zenith—Refurbish, by Bold Venture
Freedom Ring, b.c. 1963, Free America—Accroche Coeur, by Beau Max
Draft Card, ch.c. 1965, *Gallant Man—Dotted Line, by *Princequillo
Heartland, dk.b/br.f. 1965, Bold Ruler—Equal Venture, by Bold Venture
Out of the Way, dk.b/br.c. 1965, Mamboreta—Way Out, by *Alibhai
Gallant Bloom, b.f. 1966, *Gallant Man—Multiflora, by Beau Max
Kerry's Time, ch.g. 1966, Kerry Chief—Airosa, by *Heliopolis
Never Confuse, dk.b/br.g. 1966, Never Bend—Bemuse, by *Princequillo
Rio Bravo, ch.c. 1966, Saidam—Riverina, by *Princequillo
Heed the Call, b.c. 1969, To Market—Get Away Day, by Beau Max
Prove Out, ch.c. 1969, Graustark—Equal Venture, by Bold Venture
Free Hand, dk.b/br.c. 1970, *Gallant Man—Green Finger, by Better Self
Gentleman's Word, ch.c. 1970, *Gallant Man—Dotted Line, by *Princequillo
No Bias, ch.g. 1970, Jacinto—Fairness, by Cavan
Portentous, dk.b/br.c. 1970, Brambles—Turn Out, by *Turn-to

Distant Land, b.c. 1972, Graustark—Heartland, by Bold Ruler

Pakua, b.f. 1973, Zenith—On Tap, by Better Self

Pressing Date, dk.b/br.f. 1974, Never Bend—Monade, by Klairon

Harry Hastings (stp.), dk.b/br.c. 1979, *Vaguely Noble—Country Dream, by *Ribot (SW in England)

Too Chic, b.f. 1979, Blushing Groom (Fr)—Remedia, by Dr. Fager

Cryptic, b.f. 1980, Roberto—Befuddled, by Tom Fool

Solford, b.c. 1980, Nijinsky II—Fairness, by Cavan

Fiesta, b.f. 1981, Graustark—Fiesta Libre, by Damascus

If I Had a Hammer, dk.b/br.g. 1982, Cox's Ridge—Calyptra, by *Le Fabuleux

Blandford Park, b.c. 1983, Little Current—Green Finger, by Better Self

Sadeem, ch.c. 1983, *Forli—Miss Mazepah, by Nijinsky II (SW in England)

Art's Prospector, b.f. 1984, Mr. Prospector—No Duplicate, by Arts and Letters

Jungle Rage, ch.c. 1984, Jungle Savage—Calyptra, by *Le Fabuleux

Midyan, b.c. 1984, Miswaki—Country Dream, by *Ribot (SW in England)

Spur Wing, ch.f. 1984, Storm Bird—Equal Change, by Arts and Letters

Aquaba, b.f. 1985, Damascus—Elect, by *Vaguely Noble

Lake Valley, dk.b/br.f. 1985, Mr. Prospector—La Vue, by Reviewer

Make Change, ch.f. 1985, Roberto—Equal Change, by Arts and Letters

Ensconse, b.f. 1986, Lyphard—Carefully Hidden, by Caro (Ire) (SW in England/Ireland)

Robertet, b.f. 1986, Roberto—Ethics, by Nijinsky II (SW in France)

Lord Florey, b.c. 1987, Blushing Groom (Fr)—Remedia, by Dr. Fager (SW in England)

Malmsey, b.c. 1988, Topsider—Rebut, by Graustark (SW in Germany)

Misty Valley, ch.g. 1988, Majestic Light—Orchid Vale, by *Gallant Man (SW in Ireland)

Helen Alexander

Auto Dial, ch.f. 1988, Phone Trick—Here and Gone, by Royal Ski

Tenacious Tiffany, b.f. 1990, Star de Naskra—Pudding Pop, by It's Freezing

Vashon, b.f. 1993, Seattle Dancer—Go for Bold, by Bold Forbes

Pantufla, b.f. 1995, Rubiano—Rebut, by Graustark

Connected, ch.g. 1997, Twining—Auto Dial, by Phone Trick

Pie N Burger, ch.g. 1998, Twining—Abarca, by Topsider

Helen Alexander and Helen Groves

Modest, dk.b/br.f. 1981, Blushing Groom (Fr)—Aesculapian, by Dr. Fager (SW in Ireland)

Electric Flash, b.c. 1986, Fappiano—Perla Fina, by *Gallant Man

Twining, ch.c. 1991, Forty Niner—Courtly Dee, by Never Bend

Fine n' Majestic, dk.b/br.c. 1992, Majestic Light—Perla Fina, by *Gallant Man

Yamanin Paradise, b.f. 1992, Danzig—Althea, by Alydar (SW in Japan)

Defacto, b.c. 1993, Diesis (GB)—Maidee, by Roberto

Alyssum, ch.f 1994, Storm Cat—Althea, by Alydar

Arch, dk.b/br.c. 1995, Kris S.—Aurora, by Danzig

Alisios, b.c. 1996, Kris S.—Aurora, by Danzig

Festival of Light, b.c. 1997, A.P. Indy—Aurora, by Danzig (SW in UAE)

Helen Alexander, David Aykroyd, and Helen Groves

Althea, ch.f. 1981, Alydar—Courtly Dee, by Never Bend

Ketoh, ch.c. 1983, Exclusive Native—Courtly Dee, by Never Bend

Destiny Dance, b.f. 1986, Nijinsky II—Althea, by Alydar

Aishah, ch.f. 1987, Alydar—Courtly Dee, by Never Bend

Aurora, b.f. 1988, Danzig—Althea, by Alydar

Aquilegia, ch.f. 1989, Alydar—Courtly Dee, by Never Bend

Namaqualand, dk.b/br.c. 1990, Mr. Prospector—Namaqua, by Storm Bird

Helen Alexander, Helen Groves, & Dorothy (Alexander) Matz

Aldiza, ch.f. 1994, Storm Cat—Aishah, by Alydar

Bertolini, b.c. 1996, Danzig—Aquilegia, by Alydar (SW in England)

Atelier, dk.b/br.f. 1997, Deputy Minister—Aishah, by Alydar

Amelia, ch.f. 1998, Dixieland Band—Aquilegia, by Alydar

Alchemilla, b.f. 1999, Deputy Minister—Aquilegia, by Alydar

Arabis, ch.f. 1999, Deputy Minister—Aishah, by Alydar

Helen K. Groves

Serape, b.f. 1988, Fappiano—Mochila, by In Reality

Avanico (stp.), dk.b/br.g. 1991, Manila—Mochila, by In Reality

Cottage Garden, ch.f. 1994, Strawberry Road (Aust)—Ranchera, by Raise a Native

Batique, dk.b/br.f. 1996, Storm Cat—Serape, by Fappiano

My Golden Son, b.c. 1998, Twining—Snow Blossom, by The Minstrel

Emory (Alexander) Hamilton and Caroline Alexander

Serapide, b.f. 1988, Miswaki—Simply Divine, by Danzig (SW in Italy)

Stuck, dk.b/br.c. 1989, Conquistador Cielo—Simply Divine, by Danzig (SW in Italy)

Simply Tricky, dk.b/br.c. 1991, Clever Trick—Simply Divine, by Danzig (SW in France)

Dorothy (Alexander) Matz

Camella, dk.b/br.f. 1995, Housebuster—Have It Out, by Roberto

Offlee Wild, dk.b/br.c. 2000, Wild Again—Alvear, by Seattle Slew

Emory (Alexander) Hamilton

Chic Shirine, b.f. 1984, Mr. Prospector—Too Chic, by Blushing Groom (Fr)

Queena, b.f. 1986, Mr. Prospector—Too Chic, by Blushing Groom (Fr)

Tara Roma, b.f. 1990, Lyphard—Chic Shirine, by Mr. Prospector

Waldoboro, dk.b/br. 1991, Lyphard—Chic Shirine, by Mr. Prospector

Brahms, dk.b/br.c. 1997, Danzig—Queena, by Mr. Prospector

Serra Lake, b.f. 1997, Seattle Slew—Tara Roma, by Lyphard

Cappuchino, b.c. 1999, Capote—Tara Roma, by Lyphard

La Reina, ch.f. 2001, A.P. Indy—Queena, by Mr. Prospector

Henrietta Alexander

Wheelaway, gr/ro.c. 1997, Unbridled—Cuidado, by Damascus

Blue Springs, gr/ro f. 1998, Wekiva Springs—Via Dei Portici, by It's Freezing

Sources: *The Blood-Horse* Archives and The Jockey Club

CHAPTER

4

John W. Galbreath and Darby Dan Farm

John W. Galbreath had been involved with breeding and racing Thoroughbreds for about twenty years when he decided he wanted to play at the top of the game. He didn't have much in the way of a master pedigree strategy, only the willingness to go for the best whenever it was available, with the classic races as the goal. Swaps, a great horse in the middle 1950s, was owned by a fellow who could not afford to turn down major money, so Galbreath bought the 1955 Kentucky Derby winner and 1956 Horse of the Year. Prince Aly Khan had some glamorous European mares he was willing to hand off to a rich American, so Galbreath struck a deal. *Ribot dominated his rivals as the best horse in Europe, so Galbreath leased him. When the owner of Brookmeade Stable died, her fashionable broodmares came onto the market, so Galbreath bought them. A few years later, *Sea-Bird II emerged as the next *Ribot, so Galbreath dealt for him, too.

With his cache, Galbreath won two Kentucky Derbys, two Belmonts, the English Derby, the Preakness, and the Coaching Club American Oaks, all between 1962 and 1974, and three of his homebreds bore the wreath of champion. The quest for the best could be rated a success, and he still had fourteen more years of life left to add to his distinction.

Of all the Thoroughbred racing individuals who summon the frayed but sincere phrase "American success story," Galbreath must rank with the best. We have that on the testimony of no less than Norman Vincent Peale, the theologian and philoso-

pher who spoke at his friend's funeral. Dr. Peale gave to the Galbreath tale the additional phrase "old, traditional American success story" and spoke of a "synthesis" of qualities including "creativity and genius." Still, as Dr. Peale stressed, Galbreath "never lost the ability to care about and relate to people of various strata of life."

That a distinguished theologian would speak of him at his funeral in a crowded, staunch, and comforting church in Columbus, Ohio, proffered many threads of appropriateness. John Wilmer Galbreath was the grandson of a Methodist minister and the son of a farmer. He was born in 1897 in Darby, Ohio, one of six children, and by the age of ten he had gone into the business of selling horseradish. He was raised on family values, and while attending Ohio University he hit on a way to link those bonds with commerce. Everybody loves their children, so why not set up a darkroom, roam the campus taking pictures of students, and sell the photographs to their parents?

World War I interrupted this budding businessman, and he served as a lieutenant in the field artillery. He returned from the service and graduated in 1920. By the end of that decade, he was married and had two children and was making his way in real estate. The stock-market crash and Depression stymied many a businessman, but Galbreath turned it into opportunity, not only for himself but also for the fellow citizens of his adopted city, Columbus. As described in a 1955 *Fortune* magazine article, declining real estate prices had sent commerce spiraling downward, but Galbreath convinced a number of local residents who owned land free and clear to invest in foreclosure property at its reduced value: "In the darkest year of the Depression, Galbreath sold some $7 million worth of Columbus real estate...on 5 percent commission, and the bankers and investors of Columbus couldn't find the words to thank him adequately for his ingenuity."

Columbus would hold Galbreath's heart for the rest of his life, and he gradually built a sprawling property for farm use and recreation. It was named Darby Dan, for the creek that ran through the farm and for his son Daniel.

Mr. & Mrs. John W. Galbreath

DELL HANCOCK

Despite his love of Ohio and Columbus, specifically, Galbreath's business horizons knew only the bounds of gravity and Earth. John W. Galbreath Corporation in partnership with fellow entrepreneur Peter Ruffin developed a worldwide reputation for quality and honesty as builders. Galbreath projects included building the forty-two-story Mobil Oil Building in New York, the sixty-four-story U.S. Steel Building in Pittsburgh, and the forty-story First National Bank Building in Louisville, Kentucky. Farther afield, Galbreath built housing for entire communities in Canada, one of which was named Bramalea — a handy name for a filly from the mare Rarelea. Bramalea won the 1962 Coaching Club American Oaks.

In Hong Kong, Galbreath built a housing project of one hundred buildings of fifteen stories each, sitting atop land reclaimed from the sea. Still, even late in life, interviewed about his career, he was wont to stress the satisfaction he took in handing an individual a document that made him a homeowner.

He mixed profession and passion in the construction of the new state-of-the-art Aqueduct racetrack, having become involved with Thoroughbred racing in the mid-1930s and being elected to The Jockey Club in 1955. In the 1960s he also was there to help

New York racing in the rebuilding of Belmont Park, and he was involved in the purchases of both Hialeah and Churchill Downs when economics or outside ownership seemed to threaten the traditions they embodied.

Meanwhile, Galbreath had expanded his equine horizons from an early dabbling in polo. He bought a Thoroughbred stallion, Tommy Boy, and some hundred-dollar mares with the idea of breeding polo ponies, but the offspring wound up at the local racetrack, Beulah Park. There, Martha Long captured a four-hundred-dollar race in 1935 to become his first winner. Tommy Boy sired his first two stakes winners, Darby Dean and Darby Dienst. (The latter was named for Galbreath's Columbus friend Robert J. Dienst, who two decades later would own the champion sprinter Decathlon, whom Galbreath would stand as a stallion.) In 1945 Galbreath was represented as breeder of his first Kentucky Derby entrant, Mrs. W.G. Lewis' Darby Dieppe (*Foray II—*La Croma, by Solario). The colt ran third behind Hoop, Jr.

By 1950 Galbreath had bred only a half-dozen stakes winners. However, he had sewn the seeds of an expanded operation as early as 1944, when he purchased one hundred acres in Kentucky. The property, acquired from W.P. Veal, was adjacent to Colonel E.R. Bradley's famed Idle Hour Stock Farm.

Galbreath was to have many ties to Idle Hour, one of which was standing a 1930s Idle Hour Kentucky Derby winner, Burgoo King, at Darby Dan in Ohio. Galbreath's foal crops in 1950–53 indicated the upgrade represented by the Kentucky farm purchase, for they included ten stakes winners. Several were major winners, such as Errard King, Skipper Bill, and Clear Dawn.

The stage was set for his reach for the classics. Bradley had died in 1946 and Galbreath later acquired the core property of Idle Hour, changing its name to Darby Dan and maintaining the handsome old stud barn and other buildings. He also later brought back to that site Olin Gentry, who had managed it for years for Bradley. Gentry signed on with Galbreath in 1946, left in 1949, but returned to the fold in 1956. There was only one Colonel Bradley as the ultimate hero among his bosses in Gentry's life, but over the years he developed an enormous admiration for Galbreath: "He's a little guy. Thinks he can do anything. Damn near can."

Galbreath had been widowed for nine years when he married Dorothy Bryan Firestone, the former Mrs. Russell Firestone in 1955. The new Mrs. Galbreath happened to have a high-class horse at the time, Summer Tan, who was retired to Darby Dan. Soon Rex Ellsworth's 1956 Horse of the Year Swaps joined him. The Galbreaths bought Swaps in two $1 million transactions for a half-interest at a time.

In Swaps' first crop came the first to carry the fawn-and-brown Darby Dan colors to championship honors. She was Primonetta, foaled in 1958 from Banquet Bell, a Polynesian mare Gentry had purchased as a yearling for nine thousand dollars for Galbreath before becoming his full-time employee. Primonetta won the Alabama Stakes, Delaware Oaks, and Prioress Stakes at three in 1961 but lost the division championship when beaten by Bowl of Flowers in the Spinster Stakes. Bowl of Flowers was owned by Isabel Dodge Sloane's Brookmeade Stable and was by Sailor, whom Galbreath stood at Darby Dan. Mrs. Sloane's 1959 Horse of the Year Sword Dancer also had been sent to Darby Dan for stud duty when he retired at the end of 1960.

In 1962 Primonetta hauled in the championship for the older filly and mare division. She won the Spinster that year and also took the Molly Pitcher, Falls City, and Regret handicaps. She was retired with seventeen wins in twenty-five starts and earnings of $306,690. By that time her handsome full brother Chateaugay was showing a similar pattern of maturity for trainer Jim Conway. The following spring Chateaugay secured Galbreath's first classic races.

The 1963 season brought out a high-quality three-year-old crop. Candy Spots and No Robbery both reached the Kentucky Derby undefeated, and Never Bend was also known to be a colt of high quality. Chateaugay's progress got him a narrow victory in the Blue Grass Stakes, but he was slightly overlooked at 9-1 on Derby Day. The Darby Dan contract rider, Braulio Baeza, brought the glamorous chestnut up from sixth to take command in the stretch. The stamina and classic ability Galbreath had sought were well displayed as Chateaugay won by more than a length over Never Bend, with Candy Spots third.

At Pimlico, Chateaugay had a speedy work near the day of the race and perhaps was compromised. At any rate, Candy Spots got his revenge in the Preakness. That year's running of the Belmont was the first one transferred to Aqueduct

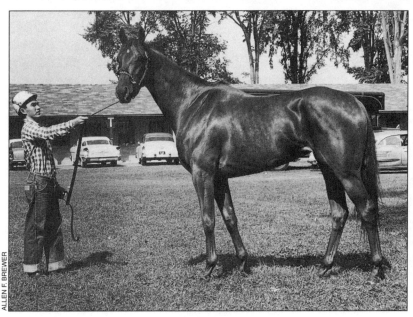

ALLEN F. BREWER

Primonetta

51

because of the condemnation and rebuilding project at Belmont Park. Prior to building the new Belmont grandstand, Galbreath had the thrill of leading in his first Belmont winner at a track he had already built! Chateaugay reigned as three-year-old champion. He had limited success as an older horse and went to stud with a record of eleven wins in twenty-four races and earnings of $360,722. His sister Primonetta became a distinguished producer, a Broodmare of the Year, while Chateaugay had limited stud success and was sent to Japan. He did leave behind, however, the good Darby Dan stakes winner True Knight, plus Chateaucreek, the dam of 1980 English Derby winner Henbit, and Chateaupavia, the second dam of 1996 Kentucky Derby winner Grindstone.

Of the important acquisitions by Galbreath, the leasing of *Ribot was probably the most internationally spectacular. *Ribot was a foal of 1952 and had authored a wondrous career in Europe, going unbeaten in sixteen races. He ventured out of his Italian base several times, winning two runnings of France's Prix de l'Arc de Triomphe as well as one edi-

tion of England's King George VI and Queen Elizabeth Stakes. Those who saw him recognized him as a great horse, perhaps the greatest Europe had ever seen.

*Ribot was one of several unbeaten public heroes bred and trained by Federico Tesio, whose others were Cavaliere d'Arpino and Nearco. The latter had become one of the most important elements in pedigrees in North America as well as Europe, but this *Ribot character had a pedigree of names largely unfamiliar to American horsemen. He was by Tenerani out of Romanella, by El Greco.

Tesio had died before *Ribot got to the races, but his partner, Marchese Mario Incisa della Rochetta, continued the stable and stud. *Ribot was sent to stand one year in England and then returned to the Italian stud, Dormello, where he stood for three years.

Galbreath was not unfamiliar with imaginative horse deals. He had entered into an unusual deal with Prince Aly Khan whereby a group of mares would be shared, to stay three years in Kentucky and shuttle back for the next three years to Ireland,

Chateaugay

THE BLOOD-HORSE

repeating the pattern for as long as they were producing. (Prince Aly was killed in an automobile accident a couple of years later and the ocean-crossing plan was scuttled.) During the time he was negotiating the Swaps deal, Ellsworth offered a plan that Swaps stand in Kentucky one season, California the next — a type of shuttling that might seem mild by today's dual hemisphere pattern but was highly unusual in the 1950s. This was rendered unnecessary when the Galbreaths bought the second half of Swaps.

With a proud European owner reluctant to let *Ribot be lost forever to his compatriots, and a proud American determined to get his hands on the great champion, a lease was arranged. Galbreath agreed to lease *Ribot for five years for $1.35 million. To put his deal into the context of its time, the first million-dollar deal for a horse had occurred only four years earlier, with complete and permanent ownership of Nashua being acquired for less, $1,251,250. *Ribot arrived in New York in June 1960 and was exhibited at Belmont Park. If Gentry and his staff had known then how irascible *Ribot could be, any idea of such a side trip would likely have been shot down. At any rate, *Ribot arrived safely at Darby Dan, and he never left. By the time the five-year lease was up, the Dormello group was convinced that it would be unsafe to move *Ribot across the ocean again. The horse had developed a reputation for wildness documented by the high hoof scars in his stall and on a tree in his paddock. Gentry used

to conjecture that he must have reared up so violently that he fell over backward and "scrambled what brains he had." *Ribot lived to age twenty and died of a twisted intestine in 1972.

Pedigree and personality notwithstanding, *Ribot proved a great sire in this country, turning out winners here as well as major winners sent back to Europe. Mr. and Mrs. Galbreath had mixed success with *Ribot. They bred only five stakes winners by the horse, but two of them were the significant full brothers Graustark and His Majesty. The brothers were out of Flower Bowl, a high-class *Alibhai mare who was part of the draft of Brookmeade Stable mares Galbreath purchased. Mrs. Sloane died in 1962, and international horseman Winston Guest bought many of her mares. A short time later Galbreath acquired a number of them from Guest.

Flower Bowl's champion daughter Bowl of Flowers had been a rival of Darby Dan's Primonetta. Now, Flower Bowl would be an ally of the Galbreaths. In the 1963 *Ribot colt Graustark, she produced the most spectacular horse of a lifetime. A huge and handsome dark chestnut, Graustark was sent to Darby Dan's Chicago stable with trainer Loyd Gentry, nephew of the farm manager. He was widely touted before his first start, and he toured through seven races unbeaten. These included only two stakes, for nagging injuries interrupted his racing career. Nevertheless, as the 1966 Derby approached, Graustark seemed to have greatness before him. Then he was outnodded at the finish of the Blue

GRAUSTARK				
*Ribot, 1952	Tenerani, 1944	Bellini, 1937	Cavaliere d'Arpino (Havresac II—Chuette)	
			Bella Minna (Bachelors Double—Santa Minna)	
		Tofanella, 1931	Apelle (Sardanapale—Angelina)	
			Try Try Again (Cylgad—Perseverance)	
	Romanella, 1943	El Greco, 1934	Pharos (Phalaris—Scapa Flow)	
			Gay Gamp (Gay Crusader—Parasol)	
		Barbara Burrini, 1937	*Papyrus (**Tracery**—Miss Matty)	
			Bucolic (Buchan—Volcanic)	
Flower Bowl, 1952	*Alibhai, 1938	Hyperion, 1930	Gainsborough (Bayardo—*Rosedrop)	
			Selene (Chaucer—Serenissima)	
		Teresina, 1920	**Tracery** (*Rock Sand—*Topiary)	
			Blue Tit (Wildfowler—Petit Bleu)	
	Flower Bed, 1946	*Beau Pere, 1927	Son-in-Law (Dark Ronald—Mother-in-Law)	
			Cinna (Polymelus—Baroness La Fleche)	
		*Boudoir II, 1938	*Mahmoud (*Blenheim II—Mah Mahal)	
			Kampala (Clarissimus—La Soupe II)	

Grass Stakes by Abe's Hope and was diagnosed as having a fractured coffin bone. He was retired to stud soon thereafter.

In 1968 Flower Bowl had another *Ribot colt, His Majesty. Although not nearly so spectacular in appearance or performance as Graustark, His Majesty also was retained for stud after a career that included victory in the Everglades Stakes and placings in the Widener and Seminole handicaps.

Both Graustark and His Majesty became major sires, and His Majesty led the sire list in 1982. As was true of *Ribot, many of the greatest dividends benefited other breeders and owners. Paul Mellon bred *Ribot's champion son Arts and Letters; Raymond Guest bred *Ribot's champion son Tom Rolfe; Mellon bred Graustark's champion son Key to the Mint; and Thomas Mellon Evans bred His Majesty's

Derby-Preakness winner Pleasant Colony. Such other stalwarts of His Majesty's sire line as Derby winner Go for Gin and European classic sensation St. Jovite also flew other colors. The His Majesty filly Ribbon was bred by the Galbreaths but was owned by Arthur Hancock III when she produced the champion Risen Star. Mehmet was another Galbreath-bred by His Majesty that became a major winner, but he had been included in a consignment to a Keeneland sale.

Darby Dan was hardly shut out, however. The Galbreaths bred and raced Tempest Queen, a Graustark filly from Queen's Paradise, by Summer Tan, and raced her to three-year-old filly championship honors of 1978. Tempest Queen won the Spinster, Acorn, and three other stakes that year. Her full sister Steal a Kiss also won stakes. Another homebred Graustark filly was Primonetta's daughter Maud Muller, winner of the 1974 Gazelle Handicap and a division of the Ashland Stakes. Then, late in life, Galbreath bred and raced the big Graustark colt Proud Truth, out of Wake Robin, by Summer Tan. A sentimental moment of the old gentleman's final years on the Turf found him walking into the winner's circle holding jockey Jorge Velasquez by the hand after Proud Truth's 1985 Breeders' Cup Classic victory.

If Galbreath might be said to have supplied other breeders with riches through his acquisition of *Ribot and that horse's successive generations, he certainly got a bit of his own back in his patronage of Hail to Reason. This champion son of *Turn-to stood his entire career at Hagyard Farm, but he won a special place in the affections of Galbreath's succeeding generations at Darby Dan.

Foremost among the Hail

Roberto

ALEC RUSSELL

54

to Reasons bred and raced by Galbreath were the sportsman's second Kentucky Derby winner, Proud Clarion, and his Epsom Derby winner and international sire, Roberto. Proud Clarion flashed briefly on the stage, but he chose his one moment of great glory wisely. The lean colt from Breath O' Morn, by *Djeddah, flashed out of the damp and fog to win the Derby at 30-1 odds on the first Saturday of May in 1967. He won by a length from Barbs Delight, with favored Damascus third. He later won one additional stakes, the Roamer. Proud Clarion's second dam was Darby Dunedin, who, being by *Blenheim II, was one of the first higher-class mares Galbreath owned. She was the dam of Clear Dawn, a *Heliopolis filly who won the Black Helen and New Castle handicaps and Miss Woodford Stakes in the 1950s.

Roberto was also by Hail to Reason, and his name reflected another aspect of Galbreath's sporting interests. Roberto Clemente was a great baseball player for the Pittsburgh Pirates from 1955 to 1972. The Galbreaths owned the World Series-winning team, and the colt was named in tribute to Clemente. Roberto the horse traced to a mare Gentry had known in Colonel Bradley's time and illustrated how compatible Bradley's seeking of the classic horse was with Galbreath's. In 1926 Bradley imported a weanling filly by the French star Sardanapale, who was an obvious candidate to add classic stamina, having won the French Derby and Grand Prix de Paris a dozen years earlier. Sardanapale also became the leading sire in France. Coincidentally, Gentry's lifelong friend, trainer Ben A. Jones, was so impressed by the Sardanapale colt *Cotlogomor's victory in the 1927 New Orleans Handicap that he was determined to get the horse as a stallion at his Missouri farm. *Cotlogomor was a bust, failing to get horses that even the masterful Jones could keep sound, and Bradley's imported Sardanapale filly never started either.

The Sardanapale runners had shown enough quality and speed that these experiences did not convince Bradley and Gentry to shun his blood, but they put

THE BLOOD-HORSE

Graustark

into Gentry's mind the need to compensate with soundness in future pedigrees.

The unraced Sardanapale filly was named *Forteresse, and she foaled the Kentucky Derby winner Brokers Tip for Bradley. *Forteresse's last foal was the filly Bleebok, by Blue Larkspur. She was scheduled to be culled, and Gentry suggested offering her to Galbreath since the Ohio businessman had had an association with the farm. For Galbreath, Bleebok foaled Rarelea to the cover of the great sire Bull Lea. Rarelea was only a moderate winner, but her pedigree assured her a place in the Darby Dan broodmare band. Seeking to double the Sardanapale influence, but through sound representatives, Gentry and Galbreath sent Rarelea to Nashua. A horse of marvelous soundness, Nashua was out of a mare whose dam was by Sardanapale. Galbreath named his 1959 Nashua—Rarelea filly Bramalea, and she carried his colors proudly, outdueling the champion Cicada to win the 1962 Coaching Club American Oaks. The Oaks has been run at various distances, including a mile and three-eighths and a mile and one-eighth; in Bramalea's running, the distance was a mile and one-quarter. Bramalea also won the Jasmine Stakes and Gazelle Handicap among eight wins from thirty-eight starts and earned $192,396.

Since Nashua was by the Nearco stallion *Nasrullah and Hail to Reason by *Turn-to (he by the Nearco stallion *Royal Charger), the Hail to Reason—Bramalea cross also provided inbreeding to Nearco as well as to the important European mare Mumtaz Begum. Moreover, Roberto's pedigree also contained a 4x4 inbreeding to Bradley's champion Blue Larkspur. (Both female families have the same Blue Larkspur/*Teddy pattern; Roberto's sire is out of the mare Nothirdchance by a Blue Larkspur stallion and out of a Sir *Gallahad III mare while Rarelea's second dam is by Bull Lea, a son of *Sir Gallahad III's full brother *Bull Dog, out of a Blue Larkspur mare. Both Sir *Gallahad III and *Bull Dog are by *Teddy.)

Roberto was sent to Ireland and trainer Vincent O'Brien. With North American-breds Sir Ivor and Nijinsky II, O'Brien had won the Epsom Derby in 1968 and 1970, respectively, and another American-bred, Mill Reef, had won that seminal classic in 1971. Galbreath did not always micromanage his

stable, but on several occasions during Roberto's career he stepped in to make a decision about his rider. As the 1972 Epsom Derby approached, Roberto's rider, Bill Williamson, was injured in a spill, and had he been retained for the Derby, he would have gone into the race off very little recent activity. Galbreath, who stood to become the only person ever to breed and race both a Kentucky Derby winner and an Epsom Derby winner, did not like having a rider come into the race after a layoff. Williamson's take on it was that he had been cleared by physicians to ride, but he chose to take no chances, so he did not ride in any races until two days before the Derby. That same day came the announcement that Galbreath had instructed that the great jockey Lester Piggott be booked on Roberto. He stipulated that Williamson would be provided an equal amount to whatever Piggott earned in the race.

Clearly, it was exceptional luck that the great rider did not have a major mount that soon before the greatest race in England, and virtually any owner would have loved to have Piggott on his horse, but the public and press turned their sympathy toward

Proud Clarion

Williamson and against Galbreath. The running of the race, however, justified the thought that Piggott made the difference. Months after Roberto had driven home by a short head after a searing battle with Rheingold across the historic ground, the *Bloodstock Breeders' Review* conceded: "No indignation, however, can alter the basic truth. Mr. Galbreath's decision, though unpopular, was crucial. Without it, and without Piggott in the saddle, Roberto would surely have lost the Derby. It is beyond belief that any other jockey, past or present, could have forced the favorite home...Unwavering, thrusting his head out in grim determination, Roberto took all the relentless driving and answered it with superb courage. Breathtakingly locked together, the pair passed the post with barely a nostril separating them. But it was the nostril of Roberto which showed just ahead in the photo-finish print." Rheingold was ridden by Ernie Johnson, who probably bristled a bit at suggestions that Piggott would have won on either colt.

Since the Galbreaths had some horses in partnership and raced some separately, Mrs. Galbreath also had a starter in that Derby in Manitoulin, who finished in midpack. (A footnote to that 1972 Epsom Derby was the quality of future stallions involved. Roberto became an acclaimed international sire as did little Lyphard, the French-based invader who occasioned some jollity among the English by bolting so wide on the final turn that he took himself out of contention.)

So, John W. Galbreath became the first person to breed and own both a Kentucky Derby winner and an English Derby winner. This distinction would go unmatched for two decades, when Epsom Derby winner Mill Reef's owner, Paul Mellon, won the 1993 Kentucky Derby with homebred Sea Hero. (Even then there was a bit of Darby Dan, since Sea Hero's dam, Glowing Tribute, was a daughter of Graustark.)

Galbreath's role as owner/jockey's agent was instrumental in another major victory that year for Roberto. During the summer the colt was aimed at a meeting with England's darling of the day, the unbeaten four-year-old Brigadier Gerard. They were on a collision course for the first running of the Benson and Hedges Gold Cup, at one and five-sixteenths miles at York.

Brigadier Gerard needed but one more victory to match *Ribot's career mark of sixteen races unbeaten and, as an older horse, was the odds-on favorite.

Rheingold had defeated French Derby winner Hard to Beat in the Grand Prix de Saint-Cloud and was second choice, since Roberto had disappointed in the Irish Derby subsequent to his Epsom heroics. Galbreath brought over his American-based rider, Braulio Baeza, for his first-ever ride in England, a move the local press conjectured would cost Roberto ten lengths. A few years before, Baeza had ridden the American turf star Assagai in France, in the Grand Prix de Saint-Cloud and put him on the lead, a tactic that did not produce victory. Nevertheless, Baeza dashed Roberto into a four-length lead early and forced the English riders either to hand him the victory without a whimper or foresake their accustomed waiting tactics. Joe Mercer challenged aggressively on his champion, but Brigadier Gerard expended so much energy in cutting the lead to a length that he had nothing left for the final business of actually passing Roberto. The American colt extended his lead to three lengths. The Brigadier's winning streak was finished.

Roberto's great-race, failing-race pattern extended to the autumn, when similar front-running tactics in France's Prix de l'Arc de Triomphe produced a far different result. Roberto faded to seventh. The following spring, once again on the grounds of Epsom, Roberto won the Coronation Cup. He later was brought home to Darby Dan with seven wins from fourteen races and earnings of $332,272.

Only two years after Roberto presented him in the hallowed Epsom Derby winner's circle, Galbreath had another classic winner at home. This was the elegant *Sea-Bird II colt Little Current. In 1965 *Sea-Bird II had been so breathtaking in his victories in the Epsom Derby and Prix de l'Arc de Triomphe that, although he lacked *Ribot's tag line of "unbeaten," many comparisons were made. This time, Galbreath moved quickly, and by year's end had leased *Sea-Bird II from Jean Ternynck for $1.5 million for five years. *Sea-Bird II was by the French classic-placed Dan Cupid, a son of America's great Native Dancer. Dan Cupid was out of a mare by *Sickle, to whom he was inbred, and the second dam was by Gallant Fox. Thus, *Sea-Bird II was something of an American in Paris. *Sea-Bird II's dam, Sicalade, was by the grand French stallion Sicambre and represented classic French breeding a couple of generations back. It is doubtful Galbreath paid much attention to the pedigree that produced such greatness as *Sea-Bird II possessed. We can hardly see him saying "No thanks" had *Sea-Bird II's ancestry been otherwise.

*Sea-Bird II was a good sire without being a great one, and by the time Galbreath won classic races with the son Little Current, the stallion's lease was up and Sea-Bird II had been returned to France. There he would be known as the sire of the high-class Gyr and the popular filly champion Allez France, both American-breds conceived in Kentucky.

Little Current was another dividend from the mare Banquet Bell, who had produced Primonetta and Chateaugay. Three of Banquet Bell's first five foals were by Swaps, but in 1962 she was bred to the outside stallion *My Babu. The resulting foal, Luiana, was unraced but was clearly a broodmare prospect. Luiana foaled Little Current in 1971. Little Current developed late for Lou Rondinello, who was then Darby Dan's trainer, winning the Everglades Stakes in the winter of his three-year-old season. He caught the largest Kentucky Derby field in history, twenty-three starters, for the hundredth running of the event. A stretch runner by inclination, Little Current had a start-and-stop trip trying to come through from being dead last early and progressed only as far as fifth place. Two weeks later his emergence was complete, and he won the Preakness Stakes by seven lengths, giving Galbreath a victory in the one American Triple Crown race that had eluded Chateaugay. Little Current then won Galbreath's second Belmont Stakes by the same margin. Although he subsequently lost his remaining three races and was forced into retirement by injury, Little Current was the Eclipse Award winner as champion three-year-old colt. He retired to Darby Dan with four wins in sixteen starts and earnings of $354,704.

Little Current was moderate at stud. He sired thirty-five stakes winners (7 percent) and, after being moved from farm to farm, wound up a revered senior citizen at a farm in Washington state. He lived to be thirty-two, dying as a pensioner in 2003. In that season he gained distant additional distinction through a daughter, Belle of Killarney, second dam of 2003 Kentucky Derby-Preakness winner Funny Cide. There must be some hardy genes in the brood of Banquet Ball, for even living to thirty-two, Little Current missed by three years matching the longevity of his dam's half sister Primonetta, who lived to thirty-five and was cited annually at Valentine's Day birthday celebrations at

the original Darby Dan in Ohio.

Hail to Reason, sire of classic winners Proud Clarion and Roberto, continued to be used. For either John W. or his heir, Dan W. Galbreath, the stallion's additional major winners included Hail to Pirates, whom Dan named for the baseball team of which he became president. Primonetta, Broodmare of the Year in 1978, produced two major winners by Hail to Reason in 1975 Florida Derby victor Prince Thou Art and $405,207 earner Cum Laude Laurie. Cum Laude Laurie wore Dan Galbreath's colors and won the Spinster, Beldame, Ruffian, and Delaware Oaks in 1977. More than one filly champion has had a less distinguished list of victories to her credit, but Cum Laude Laurie was denied a championship by Our Mims' string of the Fantasy, CCA Oaks, Alabama, and Delaware Handicap for Calumet Farm. Cum Laude Laurie's name commemorated the high school "championship" form of Dan Galbreath's younger daughter.

The next year, Tempest Queen, bred and raced in Mrs. John Galbreath's name, won the three-year-old championship with arguably no more impressive a list of victories than Cum Laude Laurie's. Tempest Queen won the Spinster, Acorn, Poinsettia, Gazelle, and Prioress. She was produced from Queen's Paradise, a filly by Mrs. Galbreath's sterling Summer Tan. Queen's Paradise was acquired from Summer Tan's trainer, Sherrill Ward, and his brother, John Ward Sr.

While Little Current was only a qualified success at stud and was later moved, Roberto became an international influence befitting his pedigree and his racing distinction. He sired eighty-five stakes winners (17 percent) and gave rise to reference to the "Roberto sire line" instead of being lumped into the "Hail to Reason line" or the "*Turn-to line." The Robertos ranged from two-year-olds like Darby Dan's Hopeful winner Capital South, European champion juveniles such as Sookera and Critique, and the precocious Red Ransom, to the outer ranges of stamina in major flat races, to wit, the Melbourne Cup winner At Talaq. Roberto also got an English St. Leger winner in Touching Wood as well as a number of prominent middle-distance stars that have proliferated the sire line. The latter includes Brian's Time, Silver Hawk, Kris S., Dynaformer, and Lear Fan.

Although the Galbreaths did not breed a high number of Roberto stakes winners, various members of the sporting family bred at least one major horse by the stallion:

John W. Galbreath bred the Roberto colt Sunshine Forever, who emerged as Galbreath's final champion in the last year of the old gentleman's life; Mrs. Galbreath bred Capital South; Dan Galbreath bred and raced major Roberto winner Script Ohio; Galbreath's daughter, Jody, and her husband, James W. (Wally) Phillips, bred the good Roberto stakes winner Darby Creek Road (third in the Affirmed-Alydar Belmont Stakes) and the 1988 Florida Derby winner Brian's Time, who has helped extend the bloodline's international prominence with his success as a sire of champions in Japan.

John W. Galbreath died at Darby Dan Farm in Ohio, on July 20, 1988. He was ninety. Mrs. Galbreath had passed away two years before. Among those sending condolences were former President Gerald Ford and Queen Elizabeth II, for whom Galbreath had boarded mares. At the same time, folk of different distinction, such as the porter and waiters at the Columbus Club, tearfully told Dan how much they would miss his father.

John Wilmer Galbreath had remained a man of the land, and Columbus was the land of his most steadying love. He taught such to his grandchildren and expressed in his own poetry his thoughts about having reared their mother and father at Darby Dan. The son and daughter, Joan Phillips, carried on. They had also become involved with Ohio State University, which is in Columbus and of which Dan served as chairman of the board of trustees. The new equine hospital and research center bears Dan Galbreath's name.

Dan Galbreath, however, lived only seven more years before his own death from cancer. He had gloried in the emergence of Sunshine Forever as a final champion bred in his father's era. With the stable then trained by John Veitch, Sunshine Forever became the 1988 champion grass horse, when he won the vaunted Washington, D.C., International as well as the Turf Classic, Man o' War, and two lesser graded stakes. Sunshine Forever was sent home to Darby Dan after the 1989 season with eight wins from twenty-three starts and earnings of $2,084,800. Olin Gentry had often declared that breeding a stallion was far tougher than breeding a good racehorse. Roberto had been both, whereas Sunshine Forever did not materialize as a sustained commercial success at stud.

Still, a great deal of family history had been involved in this last major winner for the founder. In addition to being a son of Epsom hero Roberto, Sunshine Forever was foaled from a daughter of the vaunted Graustark. His dam was Outward Sunshine, a stakes-placed filly whose dam, Golden Trail, had been one of the Brookmeade Stable mares acquired more than two decades earlier.

Golden Trail was the foundation of a Darby Dan branch of mares that also produced Brian's Time. Both Sunshine Forever and Brian's Time were from daughters of Golden Trail, who also foaled Galbreath-bred stakes winners Gleaming Light, Sylvan Place, and Java Moon. This was also the family that produced Darby Creek Road, Andover Way, and Memories of Silver for the Darby Dan family, as well as the successful Roberto stallion Dynaformer, the champion filly Ryafan, and other stakes winners for various owners/breeders.

A Rare And Devoted Continuity

In the decade-plus since the death of John W. Galbreath, his family has been buffeted by the vicissitudes of life. As a provider extraordinaire, the patriarch might have hoped he could protect his issue from such realities, but as a man of the church he probably grasped that monetary success is not an everlasting armor against day-to-day problems. In the interim the family has seen the deaths of son-in-law Wally Phillips, as well as of Dan Galbreath, and has met head-on a tumultuous economic scenario that generated *Wall Street Journal* coverage as to the stability of the financial empire. Moreover, an uncomfortable family drama had to be confronted and resolved when Mrs. Galbreath's family made accusations relative to her will.

Through it all, several qualities have prevailed, and — however presumptuous it might seem to take unto one's self such judgment — we can only believe that John W. Galbreath would applaud. There has been a deep sense of family, and, on the Turf, an imaginative crafting of economic realities that still feeds upon the tradition of breeding for the best races. The Phillips' son, John W. Phillips, is in charge of Darby Dan, and his mother, Jody, remains deeply involved. Phillips has converted Darby Dan to embrace more commercialism, stallions selected to please the market, and he has tempered the tradition of breeding to race for the classics with breeding to sell, too. Million-dollar sales of yearlings have been added to the great moments of success on the racetrack that Darby Dan authored over the years. Horses of stout, staying blood, such as Turkoman and Proud Truth, have been recruited or retired from the racing stable, while outside horses perhaps more noted for speed, such as Meadowlake, have also been brought in as farm stallions. Sunshine Forever did not prove a major success, but Phillips has mined other veins of Roberto's ongoing influence with telling success through such as Kris S. and Silver Hawk — outside stallions, but sons of Roberto. In keeping with their heritage, each of these sons of Robert sired an Epsom Derby winner: Silver Hawk sired the 1997 winner Benny the Dip, and Kris S. sired 2003 winner Kris Kin.

For Darby Dan, Silver Hawk sired the sale graduate Grass Wonder, who earned nearly $6 million as a champion in Japan, and Silver Hawk is also the sire of the high-class Phillips homebred Memories of Silver.

The Phillips/Galbreath family has had an unusual run of recent success with fillies, including five winners of the Queen Elizabeth II Challenge Cup, a grade I stakes for three-year-old distaffers on grass at Keeneland.

A recent example of the vigor of Darby Dan stock was James D. Squires' breeding of 2001 Kentucky Derby winner Monarchos from a Darby Dan mare whose produce record was such that business realities indicated she should be sold. While this might seem a disappointment in one sense, it underscores the present Darby Dan management's adaptations without releasing the old values.

Rarely do family farms, or businesses of any sort, survive into the third generation. Various changes in the makeup of the stallion roster, the yearling sale philosophy, and the bringing in of partners have allowed Mrs. Phillips and son John to convert Darby Dan into a new era. Early in the twenty-first century, for example, Phillips acquired the Eclipse Award sprinter Aldebaran to stand at the farm.

At the time John W. Galbreath died, in 1988, he had bred ninety-one stakes winners in more than fifty years. When we take into account the record of son Dan, daughter Jody and her husband, the present-day Galbreath/Phillips Racing Partnership, and various other outreach partnerships, the total for the various family entities as of early 2003 stands at 160.

Few experiences on the Turf for a family with a tradition to guide it could be more pleasing than the Galbreath/Phillips success with Soaring Softly, the

Four winners of the Queen Elizabeth II Challenge Cup Stakes in front of the Darby Dan mansion (from left): Plenty of Grace, Tribulation, Memories of Silver, and Love You by Heart

1999 Breeders' Cup Filly and Mare Turf winner and champion turf female. Soaring Softly had Darby Dan history flowing from every pedigree portal. Her sire, Kris S., was one of the best sons of Roberto. Her dam, Wings of Grace, was by Graustark's son Key to the Mint and was from a female family dating to the unusual agreement between John W. Galbreath and Prince Aly Khan. Four decades separated that deal and the emergence of Soaring Softly. Family decisions had attended every year in between.

*Skylarking II was one of the individuals involved in the aforementioned deal whereby a number of mares were to be shunted between continents in three-year intervals. Prince Aly was killed in an automobile accident several years later, however, and *Skylarking II never returned to Europe. She was a daughter of Mirza II and was a stakes winner in France. Prior to importation, she had foaled the French One Thousand Guineas winner Yla and an additional stakes winner in Karim.

For Galbreath, *Skylarking II's foals included the Swaps filly Soaring, a 1960 foal who placed but failed to win. Soaring's first foal was Miss Swapsco, by Cohoes. Miss Swapsco was sold as a yearling and her influence has been felt around the world to the benefit of other breeders and owners. She is the dam of Ballade, who produced the champion colt and major sire Devil's Bag, the distaff champion Glorious Song, and the additional major stallion Saint Ballado. Glorious Song, in turn, produced the international grade/group I winner Singspiel and the crack miler and high-class sire Rahy.

This might smack of letting the best of a broodmare's potential get away, but Darby Dan bred and raced various of Soaring's other foals. These included Far Beyond, a daughter of English Triple Crown winner Nijinsky II. Far Beyond won only once, but bred to Key to the Mint, she produced the elegantly named Wings of Grace. Wings of Grace was bred in the names of Mr. and Mrs. John W. Galbreath and won the 1981 Boiling Springs Handicap on the grass. Her foals include Plenty of Grace, by good old Roberto. Plenty of Grace, a foal of 1987, won the Yellow Ribbon, Queen Elizabeth II Challenge Cup, and three other stakes and earned $744,499.

In 1995 Wings of Grace foaled the Kris S. filly Soaring Softly. Trained by J.J. Toner, Soaring Softly was lightly raced at two and three and then emerged as another champion for the family. Prior to the Breeders' Cup Filly and Mare Turf, she had won four other stakes in 1999. One of them was the grade I Flower Bowl, named for the dam of her great-grandsire Graustark — one more stanza in the epic poem that is Darby Dan.

Champions Bred

Primonetta
1962 champion handicap female

Chateaugay
1963 champion three-year-old colt

Little Current
1974 champion three-year-old colt

Tempest Queen
1978 champion three-year-old filly

Sunshine Forever
1988 champion turf male

In Europe:

Roberto
1971 Irish champion two-year-old colt
1972 English champion three-year-old colt

Galbreath/Phillips Racing Partnership

Soaring Softly
1999 champion turf female

Stakes Winners Bred by John W. Galbreath and Family

Darby Dean, br.c. 1936, Tommy Boy—Fair Arrow, by For Fair

Darby Dienst, ch.c. 1937, Tommy Boy—Sun Lily, by *Sun God II

Darby Dimout, b.c. 1941, Flares—Sun Lily, by *Sun God II

Darby Dieppe, gr.c. 1942, *Foray II—*La Croma, by Solario

Argyle, b.c. 1948, Best Seller—*Bosnia, by Bosworth

Service, br.c. 1949, Bless Me—Lady Diver, by *Sir Gallahad III

Sharbot, ch.f. 1950, Pavot—Bright Blue, by Burgoo King

Skipper Bill, ch.c. 1950, Errard—Nipmenow, by *Bull Dog

Clear Dawn, ch.f. 1951, *Heliopolis—Darby Dunedin, by *Blenheim II

Errard King, b.c. 1951, Errard—Darby Dover, by Burgoo King

Lady Bouncer, b.f. 1951, Roman—Bum's Rush, by Blue Larkspur

My Fault, b.g. 1951, Faultless—*Valdina Spirea, by Canon Law

Craigwood, b.c. 1952, *Djeddah—Bum's Rush, by Blue Larkspur

Errasina, b.f. 1952, Errard—Lesina, by *Sir Gallahad III

Nirdar, b.c. 1953, *Nirgal—Darby Dingo, by Fighting Fox

Evening Time, br.f. 1954, Citation—Darby Dunedin, by *Blenheim II

Big Klu, gr.c. 1957, Olympia—Gay Mood, by *Mahmoud

Editorialist, b.g. 1958, Olympia—Stratafire, by *Mahmoud

Memories of Silver

Flavia, b.f. 1958, Roman—*Mbale, by Big Game

Primonetta, ch.f. 1958, Swaps—Banquet Bell, by Polynesian

Sailor Beware, dk.b.c. 1958, Sailor—*Trimbeth, by Big Game

Black Beard, br.c. 1959, Swaps—Darby Dunedin, by *Blenheim II

Bramalea, dk.b/br.f. 1959, Nashua—Rarelea, by Bull Lea

Summer Savory, ch.g. 1959, Summer Tan—Darby Damozel, by War Admiral

Chateaugay, ch.c. 1960, Swaps—Banquet Bell, by Polynesian

Ornamento, ch.g. 1960, Bold Ruler—War Ribbon, by Bimelech

Big Darby, blk.c. 1962, Swaps—Darby Dunedin, by *Blenheim II

Candalita, b.f. 1962, Olympia—Bravura, by Niccolo Dell'Arca

Country Friend, b.g 1962, Olympia—Molly Barker, by Errard

Happy Hull, ch.g. 1963, Summer Tan—Orientation, by Questionnaire

Rego, b.g. 1963, *Amerigo—Hegira II, by *Heliopolis

Proud Clarion, b.c. 1964, Hail to Reason—Breath O'Morn, by *Djeddah

Metric Mile, b.c. 1965, Olympia—Masindi, by Nashua

Gleaming Light, b.g. 1966, Never Bend—Golden Trail, by Hasty Road

Onandaga, ch.c. 1966, *Ribot—Red Pippin, by Royal Charger

Summer Cottage, ch.c. 1966, Chateaugay—Happy Summer, by Summer Tan

Summer Air, ch.c. 1967, Chateaugay—Happy Summer, by Summer Tan

Grenfall (Ire), ch.c. 1968, Graustark—Primonetta, by Swaps

Roberto (Eng), b.c. 1969, Hail to Reason—Bramalea, by Nashua

True Knight, dk.b/br.c. 1969, Chateaugay—Stealaway, by Olympia

Infuriator, b.c. 1970, Chateaugay—Flavia, by Roman

Java Moon, b.f. 1970, Graustark—Golden Trail, by Hasty Road

Prince Dantan, ch.c 1970, Graustark—Lobelia, by Bold Ruler

Little Current, ch.c 1971, Sea-Bird—Luiana, by *My Babu

Lover John, ch.c. 1971, Damascus—Evening Primrose, by Nashua

Maud Muller, ch. f. 1971, Graustark—Primonetta, by Swaps

Stage Door Betty, ch.f. 1971, Stage Door Johnny—Lobelia, by Bold Ruler

Prince Thou Art, dk.b/br.c. 1972, Hail to Reason—Primonetta, by Swaps

Steady Stride, b.g. 1972, Stage Door Johnny—Ya Es Hora, by Nashua

Sylvan Place, dk.b/br.c, 1972, Graustark—Golden Trail, by Hasty Road

Olympic Circuit, b.g. 1973, Banquet Circuit—Candalita, by Olympia

Cum Laude Laurie, dk.b/br.f, 1974, Hail to Reason—Primonetta, by Swaps

Bold Voyager, b.c. 1975, His Majesty—Meadow Stream, by Native Dancer

Fool's Prayer, b.c. 1975, Roberto—Beautiful Morning, by Graustark

Ribbon, b.f. 1977, His Majesty—Break Through, by Hail to Reason

Sunshine Always, ch.f. 1984, Arts and Letters—Inward Joy, by Raise a Native

Blossoming Beauty, b.f. 1985, Little Current—Crab Apple Lane, by Graustark

Sunshine Forever, b.c. 1985, Roberto—Outward Sunshine, by Graustark

Matthew's Moment, b.c. 1986, Little Current—Captured Moment, by Graustark

Memories, dk.b/br. f. 1986, Hail the Pirates—All My Memories, by Little Current

Pichy Nany, b.f. 1986, Gran Zar—True Esteem, by Roberto

Apple Current, dk.b/br.c. 1987, Little Current—Crab Apple Lane, by Graustark

Carverali, b.g. 1987, Roberto—Far Beyond, by Nijinsky II

Memories of Pam, dk.b/br.f. 1987, Graustark—Memories We Share, by Roberto

Plenty of Grace, b.f. 1987, Roberto—Wings of Grace, by Key to the Mint

Apple Creek, b.c. 1988, Darby Creek Road—Crab Apple Lane, by Graustark

Mr. and Mrs. John W. Galbreath

Graustark, ch.c. 1963, *Ribot—Flower Bowl, by *Alibhai

Miss Swapsco, dk.b/br.f. 1965, Cohoes—Soaring, by Swaps

His Majesty, b.c. 1968, *Ribot—Flower Bowl, by *Alibhai

Swapsann, ch.f. 1968, Swaps—La Bonne Mouche, by Bold Ruler

Ben Adhem, b.c. 1972, *Ribot—Idun, by Royal Charger
Mehmet, ch.c. 1978, His Majesty—Soaring, by Swaps
Wings of Grace, b.f. 1978, Key to the Mint—Far Beyond, by Nijinsky II
Buckeye Gal, b.f. 1982, Good Counsel—Chanderelle, by Graustark
Dawn's Curtsey, ch.f. 1982, Far North—Liberty Spirit, by Graustark
Grand Exchange, b.g. 1983, Gran Zar—Chanderelle, by Graustark

Mr. and Mrs. John W. Galbreath and Daniel M. Galbreath
Ribellor, dk.b/br.c. 1968, *Ribot—Toute Belle II, by Sicambre

Mrs. John W. Galbreath
Clover Leaf, ch.c 1959, Swaps—Miss Traffic, by Boxthorn
Macedonia, ch.g. 1960, Olympia—Pandora, by Unbreakable
Summer Sorrow, b.f. 1965, Summer Tan—*Sauchrie, by Le Lavandou
Ocean Roar, ch.c. 1966, Swaps—Summer Serenade, by Summer Tan
Power Ruler, dk.b/br.c. 1966, Bold Ruler—Polylady, by Polynesian
Great Heron, b.c. 1967, *Sea-Bird—Big Effort, by Endeavour II
Jogging, ch.c. 1967, Swaps—Doricharger, by Royal Charger
King Rail, ch.c. 1967, *Sea-Bird—Real Effort, by Sailor
Villager, ch.c. 1967, Chateaugay—Fiorita, by *Princequillo
Good Counsel, b.c. 1968, Hail to Reason—Polylady, by Polynesian
Shearwater, dk.b/br.f. 1969, *Sea-Bird—*Sauchrie, by Le Lavandou
Arum Lily, dk.b/br.f. 1970, Bold Lad—Wake Robin, by Summer Tan
Barrydown, b.c. 1970, Intentionally—Foxy Quilla, by *Princequillo
Liberal, ch.c. 1973, Arts and Letters—Fiorita, by *Princequillo
Stark Winter, ch.f. 1973, Graustark—Winter Wren, by *Princequillo
Salesian, b.c. 1974, Proud Clarion-Tender Promise, by Sailor
Tempest Queen, b.f. 1975, Graustark—Queen's Paradise, by Summer Tan
Erin's Word, b.f. 1977, Good Counsel—Machree, by Graustark
Capitol South, b.c. 1981, Roberto—Polylady, by *Polynesian
Bloomin' Bypass, b.f. 1982, Squire—Peach Bloom, by *Ribot
Plairoh, ch.c. 1982, True Knight—Summer Magic, by Roberto
Proud Truth, ch.c. 1982, Graustark—Wake Robin, by Summer Tan
Steal a Kiss, b.f. 1983, Graustark—Queen's Paradise, by Summer Tan
Old Exclusive, b.c. 1985, His Majesty—Old Grenada, by Pronto

Mr. and Mrs. Daniel M. Galbreath
Hail the Pirates (Ire), b.c. 1970, Hail to Reason—Bravura, by Niccolo Dell'Arca
Fleetferd, b.g. 1971, Banquet Circuit—Laurie Lind, by Ribot

Daniel M. Galbreath
Robin's Song, b.c. 1975, Northern Dancer—Little Red Robin, by Ribot
Trained Nurse, ch.f. 1975, Graustark—Bold Nurse, by Bold Ruler
Abbey Leix, ch.f. 1978, Banquet Circle—Lady Donna, by Graustark
Our Captain Willie, b.c. 1978, Little Current—Little Red Robin, by Ribot
Prayers 'n Promises, b.f. 1978, Foolish Pleasure—Luiana, by *My Babu
Country Pine, b.c. 1980, His Majesty—Mountain Sunshine, by *Vaguely Noble
Aspen Rose, ch.f. 1980, Little Current—Lady Donna, by Graustark
Script Ohio, b.c. 1982, Roberto—Grandma Lind, by Never Bend
Love You by Heart, b.f. 1985, Nijinsky II—Queen's Paradise, by Summer Tan
Kid Russell, b.g. 1986, Topsider—Queen's Paradise, by Summer Tan
Dance O'My Life, b.f. 1988, Sovereign Dancer—Sunshine O'My Life, by Graustark

Kyle's Our Man, ch.c. 1988, In Reality—Lady Donna, by Graustark
Young Daniel, b.c. 1988, Turkoman—Grandma Lind, by Never Bend
Low Tolerance, dk.b.f. 1989, Proud Truth—Glorious Spring, by Hail to Reason

Daniel M. Galbreath, John W. Galbreath II, L. Nichols, L. Megrue
Turkish Tryst, ch.f. 1991, Turkoman—Darbyvail, by Roberto

Daniel M. Galbreath and Mrs. James W. Phillips
Magick Top, ch.f. 1989, Topsider—Coming Spring, by Graustark

Mr. and Mrs. James W. Phillips
Regal Road, b.f. 1971, Graustark—On the Trail, by Olympia
Darby Creek Road, dk.b/br.c. 1975, Roberto—On the Trail, by Olympia
Andover Way, dk.b.f. 1978, His Majesty—On the Trail, by Olympia
Killarney Road, ch.c. 1981, His Majesty—Tremont Road, by Damascus
Graceful Darby, b.f. 1984, Darby Creek Road—Graceful Touch, by His Majesty
Brian's Time, dk.b.c. 1985, Roberto—Kelley's Day, by Graustark
Legal Coin, dk.b.f. 1985, Key to the Mint—Regal Roberta, by Roberto
Rampart Road, b.c. 1986, Dixieland Band—Regal Road, by Graustark
Cobra Classic, b.g. 1987, Graustark—Darby Trail, by Roberto
Fappiano Road, b.c. 1989, Fappiano—Regal Road, by Graustark
Devoted Brass, b.g. 1990, Dixieland Band—Royal Devotion, by His Majesty
Bound by Honor, b.c. 1991, Woodman—Jody G., by Roberto
Jambalaya Jazz, ch.c. 1992, Dixieland Band—Glorious Morning, by Graustark

Mr. and Mrs. James W. Phillips, G. Arthur Seelbinder, Dr. Ted Classen
Tribulation, b.f. 1990, Danzig—Graceful Touch, by His Majesty

Galbreath/Phillips Racing Partnership
Triumph At Dawn, ch.f. 1990, Alydar—Dawn's Curtsey, by Far North
Bahamian Sunshine, ch.g. 1991, Sunshine Forever—Pride of Darby, by Danzig
Don't Read My Lips, ch.f. 1991, Turkoman—Our Dear Sue, by Roberto
Too Cool to Fool, b.f. 1991, Foolish Pleasure—Our Tina Marie, by Nijinsky II
Majestic Dream, ch.f. 1993, Majestic Light—Dream Creek, by The Minstrel
Memories of Silver, b.f. 1993, Silver Hawk—All My Memories, by Little Current
Merit Wings, b.f. 1993, Silver Hawk—Our Tina Marie, by Nijinsky II
Aavelord, br.g. 1994, Lord Avie—Sunshine Above, by Turkoman
Buckeye Search, ch.f. 1994, Meadowlake—Pride of Darby, by Danzig
Makethemostofit, b.f. 1994, Easy Goer—Love You by Heart, by Nijinsky II
Memories of Gold, dk.b.f. 1995, Time for a Change—All My Memories, by Little Current
Misty Hour, b.f. 1995, Miswaki—Our Tina Marie, by Nijinsky II
Soaring Softly, ch.f. 1995, Kris S.—Wings of Grace, by Key to the Mint

Galbreath-Phillips, Foxfield, and Pope McLean
Sizzlin Sunshine, b.f. 1992, Sunshine Forever—North Flow, by Upper Nile

Phillips Racing Partnership
Duality, b.c. 1998, Seeking the Gold—Jody G., by Roberto
Ellie's Moment, b.f. 1998, Kris S.—Kelley's Day, by Graustark

New Real Deal, b.f. 1998, Roy—Newhall Road, by Dixieland Band

Coshocton, ch.c. 1999, Silver Hawk—Tribulation, by Danzig

Best Minister, b.c. 2000, Deputy Minister—Best of Memories, by Halo

Hippogator, b.f. 2000, Dixieland Band—Gastronomical, by Sunshine Forever

Phillips Racing Partnership and John Phillips

Private Joe, b.c. 1994, Private Account—Jody G., by Roberto

Grass Wonder, ch.c. 1995, Silver Hawk—Ameriflora, by Danzig

Wonder Again, b.f. 1999, Silver Hawk—Ameriflora, by Danzig

War Trace, b.c. 2001, Storm Cat—Memories of Silver, by Silver Hawk

John Phillips and Pam Gartin

Forest Shadows, ch.f. 2000, Woodman—Dance of Sunshine, by Sunshine Forever

Darby Dan Stables

Robin de Nest, b.c. 1997, Robin des Pins—Love of Ireland, by Irish River (FR)

Love n' Kiss S., dk.b.f. 1998, Kris S.—Key to My Heart, by Mr. Prospector

Darby Dan Farm and Mr. and Mrs. Theodore Kuster

Dr. Pain, b.g. 1989, Proud Truth—Dance Or Prance, by Bold Reasoning

Darby Dan Farm and Paul Tackett

Whomsoever Proud, dk.b.g. 1990, Proud Truth—Whatonamer, by Verbatim

Sources: *The Blood-Horse* Archives and The Jockey Club

Fred W. Hooper

In recent years The Jockey Club has invited individuals of a wide range of stripe in the Thoroughbred industry into its membership. For much of its history, however, The Jockey Club tended toward homogeneity. A member was likely to have attended a good prep school, then one of the leading Ivy League institutions, perhaps business or law school, and then have gone into banking or investments, developing comfort with pinstripes. This pattern contrasts to the background of one Fred W. Hooper, whose resume read Moler Barber College, after which he went not into banking but potato farming, construction work, thence into highway construction, with blue jeans or even overalls coming into play. Nevertheless, Hooper was such an achiever and leader in Thoroughbred racing that by 1975 he was elected to The Jockey Club.

In a sport in which the American success story is an interwoven theme era after era, Hooper's is among the best. He was a large man with a capacity for hard work along with the vision of an entrepreneur. The Georgia farm boy, born in 1897, got an early introduction to horses by dealing with mustangs.

"We raised cotton on the farm, but when I was fourteen years old I decided to bring a carload of western horses from Montana to Georgia," he told us in an interview for *The Blood-Horse* in 1973. (We had been sent to interview him as part of a series to record individuals' careers "before it is too late." In Hooper's case there would be nearly three more decades of opportunity for such an article!)

The horses "were wild to begin with — had to lasso them out in the open. They would fight you with their front feet. I broke them all."

Hooper's early jobs included a brief stint cutting men's hair after leaving barber school. He also took a turn at teaching school and then moved on to Muscle Shoals, Alabama, where he took on the most dangerous of construction jobs because the crews working high up with steel got paid the most. He boxed a bit on the side.

A call to service interrupted this vigorous life, but in a handy bit of timing, his arrival at Camp Gordon, Georgia, coincided with word of the armistice of World War I, "so they just said to go on home."

After Hooper's mother died, he moved to Florida and went into potato farming, making money for a couple of years before a fungus wiped out his crop and left him in debt. Operating on credit, he put together a crew of a couple hundred laborers and bid on a job to lay some road bases. Figuring out a way to use local quarry materials, he completed the job. Suddenly, in the mid-1920s as Florida boomed, he was a contractor. With no partners or shareholders, he flourished and eventually merged Hooper Construction with General Development Company.

Hooper also ventured into agriculture, raising Herefords, some of them champions, on five thousand acres he acquired back in Alabama. By that time his father was about sixty-five and was working for the son. An early Florida Thoroughbred breeder, Carl Rose, sold them a half-bred horse named

Prince, who proved so fast and adept at turning cattle that the senior Hooper suggested they find some match races for the horse.

"We ran him fifty-five times and he won forty-nine races, in South Carolina, Georgia, and Florida, so he was what actually got me into the racing business," Fred Hooper said years later.

Hooper famously purchased a Kentucky Derby winner as his first Thoroughbred. He made that initial venture into the game in 1943, when the next world war had caused enough travel restrictions

Fred Hooper with Laffit Pincay

that Kentucky breeders scrambled to create a local sale instead of shipping yearlings to the sale in Saratoga Springs, New York. Hooper knew enough about Thoroughbred pedigrees to know that he wanted *Sir Gallahad III blood, and he had in mind buying a few fillies by the stallion.

"Well, I saw this *Sir Gallahad III colt that didn't have much flesh on him," Hooper recalled. The colt was bred by Robert A. Fairbairn, one of the investors in *Sir Gallahad III. "I just liked his walk, and his looks, and the smartness of his eye and all, so I bought the horse for $10,200." He named the colt for his son, Hoop, Jr., and won the Kentucky Derby with him in 1945.

The foregoing has been oft repeated. As a stand-alone story, it suggests that Hooper was amazingly lucky and entered the sport with one purchase, not a cheap price for the time but not a major investment. In fact, Hooper waded deeply into the waters at his first yearling sale. Joe Palmer wrote in *American Race Horses of 1945* that, "As far as Thoroughbred breeders were concerned, Mr. Hooper had dropped from the clouds while they were bewailing the retirement of Mrs. Ethel V. Mars (a major yearling buyer), and he was one of those…who turned the yearling market upward in 1943."

Hooper bought not just Hoop, Jr. but also four other yearlings, spending a total of $39,400. He was

game to spend a good deal more, for he bid up to $50,000 for Pericles, who topped the sale at $66,000.

In addition to Hoop, Jr., the yearlings he bought included a colt named for one of his adopted states, Alabama. A son of *Mahmoud—Gala Belle, by *Sir Gallahad III, Alabama cost $17,000. He defeated Polynesian in the 1944 Ral Parr Stakes and earned $28,055.

Hoop, Jr., who was trained by Ivan Parke, won the Wood Memorial as well as the Kentucky Derby, but he broke down finishing second in the Preakness and retired with earnings of $99,290. Hooper was soon in the breeding business. He bred horses in Alabama for a time and also participated in the Illinois breeding program, but from 1966 until the end of his life he was identified most closely with the fresh and vigorous Florida Thoroughbred breeding industry.

"I like it here," he said of Ocala. "It gets cold a little, but not really bad, and it gets hot, but I never suffered more from heat in my life than I did up there in New York. When we started the farm, 75 percent of the land was in woods. But that's what I wanted — virgin land. Cattle grazed on some of the land, but some of it the owners had not farmed. They were getting concessions from the government not to grow anything."

Before Hooper's breeding operation got rolling, he continued to build his stable through the yearling market. Hooper went back the next year to buy ten horses for $101,300, and in 1945 he bought five for $66,100. Education was in the latter group and raced officially in the name of Hooper's wife, Laura.

Education was a horse that Bull Hancock of Claiborne Farm thought was as fast a two-year-old as he had ever seen. In fact, Hancock's determination to stand Double Jay was prompted by that colt's defeating Education in the Kentucky Jockey Club Stakes. Education was by the aging speed sire Ariel out of Faculty, by Swift and Sure. He was bred by Walter Salmon's historic Mereworth Farm and purchased for $6,600. At two he won ten of sixteen races and earned $164,473, so that monetarily he was already a bigger bargain than Hoop, Jr. in a cost-to-earnings ratio. His major scores came in the 1946 Washington Park Futurity, Breeders' Futurity, Elementary Stakes, and Hawthorne Juvenile. He shared juvenile championship status with Double Jay. Education had a career mark of fifteen wins in twenty-nine starts and earnings of $188,698.

An even more significant acquisition than his Derby winner or Education insofar as the future breeding program was Hooper's purchase of Olympia. Ivan Parke, a former star jockey who had been training for Hooper, also bred some horses on the side. He owned a mare named Miss Dolphin, by Stimulus. She was a seven hundred dollar Saratoga yearling who had won a couple of stakes at Woodbine and New Orleans before Parke acquired her. In 1945 Parke boarded her at Coldstream Stud near Lexington, Kentucky, and bred her to the farm's high-class Hyperion stallion *Heliopolis.

Hooper bought the *Heliopolis—Miss Dolphin colt, named Olympia, and Parke continued his association as the trainer. In the winter and spring of 1949, Olympia was so brilliant in winning from coast to coast, taking the San Felipe, Flamingo, Wood Memorial, and Derby Trial, that he was the odds-on post-time favorite to become his owner's second Kentucky Derby winner. He also had taken time out from official racing to prevail in one of the nicer bits of Thoroughbred lore. Perhaps harking back to his days with Prince, Hooper agreed to the sporting proposition of matching Olympia against a brilliant Quarter Horse mare named Stella Moore. The race took place on January 5, 1949, Olympia already having won the Breeders' Futurity and two other stakes at two. With a purse of $25,000 and about $90,000 in covered bets on the line, Hooper spotted a flaw in the scenario.

"We ran between races at Tropical Park. The finish line was seventy-three feet short of a quarter-mile, when the gate was put in the chute. I changed the finish and made them run the full quarter. I wasn't going to take any of the worst of it…Olympia and Stella Moore broke nearly even. At the eighth pole, Stella Moore was about two lengths in front, but when they got to the finish Olympia was there first."

Not so in the Derby, in which Olympia faded after leading for a mile and finished sixth. He later added the mile Withers Stakes, but the gloss was off, as

			Bayardo (Bay Ronald—Galicia)
		Gainsborough, 1915	*Rosedrop (St. Frusquin—Rosaline)
	Hyperion, 1930		Chaucer (**St. Simon—Canterbury Pilgrim**)
		Selene, 1919	Serenissima (Minoru—Gondolette)
*Heliopolis, 1936			John o' Gaunt (**Isinglass**—La Fleche)
		Swynford, 1907	**Canterbury Pilgrim** (Tristan—Pilgrimage)
	Drift, 1926		Neil Gow (Marco—Chelandry)
		Santa Cruz, 1916	**Santa Brigida** (**St. Simon**—Brigid)
OLYMPIA			Commando (**Domino**—Emma C.)
		Ultimus, 1906	Running Stream (**Domino**—*Dancing Water)
	Stimulus, 1922		Uncle (*Star Shoot—The Niece)
		Hurakan, 1911	The Hoyden (*Esher—The Maid)
Miss Dolphin, 1934			Picton (Orvieto—Hecuba)
		*Light Brigade, 1910	Bridge of Sighs (**Isinglass—Santa Brigida**)
	Tinamou, 1922		Plaudit (Himyar—*Cinderella)
		Casuarina, 1911	Nuns Cloth (Melton—St. Odille)

tends to be the case when a Derby favorite disappoints. Olympia came up short at one and a quarter miles, but he had the blazing speed that American breeders respond to, and Hooper had a key cog in the future of his breeding game. After winning three stakes at four, Olympia retired with fifteen wins in forty-one starts and earnings of $365,632.

Neither Education nor Hoop, Jr. was successful at stud, although Hooper managed a few dividends by them. The first stakes winner bred in his name was Eddie Sue, a 1950 Education filly. She was out of the Menow mare Wise Sue and won the 1952 Bay State Kindergarten Stakes and the 1953 Artful Stakes. She was one of six stakes winners (4 percent) by Education. Hoop, Jr. got only three stakes winners (3 percent), but two of them were Hooper's useful pair of full brothers Hoop Band ($233,429) and Hoop Bound ($112,443). They were out of the Pilate mare Patricia P.

Olympia, however, was the real thing. Standing initially in Kentucky before being moved to Hooper's Florida farm, Olympia sired a total of forty stakes winners (12 percent), of which the best for other breeders included champions Decathlon and Pucker

Up, plus Apatontheback, Candalita, Creme dela Creme, Pia Star, Talent Show, and Editorialist. Hooper bred ten stakes winners by Olympia and seventeen out of Olympia mares. The sons of Olympia also sired the dams of another round of Hooper's stakes winners.

The Olympias had speed, such as Greek Game, who battled the two-year-old Bold Ruler and his peers, winning the 1956 Washington Park and Arlington futurities. On the other hand, Hooper's two very good classic-placed colts Crozier and Admiral's Voyage were out of Olympia mares.

Greek Game, who was out of the Questionnaire mare Sunday Supper, won ten of fifty-five races and earned $275,120. He was an early marker in a pattern that enhanced Hooper's status as an agricultural maverick who at times went against fashion's tides and succeeded with bloodstock that did precious little for others. Greek Game's lonely pair of stakes winners (2 percent) included Hooper's homebred Miami Mood ($72,980). Moreover, Hooper bred five stakes winners out of Greek Game mares, three of them from the outstanding producer Poliniss. Her foals included Wedge Shot, who in 1973 won Hooper a running of the historic Futurity Stakes at Belmont Park.

In addition to Greek Game, Hooper's homebred Olympia stakes winners included Alhambra, Olymar, Bright Holly, and My Portrait.

Another marked example of Hooper's singular success with an otherwise obscure horse was *Quibu. In 1946, at the same time he was gathering yearlings at the American sales, Hooper the newcomer cast his eyes far afield. In Argentina, *Quibu won a series of good races at distances from a mile to a mile and three-eighths. Like many Argentine horses, he was the product of expatriate British stock of strong foundation, being by Meadow, son of England's St. Leger winner Fairway, and out of a granddaughter of leading English

Crozier

THE BLOOD-HORSE

sire Son-in-Law. Hooper bought *Quibu with the thought he would challenge the likes of Armed in this country.

"We paid the most money that had ever been paid for a horse down there," Hooper recalled in the 1973 interview. "We wanted to keep it quiet, but the next day it came out in big headlines, and then the Argentine government offered more money than we had paid, to keep him in that country. But we shipped him on up with another horse we had bought named *Colosal. *Quibu was a well-bred horse and I made a mistake not breeding a lot of mares to him."

*Quibu sired eighty-four foals, of which a half-dozen (7 percent) were stakes winners. Hooper bred five of the stakes winners and the other was bred in the name of Mrs. Hooper. Moreover, one of *Quibu's daughters, Quaze, foaled his wondrous multiple champion Susan's Girl as well as 1974 Kentucky Oaks and Alabama Stakes winner Quaze Quilt.

Specialmante was another with South American ties who succeeded for Hooper while making scant mark otherwise. A son of *Nigromante, Specialmante sired only fifty foals, and, again, Hooper bred all five of his stakes winners (10 percent). The best among them was the aforementioned Quaze Quilt, a filly thus with an Argentine-bred grandsire and broodmare sire. Quaze, the distinguished mare, had placed in the 1960 Kentucky Oaks and was out of Heavenly Sun, an unraced daughter of Olympia. In this genetic handiwork, the speed of Olympia was buttressed by staying influences, and one and a quarter miles was well within the scope of its products.

While a certain iconoclastic bent was evidenced in the use of *Quibu and Specialmante, Hooper also dealt in less exotic, more fashionable stock. He bred to twenty or so outside stallions in some years, and his brilliant colt Crozier resulted from sending the Olympia mare Miss Olympia to the 1948 English Two Thousand Guineas winner *My Babu, who stood at Spendthrift Farm in Kentucky. Crozier thus represented the sire line of Tourbillion, the great French stallion who sired *Djebel, in turn the sire of My Babu. On the bottom Crozier had the influence of Hyperion through Olympia's sire, *Heliopolis, while the second dam was by the Teddy stallion *Teddy's Comet. Lady Marlboro, third dam of Crozier, was by Sweep and from a King James mare.

By the time Crozier was foaled in 1958, one of Ivan Parke's brothers, Chuck Parke, was training for Hooper. After winning the Derby on Hoop, Jr., Eddie Arcaro quipped to Hooper that it would prove an expensive moment, for the owner would likely work a long time before winning another. Hooper never did get a second Derby, but Crozier got him as close as he ever would come.

After being run down in the stretch by Carry Back to lose by narrow margins in both the Flamingo Stakes and Florida Derby, Crozier had won the Derby Trial, and he was back to try again in the Kentucky Derby. He forged to the lead in the upper stretch but, again, succumbed to the late heroics of the favorite, losing by three-quarters of a length. A subsequent third behind Carry Back and Globemaster in the Preakness deepened the thought that Crozier could not stay one and a quarter miles. Two years later, at five, he broke through at that distance, winning the Santa Anita Handicap to give Hooper one of his most important triumphs.

In between, at four, Crozier also crafted a victory over the sterling Ridan at one and one-eighth miles in the Aqueduct Stakes. All told, Crozier won nine stakes among his ten wins from thirty-four starts and earned $641,733. Hooper had bred his own next resource for continuing success.

The early 1960s were a strong time for the Hooper stable. Those years also marked the emergence of jockey Braulio Baeza. One of the contributions Hooper made to the game was the discovery of young riders in Panama. In addition to Baeza, who had great success for him and later for Darby Dan Farm, the Phipps family, and Tartan Stable, Hooper brought to America other future Hall of Famers Jorge Velasquez and Laffit Pincay Jr., he of amazing longevity.

In addition to Crozier's status as one of the best horses in training in the early 1960s, Hooper in those years brought out such homebreds as My Portrait, Main Swap, and Admiral's Voyage. My Portrait (Olympia—Me, by Challenge Me), was in the same 1958 crop as Crozier. She won the Kentucky Oaks in the spring of 1961, the day before Crozier's second in the Derby. That summer, My Portrait handed the budding champion Primonetta her first defeat, after nine races, by running her down to win the Monmouth Oaks by a neck.

In 1962, in addition to Crozier's victories at four,

Hooper had another classic contender in Admiral's Voyage. This was a son of the former handicap champion Crafty Admiral and the Olympia mare Olympia Lou. Crafty Admiral was by Fighting Fox, full brother to 1930 Triple Crown winner Gallant Fox, and was out of a War Admiral mare. Again, staying influences mingled with the speed of Olympia to telling effect.

In the winter of 1962, Admiral's Voyage won a pair of sprint stakes at Santa Anita. Working his way east, he won the Louisiana Derby and then dead-heated with Sunrise County in the Wood Memorial, gaining the official victory when his eccentric rival was disqualified. In the Kentucky Derby, Admiral's Voyage faded to ninth after hitting the lead with a quarter-mile to run, but if this were taken as evidence of deficient stamina for the job, later events would prove that a false premise. Admiral's Voyage had a knack for getting involved in ding-dong battles. In the Jersey Derby, he and Crimson Satan and Jaipur came to the wire battling side by side. Crimson Satan won, but, like Sunrise County, he had fouled, and so Jaipur's nose margin for second over Admiral's Voyage turned into a nose margin for the victory.

Then, at one and a half miles in the Belmont Stakes, in the most sublime of Admiral's Voyage's duels, he again tested Jaipur to the wire before going under by a barely discernible nose. He lost the race but proved his stamina. At four and five, Admiral's Voyage won the Carter, Midwest, and San Carlos handicaps, but his demonstrated capacity for both speed and stamina did not carry over consistently in the stud. Winner of twelve races in fifty-two starts, he had earned $455,879. He was, paradoxically, a failure and a lasting influence. Admiral's Voyage got only eight stakes winners (3 percent), but one of them was Pas de Nom, dam of Danzig. So, the old potato farmer had his role in the continued harvesting from a very fertile ground. Admiral's Voyage was the broodmare sire of thirty-eight stakes winners.

Susan's Girl with John Russell and Fred Hooper

Copelan

Also in the early 1960s came Main Swap, whose sire, Swaps, was Horse of the Year in 1956 and a two-million-dollar purchase to stand at Darby Dan Farm in Kentucky. Hooper's Main Swap was a filly out of Mainpoint, an *Ambiorix mare whose dam was by *Mahmoud. This was class and fashion, too. As a juvenile in 1962, Main Swap won the Astarita Stakes and then took the rich Gardenia Stakes at one and one-sixteenth miles over co-champion juvenile filly Smart Deb.

In the meantime, breeding to an outside sire, Paul Mellon's 1964 Belmont and Travers winner, Quadrangle, produced perhaps the most accomplished of Hooper's horses. In 1968 the *Quibu mare

Quaze was sent to Quadrangle and out of that plethora of "Qu" names came a blaze-faced filly who was to be named Susan's Girl. Hooper had this foal listed as bred by his son, Fred Jr., namesake of Hoop, Jr., so that technically she was not included as a homebred. She raced for Hooper, however, and was a treasure. Early in 2003, when Xtra Heat won her twenty-fifth stakes, she was recognized as having won more stakes races than any other distaffer in modern history, and the owner of the record she broke was Susan's Girl.

Hooper was very much a hands-on owner, not insisting on running the show but insisting upon an owner's right to cruise the barn as he wished and to give guidance. Over the years he had success with a series of trainers. After the Parke brothers, his trainers included John W. Russell, Cotton Tinsley, Ross Fenstermaker, and Jimmy Picou.

Russell was on stage for the first championship season of Susan's Girl, which came when she was three in 1972. Susan's Girl had won a pair of stakes at two, and at three she emerged as a hardy and highly professional force, winning no fewer than nine stakes from California to Kentucky to Pennsylvania to New York. These included the second of Hooper's three scores in the time-honored Kentucky Oaks, as well as scores in the Acorn and the weight-for-age Beldame Stakes.

For three more years Susan's Girl came back, later under Fenstermaker, and she won a succession of the most important races for fillies and mares — two

COPELAN	Tri Jet, 1969	Jester, 1955	Tom Fool, 1949	Menow (*Pharamond II—Alcibiades) Gaga (**Bull Dog**—Alpoise)
			Golden Apple, 1945	Eight Thirty (Pilate—Dinner Time) Thorn Apple (Jamestown—Last Straw)
		Haze, 1953	**Olympia**, 1946	*****Heliopolis** (Hyperion—Drift) **Miss Dolphin** (Stimulus—Tinamou)
			Blue Castle, 1938	*****Blenheim II** (Blandford—Malva) *Blue Dust (*Gainsborough—*Golden Araby)
	Susan's Girl, 1969	Quadrangle, 1961	Cohoes, 1954	*Mahmoud (**Blenheim II**—Mah Mahal) Belle of Troy (Blue Larkspur—*La Troienne)
			Tap Day, 1947	Bull Lea (**Bull Dog**—Rose Leaves) Scurry (Diavolo—Slapdash)
		Quaze, 1957	*Quibu, 1942	Meadow (Fairway—Silver Mist) Querendona (Diadochos—Querella)
			Heavenly Sun, 1952	**Olympia** (*****Heliopolis**—**Miss Dolphin**) Daffy (The Porter—Lady Pike)

runnings of the Delaware Handicap and the Spinster Stakes, plus the Santa Margarita, the Matchmaker, another Beldame, and so on. The first distaffer to earn one million dollars racing solely in North America, Susan's Girl wove a career of twenty-nine wins from sixty-three starts with earnings of $1,251,668. She won Eclipse Awards for three-year-old filly of 1972, older filly or mare for 1973, and older filly or mare again in 1975.

Hooper's homebreds from the same crop as Susan's Girl, 1969, also included Tri Jet. Another offspring of an outside sire, Tri Jet was by George D. Widener's Futurity winner Jester. Tom Fool, sire of Jester, was by Menow, whose blood Hooper had used from early on. Tri Jet was out of Haze, a modest winner by the reliable Olympia and the daughter of a *Blenheim II mare. Tri Jet won seven stakes among his seventeen victories in forty-six starts, to earn $413,084. Key among them were the 1973 Hawthorne Gold Cup and Bel Air Handicap and the 1974 San Pasqual Handicap and Whitney Stakes. In the last-named he blazed nine furlongs at old Saratoga in 1:47, a track record that still stands. Tri

Jet's Whitney came a week before Quaze Quilt, Hooper's third Kentucky Oaks winner, added to the stable's long list of important victories by upsetting the champion Chris Evert in the one and a quarter-mile Alabama.

In 1979 Hooper bred Tri Jet to Susan's Girl, and this in-house combination produced the colt he named for distinguished veterinarian Robert Copelan. (Hooper also bred and raced a stakes winner named Tinsley for one of his trainers, and Hooper named a horse Forsythe Boy after actor John Forsythe whimsically suggested having a Hooper-bred namesake during an Eclipse Awards presentation.)

The colt named Copelan became an example to stick in the throat of those who harbor a prejudice against hard-raced, rugged mares as producers. In 1982 Susan's Girl's foal reeled off victories in three of the most revered of New York's juvenile stakes, the Hopeful, the Futurity, and the Champagne. He also took the Sanford Stakes and, out west, the Hollywood Prevue. But for Roving Boy's streak of four West Coast stakes, including the Del Mar Futurity, Norfolk, and Hollywood Futurity, Copelan

Precisionist with Wanda and Fred Hooper

might well have landed a championship of his own. He ran a poor fourth in the Young America and was fifth behind Roving Boy in the Hollywood Futurity. Although he lost in Eclipse Award balloting, Copelan shared highweight honors with Roving Boy at 126 pounds each on the Experimental Free Handicap.

The following year Copelan won the Fountain of Youth and for a time seemed in the thick of the Kentucky Derby picture with a second in the Florida Derby and a third in the Blue Grass Stakes. He later strained a tendon and did not get back to the races. He went to Hooper's farm with a record of seven wins in seventeen starts and earnings of $594,278.

Tri Jet's forty-eight stakes winners (9 percent) included thirteen bred by Hooper, including a later Tri Jet—Susan's Girl foal, Paramount Jet. Another was Tri To Watch, winner of the 1991 Champagne Stakes. Similarly, of Copelan's thirty-seven stakes winners (10 percent), Hooper bred fourteen.

Throughout the seventies Hooper's early 1960s star Crozier was providing his breeder with a number of good horses, including Futurity winners Wedge Shot (1973) and Crested Wave (1978). (Crested Wave became leading sire in New Zealand in 1991.) Hooper eventually bred nineteen stakes winners by Crozier with the top two being Journey At Sea and the champion Precisionist.

Journey At Sea was a 1979 foal from Mararjee, a Sailor mare from Hooper's own Gardenia winner Main Swap. Many of Hooper's horses excelled on various circuits. Journey At Sea was very much a favorite on the West Coast, where he won the Swaps Stakes, Silver Screen Handicap, San Bernardino Handicap, and two other stakes at three and five; he only made one start at four. He won eight of seventeen races and earned $564,950.

Precisionist was a star on both coasts, and over several years. He was a 1981 Crozier colt from Excellently, who was by the Argentine wonder *Forli, imported to race for and then stand at Claiborne Farm. Precisionist was a monument to the many years Hooper had been breeding horses, for, in addition to his being by Crozier (with all his Hooper connections), the mare Excellently was a daughter of Poliniss, a distinguished producing daughter of the Hooper homebred Greek Game. Precisionist, then, was inbred 3x4 to the early acquisition Olympia.

Precisionist had a wide range of talents. A moderate stakes winner at two, he emerged at three in 1984

to follow Journey At Sea as a winner of the Swaps Stakes, one of the most important races for three-year-olds on the West Coast. He also took the Del Mar Invitational, San Miguel, and San Rafael. At the end of that season, he won the Malibu Stakes to launch his assault on Santa Anita's Strub Series, then a highly acclaimed series won by only four other horses (Round Table, Hillsdale, Ancient Title, and Spectacular Bid). As a four-year-old, Precisionist added the San Fernando Stakes and the final race in the series, the Charles H. Strub.

Thus acclaimed early in the year as performing "excellently" at one and a quarter miles in sunny California, the rich chestnut ended the season in overcast New York, winning the six-furlong Breeders' Cup Sprint! He was voted champion sprinter, but the title gave only one side of his talents.

At five Precisionist was sent back west and back to middle distances, and he won three major stakes in California, including the Californian. Returned to New York, he defeated Horse of the Year Lady's Secret in the Woodward and ran second in the Marlboro Cup. He then headed back to the West Coast where he won a minor stakes and ended his season with a third in the mile and a quarter Breeders' Cup Classic.

A horse with the class, pedigree, and versatility to be the next chapter in Hooper's long tale of breeding success, Precisionist was soon shown to be short in one aspect: fertility. After a year and a half he was back at the races, and, at seven, won two additional stakes before becoming more or less a pensioner, a seeming vein of ore frustratingly closed. Precisionist's final career tallies were twenty wins in forty-six races, placings in fourteen others, and earnings $3,485,398. He was elected to the Racing Hall of Fame in 2003. Exhaustive veterinary work managed to produce a few foals, and one of Hooper's middle 1990s stakes winners, Draw Again, was by Roman Diplomat out of the Precisionist mare Preciseness.

The flow of stakes winners continued until Hooper's death in 2000, but few of them reached the upper grades. One of the last of the top-class horses Hooper bred was Diplomatic Jet, who contended for an Eclipse Award on the grass in 1996.

Diplomatic Jet was another result of long-term Hooper bloodlines with outside sources. His sire, Roman Diplomat, a homebred, was only an allowance winner. However, Roman Diplomat's bloodlines prompted Hooper's decision to use him at stud. Roman Diplomat was by the superb international stallion Roberto and out of Goddess Roman, whose sire, Chieftain, was a stakes winner by Bold Ruler. Bold Ruler, eight-time leading sire, figured in the sire line of only four of Hooper's stakes winners. Goddess Roman was one of five stakes winners from Hooper's prolific mare Roman Goddess, yet another useful daughter of Olympia.

Precious Jet, dam of Diplomatic Jet, was by Tri Jet out of a daughter of Beat Inflation. Beat Inflation, who won the 1977 Hollywood Express and Los Angeles handicaps, was by homebred Crozier out of the *Quibu mare Teacher's Art, a stakes winner whose broodmare sire was — are you surprised? — Olympia. One more touch: The fourth dam of Diplomatic Jet was Bright Holly, a stakes-winning homebred by Olympia. Thus, Precious Jet was inbred to Olympia with four crosses, and her son Diplomatic Jet had another cross of the old speedster.

If this sentimental combo might suggest a horse to race against the Stella Moores of the modern world, it is interesting to note that Diplomatic Jet excelled at extended distances (at least by current American racing standards) and on the grass at that. At four in 1996, he won a series of the best turf races in America, taking the Turf Classic and Man o' War in the autumn after earlier wins in the Early Times Manhattan Stakes and W.L. McKnight Handicap. He had won two previous stakes and had a career record of nine wins in fifty-one starts and earnings of $1,267,202.

Another important winner by Roman Diplomat was Roman Envoy, who in 1992 won the Kelso Handicap and had four other stakes wins. Roman Diplomat sired eight stakes winners (7 percent), one more example of Hooper's touch with bloodstock. Counting Susan's Girl, which seems just, Hooper bred 117 stakes winners. After his death his estate sold seventy-six horses at the 2000 Ocala Breeders' Sales Company's fall mixed sale. They brought $1,731,900 to average $22,788, not high by today's standards but in its own way a tribute to the practical approach he had made work so well and so often.

Fred W. Hooper was appropriately cast as one of the grand old men of American racing for quite a long time. He was honored by the Thoroughbred Club of America as its Testimonial Dinner Guest in 1981 and was voted various Eclipse Awards — for breeder in 1975 and 1982 and the Eclipse Award of Merit in 1991.

Hooper's knowledge of soils made him a sought out consultant on racetrack design, cushions, and maintenance, and he was founder of the American Thoroughbred Owners Association, one of the forerunners of the present day Thoroughbred Owners and Breeders Association. Even as he grew elderly, Hooper gave of his energy and common sense to seeking favorable legislation for racing in Florida, and he was president of the Florida Thoroughbred Breeders' Association through much of the 1970s. Calder Race Course in Florida established a graded stakes in his name during his lifetime. Hooper, briefly a schoolteacher as a Georgia stripling in a distant day, founded the Hooper Academy in Alabama to improve education in one of the states that were particularly important to him.

Late in Hooper's life he continued to be active while tended to by his second wife, Wanda. (He had been widowed many years before and had a long second marriage with Wanda.) In 1997 a festive hundredth birthday party in Ocala was presided over by the Hoopers, but to anyone there who might have thought of it as a goodbye, the sentiment was premature.

Fred Hooper Jr. — namesake of a Derby winner, breeder of record of Susan's Girl — died on July 23, 2000. Grief perhaps weakened the will of his father, who died on August 4. Mr. Hooper was 102.

Fred W. Hooper (1897–2000) lived in three centuries. This is not the same as living for three centuries, but his was a lasting highway of solid foundation all the same.

Champions Bred

Susan's Girl
1972 champion three-year-old filly
1973 champion handicap female
1975 champion handicap female

Precisionist
1985 champion sprinter

Stakes Winners Bred by Fred W. Hooper

Eddie Sue, br.f. 1950, Education—Wise Sue, by Menow

Fierce, ch.f. 1952, Pry—Fierceness, by *Bull Dog

Hoop Band, b.g. 1953, Hoop, Jr.—Patricia P., by Pilate

Ezgo, b.g. 1954, Olympia—Patricia P., by Pilate

Greek Game, br.c. 1954, Olympia—Sunday Supper, by Questionnaire

Gridiron (Stp.), br.h. 1954, Olympia—Little Lea, by Bull Lea

Alhambra, br.c. 1955, Olympia—Robins Charm, by Epithet

Olymar, b.c. 1955, Olympia—Valdina Marl, by Teddy's Comet

Confuse, dk.b/br.f. 1957, *Quibu—Pry Miss, by Pry

Dotty Kirsten, b.f. 1957, Count Fleet—Delphis, by *Heliopolis

Hoop Bound, b.g. 1957, Hoop, Jr.—Patricia P., by Pilate

Piagal, b.f. 1957, Olympia—Gallonia, by *Sir Gallahad III

Teacation, b.f. 1957, *Quibu—Teaching, by Education

Winonly, b.c. 1957, Olympia—Sickle Sun, by *Sickle

Bright Holly, ch.f. 1958, Olympia—Holiday Time, by *Nasrullah

Crozier, b.c. 1958, *My Babu—Miss Olympia, by Olympia

My Portrait, b.f. 1958, Olympia—Me, by Challenge Me

Admiral's Voyage, dk.b/br.c. 1959, Crafty Admiral—Olympia Lou, by Olympia

Main Swap, ch.f. 1960, Swaps—Mainpoint, by *Ambiorix

Sky Gem, br.c. 1960, *Quibu—Haze, by Olympia

Miami Mood, dk.b.f. 1961, Greek Game—Hoop Mood, by Hoop, Jr.

Symbolize, ch.g. 1961, Pry—Ianna, by Olympia

Tinsley, b.c. 1963, *Quibu—Haze, by Olympia

Olympia Site, ch.c. 1964, Olympia—Me, by Challenge Me

Teacher's Art, ch.f. 1964, *Quibu—Art Teacher, by Olympia

Paderoso, ch.c. 1966, Crozier—Ocean Maid, by Sailor

T. V. Doubletalk, ch.c. 1966, T. V. Lark—Bright Holly, by Olympia

Goddess Special, b.f. 1967, Specialmante—Roman Goddess, by Olympia

Authorize, b.c. 1968, Crozier—Teaching, by Education

Mia Mood, b.f. 1968, Crozier—Miami Mood, by Greek Game

Pitching Wedge, dk.b/br.c. 1968, Crozier—Poliniss, by Greek Game

Bright Bright, dk.b/br.f. 1969, Prove It—Bright Holly, by Olympia

Cute Sketch, dk.b/br.f. 1969, Gentle Art—Educationist, by Education

Selecting, b.g. 1969, Crozier—Pretty Pat, by Swaps

Special Satin, ch.f. 1969, Specialmante—Satins Joy, by Johns Joy

Tri Jet, dk.b/br.c. 1969, Jester—Haze, by Olympia

Video Reception, b.c. 1969, T. V. Lark—Maule, by Olympia

Crocation, b.c. 1970, Crozier—Teacation, by *Quibu

Gleam of Hope, b.f. 1970, Sensitivo—Molly O'Day, by *My Babu

Goddess Roman, b.f. 1970, Chieftain—Roman Goddess, by Olympia

Visualizer, b.c. 1970, Crozier—Visualize, by *Quibu

Quaze Quilt, ch.f. 1971, Specialmante—Quaze, by *Quibu

Special Goddess, b.f. 1971, Specialmante—Roman Goddess, by Olympia

Special Team, b.f. 1971, Specialmante—Teampia, by Olympia

Wedge Shot, dk.b/br.g. 1971, Crozier—Poliniss, by Greek Game

Beat Inflation, ch.c. 1973, Crozier—Teacher's Art, by *Quibu

Beau Talent, dk.b/br.c. 1973, Beau Gar—Main Talent, by Olympia

Bound for Pleasure, b.f. 1973, Alabama Bound—Pleasure Revoked, by Revoked

Henschel, b.c. 1974, Crozier—Heavenly Flight, by Bald Eagle

Joyous Ways, ch.f. 1974, Daryl's Joy—Mariways, by Maribeau

Miami Sun, dk.b/br.c. 1974, Crozier—Miami Mood, by Greek Game

Wavy Waves, ch.f. 1974, Crozier—Fading Wave, by *Quibu

B. W. Turner, b.c. 1975, Daryl's Joy—Roman Goddess, by Olympia

Eximious, b.f. 1975, Crozier—Poliniss, by Greek Game

Crested Wave, dk.b/br.c. 1976, Crozier—Fading Wave, by *Quibu

Infusive, dk.b/br.c. 1976, *Noholme II—Fuzier, by Crozier

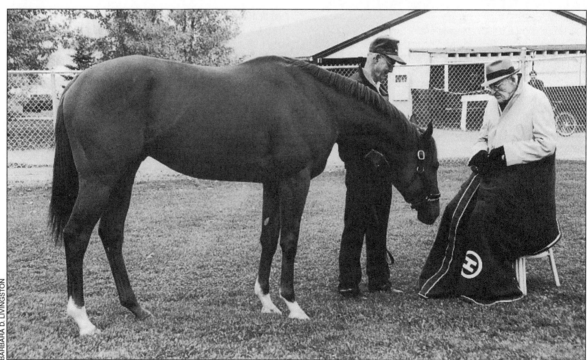

Diplomatic Jet with Fred Hooper

Noble Warrior, b.g. 1976, Roberto—Teacation, by *Quibu

If This Be So, ch.c. 1977, Secretariat—Roman Goddess, by Olympia

Sportful, b.c. 1977, Crozier—Pretty Bu, by *Quibu

Jetzier, b.c. 1978, Tri Jet—Patzier, by Crozier

Leader Jet, dk.b.g. 1978, Tri Jet—Pretend Joy, by Daryl's Joy

Motivity, b.c. 1978, Crozier—Polly N., by *Quibu

Ocean Sunset, b.f. 1978, Tri Jet—Ocean Maid, by Sailor

Pat's Joy, dk.b.f. 1978, Daryl's Joy—Queen Pat, by Crozier

The Payoff, dk.b.g. 1978, Daryl's Joy—Miss Rossean, by Intentionally

Skillful Joy, ch.f. 1979, Nodouble—Skillful Miss, by Daryl's Joy

Advance Man, ch.c. 1979, Crozier—Exchange Rate, by Maribeau

Journey At Sea, ch.c. 1979, Crozier—Mararjee, by Sailor

Copelan, b.c. 1980, Tri Jet—Susan's Girl, by Quadrangle

Flying Lassie, dk.b.f. 1980, Tri Jet—Delta Flight, by Delta Judge

Inflation Beater, ch.c. 1980, Beat Inflation—Pleasure Outing, by What a Pleasure

Chimes Keeper, b.c. 1981, Chieftain—Roman Chimes, by Crozier

Diachrony, b.f. 1981, Tri Jet—Exchange Rate, by Maribeau

Homecoming Game, b.g. 1981, Nodouble—Miami Game, by Crozier

Precisionist, ch.c. 1981, Crozier—Excellently, by *Forli

Shuttle Jet, dk.b.g. 1981, Tri Jet—Queen Pat, by Crozier

Protect Yourself, dk.b.c. 1982, Lord Rebeau—Fleeting Maid, by Crozier

Bawlmer Vision, ch.g. 1983, Visualizer—Bawlmer Queen, by Vertex

Jetting Home, b.c. 1983, Tri Jet—Mariways, by Maribeau

Lucky Rebeau, ch.c. 1983, Lord Rebeau—Petite Maid, by Quack

Contact Game, dk.b.c. 1984, Tri Jet—Miami Game, by Crozier

Ideal Change, b.f. 1984, Tri Jet—Ideal Exchange, by Crozier

Jet Pro, b.c. 1985, Tri Jet—Prozier, by Crozier

Paramount Jet, dk.b.c. 1985, Tri Jet—Susan's Girl, by Quadrangle

Basic Exchange, b.c. 1986, Copelan—Ideal Exchange, by Crozier

Classic Value, b.f. 1986, Copelan—Queen Pat, by Crozier

Copelan's Game, b.c. 1987, Copelan—Miami Game, by Crozier

Jousting Match, dk.b.g. 1987, Medieval Man—Agile Princess, by Lord Rebeau

Quality Exchange, b.f. 1987, Copelan—Ideal Exchange, by Crozier

Roman Envoy, dk.b.g. 1988, Roman Diplomat—Jetapat, by Tri Jet

Sea Art, ch.f. 1988, Journey At Sea—School of Art, by *Noholme II

Self Evident, b.g. 1988, Copelan—Apply Yourself, by Beat Inflation

Silver Duckling, gr.f. 1988, Copelan—Duck Foot, by Quack

Tri to Watch, dk.b.g. 1989, Tri Jet—I Like to Watch, by Mickey McGuire

Forsythe's Tune, ch.f. 1990, Forsythe Boy—Artistic Miss, by Quack

I Like to Win, b.c. 1990, Copelan—I Like to Watch, by Mickey McGuire

Insight to Cope, dk.b.f. 1990, Copelan—Precise Vision, by Visualizer

Prolanzier, b.g. 1990, Copelan—Prozier, by Crozier

Ambraco, dk.b.g. 1991, Copelan—Ambra Ridge, by Cox's Ridge

Carmen Copy, b.f. 1991, Copelan—Carmen C. P. R., by Bold Wizard

Co Art, b.g. 1991, Copelan—School of Art, by *Noholme II

Formatic, b.f. 1991, Forsythe Boy—Artmatic, by Tri Jet

Class Kris, dk.b.f. 1992, Kris S.—Classic Value, by Copelan

Diploman, dk.b.g. 1992, Roman Diplomat—Carmen C. P. R., by Bold Wizard

Diplomatic Jet, ch.c. 1992, Roman Diplomat—Precious Jet, by Tri Jet

On the Coast, ch.g. 1992, Florida Sunshine—Miami Game, by Crozier

Roman Rating, dk.b.g. 1992, Roman Diplomat—Jet Rating, by Tri Jet

A Point Well Made, b.g. 1993, Copelan—School of Art, by *Noholme II

Ms. Mostly, gr/ro.f. 1993, Copelan—Mostly Misty, by *Grey Dawn II

Regal Approval, gr/ro.f. 1994, With Approval—Regal Princess, by Royal and Regal

Sunshine Journey, ch.g. 1994, Florida Sunshine—Polly's Journey, by Journey At Sea

Diplomat's Reward, b.g. 1995, Roman Diplomat—Pat's Reward, by Copelan

Draw Again, ch.c. 1995, Roman Diplomat—Preciseness, by Precisionist

Recommended List, ch.c. 1995, Roman Diplomat—Waviness, by Copelan

Free of Charge, dk.b.f. 1997, Tour d'Or—Patajet, by Tri Jet

Whata Brainstorm, dk.b.c. 1997, Honor Grades—What a Future, by Roberto

Fred W. Hooper Jr.

Steel Pike, ch.g. 1963, Nadir—Quaze, by *Quibu

Pryson, ch.g. 1966, Pry—Heavenly Sun, by Olympia

Susan's Girl, b.f. 1969, Quadrangle—Quaze, by *Quibu

Source: *The Blood-Horse* Archives

6

Harry F. Guggenheim

Captain Harry F. Guggenheim bred forty-three stakes winners, a modest number when compared to most of those we have chosen to feature in recollection of leading breeders of American Thoroughbreds. What might be seen as a record that was low numerically was certainly high in quality and ultimate influence, however, for Guggenheim's stakes winners included an American Horse of the Year and three other United States champions; the internationally prominent sires Never Bend and Riverman; the broodmare sire of a Triple Crown winner; the sire of Blushing Groom; four Kentucky Oaks winners; the dam of Exceller; and a collection of horses that made him the leading owner in North America one season.

While his was an American-based operation, its influence was wide. He also bred an Irish Derby/English St. Leger winner and an Arc winner in Europe and one year was the leading breeder in England/Ireland. In addition, he bred the broodmare sire of the remarkable international stallion Sadler's Wells.

Guggenheim had a patriarchal, perhaps patronizing approach to the Turf. The man once remarked that a rabble shouting "Capitalist!" at him was among his "flatterers" and felt that racing should not be a career but a pastime. His father had once written to him that as he grew older he recognized "how much I miss in not having a hobby with which I might in time of need occupy myself." Harry filled that void in his own life with Thoroughbred racing as his partner.

"I have always believed that racing should be a man's outlet, even though it sometimes grows into big business," Guggenheim said. "This isn't to say that racing should not be taken seriously. It should be, but in my own case I wouldn't want it to be my sole or even chief concern." He appended that "I like my racing and everything to do with it," an accolade that perhaps could not have applied to something other than a sideline of life.

Guggenheim was born in West End, New Jersey, in 1890. He attended the races at Sheepshead Bay in New York as a child, and then the sparks of the Turf excited him further at England's Newmarket and in Paris during his time at Cambridge. He spent a few years with Guggenheim Brothers, in the copper division of the family enterprise, but much of his professional life for some years was involved in administering the philanthropies of the Solomon R. Guggenheim Foundation, which was named for an uncle. The foundation's affairs included supervision of the Guggenheim Museum in New York City.

With the luxury of financial security, Guggenheim also threw his energies into various forms of public service. Guggenheim was a flyer in World War I and in 1926 led the establishment of the Daniel Guggenheim Foundation (named for his father) to promote aeronautics. He was present when Charles Lindbergh took off from Curtiss Field on his historic trans-Atlantic flight to Paris in 1927. According to the National Aviation Museum, Guggenheim remarked, "When you get back from your flight, look me up," although he harbored serious doubts that

Lindbergh would survive the trip. Later, a cross-country tour by the hero Lindbergh promoted the safety of air travel and the feasibility of commercial passenger airlines.

The foundation also led to the first aviation weather reporting service and development of flying "blind" by instruments. In the 1930s Guggenheim threw the foundation's support behind the pioneering rocket research of Dr. Robert Goddard, and later created the Jet Propulsion Center at Cal Tech. Lindbergh regarded Guggenheim and Goddard as the two most important visionaries in the field of aeronautics.

Guggenheim returned to active duty during World War II, and after distinguished service in Okinawa and Sakishima, he retired in 1945 with the rank of captain.

In 1940 Guggenheim had purchased the Long Island newspaper *Newsday* to support the professional interests of his third wife, the former Alicia Patterson, and he became so interested that, although in his seventies, he took over as editor and publisher upon her death in 1963. He appointed former White House press secretary William Moyers as publisher in 1967 but maintained majority interest in *Newsday* until the year before his death. During the years the Guggenheim family controlled *Newsday*, it became America's largest suburban publication, with a circulation of more than 400,000. *Newsday* and its staffers won three Pulitzer Prizes.

Guggenheim also had an interest in governmental affairs and served under President Herbert Hoover as U.S. ambassador to Cuba. Many years later, in 1950, Guggenheim delivered a speech on inter-American relations that came to be looked upon as a virtual outline for the Organization of American States in its goals of cooperation between North American and South American nations.

Thoroughbred racing made its first official inroads into this vigorous and varied life in 1934, when Guggenheim bought, as he put it, "one very modestly priced yearling at the Saratoga auction sales." He styled his stable Falaise Stable for a time, Falaise being the name of his estate on Long Island. A later purchase, the $1,400 yearling Touch and Go, became his first stakes winner when he won the $1,000 Consolation Claiming Stakes in 1938. In the ironies implicit in the concept of a "claiming stakes," Touch and Go was duly claimed that day. It was a modest

Captain Harry Guggenheim

beginning of what would become an important chapter in the international history of the Turf.

During the 1960s Guggenheim authored an essay that he envisioned as the foreword for a book on his racing stable, an account the present author was privileged as a young man to write. (After the manuscript's completion, Guggenheim decided to maintain it only as a private history of his racing activities.)

"The racing and breeding of Thoroughbred horses is a fascinating sport and, in addition, in recent times, has become, in some instances, a business of major importance," Guggenheim wrote. "... I propose to summarize here some generalizations, for what little they may be worth, of this difficult and exciting sport and business of the Thoroughbred.

"No one has ever succeeded, although many have tried, in making of it an exact science. It is, at best, an art, and a very obscure one. Its fascinating interest lies in its unpredictability and in the major part that luck plays in every one of breeding and racing's many facets.

"After the event it is quite simple for an owner to tell how he did so and so, in his breeding plan, to produce such a magnificent animal. But ask him to repeat the performance and you will discover that his wisdom lasts over a certain limited period, and when his luck runs out, he runs out of the formula to produce champion animals...

"My own experience tells me that the most important element in breeding and racing Thoroughbreds is to have infinite patience and to keep animals sound. You must do everything right and after that, for success, Lady Luck must ride with you…"

Guggenheim early on perceived the classic races and other distance tests as the most satisfying goals, and in 1937 he got a taste of the element when his Vamoose ran third to the champion War Admiral in the Belmont Stakes. Two years later Guggenheim scored in one of the esteemed old races in New York when Red Eye won the Ladies Handicap. Red Eye had been one of ten yearlings purchased by trainer Sammy Smith at Guggenheim's behest as he was finding his way in developing a game plan for his operation. He had instructed Smith not to exceed an average of one thousand dollars for the ten, and Red Eye had been purchased for $850. Red Eye later foaled Fantan, whose son Ragusa would win the Irish Derby, English St. Leger, and King George VI and Queen Elizabeth Stakes to make Guggenheim the leading breeding in England/Ireland for 1963.

Another early purchase that had a lasting productivity for Guggenheim — and overtones still today in the breeding industry — was Lady Nicotine. Purchased as a yearling in 1937, the same year as Red Eye, Lady Nicotine was considerably more expensive. The daughter of *Sun Briar—Comixa, by Colin, was acquired from Willis Sharpe Kilmer for $5,700. Lady Nicotine was a winner who finished second to Donita M. in the 1938 Demoiselle Stakes. Lady Nicotine foaled Fighting Lady, a *Sir Gallahad III filly who in turn produced Armageddon. This was a link in the sire line that led to the Guggenheim-bred Horse of the Year Ack Ack, and thence to the present-day sire line of Broad Brush and Concern.

In 1942 Guggenheim came close to a championship when his filly Good Morning defeated Askmenow in the Matron Stakes. Askmenow, however, was then second in the Futurity and won the Selima to wrest the juvenile filly championship. Good Morning was bred by Mrs. R.A. Van Clief and was by the leading sire *Sir Gallahad III—Morning, by American Flag. Morning was from One Hour and thus, as matters transpired, became a half sister to a Kentucky Derby winner, as One Hour's *Sir Gallahad III colt Hoop, Jr. won the 1945 Derby. Good Morning, who cost $3,800 at Saratoga in 1941, won the Vineland and Hannah Dustin handicaps at four

in 1944. As was true of several of Guggenheim's earlier purchases, she paid later dividends, foaling the stakes winners Battle Morn and Victory Morn.

In 1943 Guggenheim, then in the service, changed the name of his racing operation to Cain Hoy Stable. "Cain hoy" was the pronunciation for "cane hay" in the lyrical Gullah dialect in the low country of South Carolina, where Guggenheim had purchased a large plantation. Cane hay is a native plant, whose leaf is eaten by cattle and whose stalk is used to make rattan furniture. The land was given the name Cain Hoy, and so was the racing stable.

The point of acquiring fillies was to establish a broodmare band, and the first stakes winner bred in the name of Harry F. Guggenheim was foaled in 1942. This was the Questionnaire—Lady Johren, by *Johren, gelding Reply Paid, who in 1945 won the Gallant Fox and New York handicaps. The next stakes winners were Battle Morn and Crafty Admiral, both foaled in 1948. From the aforementioned Good Morning, Battle Morn was by the leading sire *Blenheim II. Battle Morn won only one stakes, the Grand Union Hotel at Saratoga at two in 1950, but he was beaten only a head in the Wood Memorial, was third in the Derby Trial, and wound up the favorite in the Kentucky Derby. He finished sixth behind Count Turf. Battle Morn got the good stakes winner Warhead before being sold to Puerto Rico, where he was quartered in a paddock within view of the El Commandante Racetrack and became a public hero as well as the leading sire.

Crafty Admiral fell afoul of the reality that any breeder has to sell sometimes to keep numbers down. Cain Hoy had sold yearlings in the past, and in 1949 Crafty Admiral was entered in a paddock sale at Belmont Park. He was by Fighting Fox, a brother of Triple Crown winner Gallant Fox, and was out of Admiral's Lady, a moderate winner by the Triple Crown winner War Admiral. Admiral's Lady had died after producing the colt. Although a very late foal, born June 6, 1948, Crafty Admiral was impressive enough that Cain Hoy's trainer of the time, Moody Jolley, and another owner, Bull Hancock of Claiborne, bid on him. They stopped when Hugh Grant bid $6,500. Grant later resold the horse to Charles Cohen, who raced as Charfran Stable, and Crafty Admiral became 1952 champion older horse for that stable. He won the $100,000 Washington Park Handicap, the Brooklyn, and two runnings of the Gulfstream Park Handicap among eighteen wins from thirty-nine starts and earned $499,200.

Crafty Admiral was an influential stallion. He statistically got only twenty-six stakes winners (7 percent), but they included Admiral's Voyage, the broodmare sire of leading stallion Danzig, while the Crafty Admiral mare Won't Tell You foaled 1978 Triple Crown winner Affirmed.

Despite the sale of Crafty Admiral, things could have gone worse for Captain Guggenheim. He had been convinced more or less that he should sell his entire yearling crop that year, but he decided to keep two. They were Battle Morn and the filly Spring Run, later the dam of stakes winner Red God. Guggenheim put emphasis on learning from experience, and he found that selling yearlings before they had been tried as runners was something to avoid in the future.

The 1952 foal crop bred by Guggenheim included the stakes winners Flying Fury, Racing Fool, Happy Princess, and Lalun. The last-named was destined to lasting influence as the dam of Never Bend and Bold Reason. In the meantime, Guggenheim had made two purchases that would vault him into headlines that exceeded any Crafty Admiral might have gained for him had he been retained. Their names were Dark Star and *Turn-to, and in 1953 they combined to win the most famous race in America and the richest race in the world.

Dark Star was purchased for $6,500 at the 1951 Keeneland summer sale. This was not a high price for the time but was not very far under the sale average of $8,237. He was in the first crop by *Royal Gem II, an Australian horse who bore the questions about his suitability to American racing and breeding that any horse from that distant circuit presented, but who had won under as much as 149 pounds while taking twenty-three of fifty-one races. *Royal Gem II had been brought to stud at Warner L. Jones Jr.'s Hermitage Farm near Louisville. Dark Star's dam was Isolde, by leading sire *Bull Dog. Isolde was out of Fiji, who had won the Kentucky Oaks, Latonia Oaks, and Latonia Derby in 1934 and who had four foals before being killed in a van accident. Isolde won fourteen races in sixty-six starts and earned $10,865.

At two, Dark Star, like other two-year-olds of 1952, was overshadowed by the unbeaten juvenile champion Native Dancer. Dark Star placed in the Juvenile Stakes and Belmont Futurity. By Kentucky Derby Day the following spring, he had won the Derby Trial but was sent off at nearly 25-1 against the still unbeaten gray Dancer in the classic. Dark Star, trained by Eddie Hayward and ridden by Henry Moreno, led from the start. Native Dancer was buffeted early and had to go around horses in the stretch, and his late run was gamely repulsed by Dark Star, who won by a head. Many regarded the defeat of Native Dancer as a fluke, and, indeed, it was the only loss of his career. Dark Star, however, ran a courageous race. His opportunity to verify the result ended when he bowed a tendon in the Preakness. He had won six of thirteen races and earned $131,337.

Guggenheim bred six of Dark Star's twenty-five stakes winners (8 percent), and several were among the stallion's best, i.e., Suburban Handicap winner Iron Peg and the full sisters Hidden Talent (a divi-

TURFOTOS

Lalun

sion of the 1959 Kentucky Oaks) and Heavenly Body (1959 Matron Stakes).

*Turn-to was a year younger than Dark Star, and whereas the earlier colt arrived at Derby Day as a longshot, *Turn-to seemed destined to arrive the next year as the favorite. He was bred in Ireland by Major E.R. Miville and Mrs. G.L. Hastings and purchased privately on behalf of Claude Tanner by agent Frank Moore O'Ferrall. Tanner passed away, and his widow asked Bull Hancock of Claiborne Farm to sell the colt for her. By that time Guggenheim had become one of the many successful owners who were clients of Claiborne, and Hancock liked the colt well enough to recommend him for Cain Hoy. He bid $20,000 to acquire him for Guggenheim at the Keeneland summer sale.

*Turn-to was by the Nearco stallion *Royal Charger, whose standing as a source of stamina was questioned at the time. However, the colt was out of the mare *Source Sucree, whose sire, Admiral Drake, had won the one and seven-eighths mile Grand Prix de Paris and was a son of the great producer Plucky Liege (also dam of *Sir Gallahad III, *Bull Dog, and Bois Roussel). *Source Sucree was a half sister to the crack French two-year-old *Ambiorix and to the dam of English Two Thousand Guineas winner *My Babu. The *Royal Charger colt had been named Source Royal, but while Guggenheim liked the physical being and pedigree, he was not keen on the name. He often used nautical names for his horses, and he changed this one to *Turn-to, sailor jargon for something like "get to work."

Hayward brought the colt out in August at Saratoga. *Turn-to won his first start and then inherited the Saratoga Special when apparent winner Porterhouse fouled him. *Turn-to bucked his shins while running third in the Hopeful, but for the first time in a half-century having to miss the Futurity did not mean missing the climax of the juvenile season. Garden State Park in New Jersey that year had swelled the purse of the Garden State Stakes to make it the richest race in the world, and even the traditional stables had to take notice. *Turn-to was back to wear the captain's blue-and-white-block Cain Hoy silks, and at 14-1 he raced forwardly throughout and took over to win by two lengths. The winning purse was $151,282. Porterhouse was voted champion, but *Turn-to was ranked equal to him, at 126 pounds, on the Experimental Free Handicap.

The following winter *Turn-to was a dashing winner of the Flamingo Stakes, then the most important Florida race for classics contenders, and he was the reigning favorite for the Derby until he was found in his stall at Keeneland one morning with a bowed tendon. He was retired to Claiborne with six wins in eight starts and earnings of $280,032.

Although he would become known for unsoundness in a considerable percentage of his offspring, *Turn-to proved a highly influential stallion. He sired twenty-five stakes winners (7 percent), of which Guggenheim bred six. The sons of *Turn-to bred on magnificently, and the line once suspect for stamina became associated with classic winners that excelled at a mile and a half. His best sons at stud included Hail to Reason, Sir Gaylord, First Landing, Best Turn, and Cyane. (An unusual career was authored by one of Cain Hoy's homebreds by *Turn-to, the colt All Hands. Although he never won a stakes, All Hands placed in fourteen and famously pushed Kelso to the limits in the Metropolitan Handicap. All Hands wound up at stud in Mexico, but while still in the United States he sired the 1971 Belmont Stakes winner Pass Catcher.) Guggenheim's later champion Ack Ack was out of a *Turn-to mare. In 1958, when his first two-year-olds included champion First Landing, *Turn-to was syndicated for $1.4 million.

First Landing was a homebred for C.T. Chenery, who along with Guggenheim and John W. Hanes had been tapped by New York's racing leadership to address the crisis facing the state's Turf. Traditionally the leader in American racing, New York's version of the sport was declining in prestige and financial health. Different groups owned the various tracks, most in deteriorating condition. Other states, meanwhile, were gaining ground. Guggenheim might have been pleased to ship across the state line and win the richest race in the world, but he did not relish seeing New York's prestige diminished. The three entrepreneurs, members of The Jockey Club, devised the plan to close Jamaica racetrack and replace the old Aqueduct with a super new version, leaving the state three major tracks. Belmont and Saratoga were the others. This placed the ownership under the New York Racing Association, which for nearly a half-century has been the structure of the state's Turf, successful despite being thrashed from time to time by political whims, internal problems, and the difficulties in adapting to the eras of off-track betting. Later

Guggenheim was at odds with most New York racing leaders when Belmont Park was to be rebuilt. He and famed architect Frank Lloyd Wright had visions of a grandstand-clubhouse design that was a considerable departure from traditional racetracks, featuring more height to give more fans a close angle to the finish line, but they did not carry the day.

Through the remainder of the 1950s, Cain Hoy continued to produce a series of major stakes winners. Guggenheim liked the concept of a team, all loyal and seeking the same goals. He hired Kentucky horseman Woody Stephens as his private trainer, and in the glory days Cain Hoy was one of the last remaining stables to have a contract jockey, the strong and skillful Panamanian Manuel Ycaza. Cain Hoy statistically reached the top in 1959, leading the owner's list with earnings of $742,081. The stable ranked second to C.V. Whitney the next year.

In 1955 Cain Hoy won the first of its four Kentucky Oaks, with Lalun. Oaks winners Hidden Talent, Make Sail, and Sally Ship, all also homebreds, would follow. Stephens had not yet come aboard, and Lalun was trained by Loyd (Boo) Gentry. (Flying Fury and Racing Fool tried for the Derby for Cain Hoy the day after the Oaks but could not handle a field including Swaps, Nashua, and Summer Tan.) In the naming of Lalun, Guggenheim turned not to the sea, but to literature, Kipling to be specific. The dam of the filly was Be Faithful, and in the poem "With Scindia to Delhi," Lalun was the faithful lover of an Indian prince.

Be Faithful was a high-class filly from Colonel E.R. Bradley's Idle Hour Stock Farm. She was by Bimelech and out of the first Broodmare of the Year, Bloodroot, by Blue Larkspur. She had had several owners, and she was among eight mares leased by Guggenheim from the potent bloodstock holdings of John S. Phipps. She was in foal to *Djeddah, the Eclipse and Champion Stakes winner representing the sire line of the French Tourbillon, and her 1952 foal was officially credited to lessee Guggenheim as breeder.

Lalun won the Beldame Stakes and Pageant Stakes as well as the Kentucky Oaks in 1955. She escaped tragedy in the Beldame. In the fifteen-horse field, Gainsboro Girl ran up on her heels and nearly cut her down. Gainsboro Girl fell, and Rico Reto and Open Sesame then fell over her. Gainsboro Girl had to be destroyed and the former champion rider Tony DeSpirito was gravely injured, although he survived.

Lalun defeated a strong, if reduced field, including Searching, Countess Fleet, Myrtle's Jet, and Clear Dawn. Given what Lalun would accomplish as a broodmare, the fine line between her being an accident victim or the winner of the 1955 Beldame is something upon which to ponder.

In addition to her victories, Lalun finished second to High Voltage in the Coaching Club American Oaks. Cain Hoy had won that premier distaff test of stamina the year before with yet another bargain purchase, Cherokee Rose. A *Princequillo filly, Cherokee Rose was bred by Mrs. Audrey Emery and was purchased privately as a yearling by Guggenheim. Cherokee Rose was destined to be the second dam of Horse of the Year Ack Ack.

In 1958 Guggenheim's Red God won the seven-furlong Roseben Handicap. This seemed of no particular moment in the history of the Turf at the time, but it was to have its ramifications over an ocean and an era. For some years Guggenheim sent a few of his homebred prospects to England to be trained by Captain Cecil Boyd-Rochfort. Red God was one of these. A foal of 1954, he was by the leading Claiborne stallion *Nasrullah, in whom Guggenheim had five shares, and out of the Menow mare Spring Run, who had been saved out of the crop of Crafty Admiral. In the interim Spring Run had foaled the Blue Grass Stakes winner Racing Fool (by Jet Pilot) for Cain Hoy.

Red God won the Richmond Stakes and finished second in the Champagne Stakes at two in England. He was returned to America with the classics in mind. He won his first race but was injured, and the Roseben at four was the only later stakes victory of his career. Sent again across the ocean, he stood at stud in Ireland and at the age of nineteen covered the Wild Risk mare Runaway Bride and thus became the sire of Blushing Groom. A brilliant miler who stayed well enough to place in the Epsom Derby, the Aga Khan's Blushing Groom became one of the great international sires of the later years of the twentieth century. Blushing Groom's sons include the Epsom Derby winner Nashwan, the Arc winner Rainbow Quest, brilliant juvenile champion Arazi, miler Rahy, sprinter Mt. Livermore, older male champion Blushing John, and American distaff champion Sky Beauty. The sire line tracing from the 1958 Roseben winner remains prominent in the early twenty-first century.

In 1959, the year Cain Hoy led the owner's list, the star of the stable was another who had been sent first to Boyd-Rochfort. This was another *Nasrullah colt, the tall and elegant but willful Bald Eagle. The dam of Bald Eagle, Siama, was another bargain, she by *Bull Dog's Arlington and Washington futurities winner Tiger. The female family had several very strong old American elements. Siama's dam, China Face, was by the high-class Fair Play stallion Display and out of the Sweep mare Sweepilla.

Siama was bred by E.K. Thomas and initially raced for a $7,500 claiming tag. After she had placed in a couple of stakes at two, Guggenheim purchased her for $22,500. She was to prove a classy filly and had championship possibilities in the spring and summer of 1950 when she defeated eventual champion Next Move in the Acorn and beat Busanda in the Monmouth Oaks. She tailed off for a time but came back to win the Comely Handicap before her retirement the next year. Siama produced two stakes winners by *Nasrullah in the champion Bald Eagle and the good-class turf specialist One-Eyed King, and an additional stakes winner, Dead Ahead, by *Turn-to. Siama was Broodmare of the Year for 1960.

During his stay in England, Bald Eagle brought back memories of his irascible sire, *Nasrullah, who was consistently seen as an underachieving colt of virtually limitless potential. Bald Eagle won his only start at two and became viewed as a plausible candidate for the ultimate of English targets, the Epsom Derby. The following spring he won the Craven Stakes but failed stunningly in the Two Thousand Guineas. Although he rebounded to win the Dante Stakes, he was unconvincing in so doing, and his retreat from a promising position to finish twelfth in the Derby was not universally a surprise. He was then third in the St. James's Palace Stakes at a mile and was more or less seen as a disappointment as he slunk back to America.

The following summer, with Stephens having given him plenty of time to adjust to training on dirt tracks, Bald Eagle burst back into attention by winning the historic Suburban Handicap. In the autumn Guggenheim and Stephens decided to put him back on the grass, and he finished fourth in the Man o' War but was still invited to the Washington, D.C., International, a pioneering international race then only seven years old. Bald Eagle turned in a devastating performance, dominating in front from the beginning of the mile and a half event and winning by two and a half lengths over Midnight Sun.

Switched back to dirt, he then won the Gallant Fox Handicap at a mile and five-eighths.

At five in 1960, Bald Eagle became the champion older horse in American racing. His versatility was striking. The campaign included a victory in track-record time in Hialeah's Widener Handicap at a mile and a quarter, a companion victory in the Gulfstream Park Handicap at the same distance, and then a stunning mile victory in the Metropolitan Handicap in 1:33 3/5 — a track record at Aqueduct and only two-fifths of a second away from the world mark. Bald Eagle carried 128 pounds in the Metropolitan. That three-race scenario found him defeating champions Sword Dancer and First Landing as well as On-and-On and *Amerigo.

Bald Eagle was beaten under 133 pounds and 130 pounds in the Suburban and Brooklyn handicaps,

BELMONT PARK PHOTO

Siama

respectively, but won the one-mile Aqueduct Handicap over the speedy Intentionally. He then became one of the first of five years' worth of victims of the durable Kelso in the autumn weight-for-age races in New York. Back at Laurel, Bald Eagle won a second Washington, D.C., International on grass and clinched the older male championship. Bald Eagle had won twelve of twenty-nine starts and earned $692,946. He was syndicated for $1.4 million and was retired not to Claiborne, but to Leslie Combs II's Spendthrift Farm, to which Guggenheim had transferred his breeding stock after a falling out with Bull Hancock. Guggenheim also brought aboard as a pedigree adviser the worldly Humphrey S. Finney, president and later chairman of Fasig-Tipton Sales Company.

Bald Eagle was by and large a disappointing sire, getting only twelve stakes winners (5 percent), but for Guggenheim he sired the speedy filly Too Bald (Broodmare of the Year many seasons after Guggenheim's death) and the 1972 Prix de l'Arc de Triomphe winner San San (sold in Guggenheim's dispersal).

Another stable star who helped propel Cain Hoy to the top of the earnings list in 1959 was the homebred juvenile filly Heavenly Body. A daughter of Derby winner Dark Star, Heavenly Body was foaled from the *Nasrullah mare *Dangerous Dame. Heavenly Body was a year younger than her full sister Hidden Talent, who won the Kentucky Oaks and Ashland Stakes the same year (and later foaled Too Bald). The dam, *Dangerous Dame, was in one of *Nasrullah's final European-sired crops and had been imported by Guggenheim. She was out of Lady Kells, by the Hyperion stallion His Highness, and the next dam,

Anyway, had foaled the Irish Two Thousand Guineas winner Solonaway.

Heavenly Body won the Princess Pat Stakes in Chicago and then captured the historic Matron back in New York. She then tried the Gardenia, filly companion of the Garden State Stakes, but she was third behind My Dear Girl, who also won the Frizette and the juvenile filly championship. Heavenly Body won three of six races and earned $156,115. She had chips removed from a knee and did not get back to the races, but as a broodmare she foaled A Thousand Stars, a stakes winner in France and America, as well as Tobira Celeste, the dam of English champion Celestial Storm. More than thirty years after her own racing career, the name Heavenly Body popped up again as the third dam of the remarkable Sakhee, winner of the Prix de l'Arc de Triomphe and second in the Breeders' Cup Classic of 2001. (Sakhee's sire, Bahri, was by the Guggenheim-bred Riverman.)

Perhaps the greatest influence of a Cain Hoy homebred internationally was that authored by

Bald Eagle with Woody Stephens and Harry Guggenheim

Never Bend, the *Nasrullah colt that the aforementioned Lalun foaled in 1960. After Bald Eagle's retirement, Guggenheim had written to Ycaza, "I told Woody (Stephens) that one day I would give him a horse that all you have to do was put on the saddle and say 'go.' Such a horse was (Bieber-Jacobs Stable's) Hail to Reason. He was easy to train and easy to ride."

Only two years later Guggenheim made good on his rhetorical promise. The colt was Never Bend. Speed was his hallmark, and if his free-running style perhaps helped account for his defeats in the Sapling, Arlington-Washington Futurity, and Garden State, it did not cost him a championship. Never Bend won the Cowdin, Futurity, and one-mile Champagne so convincingly that he was voted the 1962 juvenile colt title. That same year Jaipur was the champion three-year-old colt, the pair raising *Nasrullah's number of American champions to nine. Guggenheim, also breeder of Bald Eagle, was the only breeder to account for more than one of those champions.

At three Never Bend won the Flamingo and prompted Guggenheim to think in international terms. The owner nominated him to the Epsom Derby as well as to the Kentucky Derby. The colt, however, just did not have the combination of stamina and style to be a classic winner in this country, finishing second in the Kentucky Derby and third in the Preakness, and was not tried in England. He later won the Yankee Handicap and was second to Mongo on grass in the United Nations and to Kelso — still around and going strong — in the Woodward. Early the next year Never Bend had recurring ankle trouble and was retired, being syndicated for $1,225,000. He went to Spendthrift with a record of thirteen wins from twenty-three races and earnings of $641,524.

If Never Bend had, in fact, been sent to Epsom, Guggenheim might have found himself running two colts he bred and owned against another that he had bred and sold. From the same crop came Ragusa, a son of the European champion *Ribot and out of Fantan, daughter of that early $850 purchase Red Eye. Fantan, by *Ambiorix, was booked in 1960 to Red God in Ireland. Sent overseas in advance, she foaled a smallish *Ribot colt. The following year, when Guggenheim asked Boyd-Rochfort if he would take the colt, the trainer asked if he could bypass the privilege since the youngster had been such a late foal and

was so small. Accordingly, the *Ribot—Fantan colt was sent to the Ballsbridge Sales, where trainer Paddy Prendergast purchased him for 3,800 guineas to race for James R. Mullion. From the same foal crop, Cain Hoy's Dark Star colt Iron Peg was sent over from America to Boyd-Rochfort as planned earlier.

Under the name Ragusa, the *Ribot—Fantan colt was third behind Relko in the Epsom Derby; Iron Peg finished fourteenth. Ragusa then won the St. Leger and the Irish Derby and defeated older horses in the King George VI and Queen Elizabeth Stakes. His earnings for the year of 114,745 pounds were a record for a European campaigner and made Harry Guggenheim the leading breeder in the British Isles. The following year, Ragusa won the Eclipse Stakes and his half sister, the Red God filly *Ela Marita, also sold at auction, won the Musidora Stakes and Fred Darling Stakes.

In 1964, Iron Peg — like Bald Eagle — was back in America for Woody Stephens to deal with him. Stephens brought Iron Peg along to score a dramatic victory over Kelso in the historic Suburban Handicap. Under 116 pounds and getting fifteen from the champion, Iron Peg (Dark Star—Hostage, by Roman) held off Kelso to win by a head, much as his sire had done when confronted with Native Dancer. Olden Times was third. Iron Peg suffered a hairline fracture a few weeks later but he had vindicated the thought that he had inherent quality.

In the autumn of 1965, Woody Stephens left Cain Hoy. By then, the trainer had enough success that he had a small breeding and racing operation in his wife's name, and racing officials deemed it a conflict of interest for his wife to run their horses, potentially in competition with Cain Hoy. Guggenheim turned to Roger Laurin. That young trainer had several stakes winners for Cain Hoy over the next couple of years, and then in 1967 scored a major triumph when the *Turn-to colt Captain's Gig won the historic Futurity Stakes. That little colt was out of Make Sail, Cain Hoy's 1960 Kentucky Oaks winner. Make Sail was by *Ambiorix and was out of Anchors Aweigh, a Devil Diver mare whom Guggenheim had purchased through an agent for five thousand dollars at the 1950 Keeneland summer sale. Anchors Aweigh had won the Jasmine Stakes, and in Make Sail she produced a very high-class filly. At two in 1959, Make Sail — in the same Cain Hoy crop as Heavenly Body — won the Schuylerville Stakes at

Saratoga, and at three she won the historic Alabama Stakes there, in addition to the Kentucky Oaks. She added the Top Flight Handicap at four and had a career record of seven wins from forty-one starts and earnings of $191,815. (The dam, Anchors Aweigh, also foaled the Never Bend colt Never Bow, winner of the 1970 Widener and 1971 Brooklyn handicaps and two other stakes.)

Captain's Gig was a free-running sort that ran himself into the ground in the 1968 Kentucky Derby. He had escaped disaster when the van bringing him from winter quarters in Columbia, South Carolina, to Kentucky caught fire, and without time for an unloading ramp, he and other horses had to be leapt to safety.

By that time Laurin had left the stable and been succeeded by Woody Stephens' brother, Bill. (Guggenheim must have been difficult to train for. While any horseman in his employ could respect his credo that sportsmanship and what was best for the horse were to guide all decisions, he was hardly the sort that trainers would praise for letting them do their job without interference. Over the years he came up with guidelines against working a horse on an off track except under special circumstances and against early racing of two-year-olds. He eventually decreed that temperatures be taken of all horses before they worked in the mornings, and he once called in a personal physician when he was convinced the trainer and veterinarians were sending an unsound horse to the racetrack.)

Captain's Gig got back on track later in the year and won the Jim Dandy Stakes. He had eight wins from fifteen starts and earned $205,312. He was a moderate success at stud.

In 1968 the Bald Eagle filly Too Bald was four and had matured late like her sire. She was out of the aforementioned Kentucky Oaks winner Hidden Talent, by Dark Star—*Dangerous Dame, by *Nasrullah. Hidden Talent earlier had foaled stakes winner Turn to Talent, by *Turn-to. The Bald Eagle filly Too Bald was a speedy sort, despite the staying capacities of her sire. She won two runnings of the Barbara Fritchie Handicap in Maryland, as well as taking the Bed o' Roses in New York and the Columbiana in Florida. Too Bald had a record of thirteen wins in twenty starts and earnings of $174,722.

The last major homebred to race for Guggenheim was Ack Ack. This colt represented a thread of success that Cain Hoy had sustained for some two decades. It had begun with Armageddon — a contradictory phrase in most contexts, but accurate here. Armageddon was a 1949 foal out of the *Sir Gallahad III filly Fighting Lady, a useful winner from the early purchase Lady Nicotine. Armageddon represented, and helped forward, a sire line that traced from the glories of James R. Keene's Domino. This particular strain of the sire line had survived repeated numerical challenges to its very existence, i.e., the early deaths of Domino and his son Commando and then the limited fertility of the latter's great son Colin.

			Phalaris (Polymelus—Bromus)
		Pharos, 1920	Scapa Flow (Chaucer—Anchora)
	Nearco, 1935		Havresac II (Rabelais—Hors Concours)
		Nogara, 1928	Catnip (Spearmint—Sibola)
*Nasrullah, 1940			Blandford (Swynford—Blanche)
		*Blenheim II, 1927	Malva (Charles O'Malley—Wild Arum)
	Mumtaz Begum, 1932		The Tetrarch (Roi Herode—Vahren)
		Mumtaz Mahal, 1921	Lady Josephine (Sundridge—Americus Girl)
NEVER BEND			Tourbillon (*Ksar—Durban)
		Djebel, 1937	Loika (Gay Crusader—Coeur a Coeur)
	*Djeddah, 1945		Asterus (***Teddy**—Astrella)
		Djezima, 1933	Heldifann (*Durbar—Banshee)
Lalun, 1952			Black Toney (Peter Pan—Belgravia)
		Bimelech, 1937	*La Troienne (***Teddy**—Helene de Troie)
	Be Faithful, 1942		Blue Larkspur (Black Servant—Blossom Time)
		Bloodroot, 1932	*Knockaney Bridge (Bridge of Earn—Sunshot)

Colin's offspring included Neddie, sire of Good Goods, he in turn the sire of Armageddon's sire, Alsab. By the time Kentucky horseman Tom Piatt submitted Alsab to the Saratoga sale ring as a yearling of 1940, the colt's lineage seemed years removed from the greatness of Domino. Moreover, he was out of a ninety-dollar mare. Alsab brought seven hundred dollars from Albert Sabbath but revived the sire line and proved one of the enduring examples of racing's great bargains, in the ranks of Stymie, John Henry, and Carry Back.

Early in his stud career the champion Alsab seemed more promising than his ultimate record proved. His early crops included the champion filly Myrtle Charm, as well as Armageddon. At two in 1951, Cain Hoy's homebred Armageddon was hit by a sharp stone in his left eye during the Champagne Stakes. It blinded the eye, but the colt persevered to win the race. Jockey Bill Shoemaker initially thought he had run greenly, for he dropped far back after the incident, before regaining his composure and closing some fifteen lengths for the victory. Armageddon also suffered from shelly feet, and he bypassed the Kentucky Derby the following spring but soon afterward defeated eventual Horse of the Year One Count in the one-mile Withers Stakes. He also won the Benjamin Franklin, Peter Pan, and Ventnor handicaps that year and was third in the classic Belmont Stakes, beaten about a dozen lengths by One Count. Armageddon won nine of fifty-one races and earned $191,700.

Alsab had gotten only 6 percent stakes winners (sixteen), and Armageddon got only 5 percent (seven). One of the Armageddons was Cain Hoy's homebred Battle Joined, whose dam was the Revoked mare Ethel Walker.

If the female family of Alsab had been lowly, it might have seemed virtually fashionable compared to the ebb in the fortunes of Battle Joined's female family. The fourth dam of Battle Joined was Milfoil (*Vulcain—Mill Maid, by Fair Play). In the Depression market of 1932, Milfoil was sold at the E.J. Tranter sales in Lexington for fifty dollars, even though she was already identifiable as the dam of a modest claiming stakes winner. Milfoil was in foal to Terry (for pedigree fans, that's Terry, not *Teddy). Presently, the offspring of Milfoil began to distinguish themselves, but not before the mare had been sold again, that time for twenty dollars. Her Time Clock (by On Watch) was a yearling at the time of

that first sale, and in 1934 he won the Florida Derby. Two later foals of Milfoil also won stakes. One of them was Mary Terry, the Terry foal she was carrying in 1932. Mary Terry won the 1935 Clipsetta Stakes, and in the next decade her Reaping Reward yearling filly, bred by Chaswil Farms, was purchased by Cain Hoy Stable for four thousand dollars at the 1948 Saratoga yearling sale. Named Ethel Terry, this filly won eight races, and she became the dam of Ethel Walker, a Revoked filly who also won for Cain Hoy. Battle Joined was Ethel Walker's second foal. In the meantime, the family had taken on airs when Mary Terry's son Inside Tract won the 1958 Jockey Club Gold Cup. There was an additional connection of this family for Guggenheim. In 1957 one of Guggenheim's daughters, Mrs. Joan Van de Maele, bought her first horse, paying $14,500 at Saratoga for the Hill Prince filly from Mary Terry's daughter Almerry (by War Admiral). Turned over to Woody Stephens, this filly, named Merry Hill, won the $50,000 Frizette Stakes two weeks after Inside Tract's Jockey Club Gold Cup.

Cain Hoy's chief standard-bearer from this mélange of genetic history, Battle Joined, was hampered by unsoundness but at two in 1961 won the Saratoga Special over Jaipur. At three he made it back to win one race, and it was a distance test, the mile and five-eighths Lawrence Realization. Battle Joined followed in his sire line's statistical mediocrity, getting thirteen stakes winners (7 percent), and by the time one of those, Ack Ack, had proven exceptional, the sire had been sent from Kentucky to stand in Colorado.

Battle Joined's knighted son Ack Ack was foaled from Fast Turn, a *Turn-to filly from Cain Hoy's 1954 CCA Oaks winner, the bargain purchase Cherokee Rose. There was depth aplenty in this female family, for Cherokee Rose was by *Princequillo and out of *The Squaw II (by *Sickle). Cherokee Rose's older full sister How also won the CCA Oaks, and her younger full sister Sequoia won the Spinaway. How, in turn, foaled the 1965 Broodmare of the Year Pocahontas, dam of champion Tom Rolfe and multiple stakes winner Chieftain. Fast Turn, dam of Ack Ack, was unraced but, given such a background, was retained for breeding.

In 1969 Frank A. (Downey) Bonsal was training for Cain Hoy, and he brought Ack Ack along to win the Derby Trial in the spring. Guggenheim coveted another Kentucky Derby but recognized his colt was hardly ready yet to take on the likes of Majestic Prince and Arts and Letters at a mile and a quarter and skipped the race. Ack Ack later won the Withers Stakes and shipped to Chicago to win the Arlington Classic.

In the autumn of 1969, Guggenheim decided it was time to cut down on his demanding activities. He dispersed virtually all the Thoroughbred stock he had gathered, nurtured, and guided over the years. He did retain Ack Ack, however, along with a few stallion shares, and he sent the colt to California to another masterful trainer, Charlie Whittingham. At four Ack Ack was kept to sprinting and won the Autumn Days Handicap and a trio of other races. On January 16, 1971, Ack Ack won the seven-furlong San Carlos Handicap in the second start of his five-year-old campaign. Guggenheim died the following week, January 22, at the age of eighty at his Falaise Estate.

A sportsman who had accomplished much, he left a continuing record of extraordinary class. The influence of Guggenheim breeding was soon to take on an even greater international aspect than it had in his lifetime.

E.E. (Buddy) Fogelson and his wife, the actress Greer Garson, purchased Ack Ack. Whittingham had discerned that Ack Ack was not necessarily limited to sprinting and had been looking to stretch him out. Ack Ack had lost his debut at five but did not lose again. By the end of the year, he had deepened his impact on American racing, taking seven races in succession, stretching to a mile and a quarter successfully while winning the Hollywood Gold Cup under 134 pounds and adding another dimension by winning the mile and one-eighth American Handicap on turf. He had a career record of nineteen wins in twenty-seven races and earnings of $636,641.

Never Bend

THE BLOOD-HORSE

Ack Ack

The Eclipse Awards were inaugurated in 1971 when the *Daily Racing Form*, Thoroughbred Racing Associations, and the National Turf Writers Association combined to vote for year-end champions. Ack Ack was the initial Horse of the Year under that championship voting scheme. He was the first horse to be voted Horse of the Year off a campaign of racing only on the West Coast, and he also was voted champion sprinter as well as older male.

From the tenuousness of the stud careers of Armageddon and Battle Joined, Ack Ack moved in with the advantage of eliciting the interest of Claiborne Farm, and he sired fifty-four stakes winners (9 percent). They included the French Derby winner and North American grass course champion Youth, who was a statistical failure at stud but did get the Epsom Derby winner Teenoso as an aberration to his overall record. (Ack Ack became the broodmare sire of another Epsom Derby winner, Benny the Dip.) The main strength in the sire line today, as handed off by Ack Ack, is through his son Broad Brush, winner of the Santa Anita and Suburban handicaps and the Wood Memorial among other important races. Broad Brush was the leading sire in North America in 1994, when his son Concern won the Breeders' Cup Classic. In 2002 Broad Brush's daughter Farda Amiga, with her distant links to Cain Hoy, won a pair of the races Guggenheim had won, the Kentucky Oaks and Alabama Stakes, and was the Eclipse Award-winning three-year-old filly. (There is an additional Cain Hoy

connection to Broad Brush, whose third dam was Hidden Talent and who is inbred 3x3 to Turn-to.)

The Guggenheim dispersal of 1969 was a blockbuster. Held at Keeneland and Belmont Park, the sale of 137 head grossed $4,751,200 to average $34,680. The gross surpassed the previous record of $4,447,650, attained over several years (1947–50) for 248 horses sold by Louis B. Mayer.

The Cain Hoy gems thus acquired by other breeders included Riverman, San San, Bold Reason — then weanlings or yearlings — and the stakes-winning young mare Too Bald. Too Bald was the top-priced mare at the dispersal, going to Charles W. Engelhard's Cragwood Estates for $225,000. Engelhard died two years later, and when the mare was returned to the auction ring in 1974, Too Bald's status had diminished markedly. Too Bald's Engelhard-bred *Vaguely Noble colt had brought only $25,000 that summer, and Gene Cashman bought the mare herself for $67,000.

Purchased by Nelson Bunker Hunt, the *Vaguely Noble—Too Bald colt was named Exceller and was sent to Europe. He won the Grand Prix de Paris and Prix Royal-Oak (French St. Leger) at three and the Grand Prix de Saint-Cloud and Coronation Cup at four. He returned to North America to win the Canadian International and the following year won seven stakes on dirt and turf, climaxed by his famous rally to edge Seattle Slew in the Jockey Club Gold Cup. Exceller earned $1,674,587.

Too Bald later foaled the 1986 Breeders' Cup Juvenile winner and champion two-year-old Capote and other stakes winners Vaguely Hidden, American Standard, and Baldski. She was voted Broodmare of the Year for 1986, following Siama (1960) as the second Guggenheim-bred mare given that honor.

The 1969 Guggenheim dispersal set individual American records for weanlings sold at auction, $110,000 for a filly (also a world record) and $50,000 for a colt. Both prices were paid for *Turn-to youngsters. Top future performers among the young stock at the dispersal, though, were Bold Reason, San San, and Riverman.

Bold Reason was by Hail to Reason and was from Never Bend's dam, Lalun, who was sixteen when she foaled him. Bold Reason brought $52,000 as a yearling in the dispersal and raced for William Levin. He matured later than his brilliant half brother but got so good for trainer Angel Penna that he won his last half-dozen races at three. They included the historic Travers Stakes as well as the American Derby on grass and the Hollywood Derby. Unlike his half brother, Bold Reason was not a particularly strong sire (twenty-one stakes winners or 6 percent), but planted his name squarely in many important pedigrees as the sire of the filly Fairy Bridge. The Claiborne Farm-bred Fairy Bridge became the dam of the remarkable, multi-time leading English/Irish sire Sadler's Wells and another important sire in Fairy King. Pedigree observers have noted that a portion of the success of Sadler's Wells has come when his bloodline was crossed with individuals bearing the blood of Cain Hoy's Never Bend, representing doubling up on the blood of a strong female (Lalun), a technique with many devotees.

Many additional European ramifications came from the Guggenheim dispersal. One of the weanlings was a Bald Eagle filly that brought only $15,000. She was purchased by Countess Margit Batthyany, owner of Haras du Bois-Roussel in France, and was named San San. The filly was from Sail Navy, whose dam, Anchors Aweigh, had foaled Make Sail. Sail Navy had been rheumatic and never raced, and she brought $25,000 at the Cain Hoy dispersal. She later was resold for only $5,500.

In 1972 the internationalist trainer Angel Penna, who had conditioned Bold Reason, had begun a lucrative stint in France and was handed another Guggenheim-bred. He developed the big, leggy San San to win the climactic Prix de l'Arc de Triomphe at three that autumn.

Another European who plied the waters of the Guggenheim dispersal was Alec Head. The noted breeder-owner-trainer in those days was concerned that the tradition of stamina in France had left too much of the national bloodstock lacking in sufficient speed to compete internationally. Although Never Bend had been seen lacking in classic stamina and had not yet erased those concerns through his offspring, he clearly had plenty of speed and class of

Champions Bred

Crafty Admiral
1952 champion handicap male

Bald Eagle
1960 champion handicap male

Never Bend
1962 champion two-year-old colt

Ack Ack
1971 champion sprinter
1971 champion handicap male
1971 Horse of the Year

In Europe:

Ragusa
1963 Irish champion three-year-old colt

San San
1972 French champion three-year-old filly

pedigree. Head bought a weanling by Never Bend—River Lady, by Prince John, for $41,000 for his long-time patrons, the Wertheimer family. River Lady had been a modest winner, but her dam, the Roman filly Nile Lily, won nine races, including the American Beauty Handicap. (Guggenheim had purchased Nile Lily for $26,000 at the 1955 Saratoga yearling sale. She was consigned by Daniel G. Van Clief's Nydrie Stud, which had purchased her from breeder Joe Metz. The next dam, Azalea, was a half sister to Walter M. Jeffords Sr.'s juvenile champion and 1945 Belmont Stakes winner Pavot).

The Never Bend—River Lady colt was named Riverman. He won a small stakes at two and the following spring won the classic Poule d'Essai des Poulains (French Two Thousand Guineas) at a mile. By that time any worries Head had harbored about the wisdom of looking to Never Bend as a source had been assuaged when Paul Mellon's Never Bend colt Mill Reef won the Epsom Derby in 1971.

Riverman won two other stakes and proceeded to a remarkable stallion career. He later was syndicated for $18 million and returned to America to stand at Gainesway Farm. Riverman sired 128 stakes winners (13 percent), including back-to-back Arc-winning fillies Detroit and Gold River, multi-millionairess Triptych, French Derby winner Policeman, the successor stallion Irish River, and the likes of Worldly Manner, Bahri (sire of Arc winner Sakhee), Lahib, Rousillon, Loup Sauvage, River Memories, and River Special. Riverman led the French sire list and ranked highly in North America and England. He segued into a leading broodmare sire, with his daughters' offspring including Epsom Derby winner Erhaab and champions Bosra Sham, Hector Protector, and Spinning World.

Despite the impressive numbers achieved by the Guggenheim-bred Never Bend stallion Riverman, it might have been Mill Reef who was the brightest jewel in the legacy of Guggenheim-bred Never Bend. Mill Reef was bred by Paul Mellon, foaled in

Virginia, and sent to England, where he demonstrated greatness by winning not only the Epsom Derby, but the Eclipse Stakes and King George VI and Queen Elizabeth Stakes, before traveling to France to win the Prix de l'Arc de Triomphe. Like his paternal granddam, Lalun, Mill Reef had a close call. At four he survived a major leg fracture followed by improvised surgery. Mellon turned down the more lucrative option of syndicating him in America and provided an opportunity to European breeders by standing Mill Reef at the English National Stud.

Mill Reef, two-time leading sire in England, sired sixty-two stakes winners (16 percent), and they included Epsom Derby winners Shirley Heights and Reference Point. Shirley Heights then sired an additional Derby winner in Slip Anchor, and another son of Shirley Heights, High Estate, took the sequence one generation further by siring the Epsom Derby winner High-Rise. Mill Reef's offspring also included French Derby winner Acamas, English One Thousand Guineas winner Fairy Footsteps, and other European group I winners such as Diamond Shoal, Glint of Gold, Wassl, and Milligram.

Among the European breeders who expressed belief in the Never Bend sire line was the Aga Khan, and he made good use of his conviction about the old Cain Hoy champion's qualities. The Aga Khan bred the Mill Reef colt Lashkari, winner of the first Breeders' Cup Turf. Another homebred by the stallion was English Two Thousand Guineas winner Doyoun, who sired Daylami, winner of the 1999 Breeders' Cup Turf.

The Aga Khan also looked to Mill Reef's son

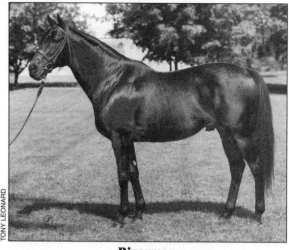

TONY LEONARD

Riverman

Shirley Heights, by whom he bred Darshaan. When Darshaan won the 1984 French Derby (Prix du Jockey-Club), it was a virtual tapestry of Cain Hoy threads, and a preview of future glories, for he defeated the distinguished horses Sadler's Wells (great-grandson of Lalun) and Rainbow Quest (grandson of Red God). Darshaan added to the Aga Khan's — and the Never Bend/Mill Reef line's — influence on the Breeders' Cup Turf when his son Kotashaan won the race in 1993, en route to an Eclipse Award as Horse of the Year. Another by Darshaan was the Aga's 2003 champion Dalakhani, winner of the Prix de l'Arc de Triomphe. Like the aeronautics industry, some of the modern Thoroughbred bloodlines took flight with a boost from the Guggenheim approach.

Stakes Winners Bred by Harry F. Guggenheim

Reply Paid, b.g. 1942, Questionnaire—Lady Johren, by *Johren

Battle Morn, b.c. 1948, *Blenheim II—Good Morning, by *Sir Gallahad III

Crafty Admiral, b.c. 1948, Fighting Fox—Admiral's Lady, by War Admiral

Armageddon, b.c. 1949, Alsab—Fighting Lady, by *Sir Gallahad III

Flying Fury, dk.b.c. 1952, *Nasrullah—Sicily, by Reaping Reward

Happy Princess, ch.f. 1952, *Princequillo—Too Sunny, by Sun Again

Lalun, b.f. 1952, *Djeddah—Be Faithful, by Bimelech

Racing Fool, b.c. 1952, Jet Pilot—Spring Run, by Menow

One-Eyed King, b.c. 1954, *Nasrullah—Siama, by Tiger

Red God, ch.c. 1954, *Nasrullah—Spring Run, by Menow

Bald Eagle, b.c. 1955, *Nasrullah—Siama, by Tiger

Victory Morn, dk.b.c. 1955, *Nasrullah—Good Morning, by *Sir Gallahad III

Amber Morn, dk.b.c. 1956, *Ambiorix—Break O' Morn, by Eight Thirty

Hidden Talent, ch.f. 1956, Dark Star—*Dangerous Dame, by *Nasrullah

Quiz Star, b.c. 1956, Dark Star—*Nut Brown Maid, by *Nasrullah

Heavenly Body, b.f. 1957, Dark Star—*Dangerous Dame, by *Nasrullah

Make Sail, br.f. 1957, *Ambiorix—Anchors Aweigh, by Devil Diver

Battle Joined, b.c. 1959, Armageddon—Ethel Walker, by Revoked

Dead Ahead, b.c. 1959, *Turn-to—Siama, by Tiger

Good Fight, b.g. 1959, Armageddon—*Nut Brown Maid, by *Nasrullah

Hoist Him Aboard (stp), ch.c. 1960, *Turn-to—*One-Eyed Queen, by *Arctic Prince

Iron Peg, b.c. 1960, Dark Star—Hostage, by Roman

Never Bend, dk.b.c. 1960, *Nasrullah—Lalun, by *Djeddah

Ragusa, b.c. 1960, *Ribot—Fantan, by *Ambiorix (bred in England; SW in England/Ireland)

Sally Ship, br.f. 1960, *Turn-to—*Nut Brown Maid, by *Nasrullah

Blinking Star, br.c. 1961, Dark Star—Peccadillo, by *Nasrullah

Pluck, b.c. 1961, Double Jay—*Nut Brown Maid, by *Nasrullah

Eagle's Top (stp), br.g. 1962, Bald Eagle—Spinning Reel, by *Alibhai

Privileged, b.f. 1962, Flying Fury—Ethel Walker, by Revoked

Sea Eagle, b.f. 1962, Bald Eagle—Rope Yarn Sunday, by *Royal Charger

Strawberry Drive, b.c. 1962, Dark Star—She Alone, by *Nasrullah

Turn to Talent, b.f. 1963, *Turn-to—Hidden Talent, by Dark Star

Too Bald, dk.b/br.f. 1964, Bald Eagle—Hidden Talent, by Dark Star

Captain's Gig, dk.b/br.c. 1965, *Turn-to—Make Sail, by *Ambiorix

Hoist Sail, dk.b/br.c. 1965, *Turn-to—*Rosaleen II, by *Jardiniere

Ack Ack, b.c. 1966, Battle Joined—Fast Turn, by *Turn-to

Never Bow, dk.b/br.c. 1966, Never Bend—Anchors Aweigh, by Devil Diver

Bold Reason, b.c. 1968, Hail to Reason—Lalun, by *Djeddah

Delay, dk.b/br.g. 1968, Decidedly—Lay Aft, by *Turn-to

Riverman, b.c. 1969, Never Bend—River Lady, by Prince John

San San, b.f. 1969, Bald Eagle—Sail Navy, by *Princequillo (SW in France)

Harry F. Guggenheim and Leslie Combs I
Wa-Wa Cy, b.c. 1959, Tom Fool—Boogooloo, by *Nasrullah

Source: *The Blood-Horse* Archives

ALLEN F. BREWER

Hidden Talent

Maxwell H. Gluck

Elmendorf Farm is one of the storied names in the history of the American Turf, and the name, intertwined with a succession of powerful men, is as resilient as the farmland that bears it is productive. Elmendorf sprawled over 10,000 acres of the Bluegrass country outside Lexington when James Ben Ali Haggin owned it in the nineteenth century. In later eras its expanse diminished as farms such as Spendthrift and Old Kenney were established on some of its parcels. In the 1940s Joseph E. Widener brought Elmendorf new glories as its owner.

In 1952 Maxwell Gluck purchased a portion of the property owned by Tinkham Veale II and S.A. Costello. The core of the property Gluck bought included handsome old barns as well as some ornamental lions and impressive columns that once guarded the entrance to a grand home constructed by Haggin. The house had been razed, but the columns became something of a Lexington landmark.

During Gluck's stewardship of some three decades, Elmendorf was the breeder of record of champions Protagonist and Talking Picture, among a series of major winners, and led the list of breeders in earnings three times — 1973, 1981, and 1982. Moreover, Elmendorf was the leading owner in earnings twice, in 1977 and 1981.

Gluck was a Texas native, but his career was somewhat at odds with the general image of the Texan. When one thinks of a Texan's clothes, it generally is not ladies apparel from an outfit carrying the corporate name of Darling stores. ("You got a rawhide belt

with a big silver buckle to cinch up that pretty little flowery frock?") This was Gluck's road to success, however. His family moved from Texas to Sharon, Pennsylvania, and established a clothing business. Gluck set his sights on New York and established the Darling brand there, producing fashionable-appearing clothes at affordable prices. Despite — or maybe because of — the timing of his first shop opening just as Wall Street crashed in 1929, Gluck built the chain into some 150 stores and became chairman of several other clothing firms.

As befits a gentleman in the clothing trade, Gluck was rather dapper, and in his later years augmented what nature had taken from his crown. He was given to natty hats and bright jackets and ties, playing to a look we tend to link with Palm Beach as well as with Del Mar, the Southern California racetrack he came to love.

By middle age, Gluck had the luxury of looking for some avocations. A Texan-cum-New Yorker, he also became taken with California, and he bought the Western Harness Racing Association track from Harry Warner. Eventually Thoroughbred racing displaced harness racing as his key hobby. In 1950, two years before he bought Elmendorf, Gluck acquired his first Thoroughbred, a $7,500 juvenile purchased at auction from Mrs. Ambrose Clark. This horse had been given the passive name of White Flag. He did not give up totally, but neither was he distinguished. The next year Gluck upped the ante to pay $17,100 for Symposium from the Samuel D. Riddle estate.

Symposium demonstrated in short order the highs and lows of the game: On an ocean voyage to Europe, Gluck received two telegrams, one announcing Symposium's victory in the Camden Handicap, the next advising him that the horse had died.

Gluck figured he needed to raise his own horses if he were to have any control of their quality, so he bought Elmendorf. During the 1950s, however, he accepted an appointment by President Dwight Eisenhower to be the U.S. ambassador to Ceylon. During this phase when that work reduced the energy he could apply to his new Thoroughbred operation, he brought on board the first in a series of excellent farm managers and advisers whose appointments exhibited his acumen at running a successful business. Lou Doherty was his manager and then was succeeded by Bob Green and later by James Brady, all three successful, hard-working, hands-on horsemen.

After he had reached the top, Gluck discussed his thoughts on his farm's success with Robert Hebert of *The Blood-Horse*: "I believe that the success of Elmendorf is based on the fact that when we were starting the farm, we were able to buy many good, quality mares at sales. At that time, it was possible to buy good mares, as some were available at almost every sale...The mares we bought were not cheap. We paid a fair price, but they were available then and they have provided a big boost to our farm...We were able to buy mares from families noted for their distance-running capabilities. We do not breed short horses. Our aim has always been for distance...The descendants of these mares, their daughters, granddaughters, and so on, are the reasons for our success now."

A key component in Elmendorf's success was Robert Bricken, who began buying for Gluck around 1953 and eventually became more than an adviser. Bricken was virtually a full-time resource. He brought to the job a cool, keen observation from a handicapping perspective, as well as a study of bloodlines and an appreciation of the relationships between performance and pedigree.

Bricken grew up amid the considerable wealth of

Maxwell Gluck

INGER DRYSDALE

his father's highly successful New York-based real estate business. He became interested in racing as a teenager on a trip to Florida with his father that was prompted in part by the young man's sinus condition. After graduating from Washington & Lee in Virginia, Bricken and his brother began attending the races on weekends, and when the brother married into the Erlanger family from New Jersey, the connection to the sport became more pronounced. Bob Bricken befriended Michael Erlanger and suggested they buy some horses as a sideline. Years later Bricken recalled that he had become an adroit handicapper by that time, the latter 1940s, but while he could read *Daily Racing Form* past performances, he was innocent of the terminology of a sale catalog. Inadvertently, he and Erlanger bought a steeplechaser, but, fortunately, Erlanger had a friend who wanted one, and they sold the horse for a quick, seven-hundred-dollar profit.

The Bricken family's earlier wealth had not funneled down in such dimension that he could go shopping for top-of-the-line yearlings, so he and Erlanger set a limit of $7,500. Bricken recalled that over some years they bought twenty-one horses, all but one winning at least one race.

Over the years the author has leaned toward the thought that quality stock and luck are the primary ingredients for sustained success in breeding. Individuals rightly seek to gain from putting into the formula other considerations, such as nicks, inbreeding or outcrossing, concentration of certain females' influence, reliance on sire lines, etc., but in the main those that succeed do so by breeding quality sires to quality female families even though they might honestly explain it differently.

Of the proponents of study, Bricken is the one who pulls us up short in our admittedly somewhat haughty generalization. In private conversations, interviews, and speeches, he was convincing that a person really *can* improve his percentages significantly by study and reflection. Gaining knowledge is not enough. The key is to understand what you have learned and how it can be applied. In that, Bricken was masterful, and in the wisdom to hire this intellect full time, Maxwell Gluck proved one of the sagest of breeders.

Gluck was somewhat reluctant to part with any dollars he did not have to. The farm was never in pristine, gleaming-white-fence shape, and he did not savor paying the highest stud fees. Placing Bricken on the payroll might have seemed an uncharacteristic extra expense, for plenty of bloodstock advisers were willing to take on new clients. It was very cost effective, however. Gluck got not just the attention of an adviser from time to time as one of many potential clients; he got the full attention of Bricken. This involved constant contact with horses at the racetrack as well as study of pedigrees and searching for prospects in catalogs.

Bricken likened a horseman's decision-making to the approach of Monroe Stahr in F. Scott Fitzgerald's *The Last Tycoon*, i.e., it was important to have a reason for doing what one did. Bricken developed a series of opinions that provided various reasons to choose one horse over another.

While astute enough to arrive at some general conclusions, he was confident enough to deviate from them when there was a reason. For example, he preferred not to breed to a stallion who still had not won after two starts. Intelligence was something he looked for in a stallion: "Horses aren't bred to win. They are bred to run, but not necessarily born to win. They have to learn how to win a race." A colt that had not figured it out or did not have the quality to win by his second race was signaling a lack. So, before breeding to a horse, Bricken preferred to know that it had won by its second start. However, with his intimate knowledge of the backstretch, he also knew if a certain trainer tended to race a horse into condition, a countering element to take into consideration. For example, he did not hold it against Hail to Reason, trained by Hirsch Jacobs, that he did not break his maiden until his sixth start.

Bricken believed in three generations of quality in a family, and he liked highly feminine fillies, but running ability was important as well. Although his assignment was to help breed distance horses, he was not unaware of the importance of speed in that pursuit. He knew that Colonel E.R. Bradley took the tack that a filly that could work very fast in the morning might be a broodmare prospect, regardless of what she did or failed to do in the afternoon. Like Bradley, Bricken often sought to evaluate a filly on her morning works, recognizing that many things could keep her from showing her true potential, not the least of which was the uneven opportunities of fillies and colts on most tracks' racing programs. Bricken also liked to make a point that Elmendorf fillies would be tried against older fillies in the stable in the mornings.

He stressed matching an individual horse with the temperament and strengths of certain trainers. For years Elmendorf ran divisions in the East and West, many of them with Johnny Campo and Ron McAnally at one stage; Walter Kelley and Farrell Jones were also among trainers Elmendorf used over the years.

In a speech at a horseman's forum organized by *The Blood-Horse* in Lexington in 1982, Bricken encouraged buyers to go to a sale with something definite in mind: "You have to have some notions, strong notions, or else it will be very difficult to function profitably at an auction. You might say to yourself, 'I want a mare by [a particular sire].' You have got to stick to that.

"You have to particularize after a while, based on your knowledge and experience. You must say to yourself, 'Even though she is a little bit big, she might have compensating qualities.' Or, 'Even though she is a little small, her mother might have been bigger.' Gathering knowledge about a family before you participate in a bidding duel is very important. You should be armed, well armed." Going to a sale that well armed is difficult for most buyers, but Bricken gave Gluck that advantage.

Some three decades after he had first begun buying for Gluck, Bricken synopsized that he had had some good luck and some bad luck. By the time of Gluck's death late in 1984, Elmendorf had bred 145 stakes winners, including foals of that year. The core of Elmendorf's success in breeding more than three dozen graded stakes winners (grading began in the early 1970s) came from fourteen mares that had been purchased over the years for an average of $12,000. They ranged in price from a $3,000 claim to a mare bought at auction in England for $22,612.

As was the case for many of the breeders included in this volume, an early acquisition had positive ramifications over the years. In 1953 Gluck and Bricken bought the filly Banta for $18,500 at a Belmont Park auction. Banta was by the Futurity winner Some Chance, and her dam was the Stimulus mare Bourtai. The importance of Bourtai was not yet clear, for Banta was the first of her stakes winners, but she was to be followed by the high-class race fillies and producers Delta, Levee, and Bayou. Elmendorf had tapped in early on a rich vein.

Banta won the Correction Handicap at five in 1954. She foaled the stakes winner Mandate, by Prince John, and her daughters set in motion a long line of success as did those of her half sisters.

Prince John represented a stroke of luck as well as purchasing cleverness that allowed Gluck to have access to a top-quality stallion without having to pay the stud fees. A son of *Princequillo—Not Afraid, by Count Fleet, Prince John was a flashy chestnut with an elongated triangle of white streaming down his face. He was bred in the name of Mrs. John D. Hertz, also breeder-owner of record of the colt's broodmare sire, 1943 Triple Crown winner Count Fleet. Elmendorf bought Prince John for $14,300 from the Keeneland sale ring in 1954. Turned over to trainer Kelley, Prince John broke his maiden in his second start and placed in the Sanford Stakes and Washington Park Futurity before finishing fourth in the Belmont Futurity. He was 24-1 in the Garden State Stakes, then the richest race in the world, but he took command turning for home and held off Career Boy and Needles, winning by a nose at a mile and one-sixteenth. Prince John was then second to Nail at the same distance in the Remsen Stakes.

Obviously a classics prospect, Prince John was sent to Florida, but he broke loose from his handlers at Hialeah, ran off, and fractured a pedal bone. He never got back to the races and entered stud in 1957 at Elmendorf with three wins in nine starts and earnings of $212,818. Prince John later was moved to Spendthrift Farm. He was an exceptional sire, his fifty-five stakes winners (10 percent) including Greentree Stable's Belmont Stakes winner Stage

			*Prince Palatine (Persimmon—Lady Lightfoot)
		Rose Prince, 1919	Eglantine (Perth—Rose De Mai)
	Prince Rose, 1928		Gay Crusader (Bayardo—Gay Laura)
		Indolence, 1920	Barrier (Grey Leg—Bar The Way)
*Princequillo, 1940			Tracery (*Rock Sand—*Topiary)
		*Papyrus, 1920	Miss Matty (Marcovil—Simonath)
	*Cosquilla, 1933		White Eagle (Gallinule—Merry Gal)
		Quick Thought, 1918	Mindful (Minoru—Noble Martha)
PRINCE JOHN			*Sunreigh (Sundridge—*Sweet Briar)
		Reigh Count, 1925	*Contessina (Count Schomberg—Pitti)
	Count Fleet, 1940		Haste (*Maintenant—Miss Malaprop)
		Quickly, 1930	Stephanie (*Stefan the Great— Malachite)
Not Afraid, 1948			Black Servant (Black Toney—*Padula)
		Blue Larkspur, 1926	Blossom Time (*North Star III—*Vaila)
	Banish Fear, 1932		*Over There (Spearmint—Summer Girl)
		Herodiade, 1923	*Herodias (The Tetrarch—Honora)

SKEETS MEADORS

Prince John

Door Johnny, the distaff champion Typecast, the stayer Jean-Pierre, and the brilliantly fast filly Deceit. Gluck made good use of Prince John, breeding twenty-two of the stallion's stakes winners, including champion Protagonist, Coaching Club American Oaks winner Magazine, Irish St. Leger winner Transworld, and the good sire Speak John.

The Prince John influence began somewhat quietly, with Mandate and Speak John, moderate stakes winners from his first crop. Speak John, however, was to prove one of those stallions that never scale the heights of fashion but beget a stream of important winners. He was an Elmendorf homebred from the imported Tornado mare *Nuit de Folies, and he won the Del Mar Derby and Tropicana Hotel of Las Vegas Handicap at three in 1961. Speak John sired only 7 percent stakes winners, but his two dozen included champion Talking Picture, $765,548-earner Text, stakes winner Thunder Puddles (sire of Travers winner Thunder Rumble), and the influential stallions Hold Your Peace and Verbatim. Speak John's classic-placed son Hold Your Peace (bred and sold by Elmendorf) sired the good stallion Meadowlake (Meadow Star, Greenwood Lake) and Breeders' Cup Juvenile winner Success Express, a successful sire in Australia and New Zealand.

In Verbatim, Speak John sired an Elmendorf homebred stakes winner from Well Kept, a non-winning daughter of Epsom Derby winner Never Say Die. Verbatim won the Gotham, Bahamas, and a division of the Bay Shore in the winter and spring of 1968 but finished thirteenth of fourteen in the Kentucky Derby. The following year he proved a very useful handicap horse, winning the Haskell, Governor Nicolls, and Whitney. Verbatim sired the Belmont Stakes winner Summing and the champion Princess Rooney, neither for Elmendorf. With so much Prince John blood in its stallion holdings and broodmare band, Elmendorf was somewhat restricted in the use of Verbatim but bred seven stakes winners by him. One of them, Alphabatim, was sold to go abroad and won the group I William Hill Futurity in 1983. Returned to this country, he won two runnings of the Hollywood Turf Cup and became a millionaire. Alphabatim's fourth dam was Oil Princess, one of the earlier mares purchased by Gluck in the 1950s and also dam of Elmendorf-bred stakes winners Oil Royalty (Beldame) and Oil Rich.

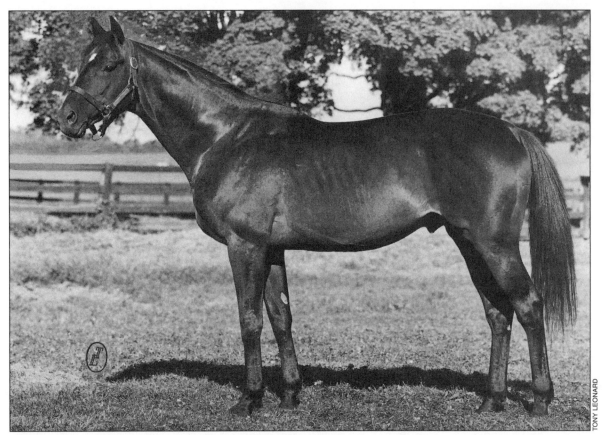

TONY LEONARD

Speak John

In a return to Prince John's early stakes winner Mandate, it will be recalled that he was from the bargain filly Banta. He was her only stakes winner, but the daughters of Banta included important producers. One was Golden Sari, by the leading sire *Ambiorix. Golden Sari, a foal of 1956, won three of twelve races. (Bricken regarded the breeding of an allowance winner on a major circuit as a success for a breeder, although the higher goal was the stakes horse, of course. "You can't breed for that three-fifths of a second difference," he said. "It's impossible.")

Golden Sari was bred to Prince John and foaled Selari, one of the Elmendorf horses that were sold, it being a regular practice to sell some yearlings at the Keeneland September sale. Racing for Lawrence Katz, Selari won the Grey Lag and was a useful sire, getting the Jockey Club Gold Cup winner Great Contractor. Silver Sari, a full sister to Selari, broke her maiden but won no other races among twelve starts. Retained for the Elmendorf broodmare band, she produced two of Gluck's best horses, Manta and Big Spruce.

The matings for Silver Sari that produced these two illustrated Bricken's willingness to go all over the map looking for stallions. Manta was by *Ben Lomond and was one of the three stakes winners by this little-known stallion. *Ben Lomond was by the English stayer and leading sire Alycidon but had been quick enough at two not only to break his maiden in his second start but also to push the high-class King Hairan to a half-length margin in the 1956 Sapling Stakes. Big Spruce, on the other hand, was by the French Derby winner *Herbager, who had been imported to stand at Claiborne Farm and was a rather fashionable number as the sire of Our Mims, Dike, *Grey Dawn II, Loud, Tiller, and so on.

Manta was a rugged little mare who won twelve stakes from three through six. Her biggest wins in California included the 1971 Santa Margarita, Santa Barbara, and Beverly Hills, and her key win in New York was the 1972 Firenze. She won eighteen of fifty-five races and earned $502,889. When John Gaines was dealing to purchase a package of Elmendorf Farm mares soon thereafter, Bricken was frank

98

enough to warn him off Manta because of her roguish behavior. (Gaines accepted her, anyway.)

The *Herbager—Silver Sari colt Big Spruce gave Gluck a pair of his more rewarding triumphs when he defeated the three-time Horse of the Year Forego in consecutive races, the Governor Stakes and Marlboro Cup, in 1974. Trained by Lefty Nickerson, Big Spruce was five at the time. Effective on both dirt and turf, he won a total of six stakes, including the San Luis Rey and two runnings of the Gallant Fox. He won nine of forty races and earned $673,117.

Syndicated to stand at Gainesway Farm, Big Spruce sired Gluck's homebred millionaire Super Moment. Super Moment was from one of the mares Bricken purchased for Gluck at Newmarket, where presumably even he had relatively little intimate inside knowledge of individual training nuances but had learned to read the catalog page effectively. Super Moment's dam was *Seductive II, for whom Elmendorf paid $22,612 in 1974. The Shantung mare had won once in three starts. Her dam, Persuader, had won the Horris Hill Stakes and had foaled a classic winner, English One Thousand Guineas victress Night Off.

Super Moment won three consecutive runnings of the Bay Meadows Handicap in the early 1980s, and his seven stakes triumphs, all on the West Coast, also included the Charles H. Strub Stakes. He won ten of forty-seven races and earned $1,017,940. (As an example of the Elmendorf stock's impact after

Gluck's death, *Seductive II, dam of Super Moment, also foaled Continental Girl, the dam of the 1990 Jockey Club Gold winner and millionaire Flying Continental.)

Yet another daughter of Banta was her 1960 filly from the last crop of Claiborne Farm's *Nasrullah. This was one example of Gluck sending a mare to the very elite in the way of a stallion. The *Nasrullah—Banta filly was named Poster Girl and was retained for breeding although she never raced — maybe it was something she did in the morning. Again giving the family a first-rung opportunity, she was bred to Round Table and foaled the Native Dancer Handicap winner Illustrious.

In 1970 Poster Girl's card was scaled back in price/quality of stallion, and she was bred to the homebred Speak John, whereupon she produced a champion. Talking Picture, the filly, was teamed with Protagonist, the colt, in a remarkable achievement for Gluck, as the breeder and owner of the 1973 juvenile champions in both divisions. Such a double was not unprecedented but placed Gluck in the company of such as Calumet Farm and the Phipps family in having achieved it.

The Speak John—Poster Girl, by *Nasrullah, filly Talking Picture won her debut but did not score again for five more races. She then won the Schuylerville at Saratoga by a nose, to start a four-race winning streak that included the Adirondack, Spinaway, and Matron. She tailed off and lost her final two starts but had done enough to win the

			Fair Play, 1905	Hastings (Spendthrift—*Cinderella)
		Chance Play, 1923		*Fairy Gold (Bend Or—Maid Masham)
	Some Chance, 1939		*Quelle Chance, 1917	Ethelbert (*Eothen—Maori)
				*Qu'elle Est Belle (*Rock Sand—Queens Bower)
			Pompey, 1923	*Sun Briar (Sundridge—*Sweet Briar)
		Some Pomp, 1931		Cleopatra (Corcyra—Gallice)
			Some More, 1920	*Polymelian (Polymelus—*Pasquita)
BANTA				*Kiss Again (Tracery—Stolen Kiss)
			Ultimus, 1906	Commando (**Domino**—Emma C.)
		Stimulus, 1922		Running Stream (**Domino**—*Dancing Water)
	Bourtai, 1942		Hurakan, 1911	Uncle (*Star Shoot—The Niece)
				The Hoyden (*Esher—The Maid)
			*Sir Gallahad III, 1920	*Teddy (Ajax—Rondeau)
		Escutcheon, 1927		Plucky Liege (Spearmint—Concertina)
			*Affection, 1914	Isidor (Amphion—Isis)
				*One I Love (Minting—The Apple)

Eclipse Award for her division. Talking Picture won a division of a little stakes, the Promise at Gulfstream Park, in her debut at three but raced with little distinction in the next few months and was retired with six wins in fifteen starts and earnings of $178,643.

Her partner in the juvenile domination was Protagonist, son of one of the astute English acquisitions. Protagonist was by Prince John and out of *Hornpipe II, a daughter of the good-class Hyperion stallion Hornbeam. Bred in England by John Jacob Astor, *Hornpipe II failed to win in three starts. Her dam, Sugar Bun, had foaled two useful stakes horses, and her second dam, Galatea II, had won the classic English Oaks of 1939. This was the sort of prospect that Gluck meant when he said they "used to be" available at most sales. Elmendorf bought *Hornpipe II for 3,500 pounds in 1967, at the end of her two-year-old season.

*Hornpipe II's first foal was the Prince John stakes winner Harbor Prince, and her third was champion Protagonist. Her fifth, foaled in 1974, was

Transworld. Bred by Elmendorf, Transworld was sold as a weanling to Gainesway Farm, which consigned him to the Keeneland yearling sales and sold him to the British Bloodstock Agency for $375,000 in 1975. He won the Irish St. Leger racing for Simon Fraser. Syndicated to return to Elmendorf as a stallion, Transworld sired the five-time American steeplechase champion Lonesome Glory among nineteen stakes winners.

Trainer Johnny Campo, who also had Talking Picture, brought Protagonist out for the first time at two on the last day of July at Saratoga. He was to be a champion, although perhaps not the sort Bricken would have chosen to breed to, for he did not win until his third start. Getting the hang of it, he was stepped immediately into top company and was third in the Futurity. Protagonist then ran off a streak similar to his stablemate's, taking a division of the Cowdin, a division of the Champagne, and then the Laurel Futurity. An Eclipse Award was his, too. Protagonist failed to run well in his three starts the

Talking Picture

next spring and was retired with four wins in ten starts and earnings of $203,995. He died from laminitis after standing one year at stud at Elmendorf.

Clearly, with a two-year-old champion who appeared to be able to stay, Gluck must have begun the year with thoughts of the Kentucky Derby with Protagonist. He tried hard to win the race, running eight horses in it from 1968 through 1984, but even good horses such as Verbatim, Big Spruce, and Super Moment were not the right horses at the right times. Gluck's best finishes in the Derby were a pair of fourths with Super Moment and Water Bank. In the other classics he got a second to Elocutionist in the Preakness with Play the Red in 1976, while in the Belmont, Gluck's Twice a Prince had the dubious honor of being closest to Secretariat (thirty-one lengths away!) at the finish. These two classic-placed colts were half brothers and, ironically, neither was a stakes winner. Twice a Prince was by Prince John—*Double Zero II, by Never Say Die, and Play the Red was by Crimson Satan.

The same year he had two juvenile champions, Gluck did win a race that has a somewhat shaky status as a classic race, though not a Triple Crown event. Some recognize the Coaching Club American Oaks as a classic for fillies, while others do not think of it in that context. In 1973 Gluck and trainer Campo took the CCA Oaks with yet another Prince John homebred, Magazine. The Oaks was the only stakes won by Magazine. Day Line, the dam of Magazine, was by the moderate stallion *Day Court (son of leading English sire Petition) and out of Fast Line, by Mr. Busher. Fast Line produced stakes winner Fairway Fun for Elmendorf, and stakes winners Filberto and White Star Line after the mare was purchased privately from Gluck by Newstead Farm.

In 1973 — the year of Protagonist, Talking Picture, and Magazine — Gluck's Elmendorf led the breeders' list with earnings of $2,128,080. He thus broke the earnings record held by Calumet Farm since 1952 and became only the second breeder to surpass $2 million in a single year. He led the breeders' list again

NYRA/BOB COGLIANESE

Protagonist

in 1981, with $2,736,029, and in 1982, with $3,049,422, just under the record set by E.P. Taylor in the interim. On the owners' list, Elmendorf led in 1977 with $2,309,200, and in 1981 with $1,928,102. The statistical leadership in these later years came from a large number of stakes winners rather than from a concentration of top-level runners as had been the case in 1973.

Gluck kept his numbers down by periodic sales, and then in 1974, he accepted the profitability of the fashion his stock had developed and sold seventeen mares and two weanlings to John R. Gaines. The mares included *Hornpipe II, Poster Girl, and Manta. The champion Talking Picture also was sold, becoming the property of Walter Haefner's Moyglare Stud in Ireland. Talking Picture produced five stakes winners, of which her 1985 Affirmed filly, Trusted Partner, won the Irish One Thousand Guineas and was champion three-year-old filly in Ireland. This family connection is not distant in history, for Trusted Partner's Danehill filly Dress To Thrill won the grade I Matriarch Stakes at Hollywood Park in the autumn of 2002.

Among the stakes winners Gluck bred that have not been mentioned above are Marry the Prince, a Prince John filly whose three stakes wins included the the Misty Isle Handicap; Sum Up, a *Day Court who caught a frozen track in the Remsen Stakes and bettered Count Fleet's record for a juvenile at a mile; High Tribute and his son, Hollywood Gold Cup winner Pay Tribute; Specious, a Prince John filly who won the Lawrence Realization; Spout, a Delta Judge filly who won the Alabama; Girl in Love, a Lucky Debonair filly who won the Mother Goose and Santa Susana; Visible, a Bold Hour colt who won the Del Mar Futurity; Road Princess, a *Gallant Man filly (from a Prince John mare) who won the Mother Goose; Top Corsage, a Topsider filly who won the Spinster Stakes; Honor Medal, an Avatar gelding who became a millionaire; and Shadow Brook, a

Champions Bred

Shadow Brook
1971 champion steeplechaser

Protagonist
1973 champion two-year-old colt

Talking Picture
1973 champion two-year-old filly

Cohoes gelding who was a steeplechase champion.

Nine years after having sold a major collection of his most successful horses, Gluck still had sufficient power in his stock to get $1,120,000 for twenty-two head (average $50,909) sold in an additional reduction at the Fasig-Tipton Kentucky fall sale.

Gluck was one masterful entrepreneur who owned Elmendorf and who eventually sold the farm to another, Jack Kent Cooke, the Canadian-born businessman and owner of the Washington Redskins and Los Angeles Lakers. Cooke contrived to buy the best farm he could, in line with the same sense of ego that sought the best in other spheres. His aim to buy Calumet Farm was a few years ahead of its time, and the next choice, Greentree, was not a go either. Cooke purchased Elmendorf and after his death and moderate success, the property was purchased by Dinwiddie Lampton, to whom can be credited about as much sentiment and historical appreciation as one could want. Louisville insurance executive Lampton, an avid coachman who somehow slipped off the pages of a Charles Dickens novel to inhabit real life, did not need the farm, had no bunch of mares to quarter there, but he just set out to save it, and so he has.

A few months before Gluck's death at eighty-five on November 21, 1984, he and Mrs. Gluck initiated a $3-million challenge. They pledged that amount to establish an equine research center at the University of Kentucky, on the condition that the industry and the state government each match the figure. The total of $9 million was rapidly achieved, and the Maxwell H. Gluck Equine Research Center was established as one of the beacons in the racing industry's willingness to fund research on behalf of the horse.

Today the name Gluck bespeaks veterinary research first, but it deserves also to remain associated with one of the smartest of Thoroughbred breeders.

Stakes Winners Bred by Maxwell H. Gluck

Cal's Choice, ch.g. 1953, Polynesian—Laurentia, by Pilate
Infantry, ch.g. 1953, Phalanx—*Miss Alesia, by Milon
My Night Out, b.g. 1953, Phalanx—*Nuit de Folies, by Tornado
Darling Adelle, b.f. 1954, Polynesian—Blue Moon, by Eight Thirty
Moon Cloud, ch.c. 1955, Nimbus—*Fair Clarissa, by Fair Trial
Oil Rich, br.f. 1955, Phalanx—Oil Princess, by Errard
Two Cent Stamp, b.f. 1955, Double Jay—Pelure, by Johnstown
Arbitrage, b.c. 1958, Prince John—*Batta, by *Nasrullah
Bombay, ch.c. 1958, Prince John—*Delightful II, by Golden Cloud
Mandate, ch.c. 1958, Prince John—Banta, by Some Chance
Oil Royalty, b.f. 1958, Greek Song—Oil Princess, by Errard
Speak John, b.c. 1958, Prince John—*Nuit de Folies, by Tornado
Royal Spirit, ch.f. 1959, *Royal Gem II—En Casserole, by War Relic
B. Major, b.c. 1960, Summer Tan—Classic Music, by Stymie
Powder 'N Paint, b.f. 1960, Gun Shot—Chanteur Star, by Chanteur II
Rash Prince, b.c. 1960, Prince John—Prompt Impulse, by Noble
    Impulse
Researcher, b.f. 1960, Dark Star—Blue Relic, by War Relic
Miss Twist, ch.f. 1961, Prince John—Night Dance, by *Cortil
Sun Coast, ch.c. 1961, Summer Tan—*Delightful II, by Golden Cloud
Fairway Fun, dk.b.f. 1962, Prince John—Fast Line, by Mr. Busher
Marry the Prince, b.f. 1962, Prince John—Rambling Mary, by
    Fighting Fox
Princess Cloud, ch.f. 1962, Prince John—*Delightful II, by Golden
    Cloud
Selari, b.c. 1962, Prince John—Golden Sari, by *Ambiorix
Sum Up, br.c. 1962, *Day Court—Toluene, by Hill Prince
Blaize Brucato, b.c. 1964, Arrogate—Tattooed Miss, by Mark-Ye-Well
Gay Pursuit, ch.g. 1964, Vertex—Debutante Kiss, by Phalanx
High Tribute, ch.c. 1964, Prince John—En Casserole, by War Relic
London Jet, b.g. 1964, Ridan—*Nuit de Folies, by Tornado
Serve Notice, dk.b/br.c. 1964, *Day Court—Kissing Belle, by Prince
    John
Shadow Brook (stp), b.c. 1964, Cohoes—*Chanteur Star, by
    Chanteur II
Illustrious, dk.b/br.c. 1965, Round Table—Poster Girl, by *Nasrullah
La Queenie, b.f. 1965, Prince John—*Batta, by *Nasrullah
Skookum, ch.c. 1965, Ridan—Silver Service, by Prince John
Verbatim, dk.b/br.c. 1965, Speak John—Well Kept, by Never Say Die
Captain Action, ch.c. 1966, Rash Prince—*Captain Tess, by Combat
Lover's Quarrel, b.f. 1966, Battle Joined—Debutante Kiss, by
    Phalanx
Manta, ch.f. 1966, *Ben Lomond—Silver Sari, by Prince John
Oil Power, b.c. 1966, Crimson Satan—Strike Oil, by Prince John
Pass Right, b.g. 1967, *Day Court—Intersection, by Prince John
Fair Test, b.c. 1968, Time Tested—Princess Fair, by Prince John
Specious, ch.f. 1968, Prince John—Sophist, by Acropolis
Big Spruce, dk.b/br.c. 1969, *Herbager—Silver Sari, by Prince John
Brad's Star, b.c. 1969, Saidam—Padrona, by Ponder
Harbor Prince, b.c. 1969, Prince John—*Hornpipe II, by Hornbeam
Hold Your Peace, b.c. 1969, Speak John—Blue Moon, by Eight Thirty
My Old Friend, dk.b/br.c. 1969, Outing Class—Rare Bouquet, by
    Prince John
Pooka, b.c. 1969, Irish Lancer—Sophist, by Acropolis (SW in
    Panama)
Tartar Chief, ch.c. 1969, Crimson Satan—My Sheer Joy, by Prince
    John

Fresh Pepper, b.f. 1970, B. Major—Rare Bouquet, by Prince John
Lilac Hill, ch.f. 1970, Prince John—Red Damask, by Jet Action
Magazine, b.f. 1970, Prince John—Day Line, by *Day Court
Proposal, b.f. 1970, Speak John—Beauty Maid, by Gun Shot
Protest, blk.f. 1970, Rash Prince—Dynamis, by Sailor
Queen's Mark, dk.b/br.f. 1970, Rash Prince—Tattooed Miss, by
    Mark-Ye-Well
Scantling, ch.c. 1970, *Ben Lomond—Minim II, by Chanteur II
Glossary, ch.c. 1971, Prince John—Sophist, by Acropolis
Protagonist, ch.c. 1971, Prince John—*Hornpipe II, by Hornbeam
Provante, b.c. 1971, Prove It—One Lane, by Prince John
Sir Jason, gr.c. 1971, *Grey Dawn II—Amber Moon, by *Ambiorix
Talking Picture, dk.b/br.f. 1971, Speak John—Poster Girl, by
    *Nasrullah
Announcer, dk.b/br.c. 1972, Verbatim—Happy Ring, by *Royal Gem II
Improviser, ch.g. 1972, Speak John—Minim II, by Chanteur II
Pay Tribute, ch.c. 1972, High Tribute—Drummer Girl, by Tompion
Podium, b.c. 1972, Prince John—Cadenza, by B. Major
Proponent, ch.c. 1972, *Gallant Man—Classicist, by *Princequillo
Rematch, b.c. 1972, Verbatim—Four Cent Stamp, by Alsab
Rouge Sang, b.c. 1972, Bold Bidder—Red Damask,, by Jet Action
    (SW in Italy)
Spout, ch.f. 1972, Delta Judge—Strike Oil, by Prince John
Sweet Old Girl, dk.b/br.f. 1972, Olden Times—Tympanist, by Tompion
Free Journey, dk.b/br.f. 1973, Verbatim—Free Ride, by Hasty Road
Girl in Love, ch.f. 1973, Lucky Debonair—Lover's Quarrel, by Battle
    Joined
Jackknife, b.c. 1973, Jacinto—Park Princess, by Prince John
National Flag, b.c. 1973, Flag Raiser—Princess East, by Prince John
Pocket Park, b.c. 1973, Verbatim—Zoo Patrol, by Jet Pilot
Queen to Be, b.f. 1973, Cornish Prince—Princess Fair, by Prince
    John
Drama Critic, b.f. 1974, Reviewer—Sophist, by Acropolis
Harvest Girl, b.f. 1974, *Herbager—Sequela, by Prince John
Make Amends, ch.c. 1974, Prince John—Lover's Quarrel, by Battle
    Joined
Postscript, b.g. 1974, Speak John—Postal Queen, by Sword Dancer
Press Notice, b.c. 1974, Crème dela Crème—Strike Oil, by Prince
    John
Replant, b.c. 1974, No Robbery—Swiss Forest, by Dotted Swiss
Rich Soil, dk.b/br.f. 1974, Verbatim—Southland, by Delta Judge
Road Princess, b.f. 1974, *Gallant Man—One Lane, by Prince John
Text, ch.c. 1974, Speak John—Sky Glory, by Gun Shot
Time Call, b.g. 1974, Bold Hour—Princess Fair, by Prince John
Transworld, ch.c. 1974, Prince John—*Hornpipe II, by Hornbeam
    (SW in Ireland)
True Statement, ch.c. 1974, Nodouble—Specious, by Prince John
Visible, ch.c. 1974, Bold Hour—See Thru, by To Market
Brash Prince, dk.b/br.c. 1975, Rash Prince—Little Blush, by Day
    Court
Holdfast, dk.b/br.c. 1975, Crewman—Everglow, by Jacinto
Josher, dk.b.g. 1975, Speak John—Little Evil, by Tompion
Miss Magnetic, b.f. 1975, Nodouble—Out Draw, by Speak John
Prince Worthy, dk.b/br.c. 1975, Twice Worthy—Park Princess, by
    Prince John
Syncopate, b.c. 1975, Marshua's Dancer—Sunny Today, by Prince
    John

Tampoy, b.g. 1975, Tumiga—Princess Fleet, by *Princequillo

Beauty Hour, ch.f. 1976, Bold Hour—Pomade, by Prince John

Double Distant, ch.c. 1976, Nodouble—Farsighted II, by Salvo

Laughing Boy, b.c. 1976, Shecky Greene—Pomade, by Prince John

Maytide, dk.b/br.f. 1976, Naskra—Melody Tree, by High Tribute

Red Crescent, ch.c. 1976, Prince John—Little Evil, by Tompion

Seascape, b.f. 1976, *Gallant Man—Sequela, by Prince John

Top Soil, dk.b/br.f. 1976, Verbatim—Southland, by Delta Judge

Berry Bush, b.f. 1977, Big Spruce—Henna, by Ridan

Major Sport, ch.c. 1977, Nodouble—Summertide, by Crewman

Super Moment, b.c. 1977, Big Spruce—Seductive II, by Shantung

Sweet Maid, dk.b.f. 1977, Proud Clarion—Honor Maid, by Prince John

Census, ch.g. 1978, Speak John—Rhinestone (GB), by Never Say Die

Maple Tree, dk.b.f. 1978, Big Spruce—Prism, by Reflected Glory

Proud Thief, b.f. 1978, Proud Clarion—No Crime, by No Robbery

Seafood, b.c. 1978, Proud Clarion—Summertide, by Crewman

Comedy Act, b.f. 1979, Shecky Greene—Everglow, by Jacinto

Girlie, dk.b/br.f. 1979, Crewman—Party Kiss, by Fleet Nasrullah

Hello Federal, dk.b/br.c. 1979, Verbatim—Henna, by Ridan (SW in Puerto Rico)

Speak of John, dk.b.c. 1979, Speak John—Kossira (Ire), by Le Levanstell

The Hague, ch.c. 1979, Transworld—Our Sue II, by Reform

Visto, b.f. 1979, Transworld—Oratorio, by Fleet Nasrullah

Water Bank, b.c. 1979, Naskra—Summertide, by Crewman

Clear Talk, b.f. 1980, Elocutionist—Pout, by Rash Prince

Continental Girl, b.f. 1980, Transworld—Seductive II, by Shantung

First Risk, dk.b/br.f. 1980, Jacinto—Party Worker, by Speak John

Red Ember, dk.b.f. 1980, Crimson Satan—Everglow, by Jacinto

Sackford, b.c. 1980, Stop the Music—Bon Fille, by Lomond

Spruce Song, b.f. 1980, Verbatim—Oratorio, by Fleet Nasrullah

Alphabatim, b.c. 1981, Verbatim—Morning Games, by *Grey Dawn II (SW in England/U.S.)

Ceylon Tea, ch.f. 1981, Giacometti (Ire)—Lady Muriel, by Prince John

Honor Medal, ch.g. 1981, Avatar—Honor Maid, by Prince John

Innamorato, ch.c. 1981, Blushing Groom (Fr)—Out Draw, by Speak John

Logjam, dk.b/br.c. 1981, Big Spruce—Brave Countess, by *Gallant Man

Ocean View, ch.c. 1981, Nodouble—Summertide, by Crewman

Office Seeker, ch.c. 1981, Sadair—Party Worker, by Speak John

Party Leader, b.c. 1981, In Reality—New Scent, by Cornish Prince

Sun Master, b.c. 1981, Foolish Pleasure—Sunny Today, by Prince John

Padua, dk.b/br.c. 1982, Caro (Ire)—Miss Magnetic, by Nodouble

Rich Earth, b.c. 1982, Transworld—Pout, by Rash Prince

Extranix, ch.c. 1983, Transworld—Little Evil, by Tompion

Free Waters, b.c. 1983, Danzig—Talk Out, by Tobin Bronze

Lover's Cross, b.c. 1983, Explodent—Girl in Love, by Lucky Debonair

New Colony, dk.b/br.c. 1983, French Colonial—Newfoundland, by Prince John

Puzzle Book, gr.f. 1983, Text—Morning Games, by *Grey Dawn II

Shotgun Wedding, b.f. 1983, Blushing Groom (Fr)—Out Draw, by Speak John

Top Corsage, dk.b.f. 1983, Topsider—Corsage, by Native Royalty

Zigbelle, dk.b/br.f. 1983, Danzig—Cornish Belle, by Cornish Prince

Barbara Sue, dk.b.f. 1984, Big Spruce—Maytide, by Naskra

Landyap, b.c. 1984, Fappiano—My Candidate, by Prince John

Momsfurrari, b.c. 1984, Elocutionist—King's Favorite, by Cornish Prince

Sail to France, b.f. 1984, French Colonial—Seapost, by Speak John

View of Royalty, dk.b/br.f. 1984, Native Royalty—Counter View, by Crewman

Source: The Blood-Horse Archives

8

E.P. Taylor

"He was the most dynamic man I ever knew," George Blackwell often said of E.P. Taylor. Blackwell, a pedigree adviser of international repute, traveled the world and dealt with moguls and commoners alike for many years. In Taylor he saw a friend and patron, someone who might send him on a mission to buy "the best mare in the Newmarket sale," and who presided over an exemplary racing circuit largely stemming from his own vision.

We knew Taylor in the latter decades of his career. He was large, round, imposing — gruff of voice, firm of handshake, quick to smile and chuckle, but thoroughly accustomed to being in charge. Some of his friends frequently referred to him as "Eddie," but this familiarity never quite seemed to fit the figure of "Mr. Taylor." Our overriding image of Taylor presents him in morning suit on Queen's Plate Day, greeting this or that "royal personage" as he called them. Thus, he peoples our memory as something of an Edwardian combination of capitalist and aristocrat. As a man of the Turf, he strode among the high and mighty, and his horses did as well. Blackwell had it about right.

Taylor had a knack for achieving expansion through constriction. This seeming contradiction worked in the Ontario brewing industry, when he bought up a series of struggling, small breweries and consolidated them into fewer brands that prospered hugely. He did the same for Ontario racetracks, consigning names such as Dufferin and Thorncliffe to

horsemen's nostalgia, but giving the sport handsome new Woodbine while keeping the old one operating, along with floral Fort Erie.

Ontario racing needed high-class horses to meet E.P. Taylor's vision, and so he bred high-class horses, often achieving an almost mystical pattern of generational upgrade. Ontario racing needed plenty of thriving owners to meet E.P. Taylor's vision, and so he devised a series of methods to provide good racing stock to other local owners while still maintaining a powerful stable himself.

Taylor came by some degree of business acumen via familial environment. His maternal grandfather had built a small empire in real estate, banking, brewing, and railways. His father, Lieutenant Colonel Plunkett Taylor, joined an investment firm after military service.

Edward Plunkett Taylor was born in Ottawa on January 29, 1901. While he was a mechanical engineering student at Montreal's McGill University, Taylor invented a type of toaster that helped pay his way through school. Barely in his twenties, Taylor, with a friend, spotted an opportunity in Ottawa, the capital of Canada, which had no meter cabs and bus service early in the 1920s. According to a profile in *Fortune* magazine years later, within a year they had put together a fleet of both types of vehicle, and then sold out at a profit.

The market crash of 1929 was not seen in all circles as an opportunity, but Taylor took the situation in hand by devising his scheme for the brewing industry. The Bank of Montreal was so skeptical of his tactics that he found it difficult to secure financing at home and so turned to a friend in England.

During World War II, Taylor was the second-youngest individual among the so called "dollar-a-year" men, wartime entrepreneurs who served their countries by providing leadership and planning for supplying, moving, and arming troops, without expectation of payment. He was sent on a mission by the minister of munitions and supplies and was aboard the *Western Prince* when it was torpedoed. Taylor and others spent eleven hours on a raft before being rescued.

Once Allied Victory was assured, Taylor resigned from wartime boards and looked ahead to a Western prosperity he sensed would be forthcoming. He and three associates formed Argus Corporation, an investment firm whose assets included all of Taylor's brewery holdings. His chief ally in Argus for many years was Bud McDougald. Taylor was so visible in many forms of business, industry, and sport that he was often mistakenly presumed to be the richest man in Canada. McDougald was less in the public eye, but it was he who was described as follows by Peter Newman's book *The Canadian Establishment*: "If the Canadian business establishment has a grand master, that all-powerful figure has to be a nearly invisible Toronto capitalist named John Angus (Bud) McDougald." (Mr. and Mrs. McDougald also became involved in racing. Typically, they had only a small stable abroad that attracted little attention, but the well-known artist Raoul Millais on occasion was commissioned to paint a modest winner or two for them. Later Mrs. McDougald quietly owned the historic Kingsclere stables, where champions from Ormonde to Mill Reef were trained.)

Taylor had less actual wealth than McDougald but could command financing, and one observer told *Fortune* Taylor also "had more of the influence of wealth than anyone else in Canada." A Taylor biog-

E.P. Taylor (right) and Horatio Luro

THE BLOOD-HORSE

rapher, Richard Rohmer, gave a similar description: Taylor, he wrote, had been called "the personification of riches gained and power wielded."

Horse racing became an avocation, then a business and a passion, for E.P. Taylor. Moving boldly — as ever an advocate of borrowing and bigness — Taylor devised and carried out a plan to build a new, modern racetrack that would give Toronto a facility to match any in the world. This was known as New Woodbine, in tribute to the original Woodbine track along the shores of Lake Ontario in downtown Toronto. The old Woodbine was home of the King's Plate, the oldest race in North America run annually without interruption. Old Woodbine was later renamed Greenwood, and the "New" was dropped from the more modern track's identity. The Plate — named King's or Queen's depending on the gender of the reigning English monarch — was moved to the new track. On numerous occasions Queen Elizabeth II or the Queen Mother presented the trophy for the splendid old event.

As far as his own horse operation is concerned, Taylor began racing in the name of Cosgrave Stable in 1936. He had become a fan of racing while at McGill. Montreal's Blue Bonnets racetrack in those days saw the likes of Exterminator and Billy Kelly on occasion. Taylor's devoted biographer, Muriel Lennox, found a commercial spin on the name Cosgrave. In a 2003 article in *Thoroughbred Times*, she recalls how Taylor years ago tried to buy Cosgrave Brewery; any recognition of a stable of that name would adroitly sidestep the prevailing law against advertising alcohol. It was in the name of Cosgrave Stable that Taylor bred his first stakes winner, Windfields.

Taylor and his wife, Winifred, also gave the name Windfields to their property in Willowdale, Ontario, in recognition of the strong breezes that swept across the Toronto suburb. Windfields, the horse, was a foal of 1943 and was by Bunty Lawless, who had defeated Cosgrave Stable's early star Mona Bell in the 1938 King's Plate.

Bunty Lawless was by Ladder—Mintwina, by Mint Briar. There was some class backing up this pedigree, for Ladder's sire, Ladkin, had won one of the international races that French invader *Epinard contested during the 1920s. Ladkin was by Fair Play, sire of Man o' War. Bunty Lawless was a local Toronto favorite who won a half-dozen stakes. He became a good regional sire who got twenty-six stakes winners (13 percent), but he was an unlikely element to figure in the inbreeding of a North American champion stallion of more than a half-century later. Yet, Bunty Lawless is there, 5x3, in the pedigree of two-time leading sire Deputy Minister.

Windfields' dam, Nandi, was by the good Kentucky stallion Stimulus—*Golden Feast, by Golden Sun. She had been claimed by Taylor's trainer, Bert Alexandra, at Pimlico in a flurry of acquisitions to build up the Cosgrave Stable. Windfields ran seventy-five times and won nineteen races, earning $83,380. He won the Victoria Stakes at two and at three took the Breeders' Stakes, one of Canada's Triple Crown races. Perhaps more important to all that would transpire later, Windfields challenged New York, with a satisfactory result when he ran second to Triple Crown winner Assault in the 1946 Dwyer Stakes. Erstwhile Kentucky Derby favorite Lord Boswell was third.

The colt, Windfields, quickly set into motion a pattern of upward mobility that was one of the recurrent themes of Taylor's remarkable record as a breeder — the other being the occasionally counterbalancing purchases at the top of the fashion and value market. Windfields sired fifteen stakes winners (10 percent), among them Bill Beasley's optimistically but accurately named Canadian Champ. Canadian Champ won Canada's Triple Crown and was its Horse of the Year in 1956.

The aforementioned Mona Bell, runner-up to Bunty Lawless in the 1938 King's Plate, was a star on the local circuit, winning the Breeders' Stakes that year, as well as the Maple Leaf Stakes and the following year's Orpen Memorial. She was a daughter of *Osiris II, a son of 1923 Epsom Derby winner *Papyrus, and was out of Belmona, by King James. Mona Bell won eleven of thirty-three starts and earned $18,120 before breaking a leg at Stamford Park and having to be euthanized. Four years later

Taylor had the opportunity to buy into the family again, paying $1,500 for Mona Bell's yearling full sister, to be named Iribelle. The second "Belle" was not as sharp a race filly as her sister, although scoring a pair of wins from thirteen starts, placing third in a Plate Trial, and earning $2,190. She, however, also keenly illustrated the upward mobility in breeding, for she foaled three stakes winners. The best of these was Canadiana, who broke through in New York by winning the Test Stakes in 1953.

By then, Taylor's involvement in breeding and racing had grown considerably. Several years earlier he had purchased the Parkwood Farm of Colonel R.S. McLaughlin, head of the Canadian division of General Motors. This farm was some sixty miles from Toronto, in Oshawa. It was known for years as the National Stud Farm, until Taylor decided to rename it Windfields, too. (The original Windfields in Willowdale outside Toronto was the site of the Taylor's handsome home, but as the city grew around that farm, it became impractical to maintain as a horse farm. There was some charm to the photo of future English Derby winner Nijinsky II taken by Peter Winants in a yearling paddock in 1968 with the Toronto skyline looming behind him, but one can look out and see school children petting the yearlings only so many times before it seems wisest to move on.)

During the 1950s Taylor also leased a farm in Kentucky, but he observed that his Oshawa-bred yearlings became as advanced as his Kentucky foals after a time, so he ceased the U.S. operation. The conviction that Ontario could raise top-class horses was a personal as well as a patriotic matter — but it had to be proven.

Canadiana returned to New York at four to win the Vagrancy Handicap. All told, this Canadian Horse of the Year won eleven stakes, defeating colts in the preeminent Queen's Plate and in Canada's top juvenile tests, the Coronation Futurity and Cup and Saucer Stakes. She had a career record of twenty wins in sixty-two starts from two through five and earned $173,116.

Canadiana was by Chop Chop, one of the key stallion acquisitions that aided the rise of Windfields Farm. Chop Chop was by the Ascot Gold Cup winner Flares, son of American Triple Crown winner Gallant Fox of the *Sir Gallahad III sire line. Chop Chop's imported dam, *Sceptical, was by the two-time Eclipse Stakes winner Buchan, who had placed in all three English colt classics and was a leading sire in England. Chop Chop was bred by Charles Thieriot and was purchased for $4,100 as a yearling in 1941 to race for Mrs. Barclay Douglas' Mill River Stable. He won the $25,000 Empire City Handicap over *Princequillo among four wins in eleven starts to earn $36,600.

Chop Chop did not race at four and was leased to Canada when he was five, to stand first at Gil Darlington's Trafalgar Farm. Darlington later became manager of Windfields, where he served a long and distinguished employment, and Chop Chop joined the Windfields stallion roster. Having sired one proven international stakes winner when bred to Iribelle, Chop Chop begot another when bred to Iribelle's last foal, Victoriana. The latter, a daughter of Windfields, injured a foot as a yearling and was unraced.

The Chop Chop—Victoriana foal was named Victoria Park and was priced at $12,500 at the pre-priced yearling sale that Taylor held for many years at National Stud-cum-Windfields. This form of sale offered a genteel and efficient way of marketing yearlings without gutting his racing stable, while at the same time putting horses in the hands of other owners to campaign at Ontario tracks. Each horse had a price tag. Invited guests wandered among the yearlings, drinks in hand, to inspect them, and either sign up to buy at the given price or not. Once half of the yearlings had been spoken for, the sale was stopped, and Taylor kept half to race in Windfields' own turquoise colors with yellow polka dots.

Victoria Park was small and toed in and was not taken. At two he led the juveniles at home, where he won the Cup and Saucer, Coronation Futurity, and Clarendon Stakes. At the end of the year, he dropped down to New York, where he won the Remsen Stakes and planted himself into the following year's Kentucky Derby scene. Victoria Park was transferred from Pete McCann's stable in Canada to Horatio Luro in the United States. At three in 1960, he mixed it up with the goodish classic crop of Venetian Way, Bally Ache, and Tompion. He did not score any breakthrough victories in a classic but was second in both the Derby and Preakness.

The Belmont conflicted with the Queen's Plate, and Taylor opted to continue the dominating record he was building in the Canadian race. Victoria Park won handily. He then returned to the United States

and that time scored an impressive victory over Tompion in the Leonard Richards Stakes at Delaware Park. Victoria Park was being pointed for the Hollywood Derby but arrived in the West with filling on both forelegs. Victoria Park was retired and paraded before the fans at Woodbine, who hardly suspected that their pride in one home country product would be overwhelmed by feelings for another only four years later. Victoria Park had won ten of nineteen races, and with earnings of $250,076 was at the time the richest Canadian-bred ever. He sired twenty-five stakes winners (7 percent), including three consecutive Queen's Plate winners in Almoner (1970), Kennedy Road (1971), and Victoria Song (1972). Kennedy Road also won the Hollywood Gold Cup in California.

The year 1960 marked the first time Taylor topped North American statistics in the list of breeders in total races won. Windfields-breds won 267 races that year. Taylor repeated atop this list for the next nine years, too, reaching a peak of 310 wins in 1966. Harbor View Farm and then Rex Ellsworth shared the honor over the next seven years, and then Taylor resumed control, leading the list nine more times (1977-85). Thus, he had been the leading breeder in races won nineteen times from 1960 to 1985. His top was 442 wins in 1978. As we shall see, he was also destined to lead many times in more of a quality measure, that of earnings.

Victoria Park's sire Chop Chop sired a total of twenty-nine stakes winners (14 percent), of which Taylor bred eighteen. One of the countless instances of Taylor's horses' key role in transforming the image of Canadian racing — from low-rent to respectability to parity — involved the Chop Chop filly Ciboulette. Her dam, Windy Answer, was by Windfields and out of Reply, an unraced Teddy Wrack mare who was in turn from an Alsab mare. Windy Answer won six Ontario stakes and then foaled three stakes winners. One of these was the Canadian champion Cool Reception, who at two won the Coronation Futurity, Cup and Saucer, and two other stakes. Racing for Green Hills Farm, Cool Reception ran second to Damascus in the following year's Belmont Stakes. The Chop Chop filly in the mix, Windy Answer's daughter Ciboulette, won four Ontario stakes for Jean-Louis Levesque, who parlayed some of his Windfields purchases into one of the most powerful stables in Canada in the 1960s and 1970s. Ciboulette in turn produced the internationally competitive Fanfreluche, who won the historic Alabama Stakes as well as major events in Canada. Fanfreluche was famed for her bizarre kidnapping from Claiborne Farm and her eventual return. She produced five stakes winners, including L'Enjoleur, Medaille d'Or, and D'Accord. The descendants of Windfields and Chop Chop were by then being bred to the likes of Northern Dancer, Nearctic, Buckpasser, and Secretariat.

In addition to Chop Chop and Bull Page, *Menetrier was another significant stallion acquisition for Windfields in the 1950s. A product of the prestigious breeding operation of Taylor's French friend Francois Dupre, *Menetrier was by the leading French sire Fair Copy, son of Fairway, he by the pivotal Phalaris. *Menetrier's dam, La Melodie, by Gold Bridge, was a French stakes winner who in turn was from the classic-placed La Souriciere.

*Menetrier won several French stakes ranging from six and a half furlongs to the Prix d'Ispahan at a mile and an eighth. Good class races at seven furlongs, such as the Prix Edmond-Blanc and Prix de la Jonchere, might have been his best venue, and he was also placed in mile and shorter stakes in England and Italy. He stood several seasons in Europe and left behind the French One Thousand Guineas winner Virgule and the good stakes winner *Bel Canto II and (later imported) *Blue Choir. *Menetrier was, in short, an example of Taylor's strategy of looking for a background of class but not in a form that commanded impracticably high investment for the racing circuit that was generally seen as the end game.

*Menetrier sired a total of nineteen stakes winners (12 percent), and several played a role in the sequence of improvements that advanced Windfields up the ladder to international fashion. The *Menetrier filly Victoria Regina, for example, was a half sister to Victoria Park (second dam was the foundation mare Iribelle) and won four stakes in Ontario. Victoria Regina then foaled two Northern Dancer colts — Viceregal, who was a brilliant champion runner and disappointing stallion, and Vice Regent, who was a compromised, unsound runner but highly significant sire (Deputy Minister.) Also, Menebora, a *Menetrier mare, foaled Canebora, who won not only the Queen's Plate but the entire Triple Crown (also include the Prince of Wales Stakes and

Breeders' Stakes) in Canada for Windfields in 1963.

The seminal moment in the rise of Windfields came in 1952. While Taylor had been successful in Ontario and was only a year removed from Canadiana's pending success in New York, he had also brushed up against international fashion and greatness. For example, he had been one of the small group that joined A.B. Hancock Jr. of Claiborne Farm in attempting to purchase *Nasrullah. (That international transaction was scuttled by an unfortunately ill-timed currency devaluation.) Now, in 1952, he asked agent George Blackwell to accompany him to the Newmarket December sale.

Pre-sale speculation was that the market would be low, and indeed it marked the first time since war-torn 1943 that the gross was under 500,000 guineas for that bellwether auction. The English and Irish horsemen sold stock, but few of them felt confident to replenish through buying. Thus, the sale was marked by international raiding, and more than 20 percent of the aggregate price represented purchases by the British Bloodstock Agency, largely for exports. Taylor, who read the nuances of international currency and finance keenly, saw an excellent opportunity. He had asked Blackwell to go to the sale to get the best mare offered. Blackwell's practiced eye was among those that focused on *Lady Angela, whose sire, Hyperion, was safely ensconced in the hierarchy of the greatest of England's racing and breeding.

In a context of feverish international purchases, it was the Canadian Taylor who attended the sale personally and topped it at 10,500 guineas for *Lady Angela. The British Bloodstock Agency acted on his behalf in outbidding English bookmaker William Hill and Humphrey Finney of America's Fasig-Tipton. *Lady Angela was eight and was out of the accomplished producer Sister Sarah, who at twenty-two was purchased at the same auction for 1,200 by none other than Winston Churchill. Sister Sarah (Abbots Trace—Sarita, by Swynford) had won twice in England at two, and by 1952 was seen as the dam of champion English juvenile filly Lady Sybil and second dam of Irish juvenile champions *The Web II and Marita. Sister Sarah's daughters later were to flower her legacy with a proliferation of descendants including Epsom Derby winner St. Paddy, Guineas runner-up and classic sire Great Nephew, and Two Thousand Guineas winner To-Agori-Mou, in addition to the wonders of *Lady Angela.

The 1952 *Bloodstock Breeders' Review* included the editorial comments that Taylor "is to be congratulated on obtaining such a high-class mare for Canada" and that "no doubt the *Review* will be recording the deeds of her Canadian offspring in years to come." This was part insight and part grandstanding, for the bloodstock sales section of that revered volume in the year in question was written by none other than G.B. Blackwell, then of the British Bloodstock Agency!

In 1956, Stanley Harrison, the essayist covering Canada gave substance to this prediction. Harrison's report in the 1956 volume of the *Review* extolled the virtues of *Lady Angela's two-year-old, Nearctic. Early in the season this Windfields homebred was being hailed as "the best ever" in the Dominion. Harrison wrote: "He made a show of his opposition in the native-bred ranks, winning his races as he pleased and taking the track and widening his margin to the wire. But, alas, as the season advanced and the distance was increased it was seen that Nearctic was strictly a sprinter." Even within that seeming limitation, he distinguished himself, winning three Canadian Stakes and the Special at Saratoga.

Nearctic was not the foal *Lady Angela was carrying when Taylor bought her. The mare was in foal at the time of sale to the unbeaten Nearco, a great racehorse and sire turned out by Federico Tesio and standing in England under ownership of Martin Benson. Taylor insisted that the mare remain in England, foal there, and be returned to Nearco.

Benson agreed, but with the proviso that the stud fee be paid in dollars.

The first Nearco—*Lady Angela foal for Taylor was disappointing, a light-boned chestnut named *Empire Day. The second, foaled at Windfields, was Nearctic. He was priced at $35,000 for Taylor's yearling sale and was not taken. While Nearctic never was exactly a stayer, the pronouncement by Harrison that he was "strictly a sprinter" proved not quite correct.

Nearctic was a strong-willed, powerful sort, and Taylor at one point sent him west to Horatio Luro to try to curb his speed craziness. The crack French jockey Rae Johnstone was visiting America between European racing seasons, and Luro induced him to help calm the horse with long, slow gallops.

As a result, Nearctic was able to stretch out a bit and won as important a race as the Michigan Mile at four in 1958. He also carried up to 130 pounds successfully as he won eight Canadian stakes that season and reigned as Canada's Horse of the Year. After one additional stakes victory at five raised his career total to fifteen stakes wins, he was retired with an overall record of twenty-one wins in forty-seven starts and earnings of $152,384.

Nearctic's arrival at the Windfields Stud barn put in place one key element for a level of success even Taylor might have regarded as a pipe dream. The other element had come from the yearling sales of 1958 in the form of a filly by the great racehorse and young stallion Native Dancer. The filly was out of Almahmoud, who at that time was already recognized

NORTHERN DANCER			
Nearctic, 1954	Nearco, 1935	Pharos, 1920	Phalaris (Polymelus—Bromus)
			Scapa Flow (Chaucer—Anchora)
		Nogara, 1928	Havresac II (Rabelais—Hors Concours)
			Catnip (Spearmint—Sibola)
	*Lady Angela, 1944	Hyperion, 1930	**Gainsborough** (Bayardo—Rosedrop)
			Selene (Chaucer—Serenissima)
		Sister Sarah, 1930	Abbots Trace (Tracery—Abbots Anne)
			Sarita (Swynford—Molly Desmond)
Natalma, 1957	Native Dancer, 1950	Polynesian, 1942	Unbreakable (*Sickle—*Blue Glass)
			Black Polly (Polymelian—Black Queen)
		Geisha, 1943	Discovery (Display—Ariadne)
			Miyako (John P. Grier—La Chica)
	Almahmoud, 1947	*Mahmoud, 1933	*Blenheim II (Blandford—Malva)
			Mah Mahal (**Gainsborough**—Mumtaz Mahal)
		Arbitrator, 1937	Peace Chance (Chance Shot—Peace)
			Mother Goose (*Chicle—Flying Witch)

as the dam of the high-class runner Cosmah, and who in later years would be the ancestress of leading sire Halo and Kentucky Derby winner Cannonade.

The Native Dancer—Almahmoud, by *Mahmoud, filly (to be named Natalma) was bred by Mrs. E.H. Augustus and her nephew Danny Van Clief and was consigned to Saratoga. "We used to take a house with my aunt every year at Saratoga. Eddie Taylor and Horatio Luro came to breakfast one morning when they had been looking at Natalma," Van Clief recalled years later. "We had these tomato preserves that our cook, Rosa Page, had made. They both gorged themselves on tomato preserves and hot bread, and I always attributed the sale to Natalma to that breakfast — I'm sure that was why they came back and looked at her one more time."

Whether the motivation was gastronomic or not, Natalma was to their taste, and Taylor paid $35,000 for her. At two Natalma beat the good Wheatley Stable filly Irish Jay in Saratoga's Spinaway Stakes but was disqualified to third. The next spring Luro was aiming for the Kentucky Oaks, but Natalma suffered a chip in a knee. With such a pedigree and proven touch of racing class, Taylor was happy to breed her rather than try to bring her back. She had won three times in seven starts and earned $16,015.

In 1961 Natalma foaled her Nearctic colt, and he was a chunky little number, a bay with a jaunty white blaze and a bit of white on three legs. Joe Thomas, who had designed — and for years admirably filled — the job as racing manager and vice president for Taylor, remarked eventually that Northern Dancer would "change the perception of what a classic horse can look like." That he did, and Thomas and Taylor liked him before the fact well enough to put a challenging price of $25,000 on him as a yearling. They escaped selling Northern Dancer, while several other horsemen were consigned to ponder forever their decision to let him go.

Northern Dancer came out running. He had speed and class and won at first asking by more than six lengths. Two races later he became a stakes winner by winning the Summer Stakes. One of Windfields' Canadian trainers, "Peaches" Fleming, had him up to then, but the lordly Luro returned from his annual summer holiday in France to take charge. Whereupon, Northern Dancer lost in the Cup and Saucer Stakes to a 44-1 shot named Grand Garcon — not a bad name for a footnote to history or the answer to a friendly cocktail party wager. Windfields, incidentally, bred and sold Grand Garcon (Censor—*Stalina, by Stalino). Northern Dancer corrected matters with victories in the Coronation Futurity and Carleton Stakes and then ventured to New York. Like Victoria Park four years earlier, Northern Dancer announced his prowess below the border with a victory in the Remsen Stakes.

Late in the little colt's juvenile season, Luro fretted over the development of a quarter crack. This seemed likely to blot out any hope of making the Kentucky Derby, but a West Coast blacksmith named Bill Bane had been working on a vulcanized rubber patch for quarter cracks and was called in. Thus treated with a Bane Patch, Northern Dancer would be a bane to the existence of other three-year-olds. He dashed home in both the Flamingo and Florida Derby with Bill Shoemaker aboard. For the Kentucky Derby the Shoe opted to stick with another three-year-old he had been riding, the Western challenger Hill Rise. Luro had won the Derby two years earlier with Decidedly for Hill Rise's owner, George Pope. The trainer turned to Bill Hartack, who had ridden that earlier Derby winner, and Hartack gently guided Northern Dancer to a half-length victory in the Blue Grass Stakes at Keeneland.

Luro regarded a mile and one quarter as beyond the actual preference of Northern Dancer but thought he could finesse it out of the horse. In the Derby the nimble little colt got a head start on Hill Rise as they worked their way up from mid pack, and he withstood the prolonged challenge of the larger colt. Northern Dancer won by a neck and set a stakes record of 2:00, breaking the mark set by Decidedly.

A decade or so after concluding he did not need a Kentucky farm, E.P. Taylor had won the Kentucky Derby with the first Canadian-bred ever to wear the roses.

In the Preakness, Northern Dancer took the lead earlier and defeated The Scoundrel by two and a quarter lengths, with Hill Rise third. No horse since Citation in 1948 had won the U.S. Triple Crown, and now that distinction loomed ahead, but it would take victory at a mile and a half in the Belmont. Northern Dancer could not achieve this, finishing third behind Quadrangle and Roman Brother, beaten six lengths. Hartack was criticized by those who by then loved the little homebred too much to be objective. The champion jockey was said to have discouraged the horse by fighting him early. Quite likely, though, a mile and a half got to the bottom of Northern Dancer, as it has so many others.

In Canada, Northern Dancer paraded happily before his proud and adoring home crowd. The parade was called the Queen's Plate, and he won by seven and a half lengths.

Later in the summer Northern Dancer strained a tendon while preparing for a trip to Chicago for the American Derby. This put him out of that race, as well as the next target, the Travers, and Taylor, Thomas, and Luro concluded that retirement was the proper call. Northern Dancer had won fourteen of eighteen races and earned $580,806, a record for his country, and he was voted the champion three-year-old colt in North America as well as Horse of the Year for Canada.

Taylor now had at hand an unusual resource, and it was not an easy fit. A horse that commanded a stud fee — $10,000 — too high for any but a few in the breeding center in which he stood, was perhaps

not ideally placed. Northern Dancer was a Canadian horse, a Canadian hero, however, and he remained a Canadian stallion for the breeding seasons of 1965-68 and attracted a goodly number of shippers from other breeding centers. Taylor then purchased a farm in Maryland to establish a Windfields division there. On October 16, 1968, Taylor announced that Northern Dancer would be moved: "It was a difficult decision," he said. "He's a Canadian hero and has done well here, but in justice to his promising future, I think we must make him more easily accessible to the finest mares in North America." The little stallion, listed at 15.2 hands for stallion registers but often said to be an inch shorter, was later syndicated for $2.4 million. Eventually, a single share would be valued at $1 million. Nearctic would also be moved to Maryland, standing at Woodstock Farm, Mrs. Richard C. du Pont Jr.'s neighboring property.

In 1968, as Northern Dancer's first runners began to appear and included the flashy Windfields homebred Viceregal, Taylor altered his marketing tactic.

Northern Dancer with Horatio Luro

Abandoning his private pre-priced sale, he instead sent a draft of yearlings to the Canadian Thoroughbred Horse Society's annual sale at its Woodbine sale pavilion. With the Northern Dancers leading the way, the sale that had been topped by a $25,500 colt the year before, suddenly took on big league status.

Still wanting to maintain a number of horses for racing, Taylor worked out a novel arrangement with Fasig-Tipton, the American auction firm that came to Toronto to conduct the Canadian Thoroughbred Horse Society sale. Windfields' top-level consignment would enter the ring with announced reserves. If no bid was forthcoming at the stated price, Taylor retained the yearling. If a bid was entered, then the auction was off and away. The top reserve of $125,000 — impressive at Keeneland and Saratoga for the time and unheard of at Woodbine — was placed on a filly whose pedigree represented two eras. She was by young Northern Dancer and out of the aging Canadiana, herself by then dam of one stakes winner. Viceregal's full brother (to be named Vice Regent) along with a colt by the international star *Ribot were each priced at $100,000. None of these sold.

However, a spectacular Canadian record for the day was attained when a representative of the top international yearling market buyer of the time, Charles W. Engelhard, bid $84,000 for a colt whose pre-sale reserve was $60,000. The Irish trainer Vincent O'Brien, who had won that year's Epsom Derby with the American-bred Sir Ivor, had come to Toronto to look at Windfields' *Ribot yearling. O'Brien was not particularly impressed by that colt, but his attention was taken by a bay colt, sired by Northern Dancer but a much larger sort than his sire. This atypical son of the promising young stallion was out of Flaming Page, and there again, the theme of breeding upward would come into play.

Bull Page, the 1948 yearling by Bull Lea—Our Page, by Blue Larkspur, had become Canadian Horse of the Year and was a solid sire, without being an outstanding one. A high-priced yearling purchase ($38,000) of 1948, Bull Page, by the leading sire Bull Lea, was Canada's Horse of the Year for Windfields in 1951. He was destined for a place of distinction in the international success that lay before Taylor. The best of his nineteen stakes winners (7 percent) were Canadian Triple Crown winner New Providence and the Queen's Plate-winning filly Flaming Page. Flaming Page was a big, rangy mare and was the dam of the record-breaking Northern Dancer yearling, who was taken to Ireland by O'Brien and given the classic name of Nijinsky. (He was not known as Nijinsky II until returned to this country for breeding; an earlier American horse had had the same name.)

Just as Northern Dancer contributed a top pedigree on the sire side, there was plenty of class on the bottom side of the record yearling, although it had not recently expressed itself in so exalted a realm as Nijinsky would operate.

Viceregal

Bull Page was not only by five-time leading sire Bull Lea, but was out of 1942 Spinaway Stakes winner Our Page, she by another outstanding sire and broodmare sire, Blue Larkspur. Bull Page, who won nine of twenty-one starts and earned $25,730, was one of five stakes winners from Our Page, whose others included Breeders' Futurity winner Brother Tex and Jerome Handicap winner Navy Page. Occult, his second dam, foaled three stakes winners and was of Whitney breeding, being by *Dis Donc—Bonnie Witch, by Broomstick.

Nonetheless, it seems unlikely that the presence of Bull Page himself as the broodmare sire generated much of O'Brien's enthusiasm for the Northern Dancer—Flaming Page colt. Bull Page had been passed along to a farm in western Canada before his new status as broodmare sire of an international star had been recognized.

Flaming Page also had strong back-up in her pedigree but without representing the best individuals of a good family. Her dam, Flaring Top, was by the high-class sire Menow and was a winner who produced nothing but winners — eleven in all. Four of Flaring Top's were Windfields-bred stakes winners in Top Tourn, Quintain, Flashing Top, and Flaming Page. The first two of these were among the only four stakes winners sired by the Tourbillon stallion Tournoi (well, not everything Taylor tried hit the big time), and Flashing Top was by *Menetrier.

Flaring Top was bred by A.B. Hancock Jr. of Claiborne Farm and was a half sister to Claiborne's two-time champion filly Doubledogdare. Their dam, Flaming Top, was by the Triple Crown winner Omaha and out of Firetop, by Man o' War. Firetop was the dam of the noted distaffer Columbiana.

One can imagine that Taylor, who had succeeded with Chop Chop — son of stayer Flares, by Gallant Fox — might find the stamina of Flares' full brother Omaha an appealing aspect of Flaring Top's pedigree although Omaha by and large was a failure at stud. Taylor also had bred several stakes winners by a noted stayer, St. Leger winner *Boswell, after that stallion had been sent from Claiborne Farm to Canada.

At any rate, in the Bull Page—Flaring Top filly named Flaming Page, Taylor had bred a very high-class race filly. Flaming Page not only defeated colts in her country's vaunted Queen's Plate but also won the Canadian Oaks. Her other stakes win at home came in the Shady Well Stakes. Prior to her Oaks-Plate double, Flaming Page had tested American waters in the spring of 1962 and was second, beaten three lengths by the champion Cicada, in the Kentucky Oaks. Flaming Page won four of sixteen races and earned $108,836.

The combination of Northern Dancer and Flaming Page was magic and helped sweep away the notion that the Belmont Stakes failure would predestine Northern Dancer's offspring also to fail at a mile and a half. Nijinsky swept unbeaten through his first eleven races and was champion at two and three in both Ireland and England. The centerpiece of his career was England's 1970 Triple Crown of the one-mile Two Thousand Guineas, mile and a half Derby, and mile and three-quarters St. Leger. No horse had won all three of those races for thirty-five years, since Bahram in 1935, and through 2004 almost a like span has ensued without a further winner. Nijinsky II also won the Irish Sweeps Derby and defeated older horses in the King George VI and Queen Elizabeth Stakes.

A more difficult St. Leger than was recognized, a recent skin rash, and an uncontrolled crush of flash-bulbs in the paddock at Longchamp conspired to help bring down Nijinsky II. In the Prix de l'Arc de Triomphe, he charged to the fore under jockey Lester

WINANTS BROS.

Nijinsky II, as a yearling at the suburban Toronto division then part of Windfields

Piggott but could not contain Sassafras. Engelhard, feeling he had let down his horse, wanted him to go out a winner and approved of O'Brien's running him in the Champion Stakes as an afterthought, but Nijinsky lost again.

Hailed a truly great champion abroad, Nijinsky II was syndicated for a record $5.44 million to go to stud at Claiborne Farm. E.P. Taylor, leading breeder of 1970 in England, had bred the most expensive Thoroughbred in the world, and he had produced him in Canada, using a Canadian-bred sire and dam.

Realities of space require that Northern Dancer's record be synopsized into virtual shorthand for these pages. A strong case can be made for him as the greatest sire of the century. He sired 147 stakes winners (23 percent) and led the English sire list four times. After Nijinsky II came two later Epsom Derby winners by Northern Dancer, both bred by Windfields, in The Minstrel and Secreto. A number of Northern Dancer's sons have been so successful at stud as to challenge his own status. He has proven a sire of sires, and his sons have, too. Among the Northern Dancers are the following outstanding sires:

Nijinsky II — himself sire of three Epsom Derby winners in Golden Fleece, Shahrastani, and unbeaten Lammtarra. (The Derby reached its two-hundredth running in 1979. To date, four stallions have sired four winners apiece, and ten others have sired three; Northern Dancer and Nijinsky II are both in that second category.) Nijinsky II also sired Kentucky Derby winner Ferdinand among 155 stakes winners (18 percent); also sire of Caerleon, Green Dancer, Royal Academy, Shadeed, Sky Classic, and De La Rose.

Sadler's Wells — dominant leading sire of England/Ireland for more than the last decade, record 239 stakes winners (13 percent) include Epsom Derby winners Galileo and High Chaparral; Breeders' Cup winners Northern Spur, Barathea, In the Wings, High Chaparral, and Islington; Prix de l'Arc de Triomphe winners Carnegie and Montjeu; plus Old Vic, Salsabil, El Prado, and Beat Hollow.

Danzig — twice leading North American sire; sire of 182 stakes winners (17 percent), including successful stallions Danehill and Chief's Crown; two-time Breeders' Cup Mile winner Lure; and champion Canadian and U.S. filly Dance Smartly.

Nureyev — sire of 137 stakes winners (17 percent), including Arc winner Peintre Celebre; Breeders' Cup

Turf winner Theatrical; milers Spinning World, Atticus; and sprinters Stravinsky and Fasliyev.

Lyphard — leading American sire once and French leading sire twice; sire of 115 stakes winners (14 percent), including Arc winners Dancing Brave and Three Troikas; Breeders' Cup Turf winner and North American champion Manila; Reine de Saba; Sabin; and Jolypha.

The Minstrel, Storm Bird, Vice Regent, Northern Taste, El Gran Senor, Secreto, Try My Best, and Shareef Dancer were among the distinguished Northern Dancers bred by Windfields, in addition to Nijinsky II.

The year after Nijinsky II was sold at Woodbine, his full brother, a bright, powerfully built chestnut with plenty of white, topped the sale at a new record, $140,000. Engelhard was again the buyer. Having named one for a classic ballet master, Engelhard looked humorously toward the different form of dancing associated with an old burlesque house for a rhyming name, Minsky. (*The Night They Raided Minsky's* was a feature film of the era.)

Like his brother, Minsky was the champion two-year-old of Ireland, where he won the Railway and Beresford stakes; at three he won the Tetrarch and Gladness stakes. Returned to this country and to the management of Windfields, Minsky failed to live up to the hype generated by *Daily Racing Form* writer Charlie Hatton, who felt he was seeing a star in the making. Minsky wound up winning Toronto's Durham Cup twice and then was exported to Japan in 1973.

The combination of Northern Dancer and Nijinsky II's female family had another champion on the horizon in The Minstrel. The Minstrel was another well-decorated chestnut, and he was by Northern Dancer and out of Fleur, who was by Chop Chop's son Victoria Park and out of Flaming Page. By the time this colt came to the yearling market, the international success of Northern Dancer had elevated Windfields Farm to exalted status. Taylor could place his yearlings where he chose. John Finney, of Fasig-Tipton, always felt that Taylor's conclusion that Keeneland embodied the height of prestige made it so, rather than responding to its already being so. Windfields sold at Woodbine still and sent some to Saratoga, but by assigning its elite consignments to Keeneland during a time when Northern Dancer had begun attracting a following from Europe as well as

America, Taylor aided Keeneland's summer sale in its climb to become the key auction in the world. Windfields was the leading consignor by average at Keeneland in 1977, 1981, and 1983. At that level Taylor's consigning technique was without the devices to maintain half of the horses. (He had once tried an approach at Woodbine of having two paired horses in the ring, with the winning bidder choosing which one he/she would take. This was fair and open but did not catch on very well.)

By the time of the 1977 sale, Engelhard had passed away, and Robert Sangster of England and his team, including adviser/trainer O'Brien, had become for a time the most visible and successful buyers. At Keeneland in 1975, Sangster's group bought The Minstrel for $200,000. A high-class stakes winner in both England and Ireland at two, The Minstrel won the 1977 English and Irish derbys at three, then, still tracking the three-year-old season of Nijinsky II, added the King George VI and Queen Elizabeth Stakes. He was rushed to America later that year to beat a pending importation ban prompted by the fears of spread of contagious equine metritis.

The Minstrel stood at Windfields' Maryland division and later at W.T. Young's Overbrook in Lexington, Kentucky. His fifty-eight stakes winners (11 percent) included the Breeders' Cup Mile winner Opening Verse and the international winner Palace Music, he in turn the sire of the two-time American Horse of the Year Cigar.

The year The Minstrel was three, 1977, also was the year that Taylor's total of stakes winners bred reached 192, giving him the all-time record surpassing the 191 bred decades earlier by Harry Payne Whitney.

In 1984 Taylor was represented by his third Epsom Derby winner as a breeder. As two horses battled for the wire, any observer with Windfields connections could relax. Both were bred by the farm! The winner was Secreto, a $340,000 Keeneland summer sale yearling, who edged El Gran Senor. El Gran Senor had already won the classic Two Thousand Guineas in England and later added the Irish Sweeps Derby. He was out of the Buckpasser mare Sex Appeal, Windfields having bought into the latter's female family; Sex Appeal's dam was Best in Show, dam of four stakes winners.

Secreto made little mark at stud. Oddly, for a sire line of such prevalence, there was a repeating ten-

dency toward decreased fertility among some of the Northern Dancer descendants — the sterile Cigar most famously — and El Gran Senor required careful management in that regard. El Gran Senor, however, has sired fifty-five stakes winners (13 percent), including Breeders' Cup Sprint winner Lit de Justice, English Two Thousand Guineas winner Rodrigo de Triano, and Strub Stakes winner Helmsman. El Gran Senor's daughter Toussaud, a grade I winner, reached exalted status as the Broodmare of the Year for 2002 and dam of the 2003 Belmont Stakes winner Empire Maker.

Vice Regent, a horse of apparent brilliance but hampered by unsoundness, was a full brother to the

The Minstrel

first Northern Dancer champion, Viceregal. After Viceregal's retirement following injury in his first start at three — and before the emergence of Nijinsky II — Taylor told the author that he thought Viceregal had been destined to be the best horse he ever bred. The year-younger brother, Vice Regent, unsold as a yearling for $100,000, inherited a bit of the mantle of local stardom when their sire Northern Dancer was sent to Maryland. Vice Regent and Viceregal were foaled from Victoria Regina, that *Menetrier filly from unraced Victoriana. The next generation back was made up of the foundation influences Windfields and Iribelle. Vice Regent sired 105 stakes winners (15 percent). His star son at stud is Eclipse Award winner and two-time leading sire Deputy Minister (Open Mind, Go for Wand, Dehere,

Deputy Commander, Touch Gold, Awesome Again, and Silver Deputy). Small Ontario breeders Morton and Marjoh Levy bred Deputy Minister.

Storm Bird was a Windfields yearling who brought $1 million from Sangster and his team at the 1979 Keeneland summer sale. He was a Northern Dancer colt whose female family presented sentiment as well as achievement. His dam, South Ocean, had won the 1970 Canadian Oaks as a first major winner in the colors of Taylor's son, Charles. Moreover, South Ocean was by New Providence, a Windfields homebred Canadian Triple Crown winner who joined Flaming Page as a star among the offspring of Bull Page.

Storm Bird

Storm Bird was delivered to master trainer O'Brien in Ireland and was the unbeaten champion two-year-old of both England and Ireland in 1980. The following year a disgruntled former employee broke into O'Brien's Ballydoyle compound and hacked at Storm Bird's mane and tail. The horse raced but once at three and was beaten, but he had been syndicated on an announced valuation of $30 million — a tribute to his sire's worldwide reputation and perhaps to creative accounting/promotion as well — and was returned to America to stand at Ashford Stud. Storm Bird sired the Preakness winner Summer Squall (in turn sire of Derby/Preakness winner and Horse of the Year Charismatic and juvenile filly champion Storm Song) as well as the English Oaks/Irish Derby winner Balanchine and Indian Skimmer. Storm Bird's signa-

ture contribution to the ongoing legacy of his sire, Northern Dancer, is a son, Storm Cat, twice-leading American stallion and sire of champions Storm Flag Flying and Giant's Causeway, Preakness/Belmont winner Tabasco Cat, and Breeders' Cup Classic winner Cat Thief.

Windfields-bred Northern Taste was a 1971 Northern Dancer colt from Lady Victoria, she by Victoria Park and out of Nearctic's dam, *Lady Angela. This close inbreeding produced a French group I winner who became a ten-time leading sire in Japan.

Windfields-bred Shareef Dancer was almost a record-setting yearling for his era. He was sold for $3.3 million at Keeneland in the summer of 1981 to the Maktoum family of Dubai, but a Windfields son of Northern Dancer—South Ocean had brought $3.5 million three hip numbers before. The year before, Taylor had suffered a stroke, and his son Charles had stepped to the fore as key leader for Windfields. Shareef Dancer represented the continual acquisition of outside bloodlines that E.P. Taylor had pursued for the Windfields racing stable and breeding operation. The dam, the Sir Ivor filly Sweet Alliance, had been a $65,000 yearling purchase by Windfields. She was out of Newstead Farm's Mrs. Peterkin, a Tom Fool mare whose own dam was the Newstead foundation matron Legendra, dam of other stakes winners Rich Tradition, Sky Clipper, and Hasty Doll. Sweet Alliance had won the 1977 Kentucky Oaks for Windfields. Unlike many other spectacularly priced yearlings, Shareef Dancer became a major success, winning the Irish Derby in 1983.

Try My Best was yet another Northern Dancer bred by Windfields. He was out of Sex Appeal and was thus a full brother to El Gran Senor. Try My Best was a foal of 1975 and was a $185,000 Keeneland yearling purchased by Sangster. Try My Best won the Dewhurst Stakes at two and was acknowledged as the champion two-year-old in England in 1977 but failed as a heavy favorite the following year in the Two Thousand Guineas. Try My Best's thirty-two stakes winners (8 percent) included Last Tycoon, winner of the 1986 Breeders' Cup Mile.

A recurrent theme of Windfields was the supplying, through sales and by standing stallions, of important horses to other breeders. Over the years (in addition to Levesque), Bud Willmot of Kinghaven Farm, Conn Smythe, Douglas Banks, George

Gardiner, Sam-Son Farm, and others benefited from the bloodlines and/or yearlings of Windfields.

Nearctic was to be overshadowed internationally by his son, Northern Dancer, but was a dominant sire of Canadian-breds and had a number of major other outside horses as well. Nearctic eventually sired forty-nine stakes winners (14 percent), and they included English Two Thousand Guineas winner Nonoalco, the important sires Icecapade and Explodent, Matron winner Cold Comfort, and Belmont Stakes runner-up Cool Reception.

Natalma, too, became a source of major success for other breeders. She produced four stakes winners, but aside from Northern Dancer her enduring influence came from fillies that were not major winners themselves. One was Arctic Dancer, a full sister to Northern Dancer. Arctic Dancer became the dam of Levesque's homebred La Prevoyante, unbeaten in twelve starts at two in 1972 when she was the Eclipse Award winner in the two-year-old division as well as Canadian Horse of the Year. Another daughter of Natalma was Raise the Standard, an unraced filly by Hoist the Flag. She produced the French stakes winner Coup de Folie and thus is a tail-female antecedent of the international sire Machiavellian, plus Coup de Genie, Exit to Nowhere, etc.

By 1975 Danny Van Clief, who had sold Natalma to Taylor for $35,000, had to go to $260,000 to buy the mare's Buckpasser filly at the Saratoga yearling sale. Named Spring Adieu, the filly won three of seven races and earned $13,757. More importantly, she became a key resource for the Van Clief family's Nydrie Stud. In 1982 Nydrie sent Spring Adieu's His Majesty filly to take her turn in her family's cameo roles in the Saratoga sale ring. This filly brought $350,000 (ten times Natalma's price). She was purchased by James Delahooke on behalf of Prince Khalid Abdullah, the Saudi Arabian aristocrat who was building Juddmonte Farms into one of the most important international breeding and racing operations of the late twentieth century-early twenty-first century era.

The His Majesty—Spring Adieu filly was named Razyana. She was sent to England and never won in three starts. Returned to the United States, Razyana was bred to the Northern Dancer horse Danzig and thus her colt foal of 1986 was inbred to Natalma 3x3. The colt was named Danehill, and he won the Ladbroke Sprint Cup Stakes, Moorestyle Convivial, and Cork & Orrery Stakes in England and Ireland. Danehill was also third behind champion Nashwan in England's Two Thousand Guineas of 1989.

The Coolmore operation headed by John Magnier gambled a reported four million pounds to acquire Danehill from Abdullah. Standing various seasons in both Ireland in the Northern Hemisphere breeding season and Australia in the Southern Hemisphere breeding season, Danehill added his name to the top echelon of the sire line of Northern Dancer. He died at seventeen in a paddock accident in the spring of 2003. Through mid-2004 he had been represented by a total of 256 stakes winners. Danehill's best included fifty-five grade/group stakes winners, of which thirty-three were Australian. He has led the Australian sire list eight times.

The Nearctic-Northern Dancer blood became the dominant

Kamar and foal

ingredient at Windfields but, of course, also necessitated the need for outcrosses. Taylor and son Charles bred a number of major horses with outside blood as well, and some with a combination. The fruit of these combinations outlined below was plucked by the avid buyers of Windfields yearlings.

The Nearctic mare Nangela produced the Quadrangle filly Square Angel, winner of the Canadian Oaks and a champion at three. Square Angel in turn foaled Kamar to the cover of Key to the Mint. Kamar became a Broodmare of the Year for Three Chimneys Farm. Another Key to the Mint filly from Square Angel was grade I Apple Blossom Stakes winner Love Smitten, dam of Swain, winner of two runnings of the King George VI and Queen Elizabeth Stakes in the late 1990s and earner of nearly $4 million.

In another instance, Taylor bred the stakes-winning Northern Dancer filly Drama School to Ogden Phipps' champion Buckpasser and got Norcliffe. Sold to Lieutenant Colonel Bud Baker, Taylor's successor as chairman of the Ontario Jockey Club, Norcliffe won the Queen's Plate and sired At the Threshold, he in turn the sire of 1992 Kentucky Derby winner Lil E. Tee.

Taylor had bought into the family of Newstead Farm's Legendra, and he also bought into the productive family of Darby Dan Farm's Soaring in the form of the *Herbager filly Ballade. Ballade's bloodline became a source of championships both with and without Northern Dancer's influence. To the cover of Halo, Ballade foaled two champions in

Glorious Song and Devil's Bag. Glorious Song, bred by Windfields and raced in partnership by Frank Stronach and Nelson Bunker Hunt, was Horse of the Year in Canada and also an Eclipse Award winner after victories in the Spinster Stakes, Santa Margarita Invitational Handicap, and other grade I races. Glorious Song produced the high-class miler and international sire Rahy to the cover of Blushing Groom, again outside the Northern Dancer blood. Glorious Song's 1996 North American Turf male champion Singspiel did have the Northern Dancer connection. His sire, In the Wings, was by Northern Dancer's son Sadler's Wells. Again the influence crossed into the next century, for Singspiel is the sire of 2003 Dubai World Cup winner Moon Ballad.

Ballade's second champion by Halo was Devil's Bag, a $325,000 Windfields-bred who raced for Mr. and Mrs. James P. Mills Hickory Tree Stable. Devil's Bag was a spectacular, unbeaten two-year-old champion whom trainer Woody Stephens was convinced was destined for greatness. The colt did not reach those heights but was syndicated for some $36 million to stand at Claiborne Farm, where he has been a solid success. Devil's Bag is the sire of Devil His Due, Twilight Agenda, Taiki Shuttle, etc.

Yet another Halo—Ballade foal bred by Windfields was the good stakes winner and successful sire Saint Ballado.

A personal favorite for the author among the many continuing sagas of Windfields Farm families stems

*Herbager, 1956	Vandale, 1943	Plassy, 1932	Bosworth (Son-in-Law—Serenissima)
			Pladda (Phalaris—Rothesay Bay)
		Vanille, 1929	La Farina (Sans Souci—Malatesta)
			Vaya (Beppo—Waterhen)
	Flagette, 1951	Escamillo, 1939	**Firdaussi (Pharos—Brownhylda)**
			Estoril (Solario—Appleby)
		Fidgette, 1939	**Firdaussi (Pharos—Brownhylda)**
			Boxeuse (*Teddy—Spicebox)
BALLADE	Cohoes, 1954	*Mahmoud, 1933	*Blenheim II (Blandford—Malva)
			Mah Mahal (Gainsborough—**Mumtaz Mahal**)
Miss Swapsco, 1965		Belle of Troy, 1947	Blue Larkspur (Black Servant—Blossom Time)
			*La Troienne (*Teddy—Helene de Troie)
	Soaring, 1960	Swaps, 1952	*Khaled (Hyperion—Eclair)
			Iron Reward (*Beau Pere—Iron Maiden)
		*Skylarking II, 1947	Mirza (*Blenheim II—**Mumtaz Mahal**)
			Jennie (Apelle—Lindos Ojos)

from another purchase at Newmarket. In 1955 Taylor bought *Queen's Statute as a yearling at the September Sale there for $10,185. She was by the sprinting Djebel stallion Le Lavandou, who was a half brother to 1956 Epsom Derby winner Lavandin. Despite his own racing tendencies, Le Lavandou turned out to be a source of stamina, but *Queen's Statute never got to the races.

Absorbed into the Windfields broodmare band, *Queen's Statute produced six stakes winners. For the most part, these were solid, Ontario campaigners, including the Canadian Oaks winner Menedict, by *Menetrier, and the Canadian champion Dance Act, by Northern Dancer. It was another of *Queen's Statute's foals who was the stepping stone toward the pinnacles to which this family would climb. Her Northern Dancer filly Royal Statute won only once in eight starts but then foaled the stakes winner Akureyri, by Buckpasser. Then, in 1979, Royal Statute had a filly by *Snow Knight.

The 50-1 winner of the Epsom Derby in 1974, *Snow Knight was a sort of horse to intrigue E.P. Taylor. *Snow Knight was owned by Mr. and Mrs. Neil Phillips of Montreal when he won the Derby, and Taylor later purchased and imported him. *Snow Knight was turned over to trainer MacKenzie Miller, who overcame the horse's roguish manners and campaigned him to an Eclipse Award as the champion turf horse of 1975.

*Snow Knight was retained for the stud by Taylor, but the Derby winner had a somewhat obscure pedigree (Firestreak—Snow Blossom, by Flush Royal) and was by and large a failure. Snow Knight did get one important filly, however, and that was his daughter out of Royal Statute. Royal Statute had also foaled a One Thousand Guineas runner-up, Konafa, by the time her *Snow Knight yearling of 1980 was consigned to the Fasig-Tipton Kentucky summer sale, which Windfields also patronized. The filly caught the eye of a new group of buyers, the Maktoum brothers of Dubai, who that summer introduced to the international Turf a family that would be one of the most important ever to express its wealth and sportsmanship via the world's racecourses, sale rings, and breeding farms.

Sheikh Mohammed bin Rashid Al Maktoum purchased the *Snow Knight—Royal Statute filly for

Glorious Song

$325,000. Given the exotic name Awaasif, she was sent abroad and won the Yorkshire Oaks in England and the Gran Premio del Jockey Club in Italy. In the Prix de l'Arc de Triomphe of 1982, the four-year-old Awaasif came close to an arithmetically more startling upset than her father's 50-1 Derby lark. At 90-1, she finished third, beaten barely a half-length by winner Akiyda.

In 1986 Awaasif foaled a Blushing Groom filly who was named Snow Bride and was bred in the name of one of the Maktoum entities, Darley Stud Management. Snow Bride officially became a classic winner when Aliysa, who had beaten her, was disqualified from victory in the Epsom Oaks after testing positive for metabolites of a forbidden substance.

Completing the circle back to Windfields, in 1992 Snow Bride foaled a colt by Northern Dancer's first great son, Nijinsky II. Named Lammtarra, this colt had a short career of only four races, but a remarkable one. He won all four, and three of those races were the Epsom Derby, King George VI and Queen Elizabeth Stakes, and Prix de l'Arc de Triomphe! The Derby and Oaks have been run for well over two hundred years, but that was the first instance of winners of the two races (Nijinsky II and Snow Bride) combining to produce another Derby winner. Lammtarra was sold to Japan for a reported valuation of $38.4 million.

Yet another aspect of the *Queen's Statute/Royal

Statute tale unfolded on the international stage. Royal Statute's classic-placed daughter Konafa foaled several nice stakes horses, including Korveya, by Riverman. Korveya foaled the French two-year-old champion Hector Protector, the English filly champion Bosra Sham, and French Two Thousand Guineas winner Shanghai. By this time the female family and it offshoots had touched several of the world's leading breeding operations, including that of the Niarchos family as well as the Maktoums'.

Gerald Leigh was co-owner of Korveya with Jayeff B Stables at the time she was consigned to the Keeneland November sale of 1998. Jayeff B had to go to a bid of $7 million to retain one half and buy out the other, that sale price matching Miss Oceana's world mark for a broodmare sold at auction.

Again the Windfields theme was upward mobility.

In stunning contrast to such tales, but with a similar outcome, was the story of the Windfields-bred *Tularia, a daughter of English classics winner *Tulyar—*Suntop II, by Dastur. A foal of 1955, *Tularia had started life as a modest stakes winner, but the circuitous route that takes horses to obscurity or great glory had chosen the former for her. She at length was part of a package of horses consigned to be sold at a sheriff's sale, and publisher Howard Sams stepped in to buy the lot for $6,000. Whereupon, *Tularia foaled 1975 juvenile champion Honest Pleasure and another grade I winner in For The Moment!

In 1974 E.P. Taylor led the list of breeders in money won for the first time, and he continued his hold on that statistical leadership seven straight years. Only Calumet Farm and Harry Payne Whitney have dominated for a longer sequence. Maxwell Gluck topped the list in 1981, stopping Taylor's streak, but Taylor returned to the top in 1983 and then for a final time in 1985, his ninth time in twelve seasons. The year 1985 also was the last season of his dominance on the list by number of winners.

Although the Kentucky Derby and Epsom Derby had become foils for Windfields-bred horses, Taylor still set great store by his home country's Queen's Plate, with its history, pageantry, and frequent royal visitors. There had been a time when Plates came his way often, and by 1978 he had bred seventeen winners of the race, of which he had raced nine. Fourteen years had passed since that last Plate victory in the home silks, however, and, he quipped, "My

wife doesn't give me any credit at all for breeding a Plate winner when someone else owns it. You know, it's pretty difficult to win when you offer everything for sale."

There had been times when the frequency of Windfields victories at Woodbine had generated carping in the Toronto press and even booing from the Woodbine stands, the implication being that "house horses" somehow had an unfair edge. When Regal Embrace broke Taylor's drought and won the Plate in Taylor's silks in 1978, however, the victory

La Lorgnette with Charles Taylor after the Queen's Plate

was seen in the sentimental light that it deserved, a subtle recognition that a "grand old man" status had been bestowed. The Canadian mogul was now seventy-seven but still quick with a quip. After the race, the Toronto columnist Dick Beddoes, a sometimes acerbic observer in the past, baited Taylor with the comment that "they cheered you today, and they didn't used to do that." Taylor shot back, "That was when they used to read you." Both laughed, and Taylor added "Touché!"

Regal Embrace was by the Northern Dancer stallion Vice Regent and out of a mare by *Nentego, a son of English Derby winner Never Say Die. Fourteen years after greeting little Northern Dancer in the Queen's Plate winner's circle, Taylor took the shank of the horse's grandson standing nearly two

Champions Bred

In the United States:

Northern Dancer
1964 champion three-year-old colt

Glorious Song
1980 champion older female

Devil's Bag
1983 champion two-year-old colt

In Canada:

Canadiana
1952 champion two-year-old filly
1952 Horse of the Year

Queen's Own
1954 Horse of the Year

Canadian Champ
1956 Horse of the Year

Nearctic
1958 Horse of the Year

Victoria Park
1959 champion two-year-old colt
1960 champion three-year-old colt
1960 Horse of the Year

Flaming Page
1962 champion three-year-old filly

Court Royal
1963 champion handicap female

Canebora
1963 champion three-year-old colt
1963 Horse of the Year

Northern Dancer
1963 champion two-year-old colt
1964 Horse of the Year

Northern Queen
1965 champion three-year-old filly

Titled Hero
1965 champion two-year-old colt
1966 champion three-year-old colt

Victorian Era
1966 champion handicap horse
1966 Horse of the Year

Cool Reception
1966 champion two-year-old colt

Arctic Blizzard
1967 champion two-year-old colt

Dance Act
1970 champion handicap horse
1971 champion handicap horse

Viceregal
1968 champion two-year-old colt
1968 Horse of the Year

Minsky
1971 champion three-year-old colt

Happy Victory
1972 champion three-year-old filly

Lord Durham
1973 champion two-year-old colt

Square Angel
1973 champion three-year-old filly

Victorian Queen
1975 champion grass horse
1975 champion handicap mare

Momigi
1975 champion three-year-old filly
1976 champion handicap mare
1977 champion grass horse

Victorian Prince
1976 champion grass horse
1976 champion handicap horse

Norcliffe
1976 champion three-year-old colt
1976 Horse of the Year
1977 champion handicap horse

Northernette
1976 champion two-year-old filly
1977 champion three-year-old filly

Sound Reason
1976 champion two-year-old colt

L'Alezane
1977 champion two-year-old filly
1977 Horse of the Year

Dance in Time
1977 champion three-year-old colt

Christy's Mount
1978 champion handicap mare

Giboulee
1978 champion handicap horse

Kamar
1979 champion three-year-old filly

Glorious Song
1980 champion handicap mare
1980 Horse of the Year
1981 champion handicap mare

Northern Blossom
1983 champion three-year-old filly

Diapason
1984 champion sprinter

Deceit Dancer
1984 champion two-year-old filly

Lake Country
1985 champion handicap mare

Imperial Choice
1985 champion three-year-old colt
1985 champion grass horse
1985 Horse of the Year

La Lorgnette
1985 champion three-year-old filly

Golden Choice
1986 champion three-year-old colt

Term Limits
1993 champion two-year-old filly

Honky Tonk Tune
1994 champion two-year-old filly

Archers Bay
1998 champoion three-year-old colt

In Europe:

Nijinsky II
1969 English champion two-year-old colt
1969 Irish champion two-year-old colt
1970 English champion three-year-old colt
1970 Irish champion three-year-old colt
1970 European Horse of the Year

Minsky
1970 Irish champion two-year-old colt

The Minstrel
1977 English Horse of the Year

Try My Best
1977 English champion two-year-old colt
1977 Irish champion two-year-old colt
1978 Irish champion miler

Storm Bird
1980 English champion two-year-old colt
1980 Irish champion two-year-old colt

Awaasif
1982 English champion three-year-old filly

Danzatore
1982 Irish champion two-year-old colt

Shareef Dancer
1983 English champion three-year-old colt
1983 Irish champion three-year-old colt

El Gran Senor
1983 English champion two-year-old colt
1983 Irish champion two-year-old colt
1984 English champion three-year-old colt
1984 English Champion miler

Street Rebel
1996 Swiss champion sprinter

In Puerto Rico:

Shake Shake Shake
1978 champion three-year-old colt

hands taller! With Regal Embrace's triumph, Taylor had bred eighteen Plate winners and raced ten of them.

The Succession

After Taylor's stroke in 1980 incapacitated him, son Charles balanced his own ambitions with those handed down by his father. As a young man, Charles had been a globetrotting journalist who covered wars in Vietnam, Nigeria, and the Middle East. He delighted in recalling the suspicions of the Chinese over cables in 1964, when he was in Peking for the Toronto *Globe & Mail* and received communiqués from the Triple Crown series, including the Preakness: "Can't you imagine a whole room of Chinese cryptographers trying to break down this ingenious Occidental code — 'Northern Dancer first The Scoundrel second Hill Rise third Love Dad.'"

Charles Taylor said as late as 1983 that "writing is the most important part of my life — it's what I am, my core. The horse thing means a great deal to me, but it's a business, a hobby, a family commitment."

He juggled the different lives. One of his five books, *Radical Tories*, examined influential Canadians' political philosophies and spoke of "compassionate conservatism" years before the phrase became a catchword for an American presidential campaign. Nevertheless, one had the feeling that the horses gradually were getting to Charles Taylor. The day he lived out his own family legacy by leading in the homebred filly La Lorgnette after the 1985 Queen's Plate, it was easy and pleasant to conjecture that, to the son of E.P. Taylor, "my core" had taken on some variety.

La Lorgnette (Val de l'Orne—The Temptress, by Nijinsky II) was identified by *The Blood-Horse* as one of forty-two horses bred by Windfields that earned divisional championships, either for Canada alone, all of North America, or European divisions. (La Lorgnette became the dam of England and Irish champion Hawk Wing and second dam of Canadian Horse of the Year Thornfield. It was also during Charles' time at the helm that two of Windfields'

three times of leading the Keeneland summer sale by consignor average were achieved.

E.P. Taylor died at eighty-eight on May 14, 1989. His compounding record in breeding stakes winners had risen to more than three hundred. His honors and positions of leadership had been many, including an Eclipse Award as Man of the Year; Thoroughbred Club of America testimonial dinner honor guest in 1959; chairmanship of The Jockey Club of Canada; membership in The Jockey Club; honorary membership in the English Jockey Club; long-time board member of the Grayson Foundation.

Charles Taylor accepted the responsibility of leadership with similar generosity and vision. He was key in running the sales and other affairs of the Canadian Thoroughbred Horse Society, a director and vice president of the Breeders' Cup, chairman and chief steward of The Jockey Club of Canada, member of The Jockey Club, director of Grayson-Jockey Club Research Foundation, director of *The Blood-Horse*'s publication oversight committee, and trustee of the E.P. Taylor Equine Research Fund and the National Museum of Racing. He was recipient of the E.P. Taylor Award of Merit for lifetime contributions to Canadian racing.

By the autumn of 1996, Charles Taylor was suffering from cancer and was unable to attend an event that was in many ways a tribute to himself, his father, and their country — the running of the Breeders' Cup at Woodbine. He was able to attend the Queen's Plate the following year and to speak with Queen Elizabeth II, but he passed away at sixty-two less than two weeks later. Under Charles' watch, the total number of stakes winners bred by E.P. Taylor and Windfields had increased to 358, an all-time record still growing. His second wife, the successful artist Noreen Taylor, had become dedicated to her husband's interest in racing and carried on the operation of Windfields Farm. By 1998, when the Windfields-bred and -sold Archers Bay won the Queen's Plate, the number of Windfields/Taylor-bred winners of that venerated event stood at twenty-two.

Stakes Winners Bred by E.P. Taylor

Windfields, br.c. 1943, Bunty Lawless—Nandi, by Stimulus
Epic, br.c. 1946, Bunty Lawless—Fairy Imp, by *Gino
Bennington, br.c. 1947, *Boswell—Iribelle, by *Osiris II
McGill, b.c. 1947, Bunty Lawless—Tinted Chick, by Tintagel
Britannia, br.f. 1948, Bunty Lawless—Iribelle, by *Osiris II
Major Factor, br.g. 1948, *Boswell—Aldwych, by Marine
Acadian, b.c. 1949, Teddy Wrack—Lady Mona, by *Osiris II
Dress Circle, br.f. 1949, *Boswell—Compensate, by Reaping Reward
Epigram, b.c. 1949, Flares—Hasty Bet, by Reigh Count
Canadiana, b.f. 1950, Chop Chop—Iribelle, by *Osiris II
Heroic Age, b.c. 1950, *Boswell—Sweet Pegotty, by Pilate
Lively Action, b.f. 1950, *Boswell—Hasty Bet, by Reigh Count
Free Trade, b.c. 1951, Windfields—Our Flare, by Flares
June Brook, b.f. 1951, Brookfield—June Bee, by Jean Valjean
My Wind, br.f. 1951, Windfields—Fairy Imp, by *Gino
Queen's Own, ch.c. 1951, War Jeep—Sea Reigh, by Reigh Count
Shine Ever, b.f. 1951, Flares—Hasty Bet, by Reigh Count
Board of Trade, b.c. 1952, Illuminable—*Mademoiselle Satan, by Sardanapale
Festivity, ch.f. 1952, *Fairaris—Bolesteo, by *Filisteo
Flirt, dk.br.f. 1952, *Menetrier—*Flora Dora II, by Deux-pour-Cent
Acushla, b.f. 1953, Bull Page—Lovely Delores, by Blue Larkspur
Air Page, br.f. 1953, Bull Page—Air Post, by Ariel
Butter Ball, dk.b.f. 1953, *Tournoi—Bernadette S., by Attention
Canadian Champ, b.c. 1953, Windfields—Bolesteo, by *Filisteo
Censor, br.c. 1953, Bull Page—Compensate, by Reaping Reward
Gracefield, br.f. 1953, Windfields—Your Grace, by Jamestown
Queen's Reigh, ch.f. 1953, Illuminable—Sea Reigh, by Reigh Count
Top Tourn, ch.c. 1953, *Tournoi—Flaring Top, by Menow
Chopadette, b.g. 1954, Chop Chop—Bernadette S., by Attention
La Belle Rose, b.f. 1954, Le Lavandou—*Missy Suntan, by Tai-Yang
Lyford Cay, b.c. 1954, Chop Chop—Famous Maid, by *Fairaris
Myanna, br.f. 1954, Chop Chop—Britannia, by Bunty Lawless
Nearctic, br.c. 1954, Nearco—*Lady Angela, by Hyperion
Our Sirdar, dk.b.c. 1954, Illuminable—Miss Rustom, by *Rustom Sirdar
Silly Lilly, b.f. 1954, Bull Page—Rustic Charm, by Reaping Reward
Chomiru, b.f. 1955, Chop Chop—Miss Rustom, by *Rustom Sirdar
Dr. Em Jay, b.g. 1955, Chop Chop—Solar Display, by Sun Again
Happy Harry, b.c. 1955, Chop Chop—Famous Maid, by *Fairaris
*Tularia, b.f. 1955, *Tulyar—*Suntop II, by Dastur (Bred in England)
Windy Answer, b.f. 1955, Windfields—Reply, by Teddy Wrack
Yummy Mummy, dk.b.f. 1955, Bull Page—Rustic Charm, by Reaping Reward
Bull Vic, b.c. 1956, Bull Page—Victoriana, by Windfields
Castleberry, dk.b.g. 1956, *Castleton—Rustic Charm, by Reaping Reward
Le Grand Rouge, ch.c. 1956, *Menetrier—*Wyndola, by Wyndham
New Providence, b.c. 1956, Bull Page—*Fair Colleen, by Precipitic
Sunday Sail, b.c. 1956, Alycidon—*Stalina, by Stalino
Willowdale Boy, b.c. 1956, Windfields—Sunny Vixen, by Gallant Fox
All Canadian, b.g. 1957, Windfields—Canadiana, by Chop Chop
Bulpamiru, b.f. 1957, Bull Page—Miss Rustom, by *Rustom Sirdar
Eltoro the Great, b.c. 1957, Chop Chop—Lap o' Luxury, by Rosemont
Menantic, dk.b.f. 1957, *Menetrier—Romantic Dream, by Reaping Reward
Men at Play, ch.c. 1957, *Menetrier—Solar Display, by Sun Again
Quintain, b.c. 1957, *Tournoi—Flaring Top, by Menow
Victoria Park, b.c. 1957, Chop Chop—Victoriana, by Windfields

Windsor Field, b.c. 1957, Windfields—*La Poloma II, by Le Pacha
Windy Ship, b.c. 1957, Windfields—Lovely Delores, by Blue Larkspur
Axeman, b.c. 1958, Chop Chop—Famous Maid, by *Fairaris
Blue Light, b.c. 1958, Chop Chop—Blen Lark, by *Blenheim II
Cut Steel, b.c. 1958, Chop Chop—*Stalina, by Stalino
Epic Queen, ch.f. 1958, Epic—*Queen's Statute, by Le Lavandou
Flashing Top, ch.f. 1958, *Menetrier—Flaring Top, by Menow
Maid o' North, dk.b.f. 1958, Bull Page—*Fair Colleen, by Precipitic
Majestic Hour, ch.c. 1958, Queen's Own—Bright Flash, by *Fairaris
Mystery Guest, b.f. 1958, *Menetrier—Consort, by *Bull Dog
Navy Wyn, b.f. 1958, Navy Page—*Wyndola, by Wyndham
Song of Even, dk.b.f. 1958, Vimy—*Evensong, by The Phoenix
Victoria Regina, ch.f. 1958, *Menetrier—Victoriana, by Windfields
Windspray, b.c. 1958, Windfields—Shipway, by Marine
Choperion, dk.b.c. 1959, Chop Chop—*Lady Angela, by Hyperion
Court Royal, b.f. 1959, Chop Chop—*Queen's Statute, by Le Lavandou
Flaming Page, b.f. 1959, Bull Page—Flaring Top, by Menow
King Gorm, b.c. 1959, Never Say Die—*White Lodge, by Casanova
Monarch Park, b.c. 1959, Chop Chop—Gai Parisienne, by *Tournoi
Peter's Chop, b.c. 1959, Chop Chop—Butter Ball, by *Tournoi
Breezy Answer, br.f. 1960, Bull Page—Windy Answer, by Windfields
Canebora, br.c. 1960, Canadian Champ—Menebora, by *Menetrier
Hop Hop, b.c. 1960, Chop Chop—*Tularia, by *Tulyar
Menedict, b.f. 1960, *Menetrier—*Queen's Statute, by Le Lavandou
Royal Maple, b.c. 1960, *Menetrier—*Wyndola, by Wyndham
Sea Service, b.c. 1960, Navy Page—*Tudorette, by *Tudor Minstrel
Son Blue, br.g. 1960, Blue Man—My Wind, by Windfields
Avec Vous, ch.f. 1961, Ace Marine—Jewel Case, by Jet Pilot
Brockton Boy, b.c. 1961, *Menetrier—*Tudorette, by *Tudor Minstrel
Ciboulette, dk.b.f. 1961, Chop Chop—Windy Answer, by Windfields
French Wind, b.f. 1961, *Menetrier—Flaming Wind, by Windfields
Grand Garcon, b.c. 1961, Censor—*Stalina, by Stalino
Northern Dancer, b.c. 1961, Nearctic—Natalma, by Native Dancer
Pierlou, b.c. 1961, Nearctic—Windka, by Windfields
Des Erables, b.c. 1962, Canadian Champ—Acacia, by *Tournoi
Lady Victoria, dk.b.f. 1962, Victoria Park—*Lady Angela, by Hyperion
Native Victor, dk.b.c. 1962, Victoria Park—Natalma, by Native Dancer
Northern Queen, dk.b.f. 1962, Nearctic—Victoriana, by Windfields
Victorian Era, b.c. 1962, Victoria Park—Ivy, by *Nasrullah
Bright Monarch, gr.g. 1963, *Grey Monarch—Bright Flash, by *Fairaris
Cangai, b.f. 1963, Canadian Champ—Gai Parisienne, by *Tournoi
Gay North, gr.f. 1963, Nearctic—Gai Gai, by Bull Page
Northern Minx, dk.b/br.f. 1963, Nearctic—Sphinxlike, by *Princequillo
Pryority D., dk.b/br.f. 1963, Canadian Champ—Merveilleux, by *Menetrier
Snow Park, dk.b/br.f. 1963, Nearctic—Question Time, by Polynesian
Solar Park, b.f. 1963, Victoria Park—Solar Display, by Sun Again
Titled Hero, blk.c. 1963, Canadian Champ—Countess Angela, by Bull Page
Battling, dk.b/br.c. 1964, Nearctic—Lyford Cottage, by Battlefield
Boot Hill, dk.b/br.g. 1964, New Providence—Miss Windsor, by Windfields
Cool Reception, ch.c. 1964, Nearctic—Windy Answer, by Windfields
Miss Snow Goose, b.f. 1964, Nearctic—Willow Lake, by Windfields
Regal Dancer, gr.g. 1964, *Grey Monarch—Natalma, by Native Dancer

Arctic Blizzard, dk.b/br.c. 1965, Nearctic—Breezy Answer, by Bull Page

Icy Note, b.c. 1965, Nearctic—Allegro, by Chop Chop

Nangela, b.f. 1965, Nearctic—*Angela's Niece, by Tim Tam

No Parando, b.c. 1965, Victoria Park—Ribola, by *Ribot

Solometeor, b.f. 1965, Victoria Park—Solar Display, by Sun Again

Accumuli, ch.f. 1966, *Stratus—Mentores, by *Menetrier

Dance Act, ch.c. 1966, Northern Dancer—*Queen's Statute, by Le Lavandou

Dobbinton, b.f. 1966, New Providence—Flaming Issue, by Ace Marine

Drama School, ch.f. 1966, Northern Dancer—*Stalina, by Stalino

Fanfaron, b.c. 1966, Victoria Park—Merry and Bright, by *Menetrier

Fire N Desire, dk.b/br.c. 1966, Nearctic—Gai Parisienne, by *Tournoi

Ice Palace, ch.c. 1966, Nearctic—Own Colleen, by Queen's Own

Viceregal, ch.c. 1966, Northern Dancer—Victoria Regina, by *Menetrier

Artic Feather, dk.b/br.c. 1967, Nearctic—*Gold Quill, by Sunny Boy III

Canadian Jerry, b.g. 1967, New Providence—Windy Response, by Windfields

Nijinsky II, b.c. 1967, Northern Dancer—Flaming Page, by Bull Page (SW in England/Ireland)

South Ocean, b.f. 1967, New Providence—Shining Sun, by Chop Chop

Swinging Apache, b.g. 1967, Northern Dancer—Allegro, by Chop Chop

Arctic Actress, dk.b/br.f. 1968, Nearctic—Lucienne C., by Primate

Canadian Victory, b.c. 1968, Canadian Champ—Lady Victoria, by Victoria Park

Great Gabe, ro.c. 1968, Langcrest—Gai Gai, by Bull Page

Minsky, ch.c. 1968, Northern Dancer—Flaming Page, by Bull Page (SW in Ireland/Canada)

New Pro Escar, dk.b/br.c. 1968, New Providence—Orchestrina, by Nearctic

New Tune, b.f. 1968, New Providence—Song of Victory, by Victoria Park

Speedy Zephyr, ch.c. 1968, Restless Wind—Speediness, by Nearctic

Victego, b.c. 1968, Nentego—Floral Victory, by Victoria Park

Winlord, ch.c. 1968, Canebora—Own Colleen, by Queen's Own

Buckstopper, ch.c. 1969, Buckpasser—Northern Queen, by Nearctic (SW in Ireland)

Flamme d'Or, b.f. 1969, Champlain—Flaming Issue, by Ace Marine

Gambier, dk.b/br.c. 1969, New Providence—No Vacation, by Nearctic

Happy Victory, b.f. 1969, New Providence—Floral Victory, by Victoria Park

Presidial, b.c. 1969, Psidium—Rose of North, by Nearctic

Takaring, ch.f. 1969, Takeawalk—Ring the Chimes, by Buisson Ardent

Victoria Song, b.g. 1969, Victoria Park—Arctic Song, by Nearctic

Down North, b.c. 1970, Victoria Park—*Queen's Statute, by Le Levandou

Impressive Lady, dk.b/br.f. 1970, Impressive—Chilly, by Nearctic

Northern Fling, b.c. 1970, Northern Dancer—Impetuous Lady, by Hasty Road

Perfect Sonnet, b.f. 1970, Right Combination—Tudor Sonnet, by Victoria Park

Square Angel, b.f. 1970, Quadrangle—Nangela, by Nearctic

Vickie's Champ, b.f. 1970, Victorian Era—Canalu, by Canadian Champ

Victorianette, b.f. 1970, Victoria Park—Lachute, by Match II

Victorian Prince, dk.b/br.c. 1970, Victorian Era—Willow Lake, by Windfields

Backstretch, b.c. 1971, Northern Answer—Queen's Song, by Queen's Own

Butterbump, ch.c. 1971, Viceregal—Butter Ball, by *Tournoi

Cool Spring Park, b.c. 1971, Victoria Park—Arctic Reel, by Nearctic

Coulisse, ch.f. 1971, Stage Door Johnny—Raise the Flag, by Raise a Native

Lord Durham, ch.c. 1971, Damascus—Solar Princess, by Summer Tan

Lost Majorette, ch.f. 1971, Majestic Prince—Lost Lagoon, by Swaps

Noble Answer, b.c. 1971, Viceregal—Prize Answer, by Shoperion

Norland, b.c. 1971, Right Combination—Unperturbable, by Victoria Park

Northern Taste, ch.c. 1971, Northern Dancer—Lady Victoria, by Victoria Park (SW in France)

Sherwood Park, dk.b/br.c. 1971, Victoria Park—Dollar Queen, by Luminary II

Victorian Queen, b.f. 1971, Victoria Park—Willowfield, by Stratus

Banqueroute, ch.g. 1972, New Providence—Fair Victoria, by Victoria Park

Belonger, b.c. 1972, New Providence—Canadian Bullet, by Northern Dancer

Greek Answer, gr.c. 1972, Northern Answer—Greek Victress, by Victoria Park

Imperial March, ch.c. 1972, *Forli—Victorian Dancer, by Northern Dancer (SW in England)

May Combination, b.c. 1972, Right Combination—Northern Willow, by Northern Dancer

Meadowsweet, b.f. 1972, Victoria Park—Burning Sand II, by Buisson Ardent

North of the Law, ch.c. 1972, Northern Dancer—*Queen's Statute, by Le Lavandou

Petrus, b.f. 1972, Sir Gaylord—Tabola, by Round Table

Quick Selection, ch.f. 1972, Viceregal—Lachine II, by Grey Sovereign

Tanzor, b.c. 1972, Nijinsky II—Lady Victoria, by Victoria Park (SW in England)

Victorian Image, b.c. 1972, Dancer's Image—Victorian Answer, by Victoria Park

Against All Flags, dk.b/br.f. 1973, Hoist the Flag—Northern Queen, by Nearctic

Christy's Mount, b.f. 1973, Vice Regent—Snowmount, by Sallymount

Cool Ted, dk.b/br.c. 1973, New Providence—Chilly, by Nearctic

Deep Meadow, dk.b.f. 1973, Right Combination—Champ de Soleil, by Champlain

Far North, b.c. 1973, Northern Dancer—Fleur, by Victoria Park (SW in France)

Gay Jitterbug, b.g. 1973, Northern Dancer—Gay Meeting, by Sir Gaylord

Golden Answer, b.f. 1973, Northern Answer—Victory Songster, by Stratus

Kirkfield Park, ch.f. 1973, Vice Regent—Sweet Romance, by Gun Bow

Laissez Passer, dk.b/br.c. 1973, Northern Dancer—Lindenlea, by Double Jay

Military Bearing, b.c. 1973, Vice Regent—Midinette II, by Tantieme

Momigi, dk.b/br.c. 1973, Laugh Aloud—Hold Me Close, by Native Dancer

Norcliffe, b.c. 1973, Buckpasser—Drama School, by Northern Dancer

Nuclear Pulse, ch.c. 1973, Nijinsky II—Solometeor, by Victoria Park (SW in France)

Regal Alibi, ch.f. 1973, Viceregal—Alibi IV, by Birkhahn

Regal Gal, ch.f. 1973, Viceregal—Impetuous Lady, by Hasty Road

Canadian Regent (stp), ch.c. 1974, Vice Regent—Canadia, by
    Canebora
Crown Count, b.g. 1974, Dancing Count—Bronzed Goddess, by
    Raise a Native
Dance in Time, b.c. 1974, Northern Dancer—Allegro, by Chop Chop
Giboulee, b.c. 1974, Northern Dancer—Victory Chant, by Victoria Park
La Malchance, b.f. 1974, Viceregal—Lindenlea, by Double Jay
Nearna, b.f. 1974, Nearctic—Native Girl, by Raise a Native
Northern Ballerina, b.f. 1974, Northern Dancer—Floral Victory, by
    Victoria Park
Northernette, b.f. 1974, Northern Dancer—South Ocean, by New
    Providence
Pro Consul, ch.c. 1974, Vice Regent—Bingo Queen, by Bing
Regent Bird, b.c. 1974, Vice Regent—Setting Sun, by Sunny
Sound Reason, dk.b/br.c. 1974, Bold Reason—New Tune, by New
    Providence
Swain, b.c. 1974, Viceregal—Sweet Story, by Candy Spots
The Minstrel, ch.c. 1974, Northern Dancer—Fleur, by Victoria Park
    (SW in England/Ireland)
Winter Wonderland, b.g. 1974, Protanto—Winter Grey, by Nearctic
Canadian Bill, dk.b.g. 1975, Right Combination—Handmaiden, by
    Nearctic
Flower Princess, dk.b.f. 1975, Majestic Prince—Fleur, by Victoria
    Park
Harpoon, b.c. 1975, Northern Native—Gun Mite, by Gun Bow
Impetuous Gal, ch.f. 1975, Briartic—Impetuous Lady, by Hasty Road

Vice Regent

L'Alezane, ch.f. 1975, Dr. Fager—Northern Willow, by Northern
    Dancer
Proliferate, b.c. 1975, Pronto—Victory Chant, by Victoria Park
Regal Embrace, b.c. 1975, Vice Regent—Close Embrace, by Nentego
Right Chilly, dk.b/br.f. 1975, Right Combination—Chilly, by Nearctic
Royal Sparkle, gr.g. 1975, Ruritania—Winter Grey, by Nearctic
Ruana, ch.f. 1975, Ruritania—Sweet Heiress, by Sir Gaylord
Sandy Isle, ch.f. 1975, Viceregal—Holiday Isle, by New Providence
Shake Shake Shake, b.c. 1975, Dancing Count—Buena Notte, by
    Victoria Park
Tikvah, b.c. 1975, Rambunctious—Aquatic Ballet, by Northern Dancer
Try My Best, b.c. 1975, Northern Dancer—Sex Appeal, by
    Buckpasser (SW in England/Ireland)
Von Clausewitz, dk.b.g. 1975, Tentam—Alibi IV, by Birkhahn

Accomplice, b.c. 1976, Graustark—Deceit, by Prince John (SW in
    Ireland)
All for Victory, b.c. 1976, One for All—Flaming Victress, by Victoria
    Park
Bronze Duchess, b.f. 1976, King's Bishop—Bronzed Goddess, by
    Raise a Native
Countess North, ch.f. 1976, Northern Dancer—Impetuous Lady, by
    Hasty Road
Country Romance, ch.f. 1976, Halo—Sweet Romance, by Gun Bow
Denim King, ch.c. 1976, Halo—Sphinxlike, by *Princequillo
Durham's Theme, b.f. 1976, Lord Durham—Theme Song, by Nearctic
Feu d'Artifice, ch.f. 1976, Northern Dancer—Quadrillion, by
    Quadrangle
Glorious Song, b.f. 1976, Halo—Ballade, by *Herbager
High Voltage Sport, ro.f. 1976, High Echelon—Respond, by Canadian
    Champ
Kamar, b.f. 1976, Key to the Mint—Square Angel, by Quadrangle
Kennedy Glamour, dk.b.f. 1976, Kennedy Road—Glamour Parade, by
    Jaipur
Lover's Answer, dk.b.c. 1976, Northern Answer—Lover's Walk, by
    Never Bend
Nonparrell, b.c. 1976, Hoist the Flag—Floral Victory, by Victoria Park
Ocean's Answer, b.f. 1976, Northern Answer—South Ocean, by New
    Providence
Solar, ch.f. 1976, Halo—Sex Appeal, by Buckpasser (SW in Ireland)
Vaguely Modest, dk.b/br.f. 1976, *Vaguely Noble—Shake a Leg, by
    Raise a Native
Victorious Answer, dk.b.f. 1976, Northern Answer—Victory Chant, by
    Victoria Park
Bejilla, gr.c. 1977, Quadrangle—Dancing Angela, by Dancer's Image
Leading Witness, b.f. 1977, Mr. Leader—Friendly Witness, by
    Northern Dancer
Let's Go South, b.c. 1977, One for All—South Ocean, by New
    Providence
Magesterial, b.c. 1977, Northern Dancer—Courting Days, by Bold
    Lad (SW in England/Ireland)
Mister Country, b.g. 1977, One for All—Glamour Parade, by Jaipur
Moneda, ch.f. 1977, Halo—Fiery Dancer, by Nearctic
Solartic, ch.f. 1977, Briartic—Solometeor, by Victoria Park
Akureyri, b.c. 1978, Buckpasser—Royal Statute, by Northern Dancer
Canadian Babe, dk.b.c. 1978, Northern Answer—Canadian Sun, by
    Canebora
Cool Tania, gr.c. 1978, Ruritania—Canadian Ballet, by Northern
    Dancer
Le Promeneur, dk.b.c. 1978, Tentam—Lover's Walk, by Never Bend
No. One Bundles, ch.f. 1978, Vice Regent—Hasty Gal, by Maribeau
Stellarette, b.f. 1978, Tentam—Square Angel, by Quadrangle
Storm Bird, b.c. 1978, Northern Dancer—South Ocean, by New
    Providence (SW in Ireland)
Alofje, b.f. 1979, Lord Durham—Canadian Ballet, by Northern Dancer
Artania, gr.f. 1979, Ruritania—Arctic Actress, by Nearctic
Awaasif, b.f. 1979, Snow Knight—Royal Statute, by Northern Dancer
    (SW in England/Ireland)
Black Rule, b.c. 1979, King Emperor—Zambia, by Native Charger
Exclusive Canadian, ch.c. 1979, Exclusive Native—Bingo Queen, by
    Bing II
Le Danseur, b.c. 1979, Lord Durham—Dancing Angela, by Dancer's
    Image
Pilgrim, b.c. 1979, Northern Dancer—Fleur, by Victoria Park (SW in
    Ireland)
Snowy Dancer, dk.b.f. 1979, Snow Knight—Aquatic Ballet, by
    Northern Dancer

Son of Briartic, ch.c. 1979, Briartic—Tabola, by Round Table

Sparkling Savage, ch.f. 1979, Jungle Savage—Champagne Carol, by Bold Lark

Suptertam, dk.b.f. 1979, Tentam—Impressive Lady, by Impressive

Danzatore, b.c. 1980, Northern Dancer—Shake a Leg, by Raise a Native

Diapason, b.g. 1980, Tentam—Coulisse, by Stage Door Johnny

Dundrum Dancer, b.f. 1980, Caucasus—Willowfield, by Stratus

Halo Reply, ch.f. 1980, Halo—Cold Reply, by Northern Dancer

Lady Ice, b.f. 1980, Vice Regent—Dancing Castanet, by Tambourine

Northern Blossom, ch.f. 1980, Snow Knight—Victorian Heiress, by Northern Dancer

Rockcliffe, ch.c. 1980, Norcliffe—Rollicking Lady, by Rambunctious

Royalesse, ch.f. 1980, Vice Regent—Fun On Stage, by Stage Door Johnny

Shareef Dancer, b.c. 1980, Northern Dancer—Sweet Alliance, by Sir Ivor (SW in England/Ireland)

Stellar Performer, b.f. 1980, T. V. Commercial—Norla, by Right Combination

Ultramate, ch.c. 1980, Nijinsky II—Gala Party, by Hoist the Flag

Valuable Witness, b.g. 1980, Val de l'Orne (Fr)—Friendly Witness, by Northern Dancer

Victorious Emperor, ch.c. 1980, Vice Regent—Springlet, by Youth Emperor

Archregent, b.c. 1981, Vice Regent—Respond, by Canadian Champ

Axe T. V., ro.c. 1981, T. V. Commercial—Princess Axe, by Majestic Prince

Born a Lady, b.f. 1981, Tentam—Natalma, by Native Dancer

Concordene, ch.f. 1981, Northern Dancer—Nocturnal Spree (Ire), by Supreme Sovereign

Cool Northerner, b.c. 1981, Dom Alaric—Imperturbable Lady, Northern Dancer

Dance Flower, b.f. 1981, Northern Dancer—Flower Princess, by Majestic Prince

Devil's Bag, b.c. 1981, Halo—Ballade, by *Herbager

El Gran Senor, b.c. 1981, Northern Dancer—Sex Appeal, by Buckpasser

Futurette, b.f. 1981, Sevastopol—Midi Skirt, by Canebora

Halo's Princess, b.f. 1981, Halo—Taberet, by Viceregal

Lake Country, b.f. 1981, Caucasus—Telltale Traces, by Tentam

Love Smitten, b.f. 1981, Key to the Mint—Square Angel, by Quadrangle

Nagurski, b.c. 1981, Nijinsky II—Deceit, by Prince John

Perfect Player, dk.b.c. 1981, Dom Alaric—Arctic Actress, by Nearctic

Pied A'Terre, b.c. 1981, Caucasus—Rainbow's Edge, by Crème dela Creme

Royal Lorna, b.f. 1981, Val de l'Orne (Fr)—Royal Statute, by Northern Dancer (SW in Italy)

Secreto, b.c. 1981, Northern Dancer—Betty's Secret, by Secretariat

Snow Blossom, b.f. 1981, The Minstrel—Floral Victory, by Victoria Park

Southern Arrow, dk.b.c. 1981, Smarten—Northern Lake, by Northern Dancer (SW in Italy)

Val Dansant, b.c. 1981, Val de l'Orne (Fr)—Dancing Doris, by Northern Dancer

Coup de Folie, b.f. 1982, Halo—Raise the Standard, by Hoist the Flag

Deceit Dancer, ch.f. 1982, Vice Regent—Deceit, by Prince John

Imperial Choice, dk.b..c. 1982, Gregorian—Your My Choice, by Barachois

In My Cap, ch.f. 1982, Vice Regent—Passing Look, by Buckpasser

Kazbek, b.g. 1982, Caucasus—Zambia, by Native Charger

La Lorgnette, b.f. 1982, Val de l'Orne (Fr)—The Temptress, by Nijinsky II

Noble Regent, dk.b.c. 1982, Vice Regent—Spirited Away, by *Vaguely Noble

Quitman, ch.f. 1982, Vice Regent—Close Embrace, by Nentego (GB)

Artic Mistral, b.f. 1983, Briartic—Telltale Traces, by Tentam

Be My Master, ch.c. 1983, Master Willie (GB)—Caught in the Act, by Nijinsky II (SW in Italy)

Cool Halo, dk.b.c. 1983, Halo—Slight Deception, by Northern Dancer

Dancing On a Cloud, b.f. 1983, Nijinsky II—Square Angel, by Quadrangle

Don Diege, dk.b.c. 1983, Gregorian—Ambitioninrest, by Bold Ambition

Golden Choice, b.c. 1983, Val de l'Orne (Fr)—Your My Choice, by Barachois

Hear Music, ch.f. 1983, Master Willie (GB)—Regal Response, by Viceregal

Improyal, ch.f. 1983, Riverman—Winsome, by Sir Ivor

Madame Treasurer, b.f. 1983, Key to the Mint—Halo Dancer, by Halo

Regency Silk, ch.f. 1983, Vice Regent—Shanghai Melody, by Shantung

Tisn't, gr.c. 1983, Shergar—Zabarella, by Clouet (SW in England)

Why Not Willie, b.f. 1983, Master Willie (GB)—Theme Song, by Nearctic

Arcroyal, ch.f. 1984, Vice Regent—Arch Miss, by Mississipian

Interrex, ch.c. 1984, Vice Regent—Betty's Secret, by Secretariat

Master Treaty, b.g. 1984, Master Willie (GB)—Happy Truce, by L'Enjoleur

Misty Magic, b.f. 1984, Master Willie (GB)—Majestic Miss, by Majestic Prince

Nordavano, b.c. 1984, The Minstrel—Noble Chick, by *Vaguely Noble

Rambo Dancer, b.c. 1984, Northern Dancer—Fair Arabella, by Chateaugay

Smart Halo, b.f. 1984, Smarten—Halo Dancer, by Halo

Imperial Colony, dk.b.g. 1985, Pleasant Colony—Impressive Lady, by Impressive

King's Deputy, b.c. 1985, Deputy Minister—Majestic Miss, by Majestic Prince

Thaidah, b.f. 1985, Vice Regent—Ballade, by *Herbager (SW in England)

Zaffaran, b.c. 1985, Assert (Ire)—Sweet Alliance, by Sir Ivor (SW in England)

Diana Dance, ch.f. 1986, Northern Dancer—Deceit, by Prince John (SW in Germany)

Norquestor, dk.b.c. 1986, Conquistador Cielo—Linda North, by Northern Dancer

Princess Caveat, dk.b.f. 1986, Caveat—Majestic Miss, by Majestic Prince

Thanx, b.g. 1986, Commemorate—Rivage, by Riva Ridge

Adorned, b.f. 1987, Val de l'Orne (Fr)—Caucasienne, by Caucasus

Alexandrina, b.f. 1987, Conquistador Cielo—La Lorgnette, by Val de l'Orne (Fr)

Floral Dancer, b.c. 1987, Limbo Dancer—Floral Victory, by Victoria Park

Primetime North, b.f. 1987, Northern Dancer—Rally Around, by Hoist the Flag

Star Standing, dk.b.f. 1987, Assert (Ire)—Minstrelsy, by The Minstrel

Majesterian, dk.b.c. 1988, Pleasant Colony—Linda North, by Northern Dancer

Masake, b.f. 1988, Master Willie (GB)—Northern Lake, by Northern Dancer (In partnership with Anderson Farms & G. Austin)

Musical Respite, b.f. 1988, Gregorian—Happy Truce, by L'Enjoleur
Run Lady Run, b.f. 1988, Smarten—Musical Ride, by The Minstrel
Street Rebel, b.c. 1988, Robellino—Street Ballet, by Nijinsky II (SW in Ireland)
Comarctic, dk.b.g. 1989, Commemorate—Arctic Fling, by Northern Dancer
Great Regent, ch.c. 1989, Vice Regent—Show Lady, by Sir Ivor
I'm Reckless, ch.c. 1989, Two Punch—Lively Affair, by Caro (Ire)
Keen Falcon, b.c. 1989, Imperial Falcon—Fun On Stage, by Stage Door Johnny
King's College, b.g. 1989, Vice Regent—Choral Group, by Lord Durham
Meadow Pipit, ch.f. 1989, Meadowlake—Delta Slew, by Seattle Slew (SW in England)
Ponche, ro.c. 1989, Two Punch—Street Ballet, by Nijinsky II
Saint Ballado, dk.b.c. 1989, Halo—Ballade, by *Herbager
Starry Val, b.f. 1989, Val de l'Orne (Fr)—Starstruck Gal, by Stage Door Johnny
Flag Down, b.c. 1990, Deputy Minister—Sharp Call, by Sharpen Up (SW in France/U.S.)
Glenbarra, b.g. 1990, Vice Regent—Supreme Excellence, by Providential
Housebound, b.g. 1991, Pancho Villa—Truly Bound, by In Reality
Norfolk Lavender, gr/ro.f. 1991, Ascot Knight—Nocturnal Spree (Ire), by Supreme Sovereign
Term Limits, ch.f. 1991, Time for a Change—Dancing Doris, by Northern Dancer

War Deputy, dk.b.c. 1991, Deputy Minister—Sweet Alliance, by Sir Ivor
Better Banker, dk.b.c. 1992, Ascot Knight—Really Taken, by In Reality (SW in Brazil)
Honky Tonk Tune, b.f. 1992, Cure the Blues—Starita, by Star de Naskra
Prospect Bay, b.c. 1992, Crafty Prospector—Baltic Sea, by Danzig
Tamayaz, dk.b.c. 1992, Gone West—Minstrelsey, by The Minstrel (SW in England/United Arab Emirates)
Classy n' Sassy, b.f. 1993, Regal Classic—Nabora, by Naskra
Gambling Girl, b.g. 1993, Secret Claim—Dawn's Deputy, by Deputy Minister
Special Deputy, b.g. 1994, Silver Deputy—Veridian, by Green Dancer
Archers Bay, b.c. 1995, Silver Deputy—Adorned, by Val de l'Orne (Fr)
Hedonist, dk.b.f. 1995, Alydeed—Play all Day, by Steady Growth
Regal Angela, gr/ro.f. 1995, Regal Intention—Dancing Angela, by Dancer's Image
Social Director, b.f. 1995, Deputy Minister—Health Farm, by Pleasant Colony
Solarity, ch.f. 1995, Ascot Knight—Dawn's Deputy, by Deputy Minister
Zaha, ch.c. 1996, Kingmambo—Play all Day, by Steady Growth
Dream About, dk.b.f. 2001, Cherokee Run—Social Director, by Deputy Minister

Sources: *The Blood-Horse* Archives and The Jockey Club

Leslie Combs II of Spendthrift Farm

Leslie Combs II combined so many facets that it was possible to overlook his bedrock horsemanship. He was a garrulous showman with a faux Southern gentleman surface charm, a foul-mouthed bully when that suited the situation better, a gambler/investor with his own money, and a high roller with others'. He built Spendthrift Farm into one of the most famous of Bluegrass horse farms and brought in a sequence of important stallions, raising the bar in syndication prices several notches.

If there was a sense of mass production in his operation, as opposed to sage matings, "Cuzin Leslie" turned out herds of stakes winners to match it: 254 in all, including those with a myriad of partners. Many of these horses brought high prices as yearlings, although sometimes what was presented as a straightforward sale was something more contrived and complicated.

Still, the horsemanship was undeniable. Combs bred one Kentucky Derby winner, selected another for a client, and the man who acquired so many stallions was the breeder of one of the century's great sires, Mr. Prospector. As judge of others' yearlings, he once cut a swath through the market, making Elizabeth Arden Graham America's leading owner with horses he advised her to buy.

When he achieved status, he did not shirk leadership. He served more than twenty years on racing commissions, in Kentucky and West Virginia; was chairman of the Kentucky commission for eight years; and had a term as president of the National Association of State Racing Commissioners. He was a member of The Jockey Club and a trustee of the Thoroughbred Owners and Breeders Association.

Thoroughbreds were large in Combs' heritage. He was a great-grandson of Daniel Swigert, a key nineteenth-century horseman and breeder of Hindoo, Firenze, and Tremont. Swigert's daughter, Mary, married Leslie Combs Jr., the U.S. ambassador to Peru whose own father, General Leslie Combs Sr., had been president of the Kentucky Association Track in Lexington. Leslie Combs Jr. was also breeder of champion El Chico and of Miyako, the second dam of Native Dancer. The children of Leslie Jr. and Mary Swigert Combs included horsemen Lucas Combs (a trustee of Keeneland) and Brownell Combs (breeder of Myrtlewood). "Uncle Brownell" was partner with Leslie II in the breeding of fourteen stakes winners, including his first champion, Myrtle Charm, and Gold Digger, the dam of Mr. Prospector.

Leslie Combs II was born in 1901 and was so vocal in calling himself "just a country boy from Coaltown (Kentucky)" that longtime friend Warren Wright Sr. of Calumet Farm named the subsequent champion Coaltown in his honor. When Combs was fourteen, his father, Daniel, died, and the youth went to live with his grandparents. He later enrolled at Centre College in Danville, Kentucky, in 1921, rooming with George Swinebroad. Given the bonhomie and sense of mischief that attended the scene years later when Swinebroad wielded the auctioneer's hammer over high-priced Combs yearlings at Keeneland, one suspects that this was a lively pair as collegians. Combs later spent a year in Guatemala with the South American Plantation Company, contracting malaria in the process. He returned to Kentucky to work for American Rolling Mills Company in Ashland.

In 1924 Combs, then in his early twenties, married Dorothy Enslow. The passions of young love notwithstanding, this was a good deal from another angle, too, for she was the daughter of the wealthy founder of Columbia Gas and Electric Company. The couple moved across the state line to Huntington, West Virginia, where Combs opened Combs-Ritter Insurance Company. He had some contact with racing, as a member of the West Virginia Racing Commission, some distance below the glamorous life of the Turf that awaited.

His inheritance from his grandmother and some assistance from Uncle Brownell helped Combs move to Kentucky and go into the horse business. In 1937 he purchased 127 acres once owned by Swigert, and he named it Spendthrift Farm in honor of that ancestor's 1879 Belmont Stakes winner.

Combs credited Uncle Brownell with much of what he learned about Thoroughbreds, and the pair hit it big as breeders with Myrtle Charm. A foal of 1946, Myrtle Charm was by the champion Alsab, then a young stallion, and out of Crepe Myrtle, by Equipoise. The operative word here is "Myrtle," for Crepe Myrtle was a daughter of Brownell's marvelous race mare Myrtlewood.

Foaled in 1932, Myrtlewood was by Blue Larkspur—*Frizeur, by Sweeper. The second dam was the distinguished Frizette. Brownell Combs had purchased *Frizeur from John E. Madden in the 1920s. Myrtlewood was beloved in Kentucky, a star in the fledgling days of the Keeneland track that replaced the Kentucky Association Track. A brilliant sprinter and the champion older female in 1936, she won fifteen of twenty-two races. Her battles with the high-class sprinter Clang were central to her standing. She beat him four of five times, and the one time he bettered her, Clang had to set a world record for six furlongs to beat her a nose. At one point Myrtlewood held the American record for females at both six furlongs and one mile, and she set track records at five Midwestern tracks.

If poesy and vision had coincided, Spendthrift Farm would have been named Myrtlewood Farm, for the mare became the foundation of so much that

Dorothy and Leslie Combs II

would catapult Leslie Combs II to the forefront of the commercial breeding scene. Her stakes-winning foals were 1942 Kentucky Oaks winner Miss Dogwood and 1943 juvenile filly champion Durazna. More than 250 stakes winners descend from her and her daughters, taking into account the female lineage only. This in addition to the gaudy records of male descendants, the most conspicuous of whom would be Mr. Prospector (180 stakes winners) and Seattle Slew (109 stakes winners).

Years later, when honored by the Thoroughbred Club of America, Combs closed his remarks with, "I want to take this opportunity to pay tribute to... Uncle Brownell, who has taught me what I know and gave me a start with the great blood of Myrtlewood."

Myrtle Charm, as stated, was a daughter of Myrtlewood's first foal and was bred by Brownell Combs and Leslie Combs II. She was ostensibly purchased as a yearling by Lester Manor Stable for $27,000. It developed later that Lester Manor was buying on behalf of Elizabeth Arden Graham, the cosmetics queen who established Maine Chance Farm. Myrtle Charm became the champion two-year-old filly of 1948 for Mrs. Graham. Some years

130

later her daughter Fair Charmer foaled My Charmer, revered today as the dam of 1977 Triple Crown winner and major sire Seattle Slew.

While he had not bred a major winner in his own name prior to Myrtle Charm, Leslie Combs II had established a presence in the upper echelons of the racing world well before that filly's championship status. In 1943 and 1944 he assisted Mrs. Graham in the purchase of a sequence of yearlings that had such success that *American Race Horses of 1945* sought to explain it by a rather tortuous suggestion, to wit: "Mr. Combs, operator of Spendthrift Farm near Lexington, is naturally in fairly constant direct contact with the other breeders for the sales, and has thus an unusual opportunity of seeing the horses in the consignments from the time they are sucklings…This was an indubitable advantage. Most trainers and most owners do not see the market yearlings until these have been shined and polished and — we must blush to add — fattened, for the sale."

Whether he had an insider's edge or not, yearlings selected for Mrs. Graham by Combs in 1944 included both the next year's juvenile champions, Star Pilot among colts and Beaugay among fillies. Maine Chance led the owners' standings with $589,170, drawing away late in the year from Hollywood producer Louis B. Mayer, whose $533,150 total was buttressed by Horse of the Year Busher.

In 1945 Combs was still at Mrs. Graham's side in the sale rings, and they purchased a *Blenheim II colt at Keeneland for $41,000. Named Jet Pilot, that colt was successful enough at two to be hailed as the most expensive yearling ever to win back his purchase price, and at three he won the Kentucky Derby. (Coincidental threads of these various sequences were connected to the eventual creation of Seattle Slew. Mrs. Graham later bought Busher from Mayer and bred her to Jet Pilot, the result being Jet Action, sire of Fair Charmer, the descendant of Myrtlewood and ancestress of Slew.)

The year 1947 was also the beginning of Combs' flexing his muscles in an adventurous method of stallion acquisition. He organized a syndicate of Americans to purchase the crack Australian *Beau Pere, who had earlier been imported to stand successfully at stud in California for Mayer. The price was $100,000. *Beau Pere died before covering any mares for the syndicate, but Combs was convinced that syndication was an idea whose time had come. Actually, it was not new, although Combs would take the concept to a new level.

Most of the famous stallion importations to America at that time were either individual purchases (*Mahmoud for example) or exclusive partnerships of between four and eight investors (*Sir Gallahad III and *Blenheim II). Combs began dealing in syndicates of thirty-two to thirty-six shares, based on the prevailing attitudes of the time about what constituted the proper book size for a stallion each year. A similar format had been worked out in the 1920s for the English St. Leger winner Tracery. Bred by August Belmont II, Tracery had been raced

		Black Toney, 1911	Peter Pan (Commando—*Cinderella)
			Belgravia (**Ben Brush**—*Bonnie Gal)
	Black Servant, 1918	*Padula, 1906	Laveno (**Bend Or**—Napoli)
			Padua (Thurio—Immortelle)
Blue Larkspur, 1926		*North Star III, 1914	Sunstar (Sundridge—Doris)
			Angelic (St. Angelo—Fota)
	Blossom Time, 1920	*Vaila, 1911	Fariman (Gallinule—Bellinzona)
			Padilla (Macheath—**Padua**)
MYRTLEWOOD		Broomstick, 1901	**Ben Brush** (Bramble—Roseville)
	*Sweeper, 1909		*Elf (Galliard—*Sylva Belle)
		*Ravello II, 1896	Sir Hugo (Wisdom—Manoeuvre)
*Frizeur, 1916			Unco Guid (Uncas—Genuine)
		Hamburg, 1895	Hanover (Hindoo—Bourbon Belle)
	Frizette, 1905		Lady Reel (Fellowcraft—Mannie Gray)
		Ondulee, 1898	St. Simon (Galopin—St. Angela)
			Ornis (**Bend Or**—Shotover)

The Combses with Elizabeth Arden Graham (center)

THE BLOOD-HORSE

Along the way were plenty of milestones in both price and in turning out exceptional horses. In 1953 the Spendthrift stallion roster picked up a major addition when California attorney Neil McCarthy bought *Royal Charger for a reported $300,000 from the Irish National Stud and placed him under Combs' management. Prior to the purchase *Royal Charger sired *Turn-to, who won the Saratoga Special and Garden State Stakes for juveniles the same year McCarthy bought his sire. *Royal Charger sired fifty-seven stakes winners (16 percent), including a high-priced yearling that became a champion for Combs and partner John W. Hanes. The *Royal Charger—

abroad, stood initially in England, and then was sold to Argentina. When his runners in England enhanced his reputation, Señor S.J. Unzue agreed to sell three-fourths of the horse to an English syndicate of thirty members, whose payments of about six thousand dollars each would entitle them to send one mare to the horse in each of the first three seasons.

Only a year after the *Beau Pere experiment, Combs put together another syndicate to purchase Mayer's successful young stallion *Alibhai for a much more exalted figure, $500,000. *Alibhai stood a long and successful term at Spendthrift and sired fifty-four stakes winners (14 percent), including 1954 Kentucky Derby winner Determine, plus Your Host (sire of Kelso), Flower Bowl, Bardstown, and Traffic Judge.

As a consignor to the yearling sales, Combs served notice in 1946 when he led the Keeneland summer sale statistics for the first time, selling five horses for an average of $26,120. Then, in 1949, he launched a sixteen-year run as leading consignor in average. The final year of that streak, 1964, found the average for a half-dozen yearlings at $67,667. Three years later Combs returned to the top in average again, sending in a much larger consignment of twenty-eight yearlings and still topping the market with an average $42,196.

Tige O'Myheart, by Bull Lea, filly, named Idun, was one of fourteen stakes winners bred by that combination of horsemen. Idun was sold for $63,000 in Spendthrift's 1956 Keeneland summer consignment. The figure was a record for a yearling filly.

Hanes, who about that time was involved in a three-man committee that designed what became the eventual New York Racing Association, was a scion of the Hanes hosiery and R.J. Reynolds Tobacco family and had served as undersecretary of the U.S. Department of the Treasury. Hanes had handled the reorganization of the vast William Randolph Hearst enterprises and had devised the merger of two giant firms into the diversified industrial giant Olin-Mathieson Chemical Company. (Just as his connection with Hanes and McCarthy indicates that Combs by then was dealing with high-scale entrepreneurs, the purchaser of Idun burnished that image. She was Mrs. Charles Ulrich Bay, whose husband was a senior partner in the Wall Street firm of A.M. Kidder & Co. Bay died before the year was out, and his wife was named president of the firm. Two years later Idun was made an honorary vice president of Kidder. By that time the most expensive yearling filly in history had become the highest earning two-year-old filly in history.)

Idun was unbeaten in eight races as the 1957 juvenile filly champion, taking the Gardenia, Frizette, and Matron stakes. Her earnings of $220,955 edged her past Top Flight's long-standing record ($219,000 in 1931) for a juvenile filly. Moneymen to the core, the Kidder people who entered the resolution of Idun's honorary vice presidency noted that the record sum had been achieved in nine minutes, twenty-three and three-fifths seconds.

Idun repeated as champion three-year-old filly in 1958, and after winning four more stakes at four, she retired with seventeen wins in thirty starts and earnings of $392,490.

The year before Idun brought a record price in the sale ring, Combs had used another form of auction to thrust Spendthrift into headlines not only in the specific world of racing but on front pages of national newspapers. The dramas that attended his acquisition of Nashua could hardly have been scripted any

better for Combs. The same could be said for the opportunity of any number of other bidders on Nashua, but it was Combs who grabbed opportunity by the scruff of the neck.

Nashua was the darling of the Turf, at least in the East, after his nationally televised victory over Swaps in a Chicago match race had avenged his Kentucky Derby defeat. In the autumn of that year, 1955, after Nashua had become the obvious choice for Horse of the Year, his socially prominent owner, William Woodward Jr., was shot and killed by his wife. (It was ruled accidental.) Nashua went on the block by sealed bid. Combs scrambled to get the horse. With a syndicate of which Christopher J. Devine was a major player, he put in a bid of $1,251,250. Typical of Combs' ability to recruit major industrialists, Devine was known as the number one bonds dealer in the United States. No horse had ever been sold for as

BERT MORGAN

Leslie Combs II leads in Nashua

leading earner as well as most expensive horse in history. He retired with twenty-two wins from thirty races and record earnings of $1,288,565.

When Combs later built a low-slung, masonry and steel u-shaped barn for the Spendthrift stallions, it was dubbed "the Nashua motel."

Over the years, many farms reluctantly had to chip away at the lovely image of tourists admiring gentle mares and their frolicking foals, and standing in awe before the patriarchs of the breed. As values of horses rose, farms reduced their accessibility. Combs, though, knew that while tourists as a whole might be something of a nuisance, rich ones could admire horses and long to own them, too. Most tourists would not be potential clients, but some would! With Spendthrift's gates open, Nashua probably came as close to Man o' War's status as an attraction as any other horse before Secretariat. The farm would grow to five thousand acres, but the small portion that made up the stallion compound was the key magnet to the outside world.

THE BLOOD-HORSE

Combs with Nashua's trainer, Sunny Jim Fitzsimmons

Nashua was a high-fashion sire of sale yearlings from the start, and

much as $1 million, but Combs needed virtually every dollar of his established price, since several other groups and individuals bid into seven figures.

Here was a horse making headlines on the front page, in the business page, the sports page, and the society section. And the name of something called Spendthrift Farm was emblazoned in all of them.

Combs decided to keep Nashua in training at four and left him with the revered horseman "Sunny" Jim Fitzsimmons. That year Swaps outshone Nashua as Horse of the Year, although they did not meet again. Nevertheless, Nashua won six of ten races and had some glorious moments. He scored dramatically in the Widener and Suburban handicaps, and easily in the Monmouth Handicap and the Jockey Club Gold Cup. Soon to be ensconced in Spendthrift's fine, old stallion barn, with its high-pitched roof and handsome paneling, Nashua claimed the title as all-time

although he tailed off as he failed to get a breakout star colt, he was an exceptional stallion. "There have been nearly 300 yearlings by Nashua which were sold at public auction," Combs noted in *The New York Times* after Nashua's death many years later, in 1982, "and a lot of those were instrumental in making Spendthrift Farm the leading consignor eighteen times…in the Keeneland summer sales…

"He put (Spendthrift) on the map. Just as an example, nearly 20,000 people a year used to come to the farm to see Nashua and talk with his groom, Clem Brooks."

Nashua sired seventy-seven stakes winners (12 percent), including the hardy champion mare Shuvee and other Coaching Club American Oaks winners Bramalea and Marshua. Bramalea foaled the Epsom Derby winner and international sire Roberto. As broodmare sire also of Mr. Prospector,

Nashua planted his name for many years in the pedigrees of a wide range of the world's major racehorses and breeding stock.

The record price for Nashua did not stand for long. Only three years later, Combs was back at it, arranging a syndicate that bought a three-quarters interest in 1957 Belmont Stakes winner Gallant Man based on an evaluation of $1,333,333.

In 1961 Spendthrift discarded the old record for an American yearling at auction. Until that time, the highest price was $87,000, paid for a Hyperion colt at Saratoga in 1956. At the 1961 Keeneland summer sale, Spendthrift sent into the ring a colt by Nashua's archrival Swaps, who was about to hit the magic $100,000 — and leave it in the distance. Humphrey Finney, president of rival sale company Fasig-Tipton and a presence in private sales as well, bid up to $130,000 (for the colt). Finney was acting on behalf of John M. Olin, the industrial giant who by then had been brought into the business and had been ardently courted by Leslie Combs II. The colt, produced from the *Mahmoud mare Obedient and named Swapson, did not amount to much, but he had his moment of glory.

That same summer Spendthrift saw Idun's filly sales record fall, when it sold a Nashua filly for $70,000.

In 1963 the most sensational two-year-old in the land was Raise a Native, who was owned by Louis E. Wolfson, another major player in American business who had entered racing in a big way a few years before. Raise a Native had been a record-priced weanling when weanling selling was not as much a part of the market as it would become. He had brought $22,000 and then had been sold to Wolfson for $39,000 as a yearling.

Winner of all his four starts in stunning fashion, Raise a Native was injured on the eve of the Sapling Stakes. Combs moved quickly, and by the end of August 1963, Raise a Native was grazing idly in a paddock at Log Cabin, a piece of property Combs owned in partnership with Hanes. In this case Log Cabin was a staging area for a move across the highway to the Spendthrift stallion barn.

Sales records were still coming quickly in the Spendthrift domain. In 1964 Combs and Hanes dispersed some breeding stock they owned in partnership. One of the mares they sold at the Keeneland November sale was La Dauphine, a seven-year-old *Princequillo mare who was a half sister to Busher. La Dauphine was in foal to Bold Ruler. That young sire's enormous importance at stud was already becoming evident, and he had sired a $170,000 Keeneland colt; sold in 1964, that colt was consigned by Warner L. Jones Jr., who thereby wrested from Combs the status of holding the yearling record. La Dauphine gave the Spendthrift camp another record, when Charles H. Wacker bought her from Combs and Hanes for $177,000. This was $40,000 higher than the previous broodmare mark held by Honeys Gem.

In 1967 Combs regained the yearling record and added a new sire syndication mark. Raise a Native

Raise a Native, 1961	Native Dancer, 1950	Polynesian, 1942	Unbreakable (*Sickle—*Blue Glass)
			Black Polly (*Polymelian—Black Queen)
		Geisha, 1943	Discover (Display—Ariadne)
			Miyako (John P. Grier—La Chica)
	Raise You, 1946	Case Ace, 1934	*Teddy (Ajax—Rondeau)
			Sweetheart (Ultimus—Humanity IV)
		Lady Glory, 1934	American Flag (Man o' War—Lady Comfey)
			Beloved (Whisk Broom II—Bill and Coo)
MAJESTIC PRINCE	*Royal Charger, 1942	Nearco, 1935	Pharos (Phalaris—Scapa Flow)
			Nogara (Havresac II—Catnip)
Gay Hostess, 1957		Sun Princess, 1937	Solario (**Gainsborough**—Sun Worship)
			Mumtaz Begum (***Blenheim II**—Mumtaz Mahal)
	Your Hostess, 1949	*Alibhai, 1938	Hyperion (**Gainsborough**—Selene)
			Teresina (Tracery—Blue Tit)
		*Boudoir II, 1938	*Mahmoud (***Blenheim II**—Mah Mahal)
			Kampala (Clarissimus—La Soupe)

was a principal in both. The young sire's first runners had been impressive enough that when a bright chestnut son of surpassing handsomeness entered the Keeneland ring, a record was in the offing. This colt, it later was revealed, was owned fifty-fifty by Combs, who was listed as breeder, and western Canadian industrialist Frank McMahon, another major capitalist whom Combs had romanced. McMahon entered the winning, and record bid, of $250,000. When the cozy connection came to light later, McMahon was said to have had the edge of bidding in fifty-cent dollars. However, he was also turning down his 50 percent of the underbids.

With such impetus of young runners and marketable yearlings, Raise a Native's value soared. He was syndicated that year for $2,625,000, which returned to Spendthrift the record status it had lost the previous year when Darby Dan Farm syndicated Graustark for $2.4 million. Only a dozen years after the spectacular acquisition of Nashua, Combs was already dealing in figures more than double that price.

The first quarter-million-dollar yearling would break the maddening pattern of record-priced yearling colts being disappointments as runners. He was out of the *Royal Charger mare Gay Hostess, a granddaughter of the grand producer *Boudoir II. Combs had had to pay only $6,700 for Gay Hostess as a yearling when he bought her from the estate of the late Louis B. Mayer. The Raise a Native—Gay Hostess colt was given the exalted name of Majestic Prince, and he managed to live up to it. In one way he was Combs' crowning achievement as a breeder. Mr. Prospector was by far the greater sire, but Majestic Prince gave Combs a Kentucky Derby.

Turned over to the recently retired jockey John Longden near the launch of his training career, Majestic Prince flashed brilliance to match his looks. When he stood off Arts and Letters to win the Kentucky Derby in 1969, he became the first unbeaten Derby winner since Morvich in 1922. The magic extended through the Preakness but then unraveled. Longden made it known that he wanted to bypass

THE BLOOD-HORSE

Majestic Prince as a Spendthrift Farm stallion

the Belmont and let the colt recover from the wear and tear. There had been no Triple Crown winner for twenty-one years, and the sporting element in McMahon's makeup made it impossible for him to turn his back on that ultimate test. Majestic Prince struggled home a distant second to Arts and Letters in the Belmont and never raced again. He had won nine of his ten races and earned $414,200.

(An incident involving Longden illustrated how Combs insulated from the world those who were his guests each year as the Keeneland summer sale approached. Shows of the yearlings, elegant parties, and long evenings preceded the sales. Spendthrift Farm yearlings pretty much constituted all the news and cultural input allowed. In the late 1960s, when race riots tore through several American cities, Longden overheard a remark about how bad the situation was in Detroit. His query was, "Why? Are the Tigers on a losing streak?")

Majestic Prince, of course, returned to stand at Spendthrift. He was not a great stallion, but his thirty-three stakes winners (9 percent) included 1979 Belmont Stakes winner Coastal and the high-class Sensitive Prince and Majestic Light. The latter, a Phipps family homebred, sired seventy-two stakes winners (8 percent) and was the vehicle for the sire line returning to glory at Churchill Downs. Majestic Light's son Wavering Monarch sired Maria's Mon, he in turn the sire of 2001 Kentucky Derby winner Monarchos.

The year after Majestic Prince won the Derby,

Combs and McMahon sent his full brother into the Keeneland sale ring. The brother to the first quarter-million-dollar yearling became the first half-million-dollar yearling. This one, bought by McMahon on a $510,000 bid, was named Crowned Prince. He was sent to England, where he won the Dewhurst and Champagne stakes to be rated the champion two-year-old of 1971. Reports of a soft-palate affliction were followed by his early retirement the next year.

Although he owned his share of important European blood, Combs by and large did not attract as much of the wave of European buyers as did the Windfields Farm consignments of the 1970s. (The sire Northern Dancer became the key influence the Europeans sought, and he was a Windfields stallion.) Spendthrift was not without a presence in this sphere, however. In addition to being the breeder of record of English champion Crowned Prince, Combs also bred Caracolero, winner of the 1974 French Derby. Caracolero was by Graustark out of another daughter of Gay Hostess, the stakes-placed Prince John mare Betty Loraine.

Combs' crowning achievement as the breeder of a future stallion was Mr. Prospector. Like Majestic Prince, Mr. Prospector was by Raise a Native. His dam, Gold Digger, was very much a Spendthrift baby, being by Nashua and from the rich and beloved line of Myrtlewood. Moreover, there was the added sentiment of her having been bred in partnership with Combs' Uncle Brownell.

Foaled in 1962, Gold Digger was produced from

MR. PROSPECTOR				
Raise a Native, 1961	Native Dancer, 1950	Polynesian, 1942	Unbreakable (*Sickle—*Blue Glass)	
			Black Polly (*Polymelian—Black Queen)	
		Geisha, 1943	Discovery (Display—Ariadne)	
			Miyako (John P. Grier—La Chica)	
	Raise You, 1946	Case Ace, 1934	*Teddy (Ajax—Rondeau)	
			Sweetheart (Ultimus—*Humanity)	
		Lady Glory, 1934	American Flag (Man o' War—*Lady Comfey)	
			Beloved (Whisk Broom II—Bill and Coo)	
Gold Digger, 1962	Nashua, 1952	*Nasrullah, 1940	Nearco (Pharos—Nogara)	
			Mumtaz Begum (*Blenheim II—Mumtaz Mahal)	
		Segula, 1942	Johnstown (Jamestown—La France)	
			*Sekhmet (Sardanapale—Prosopopee)	
	Sequence, 1946	Count Fleet, 1940	Reigh Count (*Sunreigh—*Contessina)	
			Quickly (Haste—Stephanie)	
		Miss Dogwood, 1939	*Bull Dog (*Teddy—Plucky Liege)	
			Myrtlewood (Blue Larkspur—*Frizeur)	

Sequence, a stakes-winning Count Fleet mare whose dam was Miss Dogwood. Miss Dogwood, by Bull Dog, was out of Myrtlewood herself and had won the 1942 Kentucky Oaks for Brownell Combs.

To preserve the fount of his success, of course, Combs regularly retained some fillies with the Myrtlewood connection. Gold Digger was one he kept, and, for an additional fillip of sentiment, she raced in the name of Combs' wife, Dorothy. Gold Digger failed to win a Kentucky Oaks, finishing second to Amerivan in the 1965 running, but she won the Gallorette Stakes twice, plus the Columbiana and Yo Tambien handicaps and the Marigold Stakes. She had ten wins in thirty-five starts and earnings of $127,255.

For Spendthrift, Gold Digger matched the mark of her own dam and granddam in foaling three stakes winners. One of them was Mr. Prospector. Foaled in 1970, he was one of the talked-about Spendthrift yearlings as the 1971 summer sale approached. A.I. (Butch) Savin, who had established Aisco Farm in Florida, asked trainer Jimmy Croll to help him find a

colt that might make a sire prospect. With telling prescience, Croll went for Mr. Prospector. It took a bid of $220,000, highest price of the Keeneland sale, to land him, but Savin was game.

Mr. Prospector showed sensational speed, winning a six-furlong race for Savin in 1:07 4/5 at Gulfstream Park early in his three-year-old season. Various soundness problems took him off the Derby trail and restricted his career thereafter. He won seven of fourteen starts and earned back a little over half of his purchase price, $112,171. He won two stakes, the Gravesend and Whirlaway handicaps, both sprints. Mr. Prospector had a great pedigree, great speed, and had shown high class and was, indeed, welcomed into the Aisco stallion barn.

Quickly Mr. Prospector sent out runners, and he soon wound up in Kentucky, becoming a staunch member not of the Spendthrift stallion roster but of that of the rival Hancock family's Claiborne Farm. Mr. Prospector ranks as one of the great stallions of the twentieth century. His 180 stakes winners for a

Leslie Combs II with Myrtlewood and foal

time stood as a record. They came from a total of 1,195 foals, giving him a stakes winner rate of 15 percent. Mr. Prospector sired a number of champion runners, such as Conquistador Cielo, Forty Niner, Gulch, Eillo, Queena, Ravinella, and Gold Beauty. These included a range of sprinters and stayers. Late in his career, Mr. Prospector got a Kentucky Derby winner in Fusaichi Pegasus.

Mr. Prospector also proved a sire of sires. His successful sons include Woodman, Fappiano, Forty Niner, Seeking the Gold, Gulch, Kingmambo, Afleet, Crafty Prospector, Machiavellian, Gone West, Conquistador Cielo, and Miswaki. Versatility is one of the hallmarks of these sons. Sprinter Miswaki has a Prix de l'Arc de Triomphe winner in Urban Sea (dam of 2001 Epsom Derby winner Galileo); miler Fappiano sired 1990 Derby/Breeders' Cup Classic winner Unbridled (another important sire); mile classic winner Kingmambo sired 1999 Belmont Stakes winner Lemon Drop Kid; and the versatile runner Gulch sired 1995 Derby/Belmont winner Thunder Gulch (sire of 2001 Preakness/Belmont winner Point Given).

While Myrtlewood lurks far, far in the distance now, she remains a thread to a lasting quality, in the Mr. Prospector legacy as well as that of Seattle Slew and scores of other pedigrees.

Combs often boldly declared himself America's leading breeder, lumping partnership-breds with Spendthrift-breds. Actually, he did lead the breeders' list on one occasion, 1972, when horses bred in the name of Leslie Combs II earned an aggregate of $1,578,851. Meanwhile, in the sale ring other breeders had come along to lead in average, often with much smaller consignments.

In the record syndicated stallion stair-step, Raise a Native's position at the top had not lasted through the year he was syndicated, for by the end of 1967 Bull

THE BLOOD-HORSE

Brownell Combs II

Hancock had syndicated the Phipps champion Buckpasser for $4.8 million. John R. Gaines then hit $5 million with the imported *Vaguely Noble. (Gaines had also developed a connection with Combs' man, John Olin, and it was the Gainesway stallion Bold Bidder that sired Olin's 1974 Derby winner Cannonade. Bold Bidder had raced for Gaines, Olin, and none other than John W. Hanes. Such is the way of the competitive business world of the Turf.)

Five more sires attracted record syndications before Spendthrift hit the top again with Seattle Slew, valued at $12 million in 1978. This was several years after Combs had turned management over to his son Brownell, who was named president, CEO, and general manager of Spendthrift in 1974. A large man with little of his father's outgoing flamboyance, Brownell had accepted being overshadowed and not always being shown much public respect by his father. He had his own brand of savvy, however, and in addition to landing Seattle Slew, he brought to Spendthrift the Triple Crown winner Affirmed, who was bred and raced by his father's longtime associate Louis Wolfson. (Affirmed was by Exclusive Native, an early Raise a Native stakes winner Wolfson bred and who also stood at Spendthrift.)

Brownell also purchased Fall Aspen in partnership with Francis Kernan for $600,000 from the Joseph M. Roebling dispersal at Saratoga in 1980. There, he was tapping into one of the modern mares of the ilk of Myrtlewood. The Spendthrift/Kernan connection bred the first two of Fall Aspen's nine stakes winners, then cashed in further by selling the mare for $900,000.

In 1977 the Combs father and son merged the farm and other interests into Spendthrift Incorporated. With an eye toward estate matters, Combs, a widower, and his son invited a few selected partners into the firm. Then, in 1983 amid much fanfare, the farm became a public company, its shares traded on the American Stock Exchange. A horse farm proved a poor fit with the stock market, at least at that point in American commerce. The appeal to investors, though, was beguiling. As late as 1985, the international yearling market reached the fantasy of a $13 million yearling, and a year earlier Spendthrift farm manager John Williams had sent into the ring for Spendthrift and the Wolfsons a Seattle Slew colt (Amjaad) that amazed with a $6.5 million price. That same year Spendthrift sold a Seattle Slew filly (Alchaasibiyeh) for $3.75 million — still a world record for a yearling filly. The year 1984 also brought a sentimental success at the races, when Spendthrift-bred Lucky Lucky Lucky won the Kentucky Oaks in the name of Leslie Combs II and Equites Stable. The principals in Combs' partnership in the filly included J.L. Jackson and William Bricker, president and chairman, respectively, of Diamond Shamrock Company, a Dallas-based oil, gas, coal, and chemicals firm. Cuzin' Leslie, and Brownell, still had the touch for finding important business interests.

Nevertheless, downturns in the Thoroughbred market were coming, aggravated by changes in the tax laws in 1986, which discouraged some investors who had looked to the business primarily as a tax

Champions Bred

Bred by Brownell Combs
Myrtlewood
1936 champion handicap female
1936 champion sprinter

Durazna
1943 champion two-year-old filly

Bred by Leslie Combs Jr., trustee
El Chico
1938 champion two-year-old colt

Painted Veil
1941 champion three-year-old filly

Bred by Brownell & Leslie Combs II
Myrtle Charm
1948 champion two-year-old filly

Bred by Leslie Combs II & John W. Hanes
Idun
1957 champion two-year-old filly
1958 champion three-year-old filly

Bred by Spendthrift Farm & Francis Kernan
Landaluce
1982 champion two-year-old filly

shelter. By 1985 the Combses announced they were selling their shares. Leslie Combs II was brought back in the following year as chairman, proudly proclaiming "all those directors and stockholders want me to come back and be in charge. They want me to bring it back to the way it was." This bravado from an eighty-five-year-old man who had seen so much of what he had built slip away was understandable, perhaps even admirable, but a revival was not to be. The farm so identified with the high-action world of the auction would eventually be the forlorn scene of another sort of auction, as the property, home, and many of its treasures had to be sold. Spendthrift the public company filed for Chapter 11 bankruptcy in late 1988. It was sold the following year to a group of Lexington businessmen and once again became a private farm.

Leslie Combs II died on April 7, 1990, at the age of eighty-eight.

The core of Spendthrift Farm still exists. After having had a series of ownership configurations, it became a vigorous farm again under Bruce Kline's management. Then, in 2004, B. Wayne Hughes of California bought the acreage, with Kline continuing the stallion operation.

Son Brownell Combs lives in Florida but retains an active involvement in the Thoroughbred industry. He expresses pride in his role in helping Prince Khalid Abdullah (Juddmonte Farm) come into the sport and still advises a number of clients on bloodstock matters.

The Nashua motel is still a mainstay at Spendthrift, with its statue of Nashua and Clem Brooks. The monument to the horse is made of bronze; the monument to Combs is made of memory.

Stakes Winners Bred by Leslie Combs II and Family

Leslie Combs Jr., Trustee

Adequate, b.f. 1927, Pennant—Disparity, by *Singleton

Magyar, b.c. 1929, *Stefan the Great—*Royal Dispatch, by *Ambassador IV

Sweeping Light, br.c. 1929, Manna—Sweeping Glance, by Glance

Sweet Chariot, b/br.g. 1930, Black Servant—*Song Bird, by Thrush

Allen Z., b.g. 1932, *Pharamond II—La Morlaye, by Peter Pan

Gean Canach, br.c. 1933, *St. Germans—Killashandra, by *Ambassador IV

Holl Image, ch.g. 1933, *Hollister—*Val de Grace, by Corcyra

Planetoid, gr.f. 1934, Ariel—La Chica, by Sweep

Miyako, gr.f. 1935, John P. Grier—La Chica, by Sweep

El Chico, ch.c. 1936, John P. Grier—La Chica, by Sweep

Chicuelo, blk.c. 1938, Ariel—La Chica, by Sweep

Painted Veil, b.f. 1938, Blue Larkspur—Killashandra, by *Ambassador IV

Lucas B. Combs

Kinnoul, ch.c. 1917, Peter Quince—Lychee Nut, by *Sir Modred

Black Rascal, br.c. 1919, Black Toney—Whisk Broom, by Cesarion

Moon Side, br.g. 1932, Broadside—Over the Moon, by Broomstick

Roman Hero, ch.c. 1936, Pompey—Nancy Dyer, by *Archaic

Appeasement, ch.f. 1938, Peace Chance—Refine, by Ormondale

Defense, ch.g. 1939, Jamestown—Lotofus, by John P. Grier

Burnt Cork, b.c. 1940, Mr. Bones—North Wind, by *North Star III

Jack S. L., ch.g. 1940, Jack High—Burgee, by Pennant

Grant Rice, dk.b.c. 1941, *Bull Dog—Nancy Dyer, by *Archaic

Roman Miss, br.f. 1948, Roman—Nurse Boss, by Broadside

First Aid, ch.g. 1950, *Bernborough—Humane, by Broadside

Tom Turkey, ch.c. 1959, Billings—Jungle Vine, by Dark Jungle

Farmers Market, ch.g. 1959, To Market—Swing Again, by Sun Again

Grand Square, ch.g. 1965, Battle Joined—Promenade Home, by Ky. Colonel

Graceful Native, ch.c. 1967, Billings—Graceful, by War Jeep

Chiadora, b.f. 1971, Cap Size—Keene, by Solar Slipper

Brownell Combs

My Reverie, b.f. 1919, Ultimus—Reflex, by Sir Dixon

Anna M. Humphrey, ch.f. 1920, Peter Quince—Eden Hall, by *Armeath II

Sweetheart, ch.f. 1920, Ultimus—*Humanity, by *Voter

Tuskegee, b.c. 1925, Black Toney—*Humanity, by *Voter

Manta, ch.f. 1927, Cudgel—Paloma, by Golden Sun (Bred in England)

Broadway Lights, b.g. 1929, Broadway Jones—Lotus, by *Light Brigade

Pairbypair, ch.c. 1929, Noah—*Frizeur, by *Sweeper

Crowning Glory, b.g. 1930, Black Toney—*Frizeur, by *Sweeper

Corinto, b.g. 1932, Wildair—Paloma, by Golden Sun (Bred in England)

Myrtlewood, b.f. 1932, Blue Larkspur—*Frizeur, by *Sweeper

Lotofus, blk.f. 1933, John P. Grier—Lotus, by *Light Brigade

Hard Lu, ch.f. 1937, Hard Tack—Luminosa, by Blue Larkspur

Royal Archer (stp), dk.b.g. 1937, *Royal Minstrel—La Morlaye, by Peter Pan

Miss Dogwood, b.f. 1939, *Bull Dog—Myrtlewood, by Blue Larkspur

Durazna, b.f. 1941, Bull Lea—Myrtlewood, by Blue Larkspur

Sequence, dk.b.f. 1946, Count Fleet—Miss Dogwood, by *Bull Dog

Bernwood, dk.b.c. 1948, *Bernborough—Miss Dogwood, by *Bull Dog

Bella Figura, dk.b.f. 1949, Count Fleet—Miss Dogwood, by *Bull Dog

Brownell Combs and Lucas B. Combs

Devil's Thumb, b.f. 1940, Grand Slam—Daintiness, by Blue Larkspur

Dark Jungle, br.c. 1943, He Did—Dark River, by Blue Larkspur

Brownell Combs and Leslie Combs II

Tel O'Sullivan, ch.g. 1943, Chance Play—Cartela, by *Chicle

Myrtle Charm, b.f. 1946, Alsab—Crepe Myrtle, by Equipoise

Noorsaga, b.g. 1953, *Noor—Sequence, by Count Fleet

Carrier X., ro.g. 1955, Count Fleet—Amiga, by *Mahmoud

Moon Glory, b.f. 1955, *Norseman—Moonflower, by *Bull Dog

Hermod (stp), b.g. 1956, *Royal Charger—Sequence, by Count Fleet

Dedimoud, ch.c. 1959, Dedicate—Amiga, by *Mahmoud

Journalette, b.f. 1959, Summer Tan—Manzana, by Count Fleet

Lady Wayward, dk.b.f. 1961, Dedicate—Spring Tune, by Spy Song

Gold Digger, b.f. 1962, Nashua—Sequence, by Count Fleet

Masked Lady, b.f. 1964, Spy Song—Spinosa, by Count Fleet

Tumiga, ch.c. 1964, *Tudor Minstrel—Amiga, by *Mahmoud

Lady Tramp, b.f. 1965, *Sensitivo—La Morlaye, by *Hafiz

Alert Princess, gr.f. 1966, Raise a Native—Amiga, by *Mahmoud

Leslie Combs II

Alibhai Lynn, br.f. 1950, *Alibhai—Lynn, by High Time

Pegeen, b.f. 1951, *Shannon II—Nellie's Last, by *Bull Dog

Apollo, ch.c. 1952, Mr. Busher—-Jeanne's Poise, by Equipoise

Miss Ardan, ch.f. 1952, *Ardan—Impulsive, by Supremus

Beau Pilot, ch.g. 1954, Mr. Busher—Beau Jet, by Jet Pilot

Shan Pac, dk.b.c. 1954, *Shannon II—*Pacifica II, by Puro Habano

Pilot, ch.c. 1956, Jet Pilot—War Shaft, by War Admiral

Moslem Chief, b.c. 1957, *Alibhai—Up the Hill, by *Jacopo

Shuette, b.f. 1958, Nashua—Beau Jet, by Jet Pilot

Royal Attack, ch.c. 1959, *Royal Charger—Dragona, by Bull Lea

Get Around, ch.c. 1960, Citation—Lotopep, by Menow

Polybius, ch.f. 1960, Mr. Busher—Arctic Weather, by Arctic Prince (SW in England)

Clem Pac, b.c. 1961, Clem—*Pacifica II, by Puro Habano

Intentorpoise, ch.f. 1961, Intent—Lotopep, by Menow

Timbeau, b.c. 1961, Tim Tam—Beau Jet, by Jet Pilot

Crowned Queen, ch.f. 1962, *Royal Charger—*Pagan Worship, by Hyperion

Gallant Lad, b.c. 1962, *Gallant Man—Growing Up, by Maxim

Heraldette, b.f. 1962, Bald Eagle—Sudden Impulse, by *Heliopolis

Hurry Up Dear, b.c. 1962, Dark Star—Dear April, by *My Babu

Loom, b.c. 1962, Swoon's Son—Distaff by *Beau Pere

April Dawn, ch.f. 1963, *Gallant Man—Dear April, by *My Babu

Auhsan, b.c. 1963, Nashua—Tulle, by War Admiral

Lovely Gypsy, ch.f. 1963, Armageddon—Gay Hostess, by *Royal Charger

Nashula, b.c. 1963, Nashua—Venetian Love, by The Doge

Tuffit Out, b.g. 1963, Never Give In—*Dame Galante, by Brantome

Procula, b.f. 1964, Intent—Sweet Nell, by Bull Lea

Tudor Jet, dk.b/br.f. 1964, *Tudor Minstrel—Precious Lady, by Requested

All Image, dk.b/br.c. 1965, All Hands—Gerts Image, by Mr. Music

Captain Vancouver, b.c. 1965, Never Bend—Save Time, by War Admiral

Carmen Dolores, b.f. 1965, *Sensitivo—Little Sweetie, by Bolero

Hand to Hand, gr.c. 1965, Warfare—Melon, by *Heliopolis

Owe Everything, b.f. 1965, Nashua—*Ondine II, by Timor

Royal Sue, ch.f. 1965, Raise a Native—Nassue, by Nashua

Distinctive, dk.b/br.c. 1966, Never Bend—Precious Lady, by Requested

Dorothy Glynn, b.f. 1966, Northern Dancer—Save Time, by War Admiral

Governors Party, dk.b/br.c. 1966, Nashua—Sister Dikki, by *Bernborough

Indian Emerald, dk.b/br.c. 1966, Jaipur—Endearment, by *Alibhai

Majestic Prince, ch.c. 1966, Raise a Native—Gay Hostess, by *Royal Charger

Right Cross, ch.c. 1966, Nashua—Gallatia, by *Gallant Man

Towzie Tyke, ch.g. 1966, Crimson Satan—Laquesta, by War Admiral

Coaltown Cat, b.c. 1967, Nashua—Miss Ardan, by *Ardan

Jaradara, b.c. 1967, First Landing—Dear Diane, by Alsab

Loved, ch.f. 1967, Jaipur—*Cantadora II, by Canthare

Squabble, b.c. 1967, Never Bend—Village Beauty, by *My Babu

Continental Fare, ch.c. 1968, Warfare—Secret Pocket, by Never Give In

Peninsula Princess, b.f. 1968, Crewman—Ellenwood, by *Shannon II

Silent Beauty, dk.b/br.f. 1968, Crème dela Crème—Village Beauty, by *My Babu

Struck Out, ch.c. 1968, Nashua—Cantadora II, by Canthare

Triple Bend, dk.b/br.c. 1968, Never Bend—Triple Orbit, by Gun Shot

Crown The Queen, b.f. 1969, Swaps—Village Beauty, by *My Babu

Divorce Trial, dk.b/br.c. 1969, *Sensitivo—Vacanze, by Retrial

English Silver, b.f. 1969, Mongo—Tudor Jet, by *Tudor Minstrel

Fairway Flyer, b.f. 1969, Nashua—Fairway Fun, by Prince John

Great Bear Lake, ch.c. 1969, *Gallant Man—Amerala, by *Amerigo

Heisanative, dk.b/br.c. 1969, Raise a Native—Times Two, by Double Jay

Popular Demand, ch.c. 1969, *Sensitivo—La Morlaye, by Hafiz

Rock Gold, dk.b.f. 1969, Fleet Nasrullah—Cantadora II, by Canthare (SW in England)

Wild World, ch.c. 1969, Warfare—Sister Dikki, by Bernborough

Mr. Prospector, b.c. 1970, Raise a Native—Gold Digger, by Nashua

My Gallant, ch.c. 1970, *Gallant Man—Predate, by Nashua

Never Ask, dk.b/br.f. 1970, Never Bend—Lady Wayward, by Dedicate

Torsion, b.c. 1970, Never Bend—Fairway Fun, by Prince John

Who's to Know, dk.b/br.f. 1970, Fleet Nasrullah—Masked Lady, by Spy Song

Without Peer, b.f. 1970, Crème dela Crème—Inviting, by *My Babu

Camp Whip, ch.c. 1971, *Seaneen—Had My Way, by Summer Tan

Caracolero, ch.c. 1971, Graustark—Betty Loraine, by Prince John (SW in France)

Change Purse, ch.c. 1971, Mongo—Serene Highness, by Jet Pilot

Erwin Boy, b.g. 1971, Exclusive Native—Indian Call, by Warfare

Gold Standard, b.c. 1971, *Sea-Bird—Gold Digger, by Nashua

Hasty Tudor, ch.c. 1971, King of the Tudors—Hasten On, by Fleet Nasrullah

Mr. A. Z., dk.b/br.c. 1971, *Gallant Man—Fine Call, by Tom Fool

Nicest Lady, ch.f. 1971, *Forli—Royal Sue, by Raise a Native

Quick Sea, dk.b/br.f. 1971, *Seaneen—Quick Missile, by Missile

Rufarina, dk.b/br.f. 1971, Bold Hour—Card Lady, by *My Babu

Straight as a Die, b.c. 1971, Never Bend—Melon, by *Heliopolis (SW in England)

Two Timing Lass, b.f. 1971, Prince John—Times Two, by Double Jay

Unipress, ch.g. 1971, Raise a Native—Evora, by Tatan

Baby Louise, ch.f. 1972, Exclusive Native—Careful Turn, by *Tudor Minstrel

Bombay Duck, b.c. 1972, Nashua—Egret, by *Tudor Minstrel

Make Love, ch.f. 1972, Nashua—Venetian Love, by The Doge

Pac Quick, b.c. 1972, Clem Pac—Quick Missile, by Missile

Raise a Lady, dk.b.f. 1972, Raise a Native—Masked Lady, by Spy Song (SW in France)

Visier, dk.b/br.c. 1972, Hail to All—Princess Kitty, by Billings

Yale Coed, ch.f. 1972, Majestic Prince—Mellow Marsh, by *Seaneen

Native Guest, ch.c. 1972, Raise a Native—My Guest, by Mister Gus

Romolo Augusto, b.c. 1972, Prince John—Royal Patrice, by *Royal Charger (SW in Italy)

Semi Princess, dk.b/br.f. 1972, Semi-pro—Sea Myrtle, by Swoon's Son

Straight, b.f. 1972, Never Bend—Call Card, by *Alibhai

Tiltin Milton, dk.b/br.g. 1972, Crème dele Crème—Crowned Queen, by *Royal Charger

White Fir, ch.c. 1972, Swaps—Blue Medley, by First Landing

Beach Party, b.f. 1973, Proud Clarion—Beach Talk, by *Sensitivo

Cut Corners, ch.c. 1973, Good Investment—Cold Morning, by Mr. Busher

Imacornishprince, dk.b/br.c. 1973, Cornish Prince—Had My Way, by Summer Tan

Just Jazz, ch.f. 1973, Exclusive Native—Hasten On, by Fleet Nasrullah

L'Natural, ch.c. 1973, Raise a Native—Mellow Marsh, by *Seaneen

Tuff Bear, b.c. 1973, Nashua—Hello Teddy Bear, by *Court Martial

Catahoula, dk.b.c. 1974, Never Bend—Silent Beauty, by Crème dela Creme

Chatta, b.f. 1974, Native Royalty—Air Maid, by Fleet Nasrullah

Cuzwuzwrong, dk.b/br.c. 1974, Hoist the Flag—English Silver, by Mongo

Huggle Duggle, b.f. 1974, Never Bend—Crown the Queen, by Swaps

Man's Man, b.c. 1974, *Gallant Man—Rough Mood, by Dedimoud

Pocket Princess, b.f. 1974, Exclusive Native—Secret Pocket, by Never Give In

Prince Majestic, ch.g. 1974, Majestic Prince—Easter Robin, by Prince John

Savage Bunny, dk.b.f. 1974, Never Bend—Tudor Jet, by *Tudor Minstrel

Sea Royalty, dk.b.f. 1974, Native Royalty—Sea Myrtle, by Swoon's Son

Shady Lou, b.f. 1974, Crème dela Crème—Hen, by County Clare

Forever Gallant, ch.c. 1975, Crème dela Crème—Baby Now, by *Gallant Man

Lady Hawthorne, b.f. 1975, Cornish Prince—Donna II, by Farewell

Etiquette, b.f. 1986, Raja Baba—Gold Mine, by Raise a Native

TURFOTOS

Mr. Prospector

Leslie Combs II and John W. Hanes

Prince Eric, b.c. 1952, *Priam II—Alibelle, by *Alibhai
Idun, b.f. 1955, *Royal Charger—Tige O'Myheart, by Bull Lea
Court Affair, b.c. 1956, *Royal Charger—Letmenow, by Menow
Don't Alibi, ch.g. 1956, *Alibhai—Obedient, by *Mahmoud
Francis S., ch.c. 1957, *Royal Charger—Blue Eyed Momo, by War
   Admiral
Irish Lancer, b.c. 1957, *Royal Charger—Tige O'Myheart, by Bull Lea
Colfax Maid, b.f. 1958, *My Babu—Tsumani, by Cientifico
Garwol, b.g. 1958, *My Babu—Fleece, by Revoked
Water Witch, b.f. 1960, Nashua—Water Snake, by Alycidon
Bargain Package, b.c. 1961, *Royal Charger—Princess Lea, by Bull Lea
Aqua Vite, b.g. 1963, Nashua—Moon Glory, by Norseman
Francine M., b.f. 1964, Sir Gaylord—Fleece, by Revoked
Jungle Road, b.c. 1964, Warfare—La Dauphine, by *Princequillo
Vaguely Familiar, b.f. 1970, *Vaguely Noble—Rainbow Rose,
   by *Ambiorix

Leslie Combs II, John W. Hanes, and Walmac Farm

Clems Alibi, ch.f. 1963, Clem—Many Splendored, by *Alibhai (SW in
   Mexico)
Alexander D., ch.c. 1964, *Seaneen—Native Valor, by *Mahmoud
Charles Elliott, b.c. 1964, On-and-On—Gold Lance, by *Alibhai
Mr. Hingle, b.c. 1964, Bald Eagle—Querida, by *Alibhai

Leslie Combs II, John W. Hanes, and Mrs. John M. Olin

Chalina, ch.f. 1963, *Seaneen—Babuska, by *My Babu
Betty's Pride, b.f. 1964, Dark Star—*Fernley, by *Court Martial

Leslie Combs II and Mrs. John M. Olin

Prevailing, b.c. 1966, Never Bend—Breeze-A-Lea, by Bull Lea
Shooting Starlet, b.f. 1966, Raise a Native—*Starlet II, by Nearula
Talent Search, b.f. 1967, Sir Gaylord—*Ribotina, by *Ribot
Matchless Native, b.f. 1968, Raise a Native—Royal Match, by *Turn-to
Raise a Bid, b.c. 1968, Raise a Native—Plotter, by Double Jay
Water Blossom, b.f. 1968, Nashua—*Ribotina, by *Ribot
Carezza, ch.f. 1969, *Ribot—Noblesse, by Mossborough (SW in
   England)
Gallant Knave, b.c. 1970, *Gallant Man—Plotter, by Double Jay
Where You Lead, ch.f. 1970, Raise a Native—Noblesse, by
   Mossborough (SW in England)

Leslie Combs II in various other partnerships

Fighting Jodo, ch.g. 1947, Fighting Fox—Tudor Queen, by St. James
Gay Grecque, b.f. 1949, *Heliopolis—Dark Tower, by *Blenheim II
Incidentally, ro.f. 1951, Mr. Busher—Danise M., by *Epinard
Wa-Wa Cy, b.c. 1959, Tom Fool—Boogooloo, by *Nasrullah
Deceit, dk.b/br.f. 1968, Prince John—Double Agent, by Double Jay
Lady Herald, b.f. 1968, Boldnesian—Heraldette, by Bald Eagle
Crowned Prince, ch.c. 1969, Raise a Native—Gay Hostess, by *Royal
   Charger
Introductivo, b.c. 1969, *Sensitivo—Gayway, by Sir Gaylord
Mansingh, dk.b.c. 1969, Jaipur—Tutasi, by Native Dancer (SW in
   England)
El Seetu, b.c. 1971, Prince John—Risk, by Fleet Nasrullah
Fairway Fable, b.f. 1971, Never Bend—Fairway Fun, by Prince John
Clout, b.c. 1972, *Indian Chief II—Strip Poker, by Bold Bidder
Dancers Countess, b.f. 1972, Northern Dancer—Countess Belvane,
   by *Ribot
Gallina, ch.f. 1972, Raise a Native—Gallatia, by *Gallant Man (SW in
   England)

Guillaume Tell, ch.c. 1972, Nashua—La Dauphine, by *Princequillo
   (SW in England)
Fun Forever, b.f. 1973, Never Bend—Fairway Fun, by Prince John
Jevalin, b.f. 1975, Dewan—Pagan Pagan, by Native Charger
Mashteen, ch.f. 1975, Majestic Prince—Marshua, by Nashua
Unconscious Doll, b.f. 1975, Unconscious—Long Ago, by Tom Fool

Spendthrift Farm

Angel Island, dk.b/br.f. 1976, Couger II—Who's to Know, by Fleet
   Nasrullah
Carole's Tale, ch.f. 1976, Relian—Pac of Tales, by Clem Pac
Crowned Music, ch.c. 1976, List—Crowned Queen, by *Royal
   Charger (SW in France)
Dancing Gondola, b.f. 1976, Marshua's Dancer—Air Maid, by Fleet
   Nasrullah
Doing It My Way, ch.f. 1976, Exclusive Native—Had My Way, by
   Summer Tan
Double Deceit, dk.b.f. 1976, Northern Dancer—Double Agent, by
   Double Jay
Picturesque, b.c. 1976, Native Royalty—Chic Nell, by *Seaneen
Raise Your Sights, b.f. 1976, Raise a Native—My Guest, by Mister Gus
Rossi Gold, b.c. 1976, Taj Rossi—Rock Gold, by Cantadora
Royal Imp, dk.b.f. 1976, Nashua—Prim Mim, by Bold Hour
Active Voice, b.f. 1977, Pretense—Mellow Marsh, by *Seaneen
Aerostation, ch.f. 1977, Prince John—Semi Princess, by Semi-pro
Gallant Gunner, dk.b.c. 1977, *Gallant Man—Gunner's Runner, by
   Royal Gunner (SW in Dominican Republic)
Lillian Russell, ch.f. 1977, Prince John—Gold Digger, by Nashua
Great Substence, dk.b.c. 1978, Pretense—Gay Northerner, by
   Northern Dancer (SW in France and U.S.)
Intriguing Honor, dk.b.f. 1978, Sham—Inviting, by *My Babu
Shamstar, b.c. 1978, Sham—Campus Star, by Raise a Native (SW in
   Italy)
Charge My Account, ch.f. 1979, Majestic Prince—Quick Selection,
   by Viceregal
Cost Control, ch.c. 1979, Intrepid Hero—Two Timing Lass, by Prince
   John
Dancing Partner, ch.f. 1979, Exclusive Native—Indian Call, by Warfare
Exclusive Love, ch.f. 1979, Exclusive Native—Loved, by Jaipur
Muriesk, b.f. 1979, Nashua—Cosmiah, by Olympia
Polite Rebuff, b.f. 1979, Wajima—Merely, by Dr. Fager
Proud Lou, b.f. 1979, Proud Clarion—Baby Louise, by Exclusive
   Native
Sham's Princess, gr.f. 1979, Sham—Alert Princess, by Raise a
   Native (SW in France)
Smooch, ch.f. 1979, Raise a Native—Royal Bit, by Alcibiades II
Biricchina, dk.b.f. 1980, Icecapade—Sigh Sigh, by Cyane
Brorita, ch.f. 1980, Caro [Ire]—Mellow Marsh, by *Seaneen
Fleur de Printemps, gr.f. 1980, Caro [Ire]—Savage Bunny, by Never
   Bend
Interco, ch.c. 1980, Intrepid Hero—Yale Coed, by Majestic Prince
Luigi Tobin, dk.b.c. 1980, J. O. Tobin—Tudor Jet, by *Tudor Minstrel
Migola, b.f. 1980, Raise a Native—Air Maid, by Fleet Nasrullah
Princesse Rapide, ch.f. 1980, Majestic Prince—Fast Call, by Fleet
   Nasrullah
Sugar Charlotte, dk.b.f. 1980, Wajima—Silent Beauty, by Crème dela
   Crème
Al Mundhir, b.c. 1981, Seattle Slew—Huggle Duggle, by Never Bend
   (SW in Germany)
Locust Bayou, b.c. 1981, Majestic Prince—Myrtlewood Beauty, by
   Never Bend

Lucky Lucky Lucky, b.f. 1981, Chieftain—Just One More Time, by Raise a Native

Manicure Kit, b.f. 1981, J. O. Tobin—Mellow Marsh, by *Seaneen

Milord, ch.c. 1981, Raise a Native—Evening, by Up All Hands

Northern Jazz, ro.f. 1981, Northern Jove—Just Jazz, by Exclusive Native

Our Reverie, dk.b.f. 1981, J. O. Tobin—Angel Island, by Cougar II

Rascal Rascal, dk.b.f. 1981, Ack Ack—Savage Bunny, by Never Bend

Siberian Express, gr.c. 1981, Caro [Ire]—Indian Call, by Warfare

Goldenita, b.f. 1982, Golden Act—Clandenita, by Clandestine

Natania, dk.b.f. 1982, Naskra—Astania [Ger], by Arratos

Village Sass, b.f. 1982, Sassafras [Fr]—Village Beauty, by *My Babu

Arewehavingfunyet, b.f. 1983, Sham—Just Jazz, by Exclusive Native

Nymph of the Night, b.f. 1983, Magesterial—Alert Princess, by Raise a Native

Sharrood, ro.c. 1983, Caro [Ire]—Angel Island, by Cougar II

Fast Forward, b.c. 1984, Pleasant Colony—Just One More Time, by Raise a Native

Fleet Road, b.f. 1984, Magesterial—Fleet Fashion, by Jungle Road

Holst, dk.b.c. 1984, Explodent—Angel Rouge, by Crimson Satan (SW in France)

Muskrat Love, b.f. 1984, Muscovite—Phils Love, by Philately

Sanam, ch.c. 1984, Golden Act—Rose Goddess [Ire], by Sassafras [Fr] (SW in Italy and Ireland)

Sarba, gr.f. 1984, Persepolis—Felicite, by Tapioca (SW in France)

Zaizoom, dk.b.c. 1984, Al Nasr [Fr]—Plumovent, by *Turn-to (SW in Italy and U.S.)

Adam's Run, gr.f. 1985, Fairway Phantom—Thundertee, by Ye

Double Wedge, ch.f. 1985, Northern Baby—Kit's Double, by Spring Double

Icy Stare, b.c. 1985, Icecapade—Metallurgical Gal, by Prince John

Insan, dk.b.c. 1985, Our Native—Artania, by Ruritania (SW in Ireland)

Le Fantome (stp), ro.c. 1985, Fairway Phantom—Love Bunny, by Exclusive Native (SW in France)

Ohsomellow, ch.f. 1985, Sharpen Up [Eng]—La Jalouse, by Nijinsky II (SW in England)

Shaybani, b.c. 1985, Al Nasr [Fr]—Rose Goddess [Ire], by Sassafras [Fr] (SW in South Africa)

Sheesham, dk.b.f. 1985, Sham—Hypochondriac, by Sirlad [Ire]

Truly Met, b.c. 1985, Mehmet—All Too True, by Caro [Ire]

Cyphrate (stp), dk.b.c. 1986, Saint Cyrien—Euphrate [Fr], by Royal and Regal (SW in England)

Eternity's Breath, dk.b.c. 1986, Nureyev—Sham's Princess, by Sham (SW in France)

Memorable Mitch, b.f. 1986, Mehmet—My Guest, by Mister Gus

Dinner in Rio, b.g. 1989, State Dinner—Rio by Night, by Pleasant Colony

Ten Ten Ho, ch.f. 1989, Yukon—Lonely Melody, by Raise a Native

Amal Hayati, b.f. 1990, Seattle Song—Night Fire, by Cannonade

Prince of Andros, dk.b.c. 1990, Al Nasr [Fr]—Her Radiance, by Halo (SW in England and Ireland)

Spendthrift Farm and Francis Kernan

Bally Knockan, ch.f. 1979, Exclusive Native—Ferly, by Traffic Judge

Ecstatic Pride, dk.b.c. 1980, J. O. Tobin—Ferly, by Traffic Judge

Habitassa (Ire), b.f. 1980, Habitat—Sassabunda (Ire), by Sassafras (Fr)

Landaluce, dk.b.f. 1980, Seattle Slew—Strip Poker, by Bold Bidder

Saucy Bobbie, dk.b.f. 1980, Roberto—Kadesh, by Lucky Mel

Elle Seule, ch.f. 1983, Exclusive Native—Fall Aspen, by Pretense

Wavering Kite, b.g. 1985, Wavering Monarch—Arctic Kite (Ire), by North Stoke

Spendthrift Farm, Francis Kernan, and Brownell Combs II

Northern Aspen, b.f. 1982, Northern Dancer—Fall Aspen, by Pretense

Spendthrift Thoroughbred Breeding No. 1

Bint Pasha, ch.f. 1984, Affirmed—Icely Polite, by Graustark (SW in Eng/Ire/Fr)

Canango, ch.g. 1984, Caro (Ire)—Hustle On, by Raise a Native (SW in Italy/France)

Spendthrift in various other partnerships

Cherokee, ch.c. 1976, Cabildo—Isle of Beauty, by My Redbird (SW in Panama)

Exclusive One, ch.c. 1979, Exclusive Native—La Jalouse, by Nijinsky II

Lover Boy Leslie, ch.c. 1980, Raise a Native—Clairvoyance, by Round Table

Smuggly, ch.f. 1980, Caro (Ire)—Call Me Goddess, by Prince John

My Darling One, ch.f. 1981, Exclusive Native—Princess Marshua, by Prince John

Princess Claire, ch.f. 1981, Exclusive Native—Clairvoyance, by Round Table

Light of Nashua, dk.b.c. 1982, Nashua—Village Gossip, by Buckpasser (SW in France)

Cougarized, dk.b.c. 1985, Cougar II—Jolie Jolie, by Sir Ivor

Tibullo, ch.c. 1985, Affirmed—The Fog, by Irish Castle (SW in Italy)

Brownell Combs II

Tyrnville (Fr), b.c. 1983, Tyrnavos—Karenina, by Silver Shark (SW in Belgium)

Brownell Combs II and Wilhemina McEwan

Savage Creek, ch.f. 1986, Jungle Savage—Millcreek Court, by Rainy Lake

Victory's Goldie, dk.b.f. 1986, Golden Act—Title Victory, by Title Game

Source: *The Blood-Horse* Archives

CHAPTER

10

Paul Mellon

Paul Mellon spent a lifetime fulfilling and refining the notion of connoisseur. He filled his stables with well-bred Thoroughbreds, his art collection with rare treasures, his sportsmanship and citizenry with personal achievements. He also fulfilled the notion of philanthropist, sharing his priceless masters with museums, his wealth with noble endeavors, his life's experiences with those who sought his counsel.

Mellon was born in Pittsburgh on June 11, 1907, and the Pittsburgh of his youth was the dark, smoky industrial version, unlike the city of today. If his environment were dreary, he had a pleasant land to escape to in traveling from time to time to his mother's native England. Ah, England, a land in Mellon's memory of green countryside, brightly colored trains, and robust gallops across the hunt field on frosty winter days. In England, too, Mellon found the challenges, physical and mental, of studying and crewing at Cambridge, the camaraderie of delving into the port bottle with like-minded young sportsmen, and, eventually, a victory in that hallowed of hallowed horse races, the Epsom Derby.

While the Derby at Epsom is the centerpiece of English flat racing, Newmarket is the headquarters of the Turf in the land of its origin, and Newmarket was central to England's hold on Mellon's heart. Perhaps sixty years after he was introduced to the historic course, Mellon wrote gracefully of how "Newmarket beckoned me…I still hark back to those long, soft, eminently green gallops stretching to the horizon in the slanting afternoon sun, and the late October sunlight on the warm yellow stone of the old, high stands…the bright colors of the silks flashing by, the sheen of the horses' coats…"

Of England, he also wrote, "My interest in British art is part of my fascination with British life and history. From childhood and from Cambridge days I acquired a fondness for the English landscape and for the ever-changing English light…"

Mellon did not, however, choose to live in England. Instead, he fashioned a bit of England through the prism of the similarly lovely Virginia countryside as he took the core of his mother's farm and built Rokeby Farm in Upperville into a rambling vista of fields and forests, fences and foals.

He was not a man only of the land, however. For a time after Yale and then Cambridge, he did try to follow his father's wish that he pursue banking. Board rooms and ledgers were nothing compared to hunt balls and St. Legers, and he soon went his own path. Wealth in quantities difficult for most to grasp had come the young Mellon's way, not of his own making, and he was determined to pursue a life in which he used that wealth judiciously and positively. This he accomplished to such a degree, and for such a range of purposes, that by the time he published a memoir, *Reflections in a Silver Spoon*, in 1992, an appendix therein listed more than $600 million in major gifts!

Mellon supported the work of psychologist Carl Jung, and he followed the example of his father, Andrew Mellon, who gave the nation the National

Gallery of Art; son Paul donated much of his own art to that Washington, D.C., treasure as well as to the Virginia Museum of Fine Arts and the Yale Center for British Art. The last-named did not exist until Mellon financed its construction, before filling it with a delicious collection of volumes as well as paintings by George Stubbs and the like. Similarly, Mellon and his sister funded the addition known as the East Wing of the National Gallery.

Perhaps it was with a bit of resignation that Mellon found that board rooms and ledgers were still a large part of life, after all. He was active in the leadership of the National

Paul Mellon

Gallery and other charities, concerned with good management, fiscal policy, and large staffs — such as he would have faced in banking or industry.

Paul Mellon had offices and homes in Washington, D.C., and New York, plus Cape Cod, Massachusetts, and Antigua, and he had the luxury of appointing each with handsome art. Anyone fortunate enough to have some business with this combination country squire and urbane dignitary might have the delight of gazing above his desks in New York and Washington at a series of Munnings paintings or perhaps of touring "The Brick House" at Rokeby, where he kept hundreds more sporting paintings — Stubbs and Herring, Ferneley and Marshall — as well as his beloved waxes and bronzes of Degas' dancers and racers. A Wedgwood decoration or two were spotted amidst a personalized photo of Queen Elizabeth II and Mellon's venerated oar from Clare College, Cambridge.

From our personal perspective, too, is the long-cherished memory of a lunch in the rambling main house at Rokeby. This was preceded by the honor of watching the master prepare and serve the vaunted "Mellon Martini" (including both gin and vodka!), while Mrs. Mellon mused that perhaps they would have the van Gogh original over the fireplace framed, since the canvas was slightly fraying at the corners. ("Well, we all have to economize," was a mental quip,

swallowed, not said.)

It is as a Turfman, of course, that we address Paul Mellon in this volume. He has few peers at inhabiting the term "sportsman." His reserved personality and worldly influences developed a personal manner that included a pleasant accent of languid Virginia cadence plus elements of the Old World. Tall and slender, he had a physical elegance, whether wearing pin-stripes for a formal portrait or a combination of understated plaids and stripes that on him worked just right.

Mellon was a spirited youth. Wafting down over the imprecise history of hunt breakfasts is the tale of this generally mild-mannered — perhaps even shy — stripling performing a series of cartwheels down a well-laden table and attempting a rather impressive, concluding leap to a convenient chandelier. Alas, this fine old piece, while no doubt admirable in its ability to bear and display lighting fixtures, was unequal to the task of sustaining a young squire in full flight. Mellon might have found himself sputtering "remember there are worse things than a shattered chandelier," albeit with less foreboding than the modern persona of the *Phantom of the Opera* implied in his own recitation of that sentence.

Mellon also took many a leap on horseback, with better endings, both in the hunt field and in a few point-to-point races he participated in as a jockey.

"I bought my first race horse in 1933, two years after I came home from Cambridge," Mellon noted in *Reflections* (written with John Basket; William Morrow and Company). "He was an Irish Thoroughbred called Drinmore Lad, whom my trainer, Jim Ryan, had brought over to America as a 3-year-old (of 1932). Now I was an owner…he ran in a timber race at Far Hills, N.J., and won it, an encouraging start to his and my racing career."

Fellow Virginian Ambrose Clark had won the most storied of all the world's steeplechases, England's Grand National, with Kellsboro' Jack. When Mellon contrived to attempt to seek his own Grand National victory with Drinmore Lad, Clark took him to Mrs. Aubrey Hastings' training yard in England. It was the beginning of a connection that spanned the entirety of Mellon's racing career abroad. The trainer for the yard at that time was Ivor Anthony. Drinmore Lad won a number of good races and twice was an early co-favorite for the Grand National with the great Golden Miller, with whom he dead-heated in the Crawley Chase at Gatwick in 1937. Drinmore Lad never was sound and healthy enough on National Day to contest the big race, however, and was returned to Virginia when it was discovered he had developed a heart condition.

When World War II interrupted the lives of Americans, Mellon was sent first to Fort Riley, Kansas, "in the horse section. That was great fun while it lasted and was a very nice experience, but it wasn't really of much use in the war effort." Mellon chafed under some later European assignments as well, such as being in charge of England's Victory Gardens because someone had noted "owns a farm" on his record.

"That almost drove me crazy," he said years later. "Here I was in Cheltenham, and I thought, 'This is where I used to come to the races,' and to end up in the Quartermaster Corps wasn't my idea of the war."

After another posting that struck him as out of the mainstream of the torrid affairs of the time, Mellon contrived through some connections to have himself assigned to the OSS (forerunner of the CIA): "That was pretty interesting because I had all kinds of things to do. I was SO, which is Special Operations, sabotage and that sort of thing. I had mostly to do with training agents. Later, I went over to France with an OSS detachment with the Third Army, which was to coordinate information to agents in occupied France."

Although he spoke of illness that interrupted his duty and of his later return to France with nonchalance, one can only conjecture that some fairly important stuff was going on for Mellon to have risen from private to major and earn four bronze stars.

In an interview for a 1989 series of articles on his career in *The Blood-Horse*, Mellon picked up the thread of his early days as an owner of jumpers:

"I had two or three mares that I bought locally. It was just pure luck. For instance, I bred two mares, one a mare called Sunchance and then the mare I

had ridden in England called Makista. None of this was very great bloodlines. We were just lucky. I bred them both to a horse called Gino and from that we got American Way and Genancoke." Both became important winners in the Eastern hunt meetings.

Rokeby Stable was the leading owner in the old National Steeplechase and Hunt Association standings for 1948, with $76,680. As has happened frequently in the history of American racing, the steeplechase division led to development of a flat-racing operation.

"I began to realize that I had this farm with lots of acreage and lots of good pastureland, and it seemed a shame not to have more mares," Mellon recalled in 1989.

"It seemed a shame not to have more mares" — how many owners have thought this, how many have come to second-guess the concept, and how many bloodstock agents have longed to hear the sentence?

In Mellon's case the speaker was in close control, and expansion would not be allowed to become a willy-nilly explosion of numbers. Mellon was a gentleman farmer in some respects, and he was a businessman as well. Spending for the sake of spending and size of stable just to have size were never temptations.

"It would be ridiculous to close one's eyes to the fact that behind every stable, large or small, money rears its ugly head," Mellon observed in his memoir. "It may be thought undignified to point it out, but I know of no more solid proof of successful breeding and racing than large black figures on the bottom line."

Thus, his launch into purchase of horses for racing on the flat was not a profligate exercise. He had his own bounds, and he was following good counsel: "I realized, too, that if you want to breed at all you might just as well breed the best that you could. And there was no way of telling what was the best except in flat racing. I was influenced quite a lot by a great friend of mine, Jimmy Brady, Nick Brady's father, who was a classmate at Yale. He was doing some breeding and racing, too, and so it was partly on his advice that I switched from 'chasing to flat breeding."

("Jimmy" Brady was James Cox Brady, later head of the New York Racing Association, and one of his sons, Nick Brady, is a former senator and secretary of the treasury as well as a former chairman of The Jockey Club.)

Although the flat division was to dominate the identity of Rokeby, Mellon continued to revere, and campaign, jumpers, too. A hands-on horseman as well, he was still winning hundred-mile rides in Virginia while in his sixties.

Mellon had a situation of dueling trainers he had to resolve, but each of the pair made a pivotal purchase for him during the Rokeby owner's segue from jump racing to the echelons of classic breeding for the flat races:

"At one time, I had both Jack Skinner and Jim Ryan training for me," he told us some years ago. "That did not work out very well. There was a good deal of animosity between them, so eventually I gave up Jim Ryan because he was too far away and Jack Skinner was right there and in the hunting field with me, too." (On other occasions Mellon was quoted as indicating the Ryan-Skinner affair came to fisticuffs.)

Skinner purchased for Mellon the pivotal mare Blue Banner, whom he trained and who later produced Key Bridge, dam of champions Key to the Mint and Fort Marcy. Ryan bought Tap Day as a yearling and the mare *Red Ray, who became, respectively, the dam of Belmont Stakes winner Quadrangle and the third dam of the great European champion Mill Reef.

Ryan also trained Mellon's first important flat-race winner, County Delight, a son of Triple Crown winner Count Fleet purchased at the 1948 Keeneland summer yearling sale. County Delight, bred by Mrs. John D. Hertz, cost $32,000, the eighth-highest price paid at Keeneland that summer. He won the Dixie Handicap at Pimlico on one Saturday in 1951 and the Gallant Fox Handicap at Jamaica a week later. County Delight won five stakes in 1951 and 1952. He was not much of a success at stud but did get Mellon's intrepid European stakes-winning campaigner Morris Dancer and a few other stakes winners.

Over the next decade Mellon bred a sequence of stakes horses, but they landed no trophies for him in what would be regarded today as grade I fixtures. One of his best, Pardala, had been claimed from Rokeby for $8,000 and passed among several other owners before developing into a $100,000-earner for D & H Stable. For D & H, Pardala won the Matriarch, Diana, and Black Helen handicaps. Pardala was sired by the English stallion Pardal and was out of *Double Deal II, an English mare imported by Mellon in 1952.

The aforementioned yearling purchases of Tap Day and Blue Banner provided a boost and lasting influence, for Rokeby. Tap Day was purchased from James Tupper at the 1948 Keeneland summer sale for $16,500. She was daughter of Calumet Farm's Bull Lea, known by then as the sire of Triple Crown winner Citation and a brace of other champions, and she was out of the Rockingham Park Matron winner Scurry. The latter was by Diavolo, who was known as a Cup horse but had won the Tremont at two. Scurry's dam, Slapdash, by Stimulus, had won the Great American and Schuylerville stakes and one other stakes in the 1930s. In such pedigree selection, Mellon was treading along a trail of speed and stamina. The filly was named Tap Day in commemoration of a solemn ceremonial occasion at Yale, and the name of her best son, Quadrangle, bespoke a key location on campus.

Key Bridge

Tap Day was only a modest winner, but, prior to foaling Quadrangle, she produced the moderate stakes winners Cup Man and Uncle Percy, both by *Djeddah, and the champion English sprinter Secret Step, by the great Native Dancer. Secret Step, whose wins included the 1963 July Cup, was a sprinter by the proven mile-and-a-half horse Native Dancer; conversely, Tap Day's Quadrangle was a proven stayer sired by the middle-distance horse Cohoes. In 1963, on the weekend of President Kennedy's death, Quadrangle won the Pimlico Futurity, heralding his promise for the classics of 1964. The following spring he won the Wood Memorial, prompting Mellon to quip that not only had he never had a Derby candidate before, but, in a manner of speaking, he had "never had a three-year-old."

By that time, Elliott Burch, as private trainer, handled the Rokeby horses. The son of and grandson of Hall of Fame trainers Preston and W.P. Burch had taken over after the retirement of Jack Skinner. Elliott Burch had already established himself, as trainer of 1959 Horse of the Year Sword Dancer and another champion, Bowl of Flowers, for Brookmeade Stable, before the death of its founder, Isabel Dodge Sloane.

Quadrangle ran well but was unplaced behind Northern Dancer in both the Kentucky Derby and the Preakness. Between the Preakness and Belmont Stakes, Burch sent him out against older horses in the one-mile Metropolitan Handicap, a stratagem that had worked with Sword Dancer, who won the Met as well as the Belmont. Quadrangle was second to the accomplished handicap horse Olden Times in the Met Mile but then outstayed Roman Brother and Triple Crown hopeful Northern Dancer to win the mile and a half Belmont Stakes. Less than two decades after his decision to chase the best in flat racing, Paul Mellon of Rokeby had won his first classic, and with a homebred.

Quadrangle added other races of great stature, winning Rokeby's first Travers Stakes as well as the Lawrence Realization. Roman Brother and Quadrangle represented the year's three-year-old crop well by running second and third in Kelso's fifth Jockey Club Gold Cup, but none of that was enough to supplant the retired Northern Dancer as champion three-year-old colt. Quadrangle won ten of twenty-six starts and earned $559,386. Standing at the Grayson family's Blue Ridge Farm, which neighbors Rokeby, Quadrangle got only a couple of stakes winners for Mellon, although the Belmont winner sired the champion distaffers Susan's Girl and Smart Angle for other breeders.

In 1953 Skinner paid $14,500 to purchase Blue Banner for Mellon. Like County Delight, Blue Banner was bred by Mrs. Hertz, who had raced Count Fleet. Also a Keeneland summer yearling, Blue Banner was by the Triple Crown winner War Admiral and out of Risque Blue, by Blue Larkspur. The next dam, Risque, was by Stimulus, as was Tap Day's second dam. Risque was a filly A.B. Hancock Sr. of Claiborne Farm had sold to Hertz in a package, with the declaration that she would "pay for the oth-

KEY TO THE MINT				
Graustark, 1963	*Ribot, 1952	Tenerani, 1944	Bellini (Cavaliere d'Arpino—Bella Minna)
			Tofanella (Apelle—Try Try Again)
		Romanella, 1943	El Greco (Pharos—Gay Gamp)
			Barbara Burrini (*Papyrus—Bucolic)
	Flower Bowl, 1952	*Alibhai, 1938	Hyperion (Gainsborough—Selene)
			Teresina (Tracery—Blue Tit)
		Flower Bed, 1946	*Beau Pere (Son-in-Law—Cinna)
			*Boudoir II (*Mahmoud—Kampala)
Key Bridge, 1959	*Princequillo, 1940	Prince Rose, 1928	Rose Prince (*Prince Palatine—Eglantine)
			Indolence (Gay Crusader—Barrier)
		*Cosquilla, 1933	*Papyrus (Tracery—Miss Matty)
			Quick Thought (White Eagle—Mindful)
	Blue Banner, 1952	War Admiral, 1934	Man o' War (Fair Play—Mahubah)
			Brushup (Sweep—Annette K.)
		Risque Blue, 1941	Blue Larkspur (Black Servant—Blossom Time)
			Risque (Stimulus—Risky)

ers." Sure enough, Risque won the 1930 Spinaway, 1931 Alabama, and two other stakes. Risque Blue, daughter of Risque and dam of Blue Banner, was unraced.

Blue Banner helped Rokeby to a high level of achievement. She was second in the 1954 Matron to the year's juvenile filly champion, High Voltage. The next year, she won the Test Stakes at Saratoga and was second in the Alabama. At four she won the Distaff and Firenze handicaps. She retired to Rokeby with fifteen wins in forty-six starts and earnings of $121,175.

Blue Banner's 1959 filly by the champion Claiborne Farm sire *Princequillo was named Key Bridge, for a landmark in Washington, D.C. Key Bridge did not stand training enough to race but was retained for the Rokeby broodmare band. She emerged as Broodmare of the Year for 1980, by which time two of her sons, Fort Marcy and Key to the Mint, had been voted championships in American racing. Fort Marcy emerged for his first championship season in 1967, and he followed with several years of superb form, while Key to the Mint came on as the champion three-year-old colt of 1972.

Between those years emerged the best American classic horse ever bred and raced by Mellon and, overseas, the greatest of all his horses. The American was Arts and Letters; the foreign champion was Mill Reef. Moreover, Run the Gantlet emerged as a divisional champion, too, in this country in 1971, so that on either side of the Atlantic Mellon raced a home-bred champion in 1967, 1968, 1969, 1970, 1971, and 1972 — heady days indeed.

The sequence was almost shortened before it had begun. The gelding Fort Marcy had such desultory form that he was entered in a horses-in-training sale during his three-year-old season. Still, there was some hint of potential, for Burch and Mellon did not accept the final bid of $76,000 and retained the horse. Soon thereafter he launched a series of stakes scores on the grass, climaxed by his outdueling the great colt Damascus to win the 1967 Washington, D.C., International, then the most important international contest in North America. Damascus was Horse of the Year, and Fort Marcy for the first of three times was voted the grass course championship.

Most of Mellon's mares were sent to Kentucky stallions. The mares were boarded at John A. Bell III's

Fort Marcy

Jonabell Farm for some years and later placed under Henry White's supervision at Plum Lane Farm. They routinely were returned to Rokeby to foal. Mellon also patronized some stallions in Virginia, however, and one of them was *Amerigo, sire of Fort Marcy. A willful but high-class English-bred with major wins on both sides of the Atlantic, *Amerigo seemed on the verge of becoming one of the most important sons of the highly influential sire, Nearco, but was felled by an early death.

In 1968 Fort Marcy had to share turf course honors with the versatile Dr. Fager, who made one start on grass. While Rokeby was very staunchly a New York-based operation, Burch took Fort Marcy far afield, to Chicago, Maryland, Florida, New Jersey, and California. The horse loved to fly, and he loved to run, and, as a gelding, he was still far from retirement.

In 1969 Fort Marcy won the Hollywood Park Invitational Turf Handicap and three other stakes but was outvoted for the championship of the division by *Hawaii. Rokeby's champion that year was the brilliant Arts and Letters.

Arts and Letters was technically a homebred, since the owner of a mare at the time of foaling traditionally is designated as the official breeder of the horse. Arts and Letters, however, was purchased in utero, from the 1966 dispersal sale of William du Pont Jr. A fellow Virginia breeder who also had steeplechasers as well as flat runners, du Pont was a kindred spirit insofar as breeding for stamina and giving horses plenty of time to mature.

Arts and Letters' dam was All Beautiful, whose sire, Battlefield, represented the sire line of Fair Play and Man o' War and had been a champion at two as well as a major winner at middle distances. All Beautiful, winner once in two starts, was foaled from Parlo, one of the two champion fillies on the flat that du Pont bred in some forty years. Parlo, the 1954 champion three-year-old filly, was by the Hyperion horse *Heliopolis, two-time leading sire in America.

The sire of Arts and Letters was the great, unbeaten *Ribot, champion of Europe in 1955 and 1956. *Ribot had been imported to Darby Dan Farm in Kentucky. (See Chapter 4.) In foal to *Ribot, All Beautiful was purchased by Mellon on a bid by Burch for $175,000. She was seven years old at the time. The du Pont dispersal, held at Timonium, Maryland, was spectacular for its day. A total of fifty-one lots grossed $2,401,300, averaging $47,084. John

du Pont set a record in bidding $235,000 for his father's other champion filly, Berlo. All Beautiful's dam, Parlo, was fifteen at the time and barren, and she sold for $68,000.

The du Pont sale was held on Valentine's Day. That spring All Beautiful foaled a light chestnut colt to whom Mellon gave one of the many lovely names he bestowed on his horses. Arts and Letters had a campaign redolent of his heritage: debut at Saratoga, maiden race victory in his fourth start, non-winners-of-two victory, nice effort when fourth in the Pimlico-

Mellon with Quadrangle

Laurel Futurity. Add *Ribot to a colt's escutcheon and this makes for a newly turned three-year-old with the suggestion of a future. Arts and Letters, indeed, blossomed quickly enough at three in the winter of 1969 to win the mile and one-eighth Everglades Stakes. He then placed to defending champion Top Knight in both the Flamingo and Florida Derby to earn the next step toward the Kentucky Derby.

Whereas Quadrangle had won the Wood

Memorial in New York as his final classics prep, Burch sent Arts and Letters to Kentucky for the Blue Grass Stakes. In a remarkable graduation from promising to serious, he dominated throughout and won by fifteen lengths. The great rider Bill Shoemaker had become his regular rider, but Shoe was seriously injured in a paddock accident in California a few days before the Kentucky Derby, and Braulio Baeza was called in to ride Arts and Letters.

In both the Derby and Preakness, Arts and Letters battled bravely against the unbeaten Majestic Prince before going down by a neck in the Derby and a head in the Preakness. As he was wont to do, Burch threw in the Metropolitan between the Preakness and Belmont, and Arts and Letters won from the older champion Nodouble, getting eighteen pounds of actual weight for the spring test at a mile.

For the Belmont the momentum had shifted. Majestic Prince had enough wear and tear that his trainer, John Longden, let it be known that he would prefer to rest the colt and leave the Triple Crown bid alone. Arts and Letters cruised through the stretch to win Mellon's second Belmont — and Burch's third — by five and a half lengths over the erstwhile unbeaten Majestic Prince, whose owner had insisted he run.

As was true in Quadrangle's Belmont triumph, the Rokeby winner spelled the end of a Triple Crown bid. In this case, however, the Belmont winner's form afterward was so spectacular as to wrest championship honors from the vanquished Derby-Preakness winner. After the Belmont, Arts and Letters continued along a trail of perfection, laced with tradition. He won the Jim Dandy, Travers, Woodward, and Jockey Club Gold Cup. Whereas he had gotten eighteen pounds from Nodouble at a mile in May, by autumn he was at scale weights for the Gold Cup at two miles and beat the older horse by fourteen lengths. Arts and Letters was Horse of the Year, as well as three-year-old champion.

The following year Arts and Letters won the Grey Lag Handicap among three races before a bowed tendon ended his career. He had won eleven races in twenty-three starts and earned $632,404. Arts and Letters was sent to stud in Kentucky, at Greentree Stud. He was not outstanding at stud, as his thirty stakes winners represented only 6 percent of his foals. There were great moments, however, for his offspring. He got a classic winner in 1980 Preakness victor Codex, for Tartan Farms, and while Arts and Letters sired only two stakes winners bred by Mellon, one of them was the enduring and high-class Winter's Tale, about whom more will be told later. As a broodmare sire, Arts and Letters's fifty-eight stakes winners through mid-2004 included 1996 Breeders' Cup Classic winner Alphabet Soup.

The disappointment of Arts and Letters' abruptly ended campaign was assuaged by the continuing distinction of Fort Marcy and by the delicious promise of a colt dashing across the baize fields of love and lore in England — Mill Reef. In 1970 Fort Marcy, then six, added a second Washington, D.C., International as well as winning the United Nations Handicap, Man o' War Stakes, Dixie, and Bowling Green. He not only earned a third championship in the grass division but was voted Horse of the Year on the *Daily Racing Form* poll. (In that final year before the Eclipse Award format unified championship voting, Personality was Horse of the Year in the other major poll, that of the Thoroughbred Racing Associations.)

Fort Marcy, later elected to the National Museum of Racing's Hall of Fame (as was Arts and Letters), won twenty-one of seventy-five starts and earned $1,109,791 before settling in for a long and honored retirement at Rokeby.

The emerging young star of 1970, Mill Reef was a moderate-sized, racy, dark number descending from the long-ago purchase of *Red Ray, bought on Mellon's behalf by early trainer Jim Ryan. Unraced *Red Ray, by the great sire Hyperion, cost $50,400 when purchased in England for Mellon in 1949 from the dispersal of Lord Portal's stud.

Illustrating the highs and lows of breeding and racing, the same *Red Ray who became the ancestress of his greatest horse also was a victim in one of the bleakest days Mellon ever experienced at Rokeby. *Red Ray died of colic on the same day her yearling broke his leg in the field. Ryan happened to be staying with Mellon at the time. Surviving *Red Ray was her foal of that year, 1953, a filly by Count Fleet whom Mellon named Virginia Water. "We didn't race her because she had kind of suspicious-looking ankles," Mellon recalled years later, "and, anyway, she was all we had left from the family."

Virginia Water foaled Berkeley Springs, a classics-placed Rokeby runner by another Virginia-based sire, Preakness winner Hasty Road. Berkeley Springs won the Cheveley Park with a stunning rally in 1965 and the next year vindicated that suggestion of classic potential. She was beaten a neck in the One Thousand Guineas by *Glad Rags II, raced by Mrs. James P. Mills of Virginia's Hickory Tree Farm. Berkeley Springs also finished second, to *Valoris, in the mile and a half Epsom Oaks, filly counterpart to the Epsom Derby.

In 1962 Virginia Water, whom Mellon recalled as a "beautiful gray mare," foaled a filly by the two-time leading sire *Princequillo, a noted source of stamina who stood at Claiborne. The filly was named Milan Mill. The same year Milan Mill was foaled, Captain Harry Guggenheim brought out one of the last of the brilliant sons of *Nasrullah. This colt, Never Bend, was the champion two-year-old colt that season, and he continued as a brilliant runner at three, although failing to win any of the classics. Bred to Never Bend in 1967, Milan Mill (who had failed to win in her only start) foaled Mill Reef in Virginia the following spring. The colt's pedigree was free of inbreeding for five generations.

By the time Mill Reef approached the Derby, the doubts of Never Bend as a source of staying capacity overcame the reputation of his Nearco blood to many observers.

"At stud, he had sired too many precocious youngsters, and too many tearaways, to encourage confidence in any of his sons having the durability, manageable temperament, and staying power for a demanding 1 1/2-mile test like the Derby," commented the *Bloodstock Breeders' Review*. Mellon, nonetheless, had agreed to a plan that Mill Reef be sent to the English division, then in the hands of young Ian Balding.

Balding was stationed at Kingsclere, the historic English yard where John Porter had trained the unbeaten *Ormonde and other stars of the nineteenth century. Balding was also an heir to the sequence of training

Mellon set into motion many years before.

"I trained with Ivor Anthony right up until the war," Mellon recalled in 1989. "Mrs. Hastings (who owned the yard) had a son, Peter (Hastings-Bass), and he took over after the war. When he became ill with cancer, I asked Ian Balding to come in and help him. Then, Ian married Peter's daughter. So, I've really been training with the same family since 1935." (This historic Kingsclere was later sold to Canadian Bud McDougald, who became the landlord for Balding's continued occupation of the Kingsclere yard and house. After McDougald's death Mrs. McDougald continued the arrangement.)

Over the years Mellon had considerable success in sending or breeding horses abroad, the major flat runners including Silly Season (by Tom Fool), Glint of Gold, Diamond Shoal, and Forest Flower, in addition to Secret Step and Berkeley Springs. In England Mellon used black silks, with a large golden cross pattern, whereas he had gray-and-yellow silks in this country.

By and large he relied on his trainers to make estimates as to which horses might do better, on pedigree, conformation, and rate of development, one place or the other, and he came to feel it was largely a matter of luck:

Arts and Letters

153

"In the old days I used to think a horse that was fairly backward, and maybe a little bit too big for his age, would have more of a chance in England because of the type of going and because they give them more of a chance over there, anyway," Mellon recalled. "And, I suppose a horse who had a fairly long pastern might be a little bit better on the grass over there, on the soft going. But, in general, I think it is just a matter of luck…

"Maybe some of the ones we have sent to England would have done better here and vice versa, but we'll never know. I have often wondered how Mill Reef would have stood up over here. In a way, he was a slightly delicate horse." Mill Reef is the name of a club that Mellon frequented while in Antigua, and it was one of a number of names he had at the ready for use on horses. Since the dam was Milan "Mill," he used the name on her foal.

At two, Mill Reef came out a winner at Salisbury, then dashed home in the Coventry Stakes at the Royal Ascot meeting. He was beaten in France by My Swallow in the Prix Robert Papin but followed with victories in the Gimcrack, Imperial, and Dewhurst stakes. A Gimcrack victory consigns the winning owner to make a speech on racing matters to the Gimcrack Club, and Mellon took the occasion to show his poetic talents by composing an acrostic that spelled out "Gimcrack Spirit."

Balding produced Mill Reef the following spring for a victory in the Greenham Stakes, but the colt was then second in the classic Two Thousand Guineas to a rising star, the unbeaten Brigadier Gerard, with My Swallow third. Brigadier Gerard eventually won the mile and a half King George VI and Queen Elizabeth Stakes, but in the spring of his three-year-old form he was not aimed at the Derby at that testing distance. This seminal race was, of course, Mill Reef's target.

Both Mellon's and Balding's memories of Derby Day 1971 featured clogged traffic and late arrival. Balding, whose presence was of more necessity than Mellon's in the physical playing out of saddling and getting a horse to the post, had to put in a vigorous cross-country run on foot that probably made him thankful to have kept himself fit as an amateur jump jockey and avid cricket player.

Mill Reef, with regular rider Geoff Lewis aboard, negotiated all the unique turns and hills and cambers of Epsom, the descent at Tattenham Corner included, and had the race at his mercy with a quarter-mile remaining. The consummate professional, he was in position to draw out gradually, requiring no great rallying burst. Mill Reef won by two lengths from Linden Tree.

Mrs. Mellon gleefully recalled that her husband was "white as a sheet" as he undertook that most engulfing scenario of leading his winner into the small enclosure after the Epsom Derby, surrounded by morning coats and top hats, ladies and lords.

"All I could say was 'They said he couldn't stay,'" Mellon told us years later. "Everybody had said, 'Oh, he's just a miler,' and 'no horse by Never Bend has

MILL REEF				
Never Bend, 1960	*Nasrullah, 1940	Nearco, 1935	Pharos (Phalaris—Scapa Flow)	
			Nogara (Havresac II—Catnip)	
		Mumtaz Begum, 1932	*Blenheim II (Blandford—Malva)	
			Mumtaz Mahal (The Tetrarch—Lady Josephine)	
	Lalun, 1952	*Djeddah, 1945	Djebel (Toubillon—Loika)	
			Djezima (Asterus—Heldifann)	
		Be Faithful, 1942	Bimelech (Black Toney—*La Troienne)	
			Bloodroot (Blue Larkspur—*Knockaney Bridge)	
Milan Mill, 1962	*Princequillo, 1940	Prince Rose, 1928	Rose Prince (*Prince Palatine—Eglantine)	
			Indolence (Gay Crusader—Barrier)	
		*Cosquilla, 1933	*Papyrus (Tracery—Miss Matty)	
			Quick Thought (White Eagle—Mindful)	
	Virginia Water, 1953	Count Fleet, 1940	Reigh Count (*Sunreigh—*Contessina)	
			Quickly (Haste—Stephanie)	
		*Red Ray, 1947	Hyperion (Gainsborough—Selene)	
			Infra Red (Ethnarch—Black Ray)	

ever won at more than a mile.' "

In his book Mellon gave additional insight into his reaction to so singular a moment: "...after the Epsom Derby I was exhausted but still euphoric, and I felt somehow disembodied, as if I were watching myself in the movies."

Mill Reef went from strength to strength, adding victories in three more of the most important races of Europe: England's Eclipse Stakes and King George VI and Queen Elizabeth Stakes and, finally, France's climactic Prix de l'Arc de Triomphe. All three found him defeating older horses.

Mill Reef's greatness was universally acclaimed, and the famous jockey Freddy Head, rider of Arc de Triomphe runner-up Pistol Packer, stated after the Arc that the winner was "the best horse I have ever seen."

Back in North America during Mill Reef's glorious campaign of 1971, the championship sequence was buttressed by Run the Gantlet, who added yet another Rokeby triumph in the Washington, D.C., International within a month of the Arc. For the three-year-old Run the Gantlet, the International concluded a streak of five consecutive stakes victories, and he followed his senior Rokeby mate Fort Marcy as champion turf horse.

Run the Gantlet was sired by *Ribot's champion son Tom Rolfe and was out of a mare purchased at the yearling sales. Over the years Mellon was circumspect about purchasing horses in the yearling market, but when he made a move on that chessboard, it was not made with the pawns. Run the Gantlet was out of one of Mellon's spectacular yearling buys, First Feather, by First Landing. First Feather was known as "the Quill filly" when Mellon paid a yearling-filly record price of $90,000 for her at Saratoga in 1964. Her dam, the *Princequillo mare Quill, was a champion runner and wonderful producer. In addition to Run the Gantlet, First Feather foaled the stakes winners Head of the River, Music of Time, and Lightning Leap, and produce of her daughters included the grade I grass stakes winner Dance of Life. (When Mellon paid $90,000 for "the Quill filly," he upped the yearling filly record of $83,000 he had paid at Saratoga in 1962 for a filly by Swaps and out of the dam of Kentucky Derby winner *Tomy Lee. That one he named Golden Gorse. The name was the nom de plume of a writer whose works included the tales of Moorland Mousie, a pony of the lovely English hunting country around Exmoor — an obvious tug on the Mellon heart. Alas, Golden Gorse the filly failed to win, as did her only foal.)

Run the Gantlet got thirty-nine stakes winners (9 percent), of which the best were for breeders other than Mellon. They included the French group I winner and North American grass female champion April Run, whose 1982 victory in the Washington, D.C., International matched the 1981 victory of the Run the Gantlet colt Providential. The sire's major winners in Europe included Commanche Run and Ardross, the latter of whom was so revered at the historic Warren Place training yard of Henry Cecil that a rose garden there was named in his honor.

Many horses with the all-conquering aura of a Mill Reef are retired after their three-year-old seasons to cash in on a lucrative book of mares and avoid the economic diminution inherent in a potentially disappointing four-year-old campaign. Mill Reef's owner, however, decided to keep his great colt in training at four. In 1972 Mill Reef reappeared to win the Prix Ganay in dashing style, but while he got home in front in the Coronation Cup (an older horse's companion to the Derby at Epsom), he was unimpressive in the latter. Soon, a virus overtook him, and Mellon, the most sporting of owners, was forced to declare his colt from an anticipated meeting with the still unbeaten Brigadier Gerard in the Eclipse Stakes.

Balding then set his sights on returning Mill Reef for another Arc, and the colt was proceeding nicely. Then one morning the champion's rider heard the sickening crack of a fracture as Mill Reef was dashing about his training up on the Watership Down portion of the Kingsclere training gallops. Balding looked on in shock and confusion as the rider began a desperate effort to pull the colt to a halt. In such cases, plans for this race or that are immediately replaced by a panicky desire merely to save the life of the animal. In an excruciating fellowship with his training brethren of the past, Balding perhaps at that moment was situated not far from where John Porter had been when he first heard the great *Ormonde making the noise of a roarer; as was recorded in Porter's autobiography, "I was dumfounded. The idea that the horse I almost worshipped was afflicted with wind infirmity distressed me in a way I cannot describe."

Balding's plight was worse. We are not talking

about a trans-Atlantic connection to Mr. Mellon in which he would have to say, "I think it best we stop on him, and put him to stud." The grave and present danger was of having to say to that wonderful gentleman that a worshipped horse of this era was injured so badly that "we can't save him."

Balding got the horse back to his stables at Kingsclere. A triangular piece had been broken off the cannon bone, and the sesamoids had been shattered.

"I'll never forget that," Mellon recalled. "I was in New York, and I got this call about 7:30 in the morning. It was Ian, and he told me what had happened, and naturally I thought, 'Broken leg. That's the end of it.' "

Back at Kingsclere there was no acquiescence, only an urgent attempt to utilize current technology. After the horse was stabilized, a recreation room at

Kingsclere was turned into a makeshift hospital. Balding's experience pointed toward insisting that surgery be done on the spot.

"It was probably the only good decision I made," he told us some fifteen years later. "So often, a horse will survive the surgery but then hurt himself during the recovery. I wanted him to see familiar surroundings and be attended by familiar people."

The surgeon Jim Roberts was called in. A surgery table of sorts was constructed, but the anesthetist warned Mill Reef's handlers that any miscue in pulling the horse onto the table would inevitably cause him to collapse, and nothing could likely be done. It was veterinary medicine's counterpart of a jump shot from the corner with 0.3 seconds remaining — but this was not a game.

All day Dr. Roberts toiled in the surgical repair, an achievement not dissociated to his later employment as head of the C. Mahlon Kline center at the New Bolton Center of the University of Pennsylvania. Mill Reef came through the ordeal and was fitted with a leg-length cast.

Great racehorses tend to be smart horses, and Mill Reef astounded his handlers by figuring out how to lean onto the hay bales stacked around him, stick out the injured limb, and slide down into a relaxed, recumbent position. Once the colt's survival was assured, Mellon syndicated him for two million pounds, but rather than take the most lucrative route and stand him in America, he sent him to England's National Stud in Newmarket.

Having stood up for the line of Never Bend (aided by another Never Bend race horse and stallion, Riverman), Mill Reef authored a distinguished record in the stud as well. The Aga Khan became one who espoused his belief in the "Never Bend" sire line, and he bred one of the pivotal elements of that lineage in Mill Reef's grandson Darshaan. American horsemen know of Darshaan as the sire of our 1993 Horse of the Year, Kotashaan. Also, Mill Reef's son Doyoun, an English Two Thousand Guineas winner for the Aga Khan, is the sire of the champion American turf horse Daylami.

Mill Reef was England's leading sire in 1978 and 1987. In both of those years, he was represented by Epsom Derby winners in his

Mill Reef with Mellon

own likeness, Shirley Heights and then Reference Point. Mill Reef sired sixty-two stakes winners (16 percent). They also included Lashkari, the Aga Khan's winner of the first Breeders' Cup Turf in 1984.

Mill Reef's Shirley Heights won both the English and Irish derbies in 1978, and Mill Reef's Acamas won the French Derby that same year. Later, Shirley Heights, as a stallion at the Royal Stud in Sandringham, sired his own Derby winners when his son Slip Anchor won the Epsom Derby and the aforementioned Darshaan won the French Derby.

Mellon bred no Epsom Derby winners by Mill Reef, but he did have some success with the great horse. Mill Reef sired Mellon's Glint of Gold (out of Crown Treasure, by Graustark), a wandering European with whom Balding won the Italian Derby, the group I Grand Prix de Saint-Cloud and Grand Prix de Paris in France, and the group I Preis von Europa and Grosser Preis von Baden in Germany. Glint of Gold also was second in the Epsom Derby and English St. Leger. His full brother, Diamond Shoal, also won the Grand Prix de Saint-Cloud and Grosser Preis von Baden and added the group I Gran Premio di Milano in Italy.

Mill Reef died in 1986. As a broodmare sire, he has been represented by 136 stakes winners as of mid-2004.

Returning to the run of champions for Rokeby in the late 1960s and early 1970s, Key to the Mint completed the sequence in 1972, the same year Mill Reef was struggling for survival. Key to the Mint was a powerful, huge colt by the *Ribot stallion Graustark and out of Key Bridge. Nevertheless, Burch recognized that he was a colt who needed action, and he ran him ten times at two, from June 23 through November 26. Like Quadrangle, Key to the Mint graduated into the stakes-winner category in his final start at two, the platform for that status being the Remsen Stakes. At three Key to the Mint won the Derby Trial and Withers, both at a mile, skipped the Derby, placed in the Preakness, and was fourth in the Belmont. In the second half of the year, he took over his division and eventually outvoted the early star, Riva Ridge, as champion three-year-old of 1972. He reeled off victories over older horses in the Brooklyn, Whitney, and Woodward and also won the Travers Stakes. At four Key to the Mint won the Excelsior and historic Suburban Handicap, and he was retired with fourteen wins in twenty-nine starts and earnings of $576,016.

Mellon sent Key to the Mint to Greentree Stud, where Arts and Letters also stood. Key to the Mint sired forty-eight stakes winners (8 percent), a good record but not statistically in the top class. So often, good sires tend somehow to do more for outside breeders than for their own breeder, and Mellon bred only two stakes winners by Key to the Mint. These, however, included one of the best horses Mellon ever had, Java Gold, as well as Gold and Ivory, yet another with a career similar to that of Glint of Gold and Diamond Shoal. Gold and Ivory (out of the Sir Ivor mare Ivory Wand) won the Preis von Europa and the Preis von Baden and the Gran Premio del Jockey Club at Milan as well as the group II Royal Lodge Stakes in England.

For others Key to the Mint sired champions Jewel Princess and Plugged Nickle and Broodmare of the Year Kamar. The stallion died in 1996, and his 114 stakes winners through mid-2004 as a distinguished broodmare sire include the international star and multi-millionaire Swain and champion American fillies Inside Information and Soaring Softly.

After Key to the Mint there were no classic winners or champions in the next several years of Rokeby foal crops, but successes continued. Key Bridge produced two other nice stakes winners in Key to the Kingdom, the only Mellon-bred stakes winner by eight-time sire Bold Ruler, and Key to Content, by *Forli.

One of the more important acquisitions Mellon had made in the 1960s was still paying dividends in the later 1970s, and the lasting ramifications would be numerous. In 1966 the vaunted Bieber-Jacobs Stable was selling down its stock, and a most appealing lot included in a consignment at Saratoga was the broodmare prospect Admiring. This filly had become the first stakes winner sired by the international stallion Hail to Reason when she won the 1964 Arlington-Washington Lassie. Moreover, she was out of the high-class stakes winner Searching, representing one of the verdant strands of the legacy of the mare *La Troienne. Searching already had foaled the champion Affectionately and the Futurity Stakes-winning filly Priceless Gem.

From the standpoint of Bieber-Jacobs Stable and the sale company, Fasig-Tipton, the good news was that two of the biggest players in the world at the time, Mellon and Charles W. Engelhard, both wanted Admiring. The bad news from those viewpoints of commerce was that Engelhard and Mellon decided to go partners rather than bid against each other. Nevertheless, the weighty firm of Mellon & Engelhard had to pay a record price of the time for a horse sold at auction, eventually bidding $310,000 to acquire Admiring. They alternated ownership of the first few foals of the mare, and then after Engelhard's death in 1971, Mellon became sole owner.

The 1973 Rokeby crop included Glowing Tribute, a Graustark filly from Admiring. Glowing Tribute won the Diana Handicap as well as two runnings of the Sheepshead Bay Handicap. She had nine wins in twenty-four starts and earnings of $230,819.

In 1976, when Glowing Tribute was a three-year-old, Mellon faced a difficult decision: "Elliott Burch was a very good trainer and very meticulous, but he had an unfortunate illness and had to be out for about six months. I had to do something quickly." He turned from one Hall of Fame trainer to another, Mack Miller, who had established a successful public stable in the five years since the death of his earlier boss, Engelhard's Cragwood Stable. "I was impressed (earlier) that people like Mr. Mellon, or Mr. George Widener, or Mr. John Hay Whitney always made a point to speak to country boys like me," Miller said years later. Mellon "is one of the role models you have in life, and he certainly epitomizes what an owner should be...I admire him for all the marvelous things he has done for America."

Miller, a friend of Burch's, was convinced to join Rokeby as private trainer, and another long and successful relationship was established. Until his retirement and Mellon's dispersals of the majority of his horses, they were friends as well as boss and employee. Mellon had great empathy for trainers, recognizing that their day-to-day intimate connection to every detail of the stable placed a burden on them that made the emotions of setbacks and injuries perhaps even stronger than his own reaction to such vagaries. Mellon's trainers praised him for standing back and letting them do their job, with what was best for the horse in the long run the primary criteria for decision making.

Mellon explained that he never had time to study bloodlines enough to regard himself as a pedigree expert, so he relied on professional horsemen to suggest matings for his mares. For the most part he stuck with his own trainers and the managers of farms where he boarded horses when they were being bred, rather than reach out to many private pedigree advisers. Burch, and his father, Preston, had served Rokeby well with a reliance on conformation perhaps even more important than pedigree, although Mellon always dealt in high-level stallions. Henry White and Miller both stress that during their tenure with Mellon he let them do the job of filling out suggested matings and then relatively rarely made major changes. Neither of these hardboot horsemen, tried and true in the game, strayed much from the idea of good physical match-ups and quality, to ponder esoteric nicks, dosage, etc. "It was largely luck," White said of their success, although it was also important that, despite his love of stamina, Mellon had long understood the importance of speed in sires.

One occasion White recalled in which Mellon did make some major change in plans was when "he looked down the list and said, 'You only have two mares going to Key to the Mint. We need some Key to the Mint fillies.' So, we changed things around and bred, I think, nine mares to Key to the Mint. Well, we got eight colts and only one filly, but that filly was Key Witness."

Key Witness was a 1982 foal from Summer Guest, who in turn had been one of the astute purchases Mellon and Burch had made. A daughter of Native Charger, from the stakes-winning *Heliopolis mare Cee Zee, Summer Guest was a $40,000 Saratoga yearling. She flirted with championship status by winning eight major stakes from two through five, including the time-honored Coaching Club American Oaks, Alabama, and Spinster, and a pair of handicaps over males.

Summer Guest foaled no stakes winners, but Key Witness was stakes-placed. Key Witness later provided a major dividend from Mellon's insistence on breeding for Key to the Mint fillies when she foaled the Suburban Handicap winner Key Contender as well as Mellon's classy Blushing Groom filly You'd Be Surprised. The latter won the grade I Top Flight Handicap and John A. Morris Handicap as well as other stakes for Mellon and Miller late in their affiliation, in the mid-1990s.

Another of the more gratifying horses Miller trained for Mellon came along early in their connection. Winter's Tale was a foal of 1976 by the Rokeby champion Arts and Letters and out of Christmas

Wind. Burch purchased the dam for $45,000 in 1968, in the first rendition of the Canadian Thoroughbred Horse Society yearling sale at which E.P. Taylor's Windfields Farm consigned yearlings with announced reserves. (Windfields-bred Nijinsky II, the 1970 English Triple Crown winner, came from that same sale.) Christmas Wind was by the Canadian Horse of the Year Nearctic, who had hiked Windfields to the top of the international scene by siring Northern Dancer.

Winter's Tale was a gelded four-year-old before he lived up to the hopes his pedigree might have wrought. That year he won three of the most important races in New York in the Marlboro Cup, Suburban Handicap, and Brooklyn Handicap, as well as the Nassau County Handicap. A back injury then plagued him for two years, but since he was a gelding, Miller continued to work with him. Finally, at seven, Winter's Tale had his soundness and confidence back, and he won a second Suburban and a second Nassau County. He won fourteen races from twenty-nine starts and earned $888,900 before joining Fort Marcy as an honored pensioner at Rokeby. Winter's Tale's dam, Christmas Wind, also foaled Arabia, she in turn the dam of Red Ransom. A son of Roberto, Red Ransom was a brilliant two-year-old for Rokeby in 1989 but was injured before winning any stakes. He was sent to stud at the Vinery and has proven an important stallion, as the sire of distaff grass champion Perfect Sting, and English Oaks winner Casual Look.

Winter's Tale had a total of three winning races in the series long revered as the New York Handicap Triple Crown, to wit, the Metropolitan, Suburban, and Brooklyn Handicaps. In 1984 Mellon swept all three of those with one of the relatively few horses that Miller purchased for him. That was the Chieftain colt Fit to Fight whom Miller bought at Saratoga for $175,000. If the trainer had any uneasiness about buying a horse from boyhood chums (Robert Courtney and Robert Congleton), Fit to Fight's sweep of the Metropolitan, Suburban, and Brooklyn in 1984 assuaged them. The horse joined Whisk Broom II, Tom Fool, and Kelso as only the fourth horse to sweep the series. (Some years later, we posed to Mellon the question of how Fit to Fight figured in the affections of one whose career had been so involved with breeding his own. "Naturally, I'd rather have a homebred," he replied, "but, oh no, I have nothing against Fit to Fight. We bought a couple of colts this year, and if one of them turns out to be a Mill Reef or an Arts and Letters, I'll be just as glad.")

The aforementioned Java Gold came to Miller's stable as a two-year-old of 1986. Miller wintered at Aiken, South Carolina, using stables Engelhard gave him, adjacent to the Aiken Training Center. Java Gold was not officially a champion, but Miller looked back over his career after retirement and called him "as good a horse as I have ever trained." The son of Key to the Mint was out of Javamine, who had been bred by Mrs. Engelhard and was by Nijinsky II. A horse of Nijinsky II's quality would obviously speak to the soul of a Paul Mellon, inasmuch as the son of Northern Dancer had preceded Mill Reef by one season and was the only English Triple Crown winner since 1935, soon after Mellon left Cambridge. Javamine, who raced for Engelhard, was out of a Tim Tam mare who was a half sister to the New York Filly Triple Crown winner Dark Mirage.

Java Gold was a promising two-year-old who won the Remsen Stakes, but he came into full flower after the classics had already been run in 1987. He defeated Derby-Preakness winner Alysheba in a sloppy renewal of the Travers Stakes and defeated older runners in the Whitney and the Marlboro Cup. Java Gold might well have wrested the three-year-old championship from Alysheba had the Rokeby runner won his final race of the year, the Jockey Club Gold Cup, but he was second to Creme Fraiche and came out of it with a fractured coffin bone. Miller tried to bring him back at four, but he never started again and went to stud at William S. Farish's Lane's End Farm in Kentucky. Java Gold made little mark at stud.

One of the highlights of Mellon's career was the Saratoga meeting of 1987. In addition to Java Gold's winning the Whitney and Travers, another of the traditional features at the old track was delivered into the Rokeby column when Crusader Sword (Damascus—Copernica, by Nijinsky II) won the top juvenile race, the Hopeful Stakes.

In the same crop of Java Gold, the foal crop of 1984, came another Rokeby star, Mellon's last European classic winner, Forest Flower. This elegant little chestnut filly was a testament to the willingness of Mellon and his advisory team to reach out to horses that showed speed but that did not fit the general mold of classic breeding and demonstrated stamina.

Forest Flower was sired by Green Forest, a son of sprinter Shecky Green. A product of the Bell family's Jonabell Farm, Green Forest had been the champion two-year-old of France and then won the French Two Thousand Guineas. He was established as a classic winner champion, but at a mile. The dam of Forest Flower was Leap Lively, another stakes winner by Nijinsky II. A champion at two for Ian Balding, Forest Flower came back at three to win the Irish One Thousand Guineas and convince her trainer that she was his most brilliant horse since Mill Reef. Back trouble limited Forest Flower thereafter.

By the 1990s Mellon had had a long and glorious run, winning a succession of the world's most important races, several of them more than once. He had never placed the Kentucky Derby at the top of his list of ambitions and certainly was not the sort of owner who would countenance a trainer's rushing a horse just to try to make him into a Derby colt. Nevertheless, it was a coveted race, and Miller, as a Kentuckian, certainly liked the thought although he had spent most of his career training for owners who did not make it their top priority.

Owner and trainer first had thoughts of Louisville

in the fall of 1992, when Sea Hero won the Champagne Stakes, one of New York's top two-year-old races. Virtually any Champagne winner is seen as at least a potential Derby type. Sea Hero was by the Danzig horse Polish Navy, a Phipps family homebred who had won the Woodward and Champagne but had been sold to Japan after a few years at Claiborne Farm. Sea Hero's dam was Glowing Tribute, the daughter of the Engelhard-Mellon acquisition, Admiring, that had bridged the Burch and Miller training eras. Glowing Tribute had become a wonderful producer, and Sea Hero was her fifth stakes winners. The others included two by Northern Dancer, i.e., Hero's Honor and Wild Applause. The latter, winner of the Diana and Comely, had added to the productivity of the female family as the dam of Rokeby's only winner of the historic Futurity Stakes, Eastern Echo.

(Mellon bred none of Northern Dancer's best horses, but using the great international sire had its rewards. One of them was the unusual horse Topsider, whom Mellon bred in partnership with James B. Moseley. Topsider was by Northern Dancer, sire of three Epsom Derby winners, and was out of Moseley's staying mare Drumtop, by Round Table.

Glowing Tribute

Yet, Topsider became a crack sprinter, and one of his stakes-winning sons was Mellon's homebred sprinter England Expects [out of Victoria Cross, a typically elegant bit of naming.] Returning to anticipated genetic form, Topsider also sired distaff champion North Sider, and his sixty-three stakes winners included incidence of stayers abroad; his son Salse sired the English St. Leger winner Classic Cliché.)

Once Miller decided that Sea Hero might be one to aim at the Derby, he sent him to Florida for racing rather than to Aiken for a more casual winter. Sea Hero did not do well in Florida. He was returned to

Aiken, the Derby idea off the table. In the late winter, though, Sea Hero flourished so well in the brisk Carolina air that Miller asked Mellon's approval to unfold again the map to Churchill Downs. There was not much time, but Sea Hero ran encouragingly in the Blue Grass Stakes and so the great adventure of roses was on again.

Jockey Jerry Bailey brought Sea Hero diving through to an opening on the rail in the upper stretch, and as Mellon and Miller cheered like a pair of Cambridge undergrads, Sea Hero drew out to defeat eighteen others by two and a half lengths. The Derby might not have been the greatest goal of his life, but it certainly gave Paul Mellon one of his most gratifying moments. Moreover, despite the elderly gentleman's lifetime of distinctions, it increased his fame among the sporting public manyfold, as a Derby win is wont to do. For his part, Miller was cheered the following morning when he entered his little church in the town of Versailles, Kentucky.

Ironically, the previous year Mellon had been prompted by a serious battle with cancer — plus being in his eighties — to reduce his bloodstock holdings significantly. In various stages at Keeneland, New York, and Tattersalls in England, Rokeby had sold an aggregate of sixty-three horses for $9,522,624, averaging $151,153. Mellon retained a few mares.

"In March (1993), I was worried about your survival; you looked through death's door," Mrs. Mellon remarked to her husband as he was being interviewed in the autumn of that year. "Then two months later, you won the Kentucky Derby!"

Mellon joined John W. Galbreath as the only owners of Kentucky and Epsom Derby winners. He was,

Champions Bred

American Way
1948 steeplechaser

Arts and Letters
1969 champion three-year-old colt
1969 champion handicap male
1969 Horse of the Year

Fort Marcy
1967 champion grass horse
1968 champion grass horse
1970 champion grass horse
1970 champion handicap male
1970 Horse of the Year

Run the Gantlet
1971 champion grass horse

Key to the Mint
1972 champion three-year-old colt

In Europe:

Secret Step
1962 English champion sprinter
1963 English champion sprinter

Mill Reef
1971 English champion three-year-old colt
1971 French champion three-year-old colt
1971 European Horse of the Year
1972 French champion older horse

Glint of Gold
1981 German champion three-year-old colt
1981 Italina champion three-year-old colt
1982 German champion older horse

Diamond Shoal
1983 English champion older horse

Forest Flower
1986 English champion two-year-old filly
1987 Irish champion filly

and remains, singular in having raced winners of those two races plus the Prix de l'Arc de Triomphe. Mellon liked to quip that he also had an Italian Derby to his credit, although recognizing that race does not fit with the others in international acclaim. That all of these races were won with homebreds added to the pleasure.

Sea Hero failed in the Preakness and Belmont stakes, but qualified for the $1-million bonus Chrysler then offered on a points-total basis for horses that ran in all three Triple Crown races. Mellon donated the bonus to the Grayson-Jockey Club Research Foundation, which funds equine research, with the request that the foundation double match the total. Some two years later this endowment drive was complete, so that Mellon's gift of $1 million grew to $3 million.

Sea Hero won infrequently, but his timing was superb. In addition to the Champagne and Kentucky Derby, he had one more big race in him, and it came out on Travers Day. Mellon placed considerable store in winning big races again after having already scored in one, and Sea Hero caused some welcome shuffling of the trophy cases back in Virginia when he added a fifth Travers Stakes for Rokeby. (He was preceded by Quadrangle, Arts and Letters, Key to the Mint, and Java Gold.) Mellon commissioned two statues of Sea Hero, one for Rokeby and another for Saratoga, and he personally unveiled the likeness that was placed in the walking ring of the old track. (He also had given a statue of Secretariat to the National Museum of Racing, with an endowment to maintain that one as well as the version displayed in the paddock at Belmont Park.)

BARBARA D. LIVINGSTON

Sea Hero

Several years later Mellon told us that his latest test for cancer had its best result in years. "If I had thought I would be in as good a shape today as I am, I would not have had the dispersal," he said. In the meantime he had continued to breed from a small retained band of mares and also had Miller get him some "store-bought weanlings" so he had something to look out on in the Rokeby pastures.

In 1994 Miller said he planned to retire the following year. Mellon decided not to try to develop a connection with another trainer, and curtailed his stable. Nevertheless, the quality of the mares he retained was such that after his death, eleven mares were sold by his estate at the Keeneland November sale for $4,884,000, an average of $444,000. The Rokeby influence was felt that year when Silverbulletday was named champion three-year-old filly, following her juvenile filly Eclipse Award of 1998. She was foaled from Rokeby Rose, a Mellon-bred that had been sold at an earlier dispersal.

Paul Mellon died at ninety-one on February 1, 1999. He was survived by Mrs. Mellon and by a son and a daughter from his earlier marriage. He had received many honors, both in racing and in the outside world, including the National Medal of Arts and National Medal of Arts and Humanities, presented by Presidents Reagan and Clinton, respectively. He also was recipient of the Yale Medal and had been made an Honorary Knight Commander of the Most

Excellent Order of the British Empire.

On the Turf he had served as vice chairman of The Jockey Club and director of the Grayson-Jockey Club Research Foundation, as well as having key leadership and support positions in the National Museum of Racing and National Steeplechase and Hunt Association. He received an Eclipse Award of Merit, as well as two other Eclipse Awards as owner and breeder, and was also honored by the American Association of Equine Practitioners and the Association of Racing Commissioners International. He was one of five individuals designated as Exemplar of Racing by the National Museum of Racing and Hall of Fame.

Mellon's estate left generous sums to many organizations, including some of his favorite causes in racing. Among them were both the Grayson-Jockey Club Research Foundation and National Museum of Racing, to which he had given generously during his lifetime and to each of which his executors provided $2.5 million after his death. Grayson-Jockey Club, with Mrs. Mellon's blessing, created the Rokeby Circle, a designation given in his honor to all donors of $10,000 or more in a given year. Research facilities at the University of Kentucky and Newmarket also were recipients of major gifts. Mellon's will designated up to $5 million for care of retired racehorses, the specific recipient to be determined by his executors, and eventually the Thoroughbred Retirement Foundation received that amount.

Another high honor had come to Mellon in 1975, when he was named the testimonial dinner guest of the Thoroughbred Club of America. That evening he read an original poem, which expressed a sentiment that, upon his death twenty-four years later, fluttered back to the memories of many who had admired him:

> *The Day my final race is run*
> *And, win or lose, the sinking sun*
> *Tells me it's time to quit the track*
> *And Gracefully hang up my tack,*
> *I'll thank the Lord the life I've led*
> *Was always near a Thoroughbred.*

Stakes Winners Bred by Paul Mellon

American Way (stp), b.g. 1942, Gino—Sunchance, by Chance Shot

Genancoke (stp), gr.g. 1942, Gino—Makista, by Viviani

Secret Thread, ch.f. 1948, Psychic Bid—Top Gem, by Flag Pole

Cup Man, b.c. 1952, Djeddah—Tap Day, by Bull Lea

Golden Land, br.c. 1952, Bull Lea—The Golden Girl, by Hyperion

Pardala, b.f. 1953, Pardal—*Double Deal II, by Straight Deal

Retaliation, b.c. 1953, *Princequillo—Flying Fortress, by Fair Copy

Land of the Free (stp), ch.g. 1954, County Delight—Starry Banner, by Stardust

Midsummer Night, ch.c. 1957, Djeddah—Night Sound, by *Mahmoud (SW in England)

Red Tide (stp), ch.g. 1957, County Delight—Red Sea, by Fairway (SW in England)

Drinny's Double (stp), br.g. 1958, John Constable—Donomore, by Crack Brigade (SW in England)

Uncle Percy, b.g. 1958, Djeddah—Tap Day, by Bull Lea

Owl, b.c. 1959, Our Babu—Night Sound, by *Mahmoud (SW in England)

Secret Step, ro.f. 1959, Native Dancer—Tap Day, by Bull Lea (SW in England)

Third Martini, b.c. 1959, Hasty Road—The Golden Girl, by Hyperion

Early to Rise (stp), b.g. 1960, *Ribot—Night Sound, by *Mahmoud (SW in England)

Morris Dancer, b.g. 1961, County Delight—Polyanthus, by Polynesian (SW in England/France)

Quadrangle, b.c. 1961, Cohoes—Tap Day, by Bull Lea

Cloudy Symbol, ch.f. 1962, Alcide—Veritas, by Discovery (SW in England)

Green Garden, ch.g. 1962, County Delight—*Sundial II, by Hyperion (SW in France)

Silly Season, b.c. 1962, Tom Fool—*Double Deal II, by Straight Deal (SW in England)

Berkeley Springs, b.f. 1963, Hasty Road—Virginia Water, by Count Fleet (SW in England)

Pass the Brandy, b.g. 1963, Round Table—Far Pacific, by Polynesian

Red Rumour, b.c. 1963, Hail to Reason—Red Sea, by Fairway (SW in England)

Sea Castle, b.g. 1963, Summer Tan—Coral Bell, by *Khaled

Fort Marcy, b.g. 1964, *Amerigo—Key Bridge, by *Princequillo

Green Glade, ch.f. 1964, Correspondent—Polyanthus, by Polynesian

Aldie, ro.g. 1966, Goose Creek—Elegy, by Our Babu (SW in England/Belgium)

Arts and Letters, ch.c. 1966, *Ribot—All Beautiful, by Battlefield

Dream Magic (stp), ch.g. 1967, Alcide—Magic Singing, by *Alibhai

Russian Bank, b.c. 1967, *Herbager—Russian Roulette, by *Nasrullah (SW in England)

Farewell Party, b.c. 1968, Porterhouse—Hour of Parting, by Native Dancer

Mill Reef, b.c. 1968, Never Bend—Milan Mill, by *Princequillo (SW in England/France)

Red Reef, ro.g. 1968, Goose Creek—Red Sea, by Fairway (SW in Belgium)

Run the Gantlet, b.c. 1968, Tom Rolfe—First Feather, by First Landing

Head of the River, ch.c. 1969, Crewman—First Feather, by First Landing

Key to the Mint, b.c. 1969, Graustark—Key Bridge, by *Princequillo

Branford Court, b.c. 1970, Quadrangle—Blue Bannder, by War Admiral

Key to the Kingdom, dk.b/br.c. 1970, Bold Ruler—Key Bridge, by *Princequillo

Timeless Moment, ch.c. 1970, Damascus—Hour of Parting, by Native Dancer

Winds of Thought, b.c. 1970, Hail to Reason—Prides Profile, by Free America

Elegant Tern, b.f. 1971, *Sea-Bird—Prides Profile, by Free America (SW in England)

Pass the Glass, b.c. 1971, Buckpasser—Amerigo Lady, by *Amerigo

One On the Aisle, ro.g. 1972, Stage Door Johnny—Aces Swinging, by Native Dancer

Summertime Promise, b.f. 1972, Nijinsky II—Prides Promise, by Crozier

Drop of a Hat, dk.b.f. 1973, Midsummer Night—Nonsensical II, by Silly Season (SW in Belgium)

Glowing Tribute, b.f. 1973, Graustark—Admiring, by Hail to Reason

Memory Lane, dk.b.f. 1973, Never Bend—Milan Mill, by *Princequillo (SW in England)

Place Dauphine, b.f. 1973, Quadrangle—Chez Elle, by Mongo

Music of Time, ch.g. 1974, Northern Dancer—First Feather, by First Landing

Singleton, gr.g. 1974, Dewan—Secret Symbol, by Relko

Stir the Embers, b.c. 1974, Run the Gantlet—Gently Dreaming, by Indian Hemp

Winter's Tale, b.g. 1976, Arts and Letters—Christmas Wind, by Nearctic

Key to Content, b.c. 1977, *Forli—Key Bridge, by *Princequillo

Old Dominion, b.g. 1977, In Reality—Virginia Green, by Nashua (SW in Sweden)

Rokeby Rose, dk.b.f. 1977, Tom Rolfe—Rokeby Venus, by Quadrangle

Glint of Gold, b.c. 1978, Mill Reef—Crown Treasure, by Graustark (SW in Europe)

Golden Bowl, b.f. 1978, *Vaguely Noble—Rose Bowl, by Habitat (SW in England)

Leap Lively, ch.f. 1978, Nijinsky II—Quilloquick, by Graustark (SW in England)

Crusader Castle, ch.c. 1979, The Minstrel—Mille Fleurs, by Jacinto

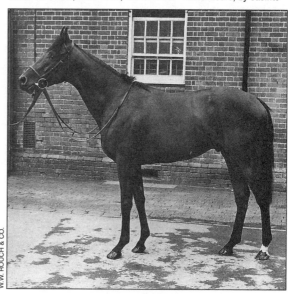

Diamond Shoal

You'd Be Surprised

(SW in Italy)

Diamond Shoal, b.c. 1979, Mill Reef—Crown Treasure, by Graustark (Bred in England; SW in Europe)

Rose Crescent, b.f. 1979, Nijinsky II—Roseliere (Fr), by Misti

Fields of Spring, ch.f. 1980, The Minstrel—Memory Lane, by Never Bend (SW in Germany)

Hero's Honor, b.c. 1980, Northern Dancer—Glowing Tribute, by Graustark

Clare Bridge, ch.f. 1981, Little Current—Gliding By, by Tom Rolfe (SW in Ireland)

Elegant Air, b.c. 1981, Shirley Heights—Elegant Tern, by *Sea-Bird (Bred in England; SW in England/Ireland)

Gold and Ivory, b.c. 1981, Key to the Mint—Ivory Wand, by Sir Ivor (SW in Europe)

Spicy Story, b.c. 1981, Blushing Groom (Fr)—Javamine, by Nijinsky II (SW in England)

Wild Applause, b.f. 1981, Northern Dancer—Glowing Tribute, by Graustark

Bobby Burns, ch.g. 1982, Roberto—Weatherwise, by Arts and Letters

Danger's Hour, ch.c. 1982, Coastal—Lively Living, by Key to the Mint

English Spring, gr.f. 1982, *Grey Dawn II—Spring Is Here, by In Reality (SW in England/Italy)

Fire of Life, b.c. 1982, Nijinsky II—Spark of Life, by Key to the Mint (SW in Italy)

Lightning Leap, ch.c. 1982, Nijinsky II—First Feather, by First Landing

River Spey, b.f. 1982, Mill Reef—Strathspey, by Jimmy Reppin (Bred/SW in England)

Dance of Life, b.c. 1983, Nijinsky II—Spring Is Here, by In Reality

Jack of Clubs, b.c. 1983, Sir Ivor—Colony Club, by Tom Rolfe

Forest Flower, ch.f. 1984, Green Forest—Leap Lively, by Nijinsky II (SW in England/Ireland)

Java Gold, b.c. 1984, Key to the Mint—Javamine, by Nijinsky II

Lights and Music, b.g. 1984, Majestic Light—Two for the Show, by Stage Door Johnny

Queen's Bridge, ch.f. 1984, Spectacular Bid—Colony Club, by Tom Rolfe (SW in Ireland)

Rose Reef, ch.c. 1984, Mill Reef—Rose Bowl, by Habitat (SW in Ireland)

Song of Sixpence, b.g. 1984, The Minstrel—Gliding By, by Tom Rolfe (SW in England)

Crusader Sword, b.c. 1985, Damascus—Copernica, by Nijinsky II

Glowing Honor, dk.b.f. 1985, Seattle Slew—Glowing Tribute, by Graustark

High Browser, b.c. 1985, Tom Rolfe—Browser, by Nijinsky II

Seattle Glow, dk.b.g. 1986, Seattle Slew—Glowing Tribute, by Graustark

Who's to Pay, ch.g. 1986, Believe It—Two for the Show, by Stage Door Johnny

Crystal Spirit (stp), b.g. 1987, Kris—Crown Treasure, by Graustark (Bred/SW in England)

Parting Moment, ch.c. 1987, The Minstrel—Farewell Letter, by Arts and Letters (SW in France/Italy)

Spinning, dk.b.g. 1987, Glint of Gold—Strathspey, by Jimmy Reppin (Bred in Ireland; SW in England)

Eastern Echo, b.c. 1988, Damascus—Wild Applause, by Northern Dancer

Fragrant Hill, b.f. 1988, Shirley Heights—English Spring, by *Grey Dawn II (Bred/SW in England)

Heart of Darkness, br.c. 1988, Glint of Gold—Land of Ivory, by The Minstrel (Bred/SW in England)

Home of the Free, b.g. 1988, Hero's Honor—Radiant, by Foolish Pleasure

Key Contender, ch.c. 1988, Fit to Fight—Key Witness, by Key to the Mint

Kiltartan Cross, b.g. 1988, Hero's Honor—Pride's Crossing, by Riva Ridge

Share the Glory, dk.b.c. 1988, Halo—Secret Sharer, by Secretariat

Blare of Trumpets, b.c. 1989, Fit to Fight—Wild Applause, by Northern Dancer

Rokeby, b.c. 1989, Lomond—Rose Bowl, by Habitat (Bred in England; SW in Germany)

Way of the World, b.f. 1989, Dance of Life—Fairy Tern (GB), by Mill Reef

You'd Be Surprised, b.f. 1989, Blushing Groom (Fr)—Key Witness, by Key to the Mint

England Expects, b.g. 1990, Topsider—Victoria Cross, by Spectacular Bid

Pure Bravado, b.c. 1990, Dancing Brave—Rose Bowl, by Habitat (Bred in England; SW in Austria)

Scoop the Gold, ch.f. 1990, Forty Niner—Leap Lively, by Nijinsky II

Sea Hero, b.c. 1990, Polish Navy—Glowing Tribute, by Graustark

Coronation Cup, b.f. 1991, Chief's Crown—Glowing Tribute, by Graustark

Kerfoot Corner, dk.b.g. 1991, Danzig—Rokeby Rose, by Tom Rolfe

Brigade of Guards (stp), b.g. 1992, Eastern Echo—Battle Drum, by Alydar

Laurie Begone, ch.f. 1998, St. Jovite—Way of the World, by Dance of Life

Paul Mellon & James B. Moseley

Topsider, b.c. 1975, Northern Dancer—Drumtop, by Round Table

War of Words, b.c. 1977, Arts and Letters—Drumtop, by Round Table

Sources: *The Blood-Horse* Archives and The Jockey Club

CHAPTER

11

Rex C. Ellsworth

For two years in the early 1960s, the leading breeder in North America was neither a hard-boot Kentucky horseman nor an industrialist with a passion for his avocation. The leading Thoroughbred breeder of 1962 and 1963 was a cowboy. He was also the leading owner in both years. Rex C. Ellsworth was a soft-spoken Mormon from Safford, Arizona, a tall, lean man with the gaunt look of Abe Lincoln and the horse savvy of Tom Mix. With his lifelong friend, Mesh Tenney, as his trainer, he scaled the heights of the game, all the while remaining a hands-on horse guy who could do his own shoeing and mix his own feed with more personal comfort than posing in the winner's circle in a suit and tie.

Readers of *The Blood-Horse* were introduced to this particular Western American icon soon after he drove a pick-up from California to Lexington in 1933 and came to the attention of editor Joseph A. Estes. Ellsworth was in town to buy some horses. In his biography of Ellsworth's *Swaps* (Eclipse Press, 2002), Barry Irwin recorded that Ellsworth had saved $625 from the $75 a month he was making as foreman of his father's farm.

Over the years readers might have found themselves wondering if it were "the same Rex Ellsworth" whose Swaps defeated the vaunted Eastern star Nashua in the Kentucky Derby, whose purchases ranged from a draft bought from the Aga Khan to exotic South American bloodstock that accounted for Preakness winner Candy Spots, and, eventually,

who wound up winning the Prix de l'Arc de Triomphe with about as fashionable a purchase as you could imagine, to wit, a son of the great European champion *Ribot. Yes, the same Rex Ellsworth accomplished all these achievements.

Swaps was one of the greatest horses of the twentieth century, and to many minds he put Ellsworth and Tenney on the map in 1955; however, the cowboy buddies from Arizona, who had pulled up stakes and moved to California, were proven commodities before then.

That first trip to Lexington for the autumn sales netted six mares and two weanlings, and a year later Ellsworth returned to look for a stallion and purchased Silver Cord. His primary experience in racing had been in the bush with Quarter Horses, but he was well read about Thoroughbreds, and with the toughness of the cattle market in the Depression, he was seeking an additional direction.

The first stakes winner attributed to Ellsworth as a breeder was Arigotal, whom he bred from a one hundred dollar eighteen-year-old mare also picked up on that second trip to Kentucky. The mare was Legotal, and she was in foal to the speed sire Ariel. Arigotal, her foal of 1935, won fourteen races, including the San Mateo Handicap at Bay Meadows for Ellsworth.

Three foals bred by Ellsworth in the foal crop of 1944 became his next stakes winners. The most important of them was by *Beau Pere, a stallion owned by Hollywood movie mogul Louis B. Mayer and one of the stud horses for whom Ellsworth had

Leslie Combs II, Rex Ellsworth, and *My Babu

bargained some cattle to pay for breeding seasons. U Time, his *Beau Pere filly, was out of Heather Time, and therein Ellsworth had struck gold, or silver, early. Heather Time had been purchased from A.E. Silver as a three-year-old in 1939. The price is not known, but given the circumstances of Ellsworth struggling to get started, it must have been modest. Heather Time was in foal at the time, but after producing an Agrarian colt, she was returned to training, winning one race out of sixteen for Ellsworth. As we shall see, she was the dam, second dam, or third dam of Ellsworth homebreds of high distinction.

U Time (*Beau Pere—Heather Time, by Time Maker) was a high-class filly. At two in 1946, she won the Hollywood Lassie and Starlet Stakes, and the next year she added the Hollywood Oaks. The two other stakes winners Ellsworth bred in 1944 were both by Arigotal. The homebred stallion sired a total of fifty foals, and Ellsworth bred all four of his stakes winners. One of them was Roman In, a 1945 foal who defeated Citation in 1:08 2/5 for six furlongs at Golden Gate Fields in 1950. Roman In was out of a mare by the leading sire *Sir Gallahad III, testimony to Ellsworth's having made considerable strides within the first dozen years of his Thoroughbred operation.

"Ellsworth has stated his own rankings of the ele-

ments important in selecting breeding stock," Estes wrote years later. "Racing class comes first, pedigree second, conformation third. (However) he began with conformation, since he couldn't afford the prices demanded by the other categories."

Over the years Ellsworth's own success seemingly tempered his reliance on the original order of importance. In an address to a 1956 meeting of the California Thoroughbred Breeders' Association, Ellsworth remarked, "Conformation is so often ignored that it is the cheapest of all measures. It isn't in books or statistics, readily available to anyone who can read. To me, it wouldn't make any difference, though, how well bred a horse was, anyone would be foolish to take a chance on him if he didn't come up to their specifications or standards of conformation. If he does, and has solid breeding, he's going to be all right."

Two pivotal purchases in Rex Ellsworth's career were *Khaled and Iron Reward. The selection, determination, and the method of dealing illustrate how a cattle rancher with horses was working his way up in the high-stakes game of Thoroughbred racing and breeding.

When he struck that early deal to breed to Mayer's stallions, he said he wanted to swap cattle for breeding rights in *Alibhai, an unraced son of England's classic winner and sire Hyperion. Ellsworth called *Alibhai "that sorrel,"employing the Western stockhorse term for the color chestnut. *Alibhai had trained brilliantly but never raced. He then defied some of the longest odds Thoroughbred breeding can generate by being an unraced horse that became an important stallion.

When Ellsworth was shopping for a stallion in Europe, another son of Hyperion was among those recommended. *Nasrullah, by Nearco, was the first choice, but Ellsworth could not make the deal. The Hyperion horse *Khaled, a nice kind of racehorse owned by the Aga Khan, was the alternative target, but the horse still strained Ellsworth's finances. Ellsworth had put up his ranch as collateral with a Denver bank to borrow $100,000, but that still left him $60,000 short of the asking price. Another loan was in order.

*Khaled had been heard to make the noise associated with being "gone in the wind" but later finished second in the English Two Thousand Guineas. The English disdained "roarers" to the point of sending even great horses off to other countries rather than breed to them, so it is not out of the question that the Aga's and son Prince Aly Khan's minions thought of themselves as taking this rustic American to the cleaners. Ellsworth, though, had had some success with the roarer Silver Cord and did not regard the condition as hereditary. In fact, *Khaled's being a roarer was something of a comfort, for it provided an explanation of why a horse bred to stay had failed to do so.

*Khaled (Hyperion—Eclair, by Ethnarch) had won the important Middle Park Stakes in England at two, and at three he won the St. James's Palace Stakes. Overall, he won five of eight races and earned $29,725. He was badly beaten in the Epsom Derby at a mile and a half and the Champion Stakes at a mile and a quarter. After being purchased by Ellsworth, *Khaled was brought to California following the 1947 season and then, oddly, was returned to training. In January of 1948, he made three starts, winning one, and then after standing at stud that spring he once again was returned to training. He finished last in a race that December, and that finally was the end of his racing.

By the time of *Khaled, Ellsworth and Tenney had moved to the ranch Ellsworth had acquired in Chino, California. As Irwin described the spread, Ellsworth designed the layout "after a cattle operation. Built for functionality and not for show, the simplistic layout allowed a single cowboy to be able to load an entire van of horses by himself, as all of the aisles and pens connected to a chute."

Ellsworth had an electric feed mill constructed, "then scoured the world in search of the finest feed and micronutrients to produce pellets that (Californian veterinarian) Jock Jocoy described as the equine equivalent of the nutritional supplements used by today's elite human athletes. A kelp imported from Norway, as an example, contained more than sixty trace minerals and vitamins."

Ellsworth was convinced California was an excellent place to raise horses. "We have much milder winters (than Kentucky), and we don't have the extreme heat of the summers," he said. "In Kentucky, young horses are kept in the barn during the harsh days of winter and the burning days of summer. They can't do much developing standing in a stall. Further, in the winter, the horse expends energy in body heat rather than growth."

In *Khaled, Ellsworth had tapped into an extraordinary resource. In his second California crop, *Khaled sired forty-five foals — a high number for the time — and when twenty-six of them won at two, he was within a single winner of the record set by *Star Shoot's juveniles of 1916. *Khaled sired sixty-one stakes winners (12 percent), and more than half, thirty-four, were bred by Ellsworth.

Epochal among the *Khaled foals was Swaps, whose dam was Iron Reward. The latter was from the powerful female family of Uncle's Lassie and Betty Derr. Purchased by Mayer, Betty Derr was sent

BOB HOPPER

*Khaled

167

KINETIC

Swaps

to Triple Crown winner War Admiral and foaled Iron Maiden. Mayer later sold Iron Maiden to Californians W.W. (Tiny) Naylor and Ellwood B. Johnston, who bred her to Mayer's *Beau Pere. Iron Maiden's foal was Iron Reward. Iron Maiden was returned to the races, won the Del Mar Handicap, and then was sold to Calumet Farm, for which she foaled 1957 Kentucky Derby winner Iron Liege.

The *Beau Pere—Iron Maiden filly, Iron Reward, was purchased by Ellsworth, who had always liked *Beau Pere. Two years before her dam distinguished herself as the producer of a Derby winner, Iron Reward had earned that distinction as the dam of Swaps.

The 1952 *Khaled—Iron Reward colt whom the Ellsworths named Swaps was a stylish, glamorous chestnut, with a bright eye and so fluid a way of going that he would set world records while seemingly under a hold. A stakes winner at two, he came on at three to win the Santa Anita Derby. It had been thirty-three years since a California-bred, Morvich, had won the Kentucky Derby. Swaps was the real deal but came with a different image. Tenney's cowboy hat contrasted to the styles of trainers at

Churchill Downs, and his tendency to sleep in the barn distinguished him further. Nevertheless, Swaps was so sensational in winning his prep at Churchill that the colt was taken seriously. The favorite was the glamorous Nashua, but Bill Shoemaker put Swaps on the lead and when Nashua came to him, Swaps repulsed the challenge. The Californian won by a length and a half.

Swaps did not contest the Preakness or Belmont but returned West, turning in a stunning performance for a three-year-old. The spring was not even over, and there he was beating older horses, including the previous year's Derby winner, Determine, at a mile and one-sixteenth in the Californian.

By late summer both Nashua and Swaps had done nothing but win after the Derby, and the racing public clamored for a rematch. In a $100,000-added, winner-take-all match race at Washington Park in Chicago, Nashua stole the lead early and ran Swaps to a six and a half-length defeat. The horses had not cooled out before word circulated claiming Swaps had been unsound, Tenney having always had to deal with the colt's tender hoof.

It was game, set, match for Nashua as Horse of the

168

Year for 1955, but Swaps was to reign the next year at four. He won eight of ten races, set or equaled five world records from one mile to a mile and five-eighths, and carried 130 pounds regularly. His wins included the American Handicap, Hollywood Gold Cup, and Washington Park Handicap. A leg fracture then threatened Swaps' life, and he stood for some time in a sling lent by Nashua's trainer, Sunny Jim Fitzsimmons. He survived.

Swaps, the clear-cut Horse of the Year for 1956, had convinced the world of his greatness with a record of nineteen wins in twenty-five starts and earnings of $848,900. Ellsworth by then had a large Thoroughbred operation, but he was not in a position to turn down millions. He sold a half-interest in Swaps to John W. Galbreath for $1 million and proposed that the horse alternate years at stud between the Ellsworth's workmanlike California ranch and Galbreath's white-fenced Darby Dan Farm in Kentucky. Eventually, Mrs. Galbreath bought the other half of Swaps, for another $1 million, and the California hero became a star stallion of the Blue Grass country.

Swaps got the Kentucky Derby winner Chateaugay and other champions Affectionately and Primonetta in early crops, but then his record waned. He sired a total of thirty-five stakes winners (8 percent). The top outfits such as Calumet and the Phipps family bred to him, and he got the dam of Ogden Phipps' champion Numbered Account and second dam of Woodman, so the name Swaps lingers in pedigrees today.

In addition to Swaps, other foals by *Khaled and out of Iron Reward included the stakes winners Like Magic and The Shoe (named for jockey Shoemaker, who rode many major winners for Ellsworth and Tenney).

Two years before Swaps an Ellsworth-bred son of *Khaled had been in the Derby picture. This was Correspondent, who was purchased to race for Mrs. Gordon Guiberson. Correspondent was out of the aforementioned Heather Time, dam of one of the earliest Ellsworth-bred stakes winners, U Time. Correspondent won the Blue Grass Stakes but ran unplaced behind Dark Star and Native Dancer in the 1953 Derby. At four he won the Hollywood Gold Cup. Although generally an unsuccessful stallion, Correspondent sired the 1961 Belmont Stakes winner Sherluck, one more bit of stamina affirmation for the roarer with whom Ellsworth had cast his lot.

The very next year after Swaps won the Derby, Ellsworth and Tenney were back at Churchill Downs with another son of *Khaled, a homebred colt named Terrang. The men were still iconoclasts in the minds of many, but iconoclasts to take seriously. Terrang was foaled from Flying Choice, whom Ellsworth had purchased in 1939 in another sojourn to the Lexington fall sales. That year Ellsworth apparently had taken a liking to Flying Heels, for he bought three yearling fillies by that sire for a total of $5,200. Flying Choice was one of them. Flying Heels, by 1925 Kentucky Derby winner Flying Ebony—Heeltaps, by Ultimus, had won thirteen stakes. Four

			Bayardo (**Bay Ronald**—Galicia)
*Khaled, 1943	Hyperion, 1930	Gainsborough, 1915	*Rosedrop (St. Frusquin—Rosaline)
		Selene, 1919	Chaucer (St. Simon—Canterbury Pilgrim)
			Serenissima (Minoru—Gondolette)
	Eclair, 1930	Ethnarch, 1922	The Tetrarch (Roi Herode—Vahren)
			Karenza (William the Third—Cassinia)
		Black Ray, 1919	Black Jester (**Polymelus**—Absurdity)
			Lady Brilliant (Sundridge—Our Lassie)
SWAPS	*Beau Pere, 1927	Son-in-Law, 1872	Dark Ronald (**Bay Ronald**—Darkie)
			Mother-in-Law (Matchmaker—Be Cannie)
Iron Reward, 1946		Cinna, 1917	**Polymelus** (Cyllene—Maid Marian)
			Baroness La Fleche (Ladas—La Fleche)
	Iron Maiden, 1941	War Admiral, 1934	Man o' War (Fair Play—Mahubah)
			Brushup (Sweep—Annette K.)
		Betty Derr, 1928	*Sir Gallahad III (*Teddy—Plucky Liege)
			Uncle's Lassie (Uncle—Planutess)

KINETIC

**Even in the aftermath of Swaps' Derby victory, trainer Mesh Tenney was
worried about the colt's soundness**

of those, including the 1929 Pimlico Futurity had come at two, and the later ones included two runnings of the Carter Handicap, plus the Manhattan and Hialeah Inaugural handicaps. Flying Heels sired thirteen stakes winners (6 percent), including Flying Choice, who won the $10,000 Starlet at two in 1940.

Terrang, the *Khaled—Flying Choice colt, won the Santa Anita Derby in 1956, as Swaps had in 1955. Terrang faded to twelfth after leading early in the Kentucky Derby, but over five seasons proved a rugged and consistent winner. He was sold to Lawrence Pollock and Roland Bond. Pollock had offered Ellsworth $255,000 for Swaps right after he won the Santa Anita Derby of 1955. Ellsworth turned down the offer but promised to keep the Texan in mind whenever he had another prospect. At the time of Terrang's retirement in 1959, his ten stakes victories at Santa Anita were unmatched at the track. His victory at six at a mile and a quarter over Hillsdale in the Santa Anita Handicap was one vehicle of proof

that the distance was within his powers. Terrang won a dozen California stakes, most of them for Pollock and Bond. He won a total of fifteen races from sixty-six starts and earned $599,285. Terrang got only seven stakes winners (6 percent), but they included the good-class Terlago and Terry's Secret.

El Drag was an Ellsworth-bred by *Khaled and out of a *Beau Pere mare that was a half sister to Terrang. El Drag, whose dam was Beauing, was foaled the year before Swaps. He won the 1955 Hollywood Premiere Handicap at six furlongs and also that year, in a seven-furlong race, set the world record of 1:20.

In the middle 1950s, after the success and the sale of Swaps, Ellsworth returned to his old ports of call in Europe and bought many mares. Some came at auction at Newmarket; some, from the Aga Khan. Among the mares thusly secured was the French mare *Djenne (Djebel—Teza, by Jock II), purchased for four thousand guineas from the Aga Khan at

Newmarket's December sale of 1957.

*Djenne was in foal to Relic, a Kentucky-bred son of War Relic. Relic had flashed impressive speed in winning the Hopeful, Hibiscus, and Bahamas stakes in Citation's crop and had been exported to France. In 1958 *Djenne foaled her Relic colt at Ellsworth's Chino ranch, and the colt was named Olden Times. He was to prove a remarkably versatile racehorse and an enduring stallion.

At two and three Ellsworth's Olden Times won five stakes in California but was hardly in the ranks of the best two-year-olds and three-year-olds. At four in 1962, he was the willing and gallant partner in a riding performance that the great jockey Bill Shoemaker always regarded as one of his best. While hardly known as a stayer, Olden Times was sent out for the mile and three-quarters San Juan Capistrano on the grass at Santa Anita. Shoemaker nursed, finessed, and coerced the distance out of Olden Times and got him home a neck in front of Juanro

while giving him ten pounds. The accomplished turf horse The Axe II was third. Two years later, at six, Olden Times won one of the top handicaps in the East, the one-mile Metropolitan Handicap on dirt, another seal on his versatility. He won a total of thirteen stakes among his seventeen wins in fifty-four starts, earning $603,875.

Olden Times was also the vehicle of one of the more permanent evidences of Ellsworth and Tenney's approach that horses were animals to be trained and managed as such, regardless of whether they were cow ponies or racehorses. At Churchill Downs the week Olden Times won the Churchill Downs Handicap in the spring of 1964 — and when Ellsworth and Tenney had the Derby candidate The Scoundrel in tow — Tenney, in cowboy hat, chaps, and boots, mounted Olden Times on a western saddle and posed for a photo. Tenney dropped the cowboy look in the afternoon at Churchill Downs, however. He was natty in a suit and snap-brim hat the

Rex Ellsworth (left) with Terrang

171

THE BLOOD-HORSE

Olden Times with Tenney aboard

Derby Week afternoon when he joined a group of reporters and mimicked taking notes from Shoemaker after the rider had won the Derby Trial on Derby favorite Hill Rise. Shoe was a rival in that Derby, and Manuel Ycaza was booked to ride The Scoundrel.

The year 1962, when Olden Times won the San Juan Capistrano, marked the first time Ellsworth rose to the top rank of American breeders and owners in earnings. Horses he bred earned a total of $1,678,769, and those he raced earned $1,154,454. In addition to Olden Times his powerful stable of homebreds that year included Candy Spots, winner of the richest race of the day; the consistent handicap horse Prove It; American Derby winner Black Sheep; and Del Mar and Bing Crosby handicaps winner Crazy Kid. This brigade represented the importation of *Khaled, as well as other purchases from Europe to South America.

Prove It was a foal of 1957, by which time *Khaled had been in Ellsworth's hands long enough that he

had producing mares. Prove It was by the accomplished Argentine racer and stallion *Endeavour II and out of the *Khaled mare Time to Khal. This was another lingering influence of the early purchase of Heather Time. Prove It's dam was a moderate runner who placed, and the second dam, Feather Time, was an unraced daughter of Heather Time.

At five in 1962, Prove It finished his campaign with six consecutive victories. They included several California races, among them the summer highlight of the Hollywood Gold Cup. He then was sent to Chicago, which was a sort of second home to the Ellsworth stable, and he won the Benjamin F. Lindheimer Handicap under 130 pounds and the Washington Park Handicap at a mile under 131.

Black Sheep, who emerged in Chicago that summer to defeat the brilliant Ridan in the American Derby, was bred by Ellsworth and sold at two for a reported $50,000 to Chase McCoy's C.R. Mac Stable. He was by *Nigromante—*Dark Justice, by Fair Trial. The dam was among mares acquired in Europe.

*Nigromante was another Ellsworth find and a more spectacular colt by him also emerged that summer in the lanky but powerful form of Candy Spots. A chestnut named because he had some white blotches on his flank and gaskin, Candy Spots was yet another Ellsworth-bred tracing to Heather Time. His dam, Candy Dish, was a *Khaled filly, an Ellsworth homebred who won the Hollywood Oaks in 1956. Candy Dish was from the unraced *Beau Pere mare, Feather Time, second dam of Prove It.

*Beau Pere, *Khaled, and Heather Time were a repeated chorus in Ellsworth's success, and *Nigromante, like the dam of Olden Times, was among the new ingredients found far away. *Nigromante was the best of the 1944 crop in Argentina, where he won six of eight races, all stakes. He was by the champion Embrujo, the latter a son of Congreve, one of the greatest of stallions in South American annals. *Nigromante had made a mark at stud, and Ellsworth imported him in time for the 1958 breeding season. The stallion died only four years later. He got seventeen stakes winners on two continents.

In that summer of 1962, Ellsworth and Tenney had their eyes on the Arlington-Washington Futurity, a new combination of two races that was to supplant the Garden State Stakes as the richest race in the world. Candy Spots, however, was not nominated. Instead, the two were eyeing the race with Space Skates and Big Kim. Then Candy Spots came to hand in his training well enough to show he could dominate that pair, and Shoemaker urged them to ante up the supplemental fee to run in the Arlington-Washington Futurity. Thus, not quite thirty years after buying eight horses for six hundred dollars, cowboy Rex Ellsworth paid $25,000 just to run one horse in one race! As Mormons, he and Tenney did not "gamble," but they certainly "invested" with a lot of heart.

The Ellsworth entry of all three horses was second choice to the leading colt of the East, Never Bend, but Candy Spots and Shoemaker battled him through the stretch and edged out to a half-length margin. The winning purse was $142,250. Never Bend had enough accomplishments back in New York that he was the champion two-year-old of 1962, but by the time they met again, Candy Spots would be the favorite.

A footnote to the 1962 season was an Ellsworth-bred named Crazy Kid. A 1958 foal with an offbeat pedigree (Krakatao—*Marmot, by Mustang), Crazy Kid had been relegated to a $2,750 claiming race at Caliente in Mexico for his first start at three in 1961. He was claimed that day by Paula Hunt. By the summer of 1962, he was in the bigger leagues, at Del Mar, where he won the Bing Crosby Handicap in 1:07 4/5, a world record for six furlongs. Thus, at the end of 1962, the world records for five key distances were held or shared by horses bred by Rex Ellsworth. Swaps held the record for one mile, a mile and one-sixteenth, a mile and five-eighths; Crazy Kid, six furlongs; and El Drag, seven furlongs. Two of Swaps'

OLDEN TIMES				
Relic, 1945	War Relic, 1938	Man o' War, 1917	Fair Play (Hastings—**Fairy Gold**)
			Mahubah (***Rock Sand**—*Merry Token)
		Friar's Carse, 1923	Friar Rock (***Rock Sand**—***Fairy Gold**)
			Problem (Superman—Query)
	Bridal Colors, 1931	Black Toney, 1911	Peter Pan (Commando—*Cinderella)
			Belgravia (Ben Brush—*Bonnie Gal)
		*Vaila, 1911	Fariman (Gallinule—Bellinzona)
			Padilla (Macheath—Padua)
*Djenne, 1950	Djebel, 1937	**Tourbillon**, 1928	***Ksar** (Bruleur—Kizil-Kourgan)
			Durban (*Durbar II— Banshee)
		Loika, 1926	Gay Crusader (Bayardo—Gay Laura)
			Coeur a Coeur (***Teddy**—***Ballantrae**)
	Teza, 1945	Jock II, 1936	Asterus (***Teddy**—Astrella)
			Naic (Gainsborough—Only One)
		Torissima, 1937	**Tourbillon** (***Ksar**—**Durban**)
			Carissima (Clarissimus—Casquetts)

other world records had been broken since 1956.

In 1963 Ellsworth repeated as the leading breeder, with yearly earnings of $1,465,069. Horses he bred made 1,468 starts that year, so that the results meant that his horses averaged about one thousand dollars every time they went to the starting gate. As leading owner a second time, Ellsworth was credited with $1,096,863. Candy Spots was his leading earner, with $604,481. Tenney was leading trainer both years, as well. The trainer had a 25 percent interest in the racing stable, but Ellsworth had full ownership of all horses returned to the breeding operation.

Tenney gave Candy Spots a cross-country preparation for the Kentucky Derby. After an allowance victory and another Santa Anita Derby triumph for the stable, Candy Spots was shipped to Florida, where he won the Florida Derby. The Santa Anita Derby victory saw him being slammed outside at a key moment and perhaps that saved him from falling

over a horrendous pile-up of tumbling horses.

Candy Spots arrived on Kentucky Derby Day unbeaten in his career of six races and was the 3-2 choice over a crack field including Never Bend, who had won the Flamingo Stakes in Florida but had not run in the Florida Derby. Shoemaker received some criticism for his ride, for Candy Spots got into trouble and was checked several times, and he finished third behind the Galbreaths' Swaps colt Chateaugay, with Never Bend second.

In the Preakness, Ellsworth and Tenney got their revenge, Candy Spots tracking the pace comfortably and then blowing by Never Bend and winning by three and a half lengths as Chateaugay came on to be second. Commented Raleigh Burroughs in the volume *American Race Horses of 1963*, "There was very little 'iffing' after the Preakness. Anyone who stood there with his eyes open was convinced that the race had been won by the best horse."

The difference in approach between Tenney and most trainers was framed in a scene after the Preakness. A standard drama of the backstretch is that the post-work or post-race bath is a matter of importance and a bit of urgency, one person holding the horse from the front while another — more often two — sponge the animal from steamy buckets of soapy water, each going about his work quickly, then rinsing, drying with cloth and scraper, and getting the horse under a blanket and walked to cool out. After the Preakness, Candy Spots' groom held Candy Spots and sprayed him with cold water, a process in which he was unaided. If any were wont to criticize the horsemanship of the stable, they would have had to walk over to the directors' room and interrupt a fellow fondling the Woodlawn Vase.

After Candy Spots had picked up the Jersey Derby, he was beaten again by Chateaugay in the Belmont Stakes. Although he later won the American Derby and Arlington Classic in Chicago, he was unable to

Candy Spots with Ellsworth and Tenney

re-establish the leadership status of the three-year-old division he had enjoyed prior to the Derby and after the Preakness. When all the best met in the Travers, outsider Crewman beat Chateaugay, Never Bend, and Candy Spots, and the Derby-Belmont winner wound up the champion.

Candy Spots won the San Pasqual Handicap two years later for his only remaining stakes victory, and he was retired with twelve wins in twenty-two starts and earnings of $824,718.

In 1963, Ellsworth's second year as leading breeder and owner, Candy Spots was backed up by Delhi Maid (*Khaled—*Bodala, by Bois Roussel), who won the Hollywood Oaks and Goose Girl Stakes, and by Dr. Kacy (*Nigromante—*Chanson Folle, by Chanteur II), who won the American Handicap.

The next year, 1964, was the year the versatile Olden Times won the Metropolitan Handicap in an Eastern invasion. Roaming farther afield, Ellsworth scored an international triumph in France's climactic autumn spectacle, the Prix de l' Arc de Triomphe. *Prince Royal II was bred by Charles Wacker III in England and was by the great *Ribot—*Pange, by King's Bench. He was sold at Newmarket for 3,600 guineas as a weanling at Newmarket and raced for an Italian partnership. Ellsworth bought him largely as a sire prospect. As described by Arthur Fitzgerald in his history of the Arc de Triomphe, the lead up to the big race was a comedy of errors that somehow resulted in a victory. Ellsworth bought *Prince Royal II for 143,000 pounds, with the agreement he would

race in the name of Dr. Guido Beradelli one more time, in the Prix Royal-Oak. His trainer instructed jockey Enrico Camici to take the horse back, which resulted in his fighting *Prince Royal II so much in the early going that he had dropped out of contention and finished last.

Ellsworth had arranged in advance that the colt be sent to the French yard of Earnie Fellowes at Chantilly after the Prix Royal-Oak, but Fellowes declined to take him after his dreadful result in that race. Ellsworth then sent him to George Bridgland. Ellsworth wanted Shoemaker to travel to France to ride, but Shoemaker declined on the basis of never having ridden at Longchamp and suggested Roger Poincelet. Despite such pre-race hubbub, *Prince Royal II came on to defeat the Epsom Derby winner Santa Claus by three-quarters of a length. Also in the field were high-class winners *Le Fabuleux, Ragusa, White Label, Nasram, and *Belle Sicambre.

As was true of Swaps, the best of the horses Ellsworth had been breeding were attractive enough to the Kentucky breeders that they eventually went to stud there. In 1965 Prove It, who had stood two years in Kentucky, and the recently retired Olden Times were sent to stand at Gainesway Farm, the Kentucky operation that John R. Gaines was developing into one of the nation's leading farms. Candy Spots joined them the next year, the three being syndicated as a $3.6 million package. Ellsworth retained about half. *Prince Royal II also went to stud in Kentucky.

Of the four, only Olden Times made a satisfactory

CANDY SPOTS			
*Nigromante, 1944	Embrujo, 1936	Congreve, 1924	Copyright (Tracery—Rectify)
			Per Noi (Perrier—My Queen)
		Encore, 1927	Your Majesty (Persimmon—Yours)
			Efilet (Let Fly—Efigie)
	Nigua, 1936	Songe, 1924	Sundari (Sunder—Gourouli)
			Salamanca (Flying Fox—Sakkara)
		Nitouche, 1926	St. Wolf (St. Frusquin—Wolf's Cry)
			Nenette (Polar Star—La Verde)
Candy Dish, 1943	*Khaled, 1943	Hyperion, 1930	Gainsborough (Bayardo—*Rosedrop)
			Selene (Chaucer—Serenissima)
		Eclair, 1930	Ethnarch (The Tetrarch—Karenza)
			Black Ray (Black Jester—Lady Brilliant)
	Feather Time, 1945	*Beau Pere, 1927	Son-in-Law (Dark Ronald—Mother-in-Law)
			Cinna (Polymelus—Baroness La Fleche)
		Heather Time, 1936	Time Maker (The Porter—Dream of Allah)
			Heatherland (Crusader—*Highland Mary)

mark from a statistical basis, although the others had a few highlights. *Prince Royal II got the Santa Anita Derby winner Unconscious, and Candy Spots got the Kentucky Derby runner-up No Le Hace and the nice filly Belle Marie. *Prince Royal II got eight stakes winners (3 percent), Candy Spots nine (4 percent), and Prove It fifteen (4 percent).

Olden Times had a vagabond stud career that saw him standing from time to time at seven different Kentucky farms. Late in that career, in 1982, he was represented by the champion juvenile colt in Roving Boy and, at twenty-four, was the oldest stallion ever to lead the American juvenile sire list. Olden Times sired fifty-nine stakes winners (10 percent), including the high-class Hagley, Dainty Dotsie, Dr. Riddick, Blue Times, and Tilt Up.

The Scoundrel, mentioned above as a Derby colt of 1964, represented another excursion into Italy. He was a son of *Toulouse Lautrec, whom Ellsworth had purchased from the famed Italian breeder Federico Tesio and imported to California in 1958.

*Toulouse Lautrec (Dante—Tokamura, by Navarro) had won the 1953 Gran Premio di Milano and Gran Premio d'Italia, but he was a big horse thought to have weak forelegs, and he had a tendency to become unbalanced when on undulating courses. His exportation by Ellsworth was not seen as a great loss to Italian breeding, but soon futile attempts were being made to recover him. His daughter Feria II won the Italian Oaks and St. Leger Italiano; another daughter, Marguerite Vernaut, was the top Italian juvenile filly of 1959; and *Toulouse Lautrec led the Italian sire list that year. In 1960 Marguerite Vernaut repeated as a champion in Italy and became the first Tesio runner ever at Newmarket, where she won the Champion Stakes.

In breeding The Scoundrel, Ellsworth crossed his Tesio stallion with Aga Khan breeding. The dam of The Scoundrel was *Malekeh, an English-bred by Stardust and out of *Majideh, the *Mahmoud mare who had produced Belmont Stakes winner *Gallant Man. On the surface The Scoundrel does not rank

The Scoundrel

with the best of Ellsworth's homebreds, but he was tantalizingly close. After winning the Haggin Stakes at two in 1963, he was prepped for the classics in Florida. He ran into Northern Dancer and was second to him twice, including the Florida Derby.

In the Kentucky Derby, The Scoundrel was prominent throughout and held on to be third, beaten about three and a half lengths as Northern Dancer edged Hill Rise in record time of 2:00. In the Preakness, The Scoundrel improved to second, splitting Northern Dancer and Hill Rise. Ellsworth then sold The Scoundrel for $500,000 to Kjell Qvale, but the colt soon was beset by tendon trouble and was retired the following year.

Thus, in ten Derby runnings from 1955 through 1964, Ellsworth had run four horses and had a win (Swaps) and two thirds (Candy Spots and The Scoundrel).

From the mid 1960s, however, fortunes waned insofar as his turning out high-class stakes winners consistently. There was still a string of stakes winners, some forty after the crop of The Scoundrel, raising Ellsworth's career total to ninety-seven, but few of the latter ones rose to national notice. *Toulouse Lautrec, for example, eventually had a career mark of twenty-nine stakes winners (9 percent), but did not sustain the quality pattern of his earlier crops.

Ellsworth had maintained a numerically large operation, some ninety in the stable at the time of Candy Spots, for example, but quality had seemed the hallmark for some years. By the 1970s he was leading

Champion Bred

Swaps
1956 champion handicap male
1956 Horse of the Year

national statistics again, but now it was in numbers. He was the leading breeder by races won four consecutive years, 1973–76. In 1974 he was the breeder of the winners of 415 races, marking the first time since John E. Madden's era (1921) that a breeder had surpassed four hundred wins in a single year. Ironically, it was during this time that Ellsworth received ugly publicity when the Society for the Prevention of Cruelty to Animals took control of his remaining horses after "a number were found dead or malnourished on his farm," recalled *The Blood-Horse* of 1997. The horses were later returned to his care, but the image of a great breeder having neglected at one time to provide sufficient funds to care for the stock was difficult to erase.

In his book on Swaps, Irwin noted that "an alliance with notorious Southern California financier C. Arnolt Smith, who went to jail for questionable banking practices, put Ellsworth on a thin financial plane. He never fully recovered." The Kentucky sales scene, seminal in Ellsworth's youthful determination to succeed, also became the stage of controversy. "Ambitious undertakings, such as building a racetrack and sales complex to compete with Keeneland, left him financially vulnerable," Irwin noted. "Nobody ever questioned his ability to develop and select horses. His business acumen, however, proved his failing. Ellsworth was more interested in the next deal then in solidifying his gains."

Ellsworth died at the age of eighty-nine in 1997. He had been an iconoclast as a newcomer, an iconoclast as a leading breeder-owner, and he remained an iconoclast to the end.

Stakes Winners Bred by Rex C. Ellsworth

Dark Air, br.c. 1944, Arigotal—Dark Heroine, by Dark Hero (SW in Mex)

Hubble Bubble, b.f. 1944, Arigotal—Spring Flower, by Hephaistos

U Time, br.f. 1944, *Bere Pere—Heather Time, by Time Maker

Roman In, b.c. 1945, Arigotal—Romanesque, by *Sir Gallahad III

Grey Spook, gr.g. 1946, Arigotal—Mad Silver, by Silver Cord

Flitting Past, br.f. 1949, Dogpatch—Flittingfeet, by Flying Heels

Goose Khal, ch.g. 1949, *Khaled—Goose Hunter, by *Hunters Moon IV

Season's Best, b.f. 1949, *Khaled—Feather Time, by *Bere Pere

Ara Time, b.f. 1950, Pere Time—Ara Cord, by Silver Cord

Correspondent, b.c. 1950, *Khaled—Heather Time, by Time Maker

Fleet Khal, b.f. 1950, *Khaled—Flittingfeet, by Flying Heels

My Heroine, br.f. 1950, Pere Time—Dark Heroine, by Dark Hero

*Se Voya, ch.g. 1950, Mirza II—Eia, by *Ortello (Bred in England)

Aunt Het, br.f. 1951, *Khaled—Feather Time, by *Beau Pere

Chorus Khal, b.f. 1951, *Khaled—Flittingfeet, by Flying Heels

El Drag, br.c. 1951, *Khaled—Beauing, by *Beau Pere

Flight Khal, blk.g. 1951, *Khaled—Flying Choice, by Flying Heels

Heather Khal, b.f. 1951, *Khaled—Heather Time, by Time Maker

Loose Shekels, b.g. 1951, *Khaled—Mad Silver, by Silver Cord

O. U. Kid, gr.f. 1951, Dogpatch—U Nada, by Silver Cord

Tussle Patch, br.c. 1951, Dogpatch—Silver Tussle, by Silver Cord

Bequeath, b.c. 1952, *Khaled—Feather Time, by *Beau Pere

Konsonet, ch.f. 1952, *Khaled—Flying Choice, by Flying Heels

New Trend, br.c. 1952, *Khaled—Pretty Pere, by Flitterpere

Swaps, ch.c. 1952, *Khaled—Iron Reward, by *Bere Pere

Ali Miss, b.f. 1953, Roman In—Ali Litt, by *Alibhai

Candy Dish, b.f. 1953, *Khaled—Feather Time, by *Beau Pere

Like Magic, b.c. 1953, *Khaled—Iron Reward, by *Bere Pere

Terrang, br.c. 1953, *Khaled—Flying Choice, by Flying Heels

Ballet Khal, b.f. 1954, *Khaled—*Peau de Balle, by Tourbillon

California Kid, b.c. 1954, *Khaled—Heather Time, by Time Maker

Straight A., b.g. 1954, *Khaled—Tall Spring, by Arigotal

Midnight Date, b.f. 1955, *Khaled—Feather Time, by *Beau Pere

The Shoe, ch.c. 1955, *Khaled—Iron Reward, by *Beau Pere

El Bandido, b.g. 1957, Tehran—*Rose of Yeroda, by Nearco

Free Copy, b.g. 1957, *Khaled—*Dama II, by Dante

Prove It, b.c. 1957, *Endeavour II—Time to Khal, by *Khaled

Small Bundle, ch.f. 1957, *Khaled—*Miriah, by Mirza II (SW in Mex)

Bushel-n-Peck, br.f. 1958, *Khaled—*Dama II, by Dante

Crazy Kid, br.c. 1958, Krakatao—*Marmot, by Mustang

Gaelic Lad, ch.g. 1958, Swaps—Scottish Miss, by *Alibhai

Light Talk, br.c. 1958, *Khaled—Flitting Past, by Dogpatch

Olden Times, b.c. 1958, Relic—*Djenne, by Djebel

Rodeo Hand, b.c. 1958, *Khaled—*Silver Lass II, by *Bahram

Black Sheep, ch.c. 1959, *Nigromante—*Dark Justice, by Fair Trial

Don't Linger, b.f. 1959, *Nigromante—So Regards, by With Regards

Dr. Kacy, ch.c. 1959, *Nigromante—*Chanson Folle, by Chanteur II

Jerri Dance, b.f. 1959, *Nigromante—Best Form, by *Khaled (SW in Mex)

Wallet Lifter, b.c. 1959, *Khaled—Lismore Liz, by Psychic Bid

Candy Spots, ch.c. 1960, *Nigromante—Candy Dish, by *Khaled

Delhi Maid, dk.b.f. 1960, *Khaled—*Bodala, by Bois Roussel

Going Abroad, dk.b.c. 1960, *Khaled—*But Beautiful, by Tehran

Quest Link, b.c. 1960, *Empire Link—Quest Cap, by *Sullivan

Space Skates, b.c. 1960, *Khaled—Chargeaway, by *Royal Charger

Close By, ch.c. 1961, *Toulouse Lautrec—Best Form, by *Khaled

Take Over, br.g. 1961, *Khaled—*Arusha, by Dante

The Scoundrel, ch.c. 1961, *Toulouse Lautrec—*Malekeh, by Stardust

Unbested, ch.c. 1961, *Nigromante—Khal n Dash, by *Khaled

I Surrender, b.f. 1962, *Toulouse Lautrec—So Regards, by With Regards

Lost Message, b.f. 1962, *Toulouse Lautrec—Khaling U., by *Khaled

Sharp Decline, b.c. 1962, *Khaled—Damp Abbey, by *King's Abbey

Toulore, ch.c. 1962, *Toulouse Lautrec—Molly Maid, by *Khaled

Embassy, b.c. 1963, *Toulouse Lautrec—So Regards, by With Regards

Feather Fan, ch.f. 1963, *Toulouse Lautrec—Glamour Babe, by *Khaled

Vague Image, dk.b/br.c. 1963, *Khaled—Flitting Past, by Dogpatch

Bargain Day, b.c. 1965, Prove It—Special Price, by *Toulouse Lautrec

Prove It Girl, b.f. 1966, Prove It—*Djenne, by Djebel

Tipping Time, b.f. 1966, *Commanding II—Tipping, by *Khaled

Mucho Loco, b.c. 1967, Crazy Kid—Latin Teacher, by *Toulouse Lautrec

Plenty Old, ch.c. 1967, Olden Times—Plenty Baby, *Khaled

Swarming Bee, ch.c. 1967, Dr. Kacy—Swapping Bee, by Swaps

Nobby Dod, b.g. 1968, Nobblns—Dody W., by Lychnus

Old Gypsy, dk.b/br.f. 1968, Olden Times—Gypsy Life, by *Khaled

Specialized, ch.g. 1968, Dr. Kacy—Artist Model, by *Toulouse Lautrec

Dollar Discount, b.c. 1969, *Prince Royal II—Special Price, by *Toulouse Lautrec

Eager Wish, b.g. 1969, Scud—Ambitious Hope, by Prove It

Magic Man, ch.c. 1969, Dr. Kacy—Sham Genie, by *Khaled

Tannyhill, dk.b/br.c. 1969, *Prince Royal II—Best Form, by *Khaled

Button Top, ch.f. 1970, The Shoe—Monissa, by *Nigromante

Pension Plan, dk.b/br.g. 1970, Olden Times—Lo May, by *Khaled

Refusal, b.g. 1970, Dr. Kacy—First Review, by *Toulouse Lautrec

Sweet Medic, b.c. 1970, Dr. Kacy—Honey Bunny, by *Khaled

Chief Spokesman, b.g. 1971, *Prince Royal II—Royal Currage, by Curragh King

Exotic Age, ch.f. 1971, Olden Times—Exotic Jungle, by *Khaled

Final K., ch.c. 1971, Dr. Kacy—Final Gesture, by *Khaled

Madam Go, ch.f. 1971, Durango—Madam Jourdain, by Valka's Boy

Speedy Quick, dk.b/br.c. 1971, Controlling—India Love, by *Khaled

Static Image, ch.g. 1971, Durango—Yasmin, by Sostenuto

Bob's Decision, dk.b/br.f. 1973, Crazy Kid—Lure, by *Toulouse Lautrec

No Ceiling, ch.g. 1973, Durango—Ballad II, by Chatsworth II

Rex Ellsworth, K.C. Ellsworth, and M.A. Tenney

Fun Sale, ch.c. 1963, Negotiation—Melifunua, by *Nigromante

Far to Reach, b.g. 1966, Binary—Queen Venus, by Nahar

Rex and Kim Ellsworth

Fairly Old, dk.b.f. 1983, Olden Times—Fancy Fair, by Prove It

California Jade, ch.c. 1985, Olden Times—Jady, by Durango

Your Hope, b.f. 1985, Mari's Book—Minimize, by Eskimo Prince

No Relaxing, dk.b.f. 1987, Dandie (Arg)—Unwinding, by Tovel

Source: *The Blood-Horse* Archives

12

Ogden Phipps and Ogden Mills Phipps

Ogden Phipps might have encouraged his mother's entry into racing, but his own interest went well beyond any concept of primogeniture. Indeed, though his mother made the first move, Phipps was "heir" only insofar as to share a common family affection for the game. After all, Ogden Phipps registered his own colors less than a decade after his mother and uncle had launched Wheatley Stable.

His mother was Mrs. Henry Carnegie (Gladys) Phipps, who founded Wheatley with her brother, Ogden Livingston Mills (see Volume I of *Legacies of the Turf*). Mrs. Phipps' twin sister, Beatrice Lady Granard, went abroad and continued the partnership in racing that their father had originated with Lord Derby.

Wheatley Stable, launched in the 1920s, was a rather new stable when Ogden Phipps joined the Turf on his own. Thus, he was emotionally involved in the success that found such as Seabiscuit, High Voltage, Misty Morn, and Bold Ruler bred from the Wheatley broodmare band, and he was to have the individual distinctions of a personal stable that would include Buckpasser, Easy Goer, and Personal Ensign.

Ogden Phipps was born in New York City in 1908, was educated at St. Paul's and Harvard, and thereafter began his professional career at the investment firm Smith Barney. In 1932 he adopted the silks of black, with cherry cap, of his grandfather, Ogden Mills, and set about establishing his own stable. Inasmuch as his mother's operation was barely out of the fledgling stage, young Phipps did not have the luxury of standing back and inheriting a ready-made source of bloodstock. He began to stock his own stable with some purchases from William Woodward's Belair Stud, which had reached an exalted level when homebred Gallant Fox won the 1930 Triple Crown.

Phipps also turned to Woodward's advisor, Arthur B. Hancock Sr. and his Claiborne Farm, which has been the wellspring of so many breeding operations. Indeed, the first stakes winner Phipps bred was foaled from a mare bred by Hancock. The mare was White Favor, she by the Claiborne import *Sir Gallahad III and a half sister to the 1927 three-year-old filly champion Nimba. Phipps did lean a bit on his mother's early success, breeding White Favor to her cup horse Diavolo and thus came up with his first homebred stakes winner, White Cockade. That colt won the Youthful Stakes at two in 1935 and added the Withers Stakes the following year.

A few years later White Cockade had segued from "the first" stakes winner bred by Phipps to "the only." As was true of many young men of the time, Phipps' personal ambitions in such matters as Wall Street and the Turf were waylaid by World War II.

"I went to Quonset in the first class," he said in an interview for *The Blood-Horse* many years later. "Then they shipped me to Miami and stranded me in Washington. I never got overseas." To have been a man of military age in a time of military urgency and to have seen no action seems always to rankle, and Phipps was hardly the only one of his age to imply regret at not having had a more active assignment. Nevertheless, he did what was asked with sufficient excellence to rise to the rank of commander and then served in the post-war Naval Reserve.

During World War II, Phipps arranged for a pivotal acquisition in his somewhat dormant bloodstock operation. He was well aware of the excellence of Colonel E.R. Bradley's imported French mare *La Troienne, and, like Bradley, perceived that crossing her blood with that of Man o' War's 1937 Triple Crown win-

Ogden Phipps with the great Buckpasser

ner War Admiral was perhaps a shortcut to the winner's circle. Phipps purchased one of *La Troienne's daughters, Selima Stakes winner Big Hurry, in time to be the breeder of record of her 1944 foal. Between assignments he might well have paused to congratulate himself when a War Admiral filly from another daughter of *La Troienne became America's Horse of the Year under the name of Busher only a year later.

Phipps sent Big Hurry to War Admiral in 1945, and the resulting foal, produced in the first post-war year, was The Admiral, who became the second stakes winner he bred. (He had acquired and raced four other stakes winners between White Cockade and The Admiral.) The Admiral won the Tremont and United States Hotel Stakes at two in 1948. By then, Big Hurry had polished her credentials further as the dam of Californian and Mexican stakes winner Be Fearless and then, far better, Bradley's 1946 champion three-year-old filly, Bridal Flower.

Bradley died in 1946, and his brother John set about disposing a collection of bloodstock excellence that was revered for having been represented by four Kentucky Derby winners and many other distinguished horses. A rare opportunity was thus presented to Thoroughbred breeders willing to step forward.

"I was staying with Jock Whitney (at Saratoga),

when Bob Kleberg (of King Ranch) spoke to me about the Bradley horses," Phipps recalled. "He thought we could buy them as a bunch." Phipps, destined to high distinction as a sportsman and horseman, also stepped forward as one of history's exemplary house guests. "I said I would go in, but I'd rather like to have Jock have a chance at it, too, since I was staying with him. Bob said that was fine. So we proceeded."

The consortium of Phipps, Kleberg, and Whitney (Greentree Stud) bought the majority of the eighty-eight horses in the Bradley holdings. The three-man syndicate retained the stallion Bimelech and sold off drafts to E.S. Moore and Charles S. Howard, while also dealing a full brother to Busher off to Elizabeth Arden Graham.

The real treasures they were interested in, the key broodmares and young stock, were retained and put into categories. As Phipps recalled a pivotal meal, the three went to lunch and celebrated the time-honored decision methodology of the drawing of matches. All won. Phipps acquired five mares, four yearlings, and three sucklings. More than four decades later, he still belied a sense of wonder — and an appreciation of his good fortune — in remarking "every one of the (seven) stakes winners I had last year except one had Bradley blood."

For the most part the verdant Ogden Phipps tale over the remaining half-century-plus was one of productive nurturing of the Bradley stock. The tale involves cluster champions, juveniles, stayers, winners of an archive of the gilded hierarchy of New York's traditional fixtures with an English classic winner thrown in, and sorties into more recently conceived headline events.

At the same time, it should not be implied that Phipps' career was merely a preordained chain of success growing from one decision. There would be the repeated luxuries of matching the best to the best, but some of this came from innovative acquisitions from outside broodmare bands and telling thrusts into the market to acquire compatible jewelry. Sometimes the sire–dam combos seemed obvious, while at other times the breedings seemed off the beaten path. So often they were either the formula for the next major stakes or a building block for one down the road. Of the more recent of his star horses, including Personal Ensign and Seeking the Gold, a number came from astute acquisitions made long after the Bradley purchase.

Phipps respected and relied upon the three generations of the Hancocks, who boarded his horses for much of his career, but he had his own opinions as well, and his interest never wavered. In the final summer of his life, Phipps was ahead of the game: his entire weanling crop had already been named.

Opinion about pedigrees is the stock-in-trade of the bloodstock breeder. In Phipps' case he used another man's opinion as a bargaining chip to exercise his own. Samuel D. Riddle, who owned Man o' War and was owner-breeder of War Admiral, had a liking for the Bradley-bred champion Bimelech. Phipps, as stated above, had ambitions toward the cross of War Admiral on *La Troienne and her daughters. As the breeding seasons of the late 1940s unfolded, he suddenly had a plenitude of *La Troienne blood in his broodmare band, and, funny thing, was part owner of *La Troienne's champion son Bimelech as well.

Phipps knew that Riddle had worked some swaps with Bradley for Bimelech seasons, and Phipps "went to do that, too, as quick as I could. War Admiral was the stallion I wanted most, because he crossed so well with La Troienne."

In addition to his nurturing of the Bradley treasure, he made some other lasting scores as well. From Woodward he acquired the mare Bellicose, by St. Leger winner *Boswell and from champion Vagrancy's dam, Valkyr, by Man o' War. From Bellicose, Phipps bred the stakes winner Torch of War, but more importantly, also bred the Menow filly Caustic. Bred to Claiborne's *Ambiorix, Caustic foaled the major winner Sarcastic, whose son, Vitriolic, was the champion two-year-old colt for Phipps in 1967. This sequence covered about a quarter-century.

Part of the magic of *La Troienne was that her issue had enormous speed and could stay, too. The two foals from Big Hurry who won stakes for Phipps were The Admiral, quick enough to win important stakes at two, and Great Captain, stout enough to win a Saratoga Cup.

With this abundance of *La Troienne blood, Phipps was in the unusual position of having to avoid "too much" of a bloodline others salivated over. Two of Big Hurry's daughters were let go after seeming to have failed as race mares, and they created several rungs of the ladder by which Bieber-Jacobs Stable climbed from a claiming outfit to be the leading breeder in the nation. Bieber-Jacobs (see Volume I, chapter 19) acquired daughters No Fiddling and Searching, and they would become ancestresses of the likes of Regal Gleam, Caerleon, Straight Deal, Affectionately, Priceless Gem, Sea Hero, and on and on.

Phipps was hardly left with no wellspring of his own. Big Hurry also foaled for him the filly Allemande, who was unraced but retained for breeding. (Phipps for many years boarded his mares with Cy White at Elsmeade Farm in Kentucky, until 1957, when White convinced him that his own health was failing and that Phipps might want to move the mares. The mares were transferred to Claiborne, where Phipps' mother had boarded her mares for some thirty years already.)

All three of the stakes winners Phipps bred from Big Hurry were by War Admiral, but the mare had succeeded with other crosses earlier. Be Fearless was by Burgoo King, and Bridal Flower by *Challenger II. Likewise, Phipps sought other influences, too, not always looking only for the top echelon. Allemande was by Counterpoint, the 1951 Horse of the Year but not an outstanding sire, and Allemande's stakes-winning daughter Marking Time was by To Market. A son of Market Wise, To Market was out of the Johnstown mare Pretty Does. To Market was a very good racehorse, numbering both the Arlington Futurity and Washington Park Futurity as well as the Arlington Handicap among his half-dozen stakes wins. He earned $387,325 in the early 1950s. To Market was a solid sire, but again, not a top one.

Allemande's To Market filly Marking Time won the 1966 Acorn Stakes and had two other wins from ten starts to earn $66,522. Bred to homebred champion Buckpasser, she produced Relaxing, who was initially sent overseas because Phipps felt her soundness was tenuous and she might fare better on grass.

When returned to this country as a mature runner, her soundness perhaps thus protected, Relaxing proved a major winner on dirt. She was one of several fillies and mares Phipps owned that competed successfully against males. At five in 1981 Relaxing won the grade I Delaware and Ruffian handicaps against other females, defeated colts in the John B. Campbell and Assault handicaps, and was beaten only a length when third to champion John Henry in the Jockey Club Gold Cup. Relaxing was voted the Eclipse Award for older mare for that season, and she retired with thirteen wins from twenty-eight races and earnings of $589,195. Bred to Calumet Farm's heroic Alydar, Relaxing produced one of the brightest stars in the Phipps history. This was Easy Goer, a glowing chestnut with a fluid stride that belied his short pasterns and less-than-perfect foot. Another Alydar—Relaxing foal was Cadillacing, a grade I winner who in turn foaled Futurity winner Strolling Along and the graded stakes winner Cat Cay, her dam's only filly. To the cover of Danzig, Relaxing foaled another major winner in Easy Now.

Easy Goer seemed stamped for glory from the time trainer Claude "Shug" McGaughey brought him out at Belmont and then went on to old Saratoga. He swept to victories in the Cowdin and Champagne with such aplomb that he was voted the 1988 two-year-old colt championship despite being upset by Is It True in the Breeders' Cup Juvenile.

At three Easy Goer won his first prep and then blazed the mile of the Gotham Stakes in 1:32 2/5, and added the Wood Memorial. He went into the Kentucky Derby the favorite, but just as Phipps' mother had been disappointed in the Derby with a pair of favorites (Bold Ruler and Bold Lad), Easy Goer found a combination he could not handle. The dashing Sunday Silence went to the front in the upper stretch and won handily as Easy Goer rallied through the mud to be second. In the Preakness the pair locked in a duel through the Pimlico stretch that immediately vaulted their rivalry almost into the Affirmed-Alydar class. Sunday Silence prevailed in a photo.

By 1989 Phipps had been winning important New York races for more than fifty years, but he had never won the Belmont Stakes, in many ways the centerpiece of the Turf on that historic circuit. Easy Goer got him that prize, drawing clear of Sunday Silence to win by eight lengths.

Thereafter, Easy Goer continued a tour of the most revered old targets, taking the Whitney (against older horses) and Travers at Saratoga, the Woodward and Jockey Club Gold Cup at Belmont. Pasterns notwithstanding, he had the look of greatness, and he ran to his looks. Such was the impression he created that Easy Goer was 1-2 against his rival Sunday Silence in the Breeders' Cup Classic. In three meetings one colt had won twice, and that form held. The nimble Sunday Silence dashed to the lead and held safe the powerful rush of Easy Goer, winning by a neck. Honors belonged to Sunday Silence.

Both were kept in training at four, but they did not meet again, being stopped in the summer by injuries. Easy Goer added one more time-honored event, the Suburban Handicap, and he was retired to fill a special stall at Claiborne, Bold Ruler and Secretariat having resided there in the past. He won fourteen of twenty races and earned $4,873,770. Easy Goer died young at age eight, but Phipps did breed another important individual by him in My Flag, winner of the 1995 Breeders' Cup Juvenile Fillies and 1996 Coaching Club American Oaks. (My Flag, in turn, is the dam of the

Easy Goer

2002 Breeders' Cup Juvenile Fillies winner and two-year-old filly champion Storm Flag Flying.)

Thus, the acquisition of Big Hurry in the early forties created and renewed its importance over a sixty-year sequence. The history of the Phipps stable is replete with these lengthy legacies of prolonged excellence, which take the storyteller traveling through the decades and then dropping back in time again to pick up another similar tale.

The sequences by which Buckpasser was a star also began with *La Troienne, to whom horses such as Easy Goer and many other Phipps runners were somewhat inbred. Among the daughters of *La Troienne that were not among her five stakes winners but were pearls as broodmares was Businesslike, a 1939 filly by Bradley's champion Blue Larkspur. Businesslike was seven when Bradley died, and she foaled for Phipps the stakes winners Auditing and Busanda. Auditing was by Count Fleet, the 1943 Triple Crown winner whom Phipps regarded as perhaps ranking with or above Citation as the best horse he had ever seen. (This evaluation was described in 1989, so that Phipps was objective enough to place those horses even above his own best.) Busanda was by War Admiral, again representing the cross Phipps was emphasizing.

Busanda was a foal of 1947, and in that same crop came her close relative Striking, she by War Admiral and out of another daughter of *La Troienne named Baby League (earlier the dam of Horse of the Year Busher).

"I was crazy about Striking," Phipps recalled. "She and Busanda ran against each other at two several times. Striking was the better two-year-old" and won the Schuylerville Stakes. Busanda came on later and at three won the historic Alabama. At four and five Busanda defeated colts in the Suburban Handicap and twice won the mile and three-quarters Saratoga Cup against males. Phipps noted that a route-loving distaffer winning that race was not unique to Busanda.

"Oddly enough, fillies won the Saratoga Cup," he recalled in 1989. "In that same period, Walter Jeffords' filly, Snow Goose, won it. Busanda could go all day, but she was stubborn. You couldn't trust her...but when she was right and wanted to run, she was good."

Busanda won ten of sixty-five starts and earned $182,460. She produced three stakes winners, two of them high-class, the last one Buckpasser. First came the Polynesian colt Bureaucracy, who won the National Stallion, Dwyer Handicap, and Providence Stakes and was second to *Gallant Man in the Travers. Then came the Double Jay colt Bupers, who won the Futurity, and finally there was Buckpasser, who was sired by Greentree Stable's great Tom Fool. Phipps had great admiration for Tom Fool and bred to him annually, but Phipps said he had no particular desire to stipulate that War Admiral mares go to the stallion. Phipps also bred the stakes-winning Tom Fool filly Funloving, whose dam, Flitabout, was by Challedon.

Alydar, 1975	Raise a Native, 1961	Native Dancer, 1950	Polynesian (Unbreakable—Black Polly)
			Geisha (Discovery—Miyako)
		Raise You, 1946	Case Ace (*Teddy—Sweetheart)
			Lady Glory (American Flag—Beloved)
	Sweet Tooth, 1965	On-and-On, 1956	*Nasrullah (Nearco—Mumtaz Begum)
			Two Lea (**Bull Lea**—Two Bob)
		Plum Cake, 1958	Ponder (Pensive—Miss Rushin)
			Real Delight (**Bull Lea**—Blue Delight)
EASY GOER	Buckpasser, 1963	Tom Fool, 1949	Menow (*Pharamond II—Alcibiades)
			Gaga (*Bull Dog—Alpoise)
Relaxing, 1976		Busanda, 1947	War Admiral (Man o' War—Brushup)
			Businesslike (Blue Larkspur—***La Troienne**)
	Marking Time, 1963	To Market, 1948	Market Wise (Brokers Tip—On Hand)
			Pretty Does (Johnstown—Creese)
		Allemande, 1955	Counterpoint (Count Fleet—Jabot)
			Big Hurry (Black Toney—***La Troienne**)

In the spring of 1965, trainer Bill Winfrey, who had succeeded Sunny Jim Fitzsimmons as trainer for the Phipps family on the latter's retirement, summoned the venerated *Daily Racing Form* writer Charlie Hatton to his barn. Winfrey, who had trained the mighty Native Dancer for A.G. Vanderbilt, wanted to show off a colt he knew was the type Hatton liked, a strong, yet elegant, bay with black points and a look of class. It was the Tom Fool—Busanda colt whom Phipps had named Buckpasser.

Years later the distinguished artist Richard Stone Reeves still refers to Buckpasser as the handsomest of horses he has studied and portrayed. Reeves recalled in *Legends* (Oxmoor House, 1989) that Buckpasser at two was "a tall, long colt, a bit on the leg then, but he was to furnish out into my ideal. Of all the horses I have painted, I believe Buckpasser comes closest to physical perfection." Reeves' notes had described a colt with "fine-crested neck and clean-cut throatlatch; deep, well-sloped shoulders; deep through the middle, good bone, straight legs; nearly perfect quarters and hind legs…I could go on and on about Buckpasser. He had an intelligent eye, a wide forehead, a finely crafted muzzle. And he was big, almost 17 hands tall. His coat was a rich bay, the color of antique furniture, dark mahogany almost."

This prettily designed vehicle was also one with a turbo-charged V-8 engine, and he hit on all cylinders — although inclined to let the foot off the throttle at awkward moments.

Buckpasser came out in the spring of 1965, winning a maiden race in his second start and quickly moving up to stakes class. He rattled off wins in the National Stallion, Tremont, Sapling, Hopeful, and Arlington-Washington Futurity. Searching's daughter Priceless Gem then gave him a defeat that smarted, a loss to a filly in the Futurity, but Buckpasser rebounded to win the Champagne to clinch the championship.

He had won nine of eleven and earned $568,096, a record for a juvenile.

As the autumn of his two-year-old days gave way to the winter of his classic season, Buckpasser was seen to have a rival in the astounding colt Graustark. Sadly, injuries to both kept them from ever meeting. After barely relenting to get down to business in time to catch Abe's Hope in a betless Flamingo Stakes, Buckpasser suffered a quarter crack. The Triple Crown (then without a winner in nearly twenty years) had beckoned but now was irrelevant.

By then Buckpasser had a new trainer in Eddie Neloy, a loquacious, self-educated combination of philosopher and wise-cracking racetracker. Despite the opportunity to train such a stable, Winfrey concluded that he had reached a time in life that he wanted less pressure, and he retired to live in California.

Neloy had Buckpasser back to the track in the summer, and the colt quickly overwhelmed the leg-up on championship honors that Kauai King had established by winning the Derby and Preakness. In the Arlington Classic, Neloy employed Phipps' Impressive as pacemaker, and at the end of six furlongs he owed that employee a bonus. Impressive got three-quarters in 1:06 4/5. Buckpasser moved in to

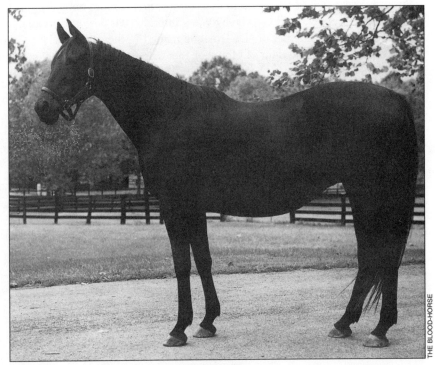

Busanda

THE BLOOD-HORSE

finish the deal, setting a world record of 1:32 3/5 for a mile. Kauai King broke down in the race.

In his debut at three, Buckpasser had dropped an allowance race to Impressive but won his remaining thirteen races that year. After the Arlington Classic these included a glorious run through many of the best New York had to offer, similar to the summer and autumn of Easy Goer two decades later. Buckpasser took the Brooklyn, Travers, Woodward, Lawrence Realization, and Jockey Club Gold Cup (then run at two miles). He spliced into this Gotham menu two more Chicago races, winning the Chicagoan and American Derby. Neither older horses, nor track conditions, nor various distances made much of an impression on Buckpasser. He was the hands down Horse of the Year as well as three-year-old champion colt for 1966. Impressive (*Court Martial—High Voltage, by *Ambiorix), bred in partnership with Phipps' mother, had won six stakes on his own, including the Fall Highweight Handicap, and was voted champion sprinter that same year.

On the last day Buckpasser was three, he came out at four. That is to say the implicit launch of his four-year-old campaign took place on December 31. He was back to sprinting, and in California, winning the Malibu at seven furlongs. After the calendar changed, he won the San Fernando before another quarter crack cost him four months on the sidelines. Neloy brought him back without a prep race, the colt carrying 130 pounds at a mile in the Metropolitan Handicap. Buckpasser won easily.

He had then won fifteen races in succession. This number came under scrutiny nearly thirty years later when Cigar won sixteen in a row. Cigar's streak matched that of Citation and was recognized as a modern American record for major racing. Had Buckpasser not had the loss to Impressive in a betless exhibition (or had the race not been recognized as an official start) in his debut at three, he, too, would have had sixteen in a row — from the 1965 Champagne through the 1967 Metropolitan.

As it was, the official streak stopped at fifteen. Phipps had the sporting inclination to take on an unusual challenge and was considering running Buckpasser in France's Grand Prix de Saint-Cloud. Since the horse never had run on grass, it was deemed prudent to give him a start on that surface at home before hying off across the Atlantic. Buckpasser was assigned a steadying 135 pounds for

the mile and five-eighths Bowling Green, giving eight pounds in his first turf start to defending grass-division champion Assagai.

Phipps was sometimes wrongly thought not to be an accepting loser, because of his tendency to leave the track immediately after seeing his horse beaten in a major stakes. One of his trainers, John Russell, defended him on that score, explaining that but for the ceremonial necessity of appearing in the winner's circle, the shy Mr. Phipps would have preferred to bolt after a victory as well!

The 1967 Bowling Green presented a unique situation, and Phipps that day was dour, but dour as a winner. Buckpasser finished third, beaten for the first time in sixteen months, but his stablemate Poker won the race. "They never saw anybody so sorry about winning a $100,000 race as I was," Phipps noted many years later.

With the winning streak stopped and the French soiree off, Buckpasser came out under 133 pounds for the Suburban Handicap, going a mile and a quarter. George D. Widener's Ring Twice was getting twenty-two pounds and, more to the point, was getting darn close to the finish line all alone before Buckpasser went full throttle for jockey Braulio Baeza and swept up to win by a half-length.

Like Wheatley Stable's Bold Ruler, Buckpasser took up 136 pounds for the Brooklyn, and like the other horse he was beaten, coming out of the race a bit less than 100 percent. This endangered the dream match-up with the scintillating three-year-olds Damascus and Dr. Fager in the Woodward Stakes. As this engulfing meeting approached, Baeza at one point advised against running Buckpasser, but one final work seemed to go all right, and so the race was set. Damascus won by ten lengths, as Buckpasser wore down Dr. Fager for second. It was a matter of full honors to the deserving winner but difficult to accept the margin as a true bill. At any rate, Buckpasser raced no more. He had won twenty-five of thirty-one races and earned $1,462,014. Damascus deserved, and received, Horse of the Year honors, while Buckpasser was voted champion older horse, a champion for the third time in a three-year career.

Bull Hancock saw to the syndication of Buckpasser, which was based on an evaluation of $4.8 million, exceeding the previous record, for Raise a Native, by more than $2 million. Buckpasser did not have a long career at stud, dying at fifteen. He sired thirty-five stakes winners (11 percent), and many of his offspring have proven of lasting importance, not just to Phipps and Claiborne but to other breeders. His son Norcliffe won the Queen's Plate for Colonel Charles Baker and then became a link in the sire line that produced a range from Kentucky Derby winner Lil E. Tee to champion sprinter Groovy. Buckpasser sired Jean-Louis Levesque's La Prevoyante, champion in Canada and the United States as an unbeaten juvenile filly with twelve wins in 1972; C.V. Whitney's Silver Buck, later the sire of Derby-Preakness winner and champion Silver Charm; and Greentree Stable's Buckaroo, later the sire of Derby winner and champion Spend a Buck and North America's leading sire in 1985.

Buckpasser never topped the sire list, but he emerged as an exceptional influence as a broodmare sire, leading that list in this country three or four times, depending on various statistical inclusions. Daughters of Buckpasser had foaled 142 stakes winners through mid-2004. Those stakes winners include such international sires as Miswaki, Woodman, and Seeking the Gold, as well as the Belmont Stakes winner and promising young sire Touch Gold. One of Buckpasser's daughters, Lassie Dear, illustrated his presence beyond one or two generations, she being in the pedigrees of A.P. Indy, Summer Squall, Charismatic, Countess Diana, Old Trieste, and Pulpit.

Another impact of Buckpasser's dam, Busanda, came through her *Tatan filly Navsup. The latter foaled Phipps' Polish Navy (by Danzig), who won the Champagne and Woodward and sired Paul Mellon's Kentucky Derby winner Sea Hero. Still another daughter of Busanda was Oak Cluster, dam of three stakes winners. Also, Harbor View Farm's 1984 juvenile filly champion Outstandingly was out of a granddaughter of Busanda.

Buckpasser was honored in the naming of a large and lovely yacht that Phipps docked in Palm Beach and used as a winter home. On the subject of names, Phipps was masterful. In the case of Busanda and Bupers, Phipps turned to acronyms from Navy days: Bureau of Supplies and Accounts, and Bureau of Personnel. More common was his weaving humor into the combinations of sires and dams with a flair and lilt and without loss of dignity in the process. The ongoing success of families from one generation to the next also was mirrored in the naming.

Thus, a Hasty Road—Flitabout filly became Broadway, whose *Court Martial filly picked up the theater district's nickname of Great White Way; Broadway's Bold Ruler foals included the champion Queen of the Stage, and another from Broadway was Will Hays, named for the arbiter of decency and taste in mass entertainment. The Runyonesque aspects of New York's Broadway were hinted at in the Buckpasser—Broadway filly Con Game, who later teamed with Mr. Prospector to produce Seeking the Gold. Con Game also foaled Stacked Pack. Yet another from Broadway was Reviewer and then the Reviewer—Bank of England colt veered into another

Tom Fool, 1949	Menow, 1935	*Pharamond II, 1925	Phalaris (Polymelus—Bromus)
			Selene (Chaucer—Serenissima)
		Alcibiades, 1927	Supremus (Ultimus—Mandy Hamilton)
			*Regal Roman (Roi Herode—Lady Cicero)
	Gaga, 1942	*Bull Dog, 1927	**Teddy (Ajax—Rondeau)**
			Plucky Liege (Spearmint—Concertina)
		Alpoise, 1937	Equipoise (Pennant—Swinging)
			Laughing Queen (*Sun Briar—Cleopatra)
BUCKPASSER	War Admiral, 1934	Man o' War, 1917	Fair Play (Hastings—*Fairy Gold)
			Mahubah (*Rock Sand—*Merry Token)
		Brushup, 1929	Sweep (Ben Brush—Pink Domino)
Busanda, 1947			Annette K. (Harry of Hereford—Bathing Girl)
	Businesslike, 1939	Blue Larkspur, 1926	Black Servant (Black Toney—Padula)
			Blossom Time (*North Star III—Vaila)
		*La Troienne, 1926	**Teddy (Ajax—Rondeau)**
			Helene de Troie (Helicon—Lady of Pedigree)

aspect of "review" to become Comptroller.

Another foal of Flitabout was Flirtatious, who foaled a pair of Bold Ruler foals delightfully named Henry the Eighth and Good Queen Bess.

A *Nasrullah—Striking filly was named So Chic, who launched a long list of clever names as well as stakes winners. So Chic's *Court Martial filly was Fashion Verdict; her Round Table colt was Beau Brummel; and her *Ribot colt was Dapper Dan.

A Tim Tam—His Duchess filly was So Social, who later foaled Snobbishness and Ward McCallister (named for the keeper of that so social list of four hundred). Fashion Verdict's Gallant Romeo colt was Sartorially Perfect.

The Dr. Fager—Cutting filly was Operating; a Bold Ruler—Sarcastic colt was champion Vitriolic; a Private Account—Grecian Banner filly was the beloved Personal Ensign, she later dam of My Flag, and thence to Storm Flag Flying (by Storm Cat).

Phipps also used other historical connections: *Ribot, named for a painter, sired the homebred St. Leger winner Boucher, named for another painter. The name of ballet's Nijinsky II as a sire tickled Phipps' naming fancy: Dancing Spree (from Blitey), Russian Roubles (from Squander), Pas de Deux, and Dancing All Night. The last-named was vehicle for a family joke when Phipps named the Private Account—Dancing All Night filly Sam's Diary, with granddaughter Samantha in mind. A personal touch also was involved in the name Cadillacing, that being a word coined by Phipps' driver, as in "now you're Cadillacing" in response to things going well.

The year Buckpasser was champion older horse in his injury-shortened campaign of 1967, Phipps had a young double of juvenile champions. His mother's Wheatley Stable had won matching juvenile titles with the Bold Ruler youngsters Bold Lad and Queen Empress in 1964, and the son matched that when the Bold Ruler homebreds Vitriolic and Queen of the Stage were juvenile colt and filly champion of 1967. If sponsorship naming had been in vogue in sports in the 1960s, the juvenile categories might have been known as the Wheatley/Phipps Division: Champions — 1963, filly Castle Forbes; 1964, Bold Lad and Queen Empress; 1966, colt Successor; 1967, Vitriolic and Queen of the Stage.

Vitriolic (Bold Ruler—Sarcastic, by *Ambiorix) won the Arlington-Washington Futurity, Pimlico-Laurel Futurity, Saratoga Special, Futurity Trial and

the Champagne. He did not train on successfully, nor was he a major sire. Vitriolic won nine of twenty-one races and earned $453,558. Queen of the Stage (Bold Ruler—Broadway, by Hasty Road) had true brilliance. She won the first seven races of her career, including some of the prerequisite major prizes for the division in the East, the Frizette and Matron, the Spinaway and the Sorority. She earlier had won the Astoria. Queen of the Stage was unplaced in the Selima and did not get back to the races for nearly a year. She won a few allowance races, but no more stakes, and retired with nine wins in fourteen starts and earnings of $316,515.

One of the yearlings in the Bradley stock acquired by Phipps was not of the *La Troienne brood but was the foundation of the female family with which Phipps bred the sequence of Broadway/Queen of the Stage. The filly was the aforementioned Flitabout, she by Challedon—Bird Flower, by Blue Larkspur. Flitabout was not a stakes winner but was second to Scattered in the 1948 CCA Oaks. In the stud Flitabout produced the Spinaway Stakes winner Flirtatious to the cover of Menow. Sent to Menow's champion son Tom Fool, Flitabout foaled Funloving, who defeated champion Bowl of Flowers in the 1961 Mother Goose Stakes and also won the Black-Eyed Susan Stakes. Funloving also was second in the CCA Oaks, Bowl of Flowers gaining revenge that day.

Flitabout foaled Broadway in 1959, and that Hasty Road filly won the Polly Drummond and placed in the Selima Stakes. To review, Broadway then foaled the champion Queen of the Stage, but it would be Broadway's less-distinguished runners that would be more influential. her other stakes winners were Reviewer, Will Hays, and Great White Way.

Reviewer was a very high-class speedster who won six stakes for Phipps, including the Sapling and Saratoga Special at two in 1968. He was not the soundest of horses, making only thirteen starts from two through four, but he won nine races, including four other stakes. In the Nassau County Handicap he stretched out to a mile and one-eighth so effectively that he set a Belmont Park track record of 1:46 4/5. Reviewer was second in the Metropolitan Handicap and two other good stakes, and he earned a total of $247,223. With this speed and pedigree, he was one of the Bold Rulers selected to stand at Claiborne.

Reviewer got nineteen stakes winners, (10 percent) before dying young, at eleven. They included the great Ruffian, who was bred and raced by Phipps' sister and brother-in-law, Mr. and Mrs. Stuart S. Janney Jr. Another champion filly, William Haggin Perry's Revidere, was by Reviewer. Phipps got only a couple of his lesser stakes winners by Reviewer, but Phipps' son, Ogden Mills Phipps, bred the stakes-winning Reviewer filly Resolver, later the dam of Kentucky Oaks and Spinster winner Dispute. Reviewer is the broodmare sire of forty-three stakes winners, including the successful stallion Mr. Greeley.

The most dramatic success in the stud barn from the current descendants of Flitabout is Seeking the Gold. A son of the Claiborne stallion Mr. Prospector, Seeking the Gold is from Con Game, an allowance winner foaled from Broadway. Con Game, sired by Buckpasser, also foaled the stakes winners Fast Play (by Seattle Slew) and Stacked Pack (by Majestic Light).

A foal of 1985, Phipps' Seeking the Gold won four stakes at three: the Super Derby, Dwyer, Peter Pan, and Swale. He just missed in several more important races, going under by a nose to Claiborne's Mr. Prospector colt Forty Niner in both the Travers and

Seeking the Gold

the Haskell. At the conclusion of his three-year-old campaign, he closed to within a half-length of Horse of the Year Alysheba in the 1988 Breeders' Cup Classic. At four he was narrowly beaten by Proper Reality in the Metropolitan Handicap, with Dancing Spree third. Seeking the Gold won eight of fifteen races and earned $2,307,000. At Claiborne he quickly established himself as one of the world's prominent stallions. Through mid-2004 he has sired fifty-nine stakes winners (9 percent), and they include the international champion Dubai Millennium, winner of the Dubai World Cup and Royal Ascot's Queen Elizabeth II Stakes; the champion Flanders, winner of the 1994 Breeders' Cup Juvenile Fillies; Cash Run, also winner of a Breeders' Cup Juvenile Fillies; Louisiana and Ohio derbys winner and promising young sire Petionville; Florida Derby winner Cape Town, and a variety of other champions and important winners around the world — Seeking the Pearl (Japan and France), Catch the Ring (Canada), Meiner Love (Japan), Secret Savings (Australia), Lujain (England).

For Ogden Phipps, Seeking the Gold sired the champion Heavenly Prize, the Matron Stakes winner Oh What a Windfall, and the grade II winner Country Hideaway.

Heavenly Prize and Oh What a Windfall were both from the Nijinsky II mare Oh What a Dance. Here we meet a family that Phipps acquired privately. Lady Pitt was a Sword Dancer filly who raced for Thomas Eazor and was the champion three-year-old filly of 1966. Bred by the Greathouse family in Kentucky, Lady Pitt offered the quality to win a CCA Oaks and classic breeding of an unusual sort. Her sire, Sword Dancer, was a Horse of the Year and sire of the champion Damascus but statistically was not a major success; the broodmare sire, Whirlaway, was a Triple Crown winner but sire of relatively few important horses. Phipps bought Lady Pitt from Eazor after she had foaled a couple of winners by Buckpasser.

For the Phippses, Lady Pitt produced two stakes winners. Emphasizing the English connotations of her name, Phipps named them Blitey (a nickname for England) and The Liberal Member. The Liberal Member, by Bold Reason, won the Brooklyn Handicap in 1979, the same year his younger half sister Blitey emerged to win three of the major stakes left untouched by Calumet Farm's dominant three-year-old filly champion Davona Dale.

Blitey was sired by Riva Ridge, the Derby-Belmont winner for the Chenery family who stood at Claiborne. She was trained by Angel Penna, as were Relaxing and numerous other stakes horses of that period. At three Blitey won the Maskette, Ballerina, and a division of the Test Stakes. She added the Twilight Tear and Imp stakes at four and had an overall record of eight wins in twenty-three starts and earnings of $297,746. She had no inbreeding through five generations and was an outcross for the Northern Dancer and Mr. Prospector blood that was rising to the top in American breeding.

To the cover of Northern Dancer's great son Nijinsky II, Blitey foaled the versatile Dancing Spree, winner of the 1989 Breeders' Cup Sprint, and Dancing All Night, winner of the 1988 Long Island Handicap. A remarkable story, Dancing Spree had once seemed so inept a prospect that he was almost donated to a university, but he came along to become a high-class and versatile racehorse. Dancing Spree also had won the mile and a quarter Suburban Handicap the year he won the Breeders' Cup Sprint. He was the middle figure in a three-year Suburban run for the stable, for Phipps won the Suburban in 1988 with Personal Flag and in 1990 with Easy Goer.

The dam, Blitey, had more to give. To the cover of Mr. Prospector, she foaled the grade I Hempstead Handicap winner Fantastic Find (subsequently the dam of grade I Acorn and Matron winner Finder's Fee, by Storm Cat of the Northern Dancer line). At the age of eighteen, Blitey foaled her fourth stakes winner, Furlough, sired by Phipps' champion Easy Goer. Furlough won the grade I Ballerina and two other stakes.

Along the way, Blitey also produced another Nijinsky II foal. Although named Oh What a Dance, this filly got to no dances at all, being unraced. Like Broadway's Con Game, though, she would earn pride of place in her family.

Again crossing this family with Mr. Prospector blood via Seeking the Gold, Phipps bred champion Heavenly Prize and stakes winner Oh What a Windfall from Oh What a Dance. At two in 1993, Heavenly Prize won her first race by nine lengths. Shug McGaughey stepped her right up to grade I company, and she won the Frizette by seven. Sent to California, she was third in the Breeders' Cup Juvenile Fillies. At three Heavenly Prize lost her first three races but burst back into the limelight with a

BARBARA D. LIVINGSTON

Heavenly Prize

seven-length victory in the mile and a quarter Alabama Stakes. By this time Ogden Phipps was an elder statesman, so that victories were seen in the context of a career that nature's realities defined as being in the latter chapters. It was his first victory in the Alabama since Busanda's in 1950 — although his mother had won the race more recently, with High Bid in 1959, and his daughter, Cynthia, had won it with Versailles Treaty in 1991. (Sorting out the Phipps family's distinctions in racing is not unlike addressing the Bach gang in music.)

Heavenly Prize added six-length victories in the Gazelle and Beldame, then got to within a neck of the older One Dreamer in the Breeders' Cup Distaff, and was the Eclipse Award winner in the three-year-old filly division for 1994. The following year, Heavenly Prize trained on powerfully, winning the Apple Blossom, Hempstead, Go for Wand, and John A. Morris handicaps, all grade I races. She seemed to have another championship salted away, but then she was beaten by more than thirteen lengths in a muddy Breeders' Cup Distaff at Belmont Park. Not to worry, the winner was the Phipps family home-bred Inside Information, and it was she who was named champion of the older distaff division. Inside

Information was bred in the name of Phipps' son Johann Christian, no, make that Dinny, and came from the family of his grandmother's great mare Grey Flight.

Heavenly Prize won nine of eighteen starts and earned $1,825,940, and she is the dam of the Storm Cat colt Pure Prize, who won the Kentucky Cup Classic in 2002.

The reader is by now well aware that recounting the history of Ogden Phipps' Thoroughbreds involves either a year-by-year recitation or repeating sequences of start-a-long-time-ago-and-bring-a-family-up-to-date, one at a time. Having by and large chosen the latter, we now address the specific detail of the *La Troienne legacy that produced Numbered Account, and much that has come after her. It will be recalled that Striking and Busanda were produced in the same Phipps-bred foal crop. Striking was by War Admiral and out of one of *La Troienne's daughters, Baby League. Striking won the Schuylerville Stakes at two in 1949, before Busanda went on to far greater distinction as a race mare and eventually foaled the great Buckpasser.

The Phipps' naming talents were brought into play by the qualities of Striking's family. Over the years several themes of names developed, one based on the word "League" and taking the word "Striking" in a baseball context. Another naming sequence looked at "Striking" in terms of appearance and led to the aforementioned chic connection, i.e., So Chic, Fashion Verdict, Garden State Stakes winner Beau Brummel, and to the *Ribot colt Dapper Dan, who never won a stakes but placed for Phipps in the 1965 Derby and Preakness and finished fourth in the Belmont. In later years this naming sequence morphed into a flag of success after William S. Farish bought into the family: hence, the Hoist the Flag filly Up the

Flagpole (out of The Garden Club, she by *Herbager—Fashion Verdict). Up the Flagpole foaled seven stakes winners, including Runup the Colors, Allied Flag, as well as one with another naming theme in Prospectors Delite (latterly dam of Horse of the Year Mineshaft among her own five stakes winners).

Striking's "baseball" foals included stakes winners Bases Full, Batter Up, and Hitting Away. The glamour side included Glamour herself, who also foaled the Bowling Green Handicap winner Poker (by Round Table), and another strain of names started with the slyly related Intriguing. Intriguing was by the California-bred champion Swaps, by Hyperion's son *Khaled, and was out of the *Nasrullah mare Glamour. Intriguing was a stakes-placed winner who then foaled the stakes winners Cunning Trick, How Curious, and Numbered Account, all by Buckpasser.

Numbered Account was in Buckpasser's first crop and carried *La Troienne strains top and bottom. Like her sire, she was a big, handsome individual who came on strongly from the beginning. In 1971 she won her first start by ten lengths and moved right into stakes company to win the Fashion Stakes.

Trainer Eddie Neloy died suddenly soon afterward and Phipps was now faced with employing a

Striking

third trainer within a decade. He turned to young Roger Laurin as Neloy's successor. Under Laurin's care, Numbered Account ran the table of most of the best filly races, taking the Fashion, Schuylerville, Spinaway, Matron, Frizette, Selima, and Gardenia, coming up short only in one race. By and large, the farther she went, the more impressive she was, and Numbered Account won the Frizette by seven lengths over Susan's Girl and the Selima by six. Finally, she defeated future champion Susan's Girl again, by two and

Numbered Account

three-quarters lengths, in the Gardenia, and Laurin was emboldened to challenge the colt champion Riva Ridge, trained for the Chenery stable by his father Lucien Laurin. Numbered Account ran well in the Garden State Stakes, but was fourth. She was the clear choice as champion two-year-old filly.

Numbered Account suffered a splint and was compromised at three, although she won the Test and then put in one more brilliant effort to win the Spinster at Keeneland. She won fourteen of twenty-two races in three seasons and earned $607,048.

Numbered Account immediately made a mark as a broodmare, and her family proliferated as well. Not all the foals even of the top mares could be retained without the numbers getting out of hand, and Phipps and Seth Hancock were adroit at selling individuals whose credentials made them exceptional prospects for others. One of these was Numbered Account's full sister, Playmate, who then produced the remarkable stallion Woodman (by Mr. Prospector). Bred by Warner L. Jones Jr. and Ed Cox, Woodman was a $3-million yearling who became a champion at two in Ireland for Robert Sangster. Woodman, who stands at Ashford Stud in Kentucky, has sired American champions Hansel and Timber Country and European champions such as Hector

Protector and Bosra Sham. Like Seeking the Gold, he has been a vehicle of the Phipps influence roaming the racing globe.

Meanwhile, Numbered Account's foals included the Northern Dancer filly Dance Number, bred by Phipps and winner of the grade I Beldame and three other stakes. Dance Number then foaled the Mr. Prospector colt Rhythm, for Ogden Mills (Dinny) Phipps' account. Rhythm was the champion two-year-old of 1989 and won the Travers the following year.

In 1975 Numbered Account was bred to Damascus, Buckpasser's erstwhile conqueror who stood at Claiborne. The following spring she foaled a colt that was named Private Account. Private Account won the Jim Dandy at three and then at four won both of the top Florida winter handicaps of the day, the Widener and Gulfstream Park. He won six of thirteen races and earned $339,396, and the combination of his success, the hint of potential unrealized, and his dam made him a prospect to stand at Claiborne. Private Account sired sixty-one stakes winners (10 percent). For other breeders they included the champion French filly East of the Moon and Wood Memorial winner Private Terms, and for the Phippses they included Dinny's distaff champion

Inside Information, the Suburban and Widener winner Personal Flag, and the prize of prizes, Phipps' unbeaten Personal Ensign.

As Ogden Phipps clocked the racing world, one mare he had spotted was the Argentine star *Dorine, a daughter of the Hyperion stallion Aristophanes (sire of the champion *Forli). "I was pretty fussy" in selecting outside mares, he told us. "One I always wanted, but could never get, until all of a sudden Wayne Murty came to me and said, 'I've got an Argentine mare that might interest you.' I said, 'What?' and he said 'Dorine.' I asked how much they wanted, and he said $125,000. I knew she was the best race filly down there and from one of the best families. She had a sister that was a champion, and She had a brother (*Dorileo) that my cousin (Michael Phipps) bought."

For Ogden Phipps, *Dorine foaled a Hoist the Flag filly named Grecian Banner, a lightly raced filly who was a modest winner. Grecian Banner foaled both Personal Flag and Personal Ensign.

Personal Ensign came to the races late in her two-year-old season in 1986. She was in the first draft of Phipps horses that Shug McGaughey had from the beginning, he having been hired in 1985. After having had one trainer, Fitzsimmons, for some three decades, the Phipps stable had a sequence of outstanding trainers in the interim, each having a number of important horses: Bill Winfrey, Eddie Neloy, Roger Laurin, John W. Russell, and Angel Penna. McGaughey has been at the helm now for twenty years.

Personal Ensign won her debut by twelve and three-quarters lengths and was so impressive that she was 3-10 when stepped up immediately to grade I company for the Frizette, which she gamely won by a head. She was to have a small margin of victory only once again.

After fifty years of major targets that in many cases had been in place since his entry into the game, Phipps had responded enthusiastically to the creation of the Breeders' Cup races in 1984. By 1989 he was speaking in terms of the Breeders' Cup as being with the top of his ambitions. Personal Ensign was set to be flown to Santa Anita, where the Breeders' Cup was held in 1986, but two days before the trip she suffered a career-threatening fracture in a workout. Dr. Larry Bramlage operated, inserting five surgical pins in the left rear leg. While McGaughey has recalled he never regarded the injury as life threat-

ening, Phipps recalls that he thought Bramlage's comment that the filly had a chance to "make it" meant a chance to live. The thought of her racing again seemed not even on the table initially, to either trainer or owner.

Nevertheless, Personal Ensign recovered to such soundness that she was returned to training and a year later McGaughey brought her back ready to win at first asking. She won two allowances, then took the Rare Perfume, and defeated older fillies and mares in a favorite old target, the Beldame. That brief campaign was not enough to wrest the three-year-old filly championship from Sacahuista, but it established such a reputation for Personal Ensign that McGaughey could not find an overnight race for her that would fill at the start of her four-year-old season.

In a bit of training that harked back to Sunny Jim's winning the Widener with Nashua off works alone, McGaughey fine-tuned Personal Ensign with the added pressure of her unbeaten status. She debuted at four in grade I company, in the mile and one-sixteenth Shuvee Handicap, seven months after the Beldame. Personal Ensign took this in stride, winning by daylight.

Next came the grade I Hempstead (renamed for Phipps after his death in 2002), and Personal Ensign floated to the wire seven lovely lengths in front. She added the Molly Pitcher and then was set on a course other Phipps fillies had addressed over history, i.e., running against high-class males. While Busanda, Numbered Account, and Relaxing had been presented with this challenge, none of them risked an unbeaten status. Personal Ensign drove through the slop at old Saratoga to run down Gulch and win the Whitney Stakes by a length and a half.

Winning Colors had become the third filly to win the Kentucky Derby that spring, and when she ran her best race she was a Hall of Fame filly on her own. In the one-mile Maskette, Winning Colors ran a powerful race on the lead, and Personal Ensign was tested severely to wear her down and win by three-quarters of a length. Another visit to the Beldame produced an easier victory, by five and a half lengths over Classic Crown.

Personal Ensign thus was unbeaten in twelve career races. American racing had not had an unbeaten horse through a major career since Colin went fifteen for fifteen in 1907-08. Phipps did not slip her into an easy retirement but allowed McGaughey to point her for the Breeders' Cup Distaff, which was run on a dark, wet, overcast day at Churchill Downs in 1988. Her ancestor, Buckpasser, had cut it close in the Flamingo and Brooklyn some twenty years earlier, but even those victories did not have the drama level of a struggle to retire unbeaten.

Personal Ensign raced so far back and seemingly struggled in the going, causing McGaughey to shake his head and conclude "not today." Phipps said to himself that it had been a mistake to run her.

Then Personal Ensign began to close. Two excellent three-year-old fillies were waging war in front of her, Winning Colors and Goodbye Halo. Under regular rider Randy Romero, Personal Ensign closed relentlessly, but she had about four lengths to make up and they were nearing mid-stretch! Somehow she got there, got there to brighten a gloomy day and win by a nose. She was thirteen for thirteen in her unbeaten career and had earned $1,679,880.

Even the most optimistic sportsman cannot be in the game sixty-five years and fail to grasp that an unbeaten horse is no realistic ambition, but Ogden Phipps and McGaughey had orchestrated that virtual impossibility. In later interviews Phipps seemed to place the 1988 Breeders' Cup Distaff atop his personal memory list — high praise from the owner of Buckpasser and Easy Goer!

Winning Colors was second in the Distaff, and Goodbye Halo was third. Goodbye Halo, a filly of championship quality that never was voted a championship, won the CCA Oaks, Kentucky Oaks, and other important tests for owners Arthur B. Hancock III and Alex Campbell. She was another dividend with a Phipps background for other owners. Goodbye Halo's dam, Pound Foolish, was bred by Dinny Phipps from the Phipps-bred Buckpasser stakes winner Squander, whose own granddam was the Wheatley Stable stakes winner Lady Be Good.

Personal Ensign was not through amazing people. Of her first seven foals, all are winners, three grade I winners, two stakes-placed, and one of those two lesser stakes horses, Our Emblem, has sired a Kentucky Derby-Preakness winner and champion in War Emblem!

Personal Ensign's first foal was Miner's Mark, by Mr. Prospector. Miner's Mark came to hand at three to win the Colin and Jim Dandy, and scored a hairbreadth victory over Colonial Affair in one of those favored Phipps races, the grade I Jockey Club Gold Cup. Had he been wont to brag, Ogden Phipps could have named a horse Oh What a Saturday in memory of that Gold Cup afternoon. The race was on a Belmont card cast as the Breeders' Cup Preview of 1993. That same day, October 16, Phipps won the grade I Frizette with Heavenly Prize and grade III Lawrence Realization with Strolling Along, and Dinny won the grade I Beldame with Dispute. McGaughey was urged by photographers to hold up a full hand to commemorate five major wins, for he also saddled Claiborne's Lure to win the grade III Kelso that astounding day.

In 1995 the Phipps stable was back in form at Belmont on Breeders' Cup Day itself. Inside Information raced off from Heavenly Prize in Dinny's colors to win the Distaff, and Phipps' My Flag won the Breeders' Cup Juvenile Fillies. My Flag

BARBARA D. LIVINGSTON

Storm Flag Flying

193

was a lengthy, darkish chestnut by the Phipps champion Easy Goer and out of the Phipps champion Personal Ensign. My Flag later added yet another win in the venerated CCA Oaks, as well as winning the Ashland, Gazelle, and Bonnie Miss. She won six of twenty starts and earned $1,557,057.

By that time the Phipps interests had been merged into the entity called the Phipps Stable, so that whose silks were worn was apparently a matter less of distinct ownership than just family decision. Dinny Phipps' children also had begun evidencing their own interests in the sport, and the stable. My Flag's second foal was the Storm Cat filly Storm Flag Flying, who emerged to win the Matron, Frizette, and Breeders' Cup Juvenile Fillies for the Phipps Stable as its most recent champion, in 2002.

Personal Ensign's third grade I winner was Traditionally, who had problems early but got it together enough to win the grade I Oaklawn Handicap in 2001.

Ogden Phipps' late wife, Lillian, also was involved individually with racing and breeding. One of her brothers, Pete Bostwick, was a distinguished steeplechase rider, and Mrs. Phipps' interest in steeplechasing was strong. Although a number of the stakes winners bred in her name were flat runners, Straight and True among them was a steeplechase champion. Mrs. Phipps raced various major jumpers that were bred by the Phipps operation and converted to that venue, but her greatest jumper was purchased, he being the immortal Neji.

Including ten stakes winners bred in partnership with his mother's Wheatley Stable (but not including those bred in his wife's name), Ogden Phipps bred 116

Champions Bred

Ogden Phipps

Ancestor
1959 champion steeplechaser

Buckpasser
1965 champion two-year-old colt
1966 champion three-year-old colt
1966 Horse of the Year
1967 champion older horse

Impressive
1966 champion sprinter
(bred in partnership with Wheatley Stable)

Mako
1966 co-champion steeplechaser
(bred in partnership with Wheatley Stable)

Queen of the Stage
1967 champion two-year-old filly

Vitriolic
1967 champion two-year-old colt

Numbered Account
1971 champion two-year-old filly

Relaxing
1981 champion older filly or mare

Easy Goer
1988 champion two-year-old colt

Personal Ensign
1988 champion older filly or mare

Heavenly Prize
1994 champion three-year-old filly

Storm Flag Flying
2002 champion two-year-old filly

Mrs. Ogden Phipps

Straight and True
1976 champion steeplechaser

---

stakes winners and twelve North American champions from 1933 through 2002. Of Phipps' champions, Buckpasser, Easy Goer, and Personal Ensign have all been elected to the Hall of Fame of the National Museum of Racing in Saratoga Springs, New York. All three were ranked in *The Blood-Horse*'s poll of the best horses of the twentieth century, Buckpasser fourteenth, Easy Goer thirty-fourth, and Personal Ensign forty-eighth. The mare Searching, whom Phipps bred and sold, also is in the Hall of Fame, giving him a total of four representatives there.

Other measures of success and recognition for Phipps include the following:

• Leading breeder in money won in 1988 and 1989, with, respectively, $6,031,305 (a record at the time) and $5,568,537;

• Leading owner in money won in 1988 and 1989, with, respectively, $5,858,168 (a record at the time) and $5,438,034;

• Breeder-owner of Broodmares of the Year Striking (1961), Grecian Banner (1988), Relaxing (1989), and Personal Ensign (1996);

• Eclipse Awards for breeder in 1988, for owner in 1988 and 1989, and Award of Merit in 2002.

Phipps accepted leadership positions that were instrumental to the sport he revered. He was elected to The Jockey Club in 1939, and in the early 1950s as vice-chairman was involved in appointing Harry Guggenheim, C.T. Chenery, and John W. Hanes as a committee of three to develop a plan to meet the challenge to the declining affairs of New York racing. The three developed the formula that created the New York Racing Association, which consolidated the state's tracks, built the new Aqueduct, and eventually rebuilt the pivotal Belmont Park. Phipps served as chairman of The Jockey Club from 1964 to 1974, and his son Dinny has been chairman since 1983. In their business careers the son also succeeded Phipps as chairman of the family's Bessemer Trust investment house.

As one who came to racing years before, via an element steeped in tradition, Phipps might have been presumed to look only at what was appreciated by the old guard and have a dismissive attitude toward hoi polloi. Such was not the case. He once gave a speech that to this day could be a platform for the National Thoroughbred Racing Association, as it seeks to reach out and serve the general public as customers.

"Who among us denies that racing long since ceased to be the Sport of Kings? Now more than ever, the sport belongs to the people," Phipps said in an address to the Kentucky Thoroughbred Breeders Association, "and John Public is the king of horse racing. As king, he is entitled to our first consideration."

The NTRA was formed in the late 1990s. Phipps made that speech in 1954!

The author had the good fortune to interview Phipps on several occasions over a number of years, and we were always struck by his appreciation of the sport, not just of his own horses, but of the whole game. He was once described to us by a family friend as "one of the nicest and shyest people you will ever meet," and there was no reason to question either. A personal recollection of Ogden Phipps that lingers, more vivid than winner's circle photos and earnings records, was of the day Polish Navy won the 1987 Woodward Stakes. Phipps was seated a distance away, in his front-row box at Belmont, as he had been so often for so many seasons, and he was surrounded by a fair portion of his then-young stable of grandchildren. In response to the latest grade I victory at beloved Belmont, this gaggle of blond heads turned to him as a unit, clapping him on the back with comfort and glee. If ever an old gent could be presumed to be "Cadillacing," it was O. Phipps that day.

When he passed away at ninety-three on April 22, 2002, that image returned, and we hoped it would not be judged blasphemous to conjecture that if heaven were just almost as good as that moment, Mr. Phipps would be fine.

Ogden Mills Phipps

Like his father, Ogden Mills (Dinny) Phipps was attracted early to what was by then a lengthy family involvement in Thoroughbred racing. He also shared his father's interest in court tennis, an ancient game at which both were champions, and powerboat racing. Dinny took a successful fling, too, in owning top-of-the-line cars on the international road racing circuits.

While Dinny Phipps has been building his own successful record of breeding and racing the sort of horses his grandmother and father bred, it has been in the organizational aspect of the business that he is often equally visible. As U.S. Treasury secretary Nicholas Brady expressed when the Thoroughbred Club of America honored Dinny Phipps in 1990, "At a time when he could have been a bystander, he has taken a leadership role." Added Ted Bassett of Keeneland, "Throughout the years, the Phipps family has provided the world of racing with a broader dimension and appreciation for classic racing conducted with an abiding respect for the integrity and finer traditions of racing."

Phipps expressed that evening that his deep involvement in the workings of racing, as opposed merely to enjoying its pleasures and challenges, was "necessary." He added that, "you cannot be only a taker in life. You must also give back."

What Phipps has given back has been hard work, vigorous leadership, and vision on many levels. A sportsman who, as Brady said, was in position to deal only with his own stable and ambitions had he chosen, Phipps instead has agreed to take on many of the grit-and-grind roles that require negotiations with individuals of various agendas, subsuming one's own preferences here and there to move the greater whole forward.

As chairman of The Jockey Club for twenty years, he has opened membership, inviting many who are like-minded in wanting to work for the game and not just be in it. In so doing, he has brought into the club individuals with a wide range of agendas that sometimes find actual members of The Jockey Club on the opposite side as a racing situation is addressed! He also spent numerous years in the difficult and often thankless task of running the New York Racing Association, a post that is just as much about prickly dealings with politicians — not always with the health of the Turf as a priority — as about breeding foals and enjoying seeing them in the fields and on the backstretch.

Phipps was also chairman of the American Horse Council, understanding its role in protecting and fostering the health of the industry in a national legislative environment that often lacks a grasp of the importance and mechanisms of the Turf. More recently he has served on the boards of the various entities of The Jockey Club, such as The Jockey Club Information Systems, and the partnership with racetracks, Equibase, which placed the records of the sport under the management of the sport. The registration procedures of The Jockey Club, which for so long sent the teeth of breeders grinding in frustration as they awaited foal certificates, have been so refined and smoothed that quick service is now the norm. Keeping up with technology, The Jockey Club has gone to improved parentage verification in steps over the years and has now committed to using DNA testing for identification.

Dinny Phipps (right) and son Ogden

Phipps plays a key role in equine research as a leader in the Grayson-Jockey Club Research Foundation, which from 1983 to 2004 provided $11 million to thirty-one universities to fund some 180 projects. Meanwhile, another foundation, The Jockey Club Foundation, provides financial relief to many involved in racing who find themselves in need of help, for treatment of maladies or for mere subsistence.

In such pivotal developments on behalf of the industry as the Breeders' Cup and National Thoroughbred Racing Association, Phipps has invested the prestige and solidity of The Jockey Club as well as his own influence.

It is, of course, as breeders of distinguished horses that individuals in this volume are addressed, and in this sphere Phipps set off on a fast pace that he has sustained.

Born on September 18, 1940, in New York City, Phipps was educated at Green Vale School and Deerfield Academy before graduation from Yale. The first stakes winner attributed to him as a breeder was foaled when Phipps was twenty-one. This was Time Tested (Better Self—Past Eight, by Eight Thirty), winner of the Great American, Dover, and James H. Connors Memorial at two in 1964. Time Tested later added five more stakes at three and four.

(Phipps told the following story on himself as a young enthusiast: "We were running a horse at Hialeah and there was national television. The announcer interviewed Mr. Fitz [trainer Sunny Jim Fitzsimmons] for some comments, and one remark he made was to the effect that he could not understand how I was at the races because 'the young man' should have been at college that day. The next day, after I had returned to New Haven, the dean, who watched the race, asked me the same question in rather greater detail!")

Dinny Phipps had some horses in partnership with his grandmother, Mrs. Henry Carnegie Phipps of Wheatley Stable, and his stock was boarded at Claiborne Farm, as were the horses of other family members. His silks resemble his father's, black with cherry cap, but with cherry cuffs to make a distinction.

"Racing has always been a part of my life," young Phipps said years ago. "I was more of a fan before Bold Ruler (Wheatley's mid-1950s champion), and then I became more involved with the business aspects of the stable. Remembering him and watching his produce grow up and race was a great enjoyment. He certainly was a landmark for the stable."

Looking back, from the vantage point of the early twenty-first century and some forty-five years of his own involvement, Phipps said that he has observed that in general his family's runners tended to be physically larger after Bold Ruler because of that great stallion's genetic influence.

The early stakes winner Time Tested's dam, Past Eight, was in the Wheatley broodmare band and was a daughter of Helvetia, who had produced for Wheatley the CCA Oaks winner Edelweiss and the Metropolitan-Suburban handicaps winner Snark. One of the other foals Past Eight produced that was bred in the name of Wheatley was Lady Be Good. Dinny Phipps is still raising major horses from this female family.

Lady Be Good was a foal of 1956 and, like Time

Tested, was by the solid King Ranch stallion Better Self. The sire was by *La Troienne's champion son Bimelech and was out of a War Admiral mare, thus combining in reverse the favored cross of Phipps' father, Ogden Phipps, who liked to breed his *La Troienne-family mares to War Admiral. A foal of the middle of the twentieth century, Lady Be Good had a pedigree resonating with key figures of many years before; she was inbred to Man o' War and Rock Sand and her sire was inbred to Black Toney.

Lady Be Good won the Colleen and National Stallion Stakes at two in 1958 and produced four stakes winners in Discipline, Disciplinarian, Full of Hope, and In Hot Pursuit. The last of the four was foaled in 1971, the year after Mrs. Henry Carnegie Phipps' death, and Lady Be Good was among the mares folded into Dinny's division, he being breeder of record of In Hot Pursuit.

In Hot Pursuit, a daughter of Bold Ruler, won the Fashion Stakes at two in 1973 and quickly took her own place in the breeding shed. In Hot Pursuit produced three stakes winners, all by the Argentine champion *Forli, who was imported to stand at Claiborne Farm. Dinny Phipps is the third generation of his family to board his mares there, and he has now been dealing for thirty years with the third generation of Hancocks (Seth) who have developed and maintained an exceptional prestige for the farm. The synergy of the two families has been fundamental to the success of each. Phipps described it as "a bond of confidence between us which would be impossible to duplicate." Several years ago Phipps and his father arranged for production of a video paying tribute to the Hancocks and celebrating the fact that Claiborne had raised 260 winners for the Phippses and Janneys.

The best of the *Forli—In Hot Pursuit foals was Posse, a 1977 colt whom Dinny Phipps raced abroad during a time that he had a racing division in England. Posse won the group I Sussex Stakes at Goodwood and the group II St. James's Palace Stakes at Royal Ascot and was second in one of the English classics, the Two Thousand Guineas. With the female family thus impressive internationally, various members of the family have been purchased to go abroad, and In Hot Pursuit's descendants include Italian champion Lonely Bird, Brazilian stakes winner Padrao Global, and Anglo-American stakes winner Hal's Pal.

In the same crop as Posse, Phipps achieved the distinction of breeding and racing a classic winner. The filly Quick as Lightning was also among those chosen to be sent abroad to English trainer John Dunlop. She was by Phipps' father's great horse Buckpasser and out of Clear Ceiling, a Bold Ruler mare, whose dam was Wheatley Stable's matriarch Grey Flight. Clear Ceiling was not one of Grey Flight's nine stakes winners, but she was a useful race filly, winning five of seventeen races. Her Buckpasser filly Quick as Lightning won the Hoover Mile in England at two in 1979 and the following spring won a classic for fillies, the One Thousand Guineas. As breeder and owner of an English classic winner, Dinny Phipps thus followed in his father's footsteps in another way, the elder Phipps having bred and raced the 1972 St. Leger winner Boucher.

A family almost as important to the Phippses' success as Grey Flight's is that of Lady Be Good. In 1971, the year after his grandmother's death, Dinny Phipps bred Lady Be Good to his own homebred Majestic Light. The resultant filly, winner Impish, became the dam of two stakes winners and second dam of several others, including the high-class runner Mining. A 1984 colt by Mr. Prospector—I Pass, by Buckpasser, Mining won six of seven races, including the grade I Vosburgh, and earned $264,030. He was sent to Japan in 1995 after several years at stud. As of mid-2004, he has sired thirty-two stakes winners (5 percent).

The aforementioned Majestic Light, sire of Lady Be Good's daughter Impish, represented success by Phipps in continuing the legacy of another of his grandmother's important mares. Majestic Light was the best son of the Spendthrift Farm stallion Majestic Prince, winner of the 1969 Kentucky Derby and Preakness. Phipps bred Majestic Light by sending to Majestic Prince the mare Irradiate, whose sire was the great *Ribot and whose dam was High Voltage, a champion at two for Wheatley and winner of the 1955 Coaching Club American Oaks. Irradiate, who placed in the 1968 Matron Stakes and won four races, was four when Mrs. Phipps died. From High Voltage, Dinny Phipps bred the stakes winners Celestial Light (by Bold Ruler) and Fluorescent Light (by *Herbager) as well as Majestic Light.

In 1976, John W. Russell, who trained for the Phipps stables for several years, unleashed a barrage of stakes winners for Phipps and his father. Majestic

Light, a pleasant development at that time, roved from coast to coast and scored on grass and dirt. He won the Swaps Stakes, Monmouth Invitational, Cinema, and Cavalcade. At four he added the Man o' War Stakes on grass, took the Amory L. Haskell on dirt, and won two other stakes. Had he won the Washington, D.C., International that fall, he probably would have been the champion grass horse, but jockey Steve Cauthen stole away to a long lead on Johnny D., who won the race and the title.

Majestic Light retired to Claiborne with a record of eleven wins in thirty-one starts and earnings of $650,158. As has been true in many cases addressed in this volume, the founder of the feast, i.e., the breeder, enjoyed few dividends. Phipps bred only a couple of stakes winners by Majestic Light. The horse, himself, however, enjoyed a long and significant career, siring seventy-two stakes winners (8 percent), including Lite Light, Solar Splendor, Lacovia, Simply Majestic, and Wavering Monarch. The last-named was a grandson of Lady Be Good and represented one of many cases of success for mares the Phippses sold to keep their operation within desired numbers. Wavering Monarch went on to sire 1995 juvenile colt champion Maria's Mon, in turn the sire in his first crop of the 2001 Kentucky Derby winner, Monarchos. Since Monarchos now stands at Claiborne, Dinny Phipps might yet tap into this sire line of Majestic Light that has been prolific for others.

Another Phipps star to rise quickly in 1976, when Majestic Light was three, was the juvenile filly Squander, who was by Buckpasser. Squander was also from the legacy of Lady Be Good. Squander's dam, Discipline, was out of Lady Be Good and by the two-time leading sire *Princequillo. Foaled in 1962, Discipline had won one important race each year at two, three, and four, i.e., the Demoiselle, Test, and Molly Pitcher.

Squander emerged in the summer of 1976 to win the grade I Sorority Stakes and the Astoria. After retiring, she produced three stakes winners and became the second dam of Goodbye Halo, an important winner for Alex Campbell and Arthur B. Hancock III. Ironically, Goodbye Halo was third behind the unbeaten filly of Dinny Phipps' father, Personal Ensign, when she won the 1988 Breeders' Cup Distaff.

Living Vicariously, the Phipps-bred that won the 1993 Brooklyn Handicap, is out of another daughter of Squander, the Alydar filly Extravagant Woman. This sequence illustrates that Dinny came along, too, as a purveyor of clever naming. Others illustrations of this acumen include Stuttering, by Ack Ack; Adjudicating, out of Resolver; Rhythm, out of Dance Number; Late as Usual, out of In Hot Pursuit; Tax Collection, by Private Account—Resolver; and the sly Inside Information, by Private Account—Pure Profit.

The influence of the Lady Be Good bottom line continued for Dinny Phipps. In 1989 he won the grade I Flamingo Stakes and American Derby with the Slew o' Gold colt Awe Inspiring, whose dam, Highest Regard, was a granddaughter of Lady Be Good. Highest Regard was by Gallant Romeo, an outside sire, and she won the Dark Mirage and Imperatrice stakes. Highest Regard also did well in races named for Phipps mares of the past, winning the High Voltage and finishing third in the Grey Flight Handicap. Bold Example, dam of Highest Regard, was by Bold Lad and was stakes-placed. She was also the dam of the Del Mar Oaks winner French Charmer, who was one of the vehicles of international outreach for this bloodline. French Charmer foaled the English Horse of the Year Zilzal. The European filly champion Culture Vulture, French champion Polish Precedent, and Italian classic winner and champion Lonely Bird also come from the Lady Be Good family.

In 1989 Phipps wryly referred to Awe Inspiring as "1-A" in recognition of the fact that his father's Easy Goer was by all odds the star three-year-old of the stable. Both contested the Kentucky Derby. Easy Goer was second, Awe Inspiring third, behind Sunday Silence. Awe Inspiring finished fourth behind Easy Goer in the Belmont Stakes.

In addition to his pair of grade I wins, Awe Inspiring won the Jersey Derby and Everglades Stakes. He won seven of twelve races, earned $994,072, and was sold to Japan as a stallion.

Yet another major horse to come along with Majestic Light and Squander in that summer of 1976 was the turf specialist Intrepid Hero. This was another *Forli colt and was out of Bold Princess, a Bold Ruler filly from Wheatley's matriarch Grey Flight. Bold Princess had been in the first crop of Bold Ruler and won the Schuylerville Stakes at two. She had also foaled Wheatley's 1969 Selima Stakes winner Predictable and Phipps' Irish stakes winner Primed. Bold Princess had an unusual pattern of having all her stakes winners when bred to South American imports standing at Claiborne. Predictable was by *Tatan, Primed by *Pronto, and Intrepid Hero by *Forli. To the cover of the world-class sire Northern Dancer, Bold Princess foaled the stakes-placed Sovereign Dancer, who was sold to France but was returned to stud in this country and sired Preakness Stakes winners Gate Dancer and Louis Quatorze. Sovereign Dancer also sired Eclipse turf champion Itsallgreektome and French champion Priolo, and was an arch example of the Phipps blood-stock's inherent, underlying potency that makes its produce attractive to breeders worldwide.

Intrepid Hero fractured both knees as a yearling, presumably from a fall, and underwent surgery by Dr. Robert Copelan before even being broken to saddle. Given plenty of time, he emerged at three to win the Hollywood Derby on the dirt and added the Secretariat Stakes and Boardwalk Stakes. He also earned a trip to Paris. Phipps' great-grandfather had won France's Prix de l'Arc de Triomphe with Kantar, owned in partnership with Lord Derby, in 1928, and Phipps took a decidedly sporting swing at the great event with Intrepid Hero in 1975. The colt ran solidly, but wound up tenth in a field of twenty-four. The distinguished fillies Ivanjica and Dahlia trailed him, small solace, but an indication of the quality of the field.

Back in the United States, Intrepid Hero won one of the top turf stakes of the time, the United Nations, as well as the Bernard Baruch. He was sent to stud at Leslie Combs II's Spendthrift Farm with nine wins in twenty starts and earnings of $405,305. He died of a heart attack after only three seasons. His best off-spring in that shortened stud career was the million-aire Interco, bred by Spendthrift.

Another potential treasure whose management passed to Dinny Phipps after his grandmother's death was Eastern Princess, a full sister to the pivotal Bold Ruler (*Nasrullah—Miss Disco, by Discovery). Stakes-placed Eastern Princess produced a husky colt by Darby Dan Farm's spectacular Graustark in 1971. Named Shady Character, the colt won three stakes, including two runnings of the Knickerbocker Handicap.

In 1989, when his father's Easy Goer trumped his own Awe Inspiring among three-year-olds in trainer Shug McGaughey's barn, Dinny Phipps accounted for the one-two punch among two-year-olds all on

his own. He was the breeder of both Adjudicating and Rhythm, and they both went into the Breeders' Cup Juvenile at Gulfstream Park that autumn.

Adjudicating was by Claiborne's champion Northern Dancer stallion Danzig and was out of the stakes winner Resolver. Also bred by Dinny Phipps, Resolver was by Reviewer—Lovely Morning, by Swaps. Lovely Morning was an unraced daughter of the Wheatley champion and exceptional broodmare Misty Morn, she from Grey Flight. A foal of 1965, Lovely Morning came along a year after another Swaps filly, Ogden Phipps' Intriguing, who in time foaled the champion Numbered Account.

Prior to foaling Adjudicating, Resolver had pro-duced for Dinny Phipps the Damascus colt Time for a Change. The highlight of this colt's career was one of those inevitable moments when the Phippses and Hancocks were on a collision course for a day. Seth Hancock had recently syndicated Mr. and Mrs. James P. Mills' spectacular Devil's Bag for $36 mil-lion. The 1984 Flamingo Stakes was the first major test at three for Devil's Bag, the 1983 juvenile colt champion, and trainer Woody Stephens seemed con-fident that Devil's Bag was truly a great prospect. It was Dinny Phipps' Time for a Change who lowered his colors in the Flamingo. (Devil's Bag never re-established leadership of his age group, but Stephens and Claiborne won the Kentucky Derby and Belmont that year with Swale.)

Time for a Change won five of nine races and earned $313,896. He was not selected to stand at Claiborne but was placed in stud at Spendthrift Farm, where he got the champion two-year-old colt Fly So Free among forty-three stakes winners (10 percent). Fly So Free, who raced for music publisher Tommy Valando, later sired the Dubai World Cup winner Captain Steve.

Time for a Change's half brother Adjudicating went into the 1989 Breeders' Cup Juvenile with a vic-tory in one of the key, traditional tests in New York, the Champagne Stakes. Rhythm had been his run-ner-up, beaten a neck. In the Breeders' Cup it was Rhythm who came on in the stretch to defeat Grand Canyon and Slavic.

Resolver also foaled a Danzig filly named Dispute. In 1993 Dispute scored in three of the most impor-tant races of the female divisions, carrying Dinny Phipps' colors to victory in the Kentucky Oaks, Gazelle Handicap, and Beldame Stakes. It was a remarkable year for three-year-old fillies, that division also including Sky Beauty and Hollywood Wildcat, and the latter secured the championship by winning the Breeders' Cup Distaff. At four Dispute added the Spinster Stakes to her tally, but prior to the 1994 Breeders' Cup Distaff she was injured and missed the race.

When Rhythm defeated Adjudicating in the 1989 Breeders' Cup Juvenile, he won not only the race, but as it later transpired, the Eclipse Award for the division. Rhythm was by another prolific Claiborne stallion, Mr. Prospector, and was out of Ogden Phipps' Beldame Stakes winner Dance Number, a Northern Dancer filly from his champion Numbered Account.

Although Easy Goer lost the Breeders' Cup Classic to nemesis Sunday Silence that afternoon, it was a bellwether day for the Phipps stables. Dancing Spree (Nijinsky II—Blitey, by Riva Ridge) won the Breeders' Cup Sprint for Dinny's father.

Rhythm has had a highly unusual career. He was generally seen as a disappointment through most of his three-year-old season, but at Saratoga he emerged again to win the Travers Stakes, in which the beaten field included that year's Belmont Stakes winner, Go and Go (Ire). The old Travers had been won the previous year by the elder Phipps' Easy Goer and also by Buckpasser many years before (1966). Rhythm also won the Colin Stakes that year and placed in the Dwyer, Woodward, and Haskell. He won six races in twenty starts and earned $1,592,532. He stood initially in Japan and also sired three crops in New Zealand before being returned to this country. He is currently standing at Diamond F Ranch in

Rhythm

Green Valley, California, seeing still another part of the world. Rhythm sired Ethereal, whose victory in the Australian national heritage race, the Melbourne Cup, is an interesting adjunct to the sire line of Mr. Prospector, once suspect as to stamina. Ethereal also won the Caulfield Cup and Queensland Oaks, among others, and earned $2,450,827. Rhythm's Japanese foals include another millionaire in Legacy Rock. Yet, as of mid-2004, he has sired only 3 percent stakes winners (seventeen) from foals.

Dance Number, dam of Rhythm, also foaled the stakes-winning Mr. Prospector filly Get Lucky, who foaled the $414,908-earner Accelerator (by A.P. Indy).

As Ogden Phipps approached his nineties, the Phipps generations' horses were merged into the entity of Phipps Stable, although horses still were bred individually in the names of Mr. Phipps and his son. Thus, when "Dinny's" Inside Information rolled away to defeat "Mr. Phipps'" Heavenly Prize by thirteen and a half lengths in the 1995 Breeders' Cup Distaff, it was more "ours," than "yours" and "mine." Inside Information was bred in Dinny's name and was by Private Account and out of Pure Profit, a filly by the outside stallion Key to the Mint (bred and raced by Paul Mellon and standing at Greentree Stud).

Pure Profit was a daughter of Clear Ceiling, whom we have met before as the dam of the 1980 English One Thousand Guineas winner Quick as Lightning. Pure Profit did not get to the races until she was four, in 1986, when she won four of five. Dinny Phipps bred three stakes winners from Pure Profit. The first two were millionaires Educated Risk and Inside Information, and the third was the more moderate winner Diamond. Educated Risk and Diamond were by Mr. Prospector, while Inside Information was by Private Account.

Educated Risk, a foal of 1990, won six stakes, including the grade I Frizette and Top Flight, and she was second in the Breeders' Cup Juvenile Fillies, Acorn, Beldame, Matron, etc. She won eleven of twenty-three starts and earned $1,163,717.

This near-championship status was one-upped by her year-younger half sister, Inside Information. In any family except one that had bred a champion filly that won thirteen of thirteen, Inside Information's record might be, well, "awe inspiring." She won fourteen of seventeen starts from two through four and earned $1,641,806. At three she took the grade I

Acorn and Ashland as well as the Bonnie Miss Stakes. At four Inside Information was presumed to be second fiddle to Ogden Phipps' Heavenly Prize, even while winning five graded stakes from a span of six starts. Her wins included the grade I Shuvee, Ruffian, and Spinster. Thus, when she won the Breeders' Cup Distaff, that win plus her previous credentials entitled her to the older filly division's Eclipse Award. She followed Rhythm as the second Eclipse Award winner bred and raced in the name of Ogden Mills (Dinny) Phipps.

In a partnership with the vice chairman of The Jockey Club, William S. Farish, Phipps bred and sold an additional champion, Storm Song. Purchased by Dogwood Stable, Storm Song won a series of major races in the autumn of 1996, climaxed by the Breeders' Cup Juvenile Fillies, and was that season's champion two-year-old filly. (In addition to his professional connections and friendship to Farish, one of Phipps' stepdaughters is married to Farish's son.)

Inside Information has had woeful luck as a producer, while the naming skills of her owner have had a fruitless run of cleverness. As of mid-2004, she has no foals to race, thus consigning to obscurity foals with the names of Devious, Scheming, and Illicit. Perhaps, the younger foals, Manipulator and Smuggler, can return Inside Information to prominence.

As of mid-2004, Ogden Mills Phipps has bred sixty stakes winners. In his early sixties, he was, as his first stakes winner might have it, "time tested," but the prospects of many more distinctions for the Phipps Stable were irresistible.

Cynthia Phipps

Although Dinny Phipps' sister, Cynthia, has maintained a numerically smaller operation, her success has been very much in keeping with family tradition. Cynthia Phipps adopted the Wheatley Stable colors of the grandmother — gold jacket, purple sleeves, purple cap. She has evidenced the family knack not only for breeding, but also for naming. Her best — in both disciplines — include:

Christmas Past (*Grey Dawn II—Yule Log, by Bold Ruler), winner of the Coaching Club American Oaks, Ruffian, Monmouth Oaks, as Eclipse Award winner among three-year-old fillies of 1982, and later winner over males of the Gulfstream Park Handicap;

Versailles Treaty (Danzig—Ten Cents a Dance, by Buckpasser), winner of the historic Alabama and the Ruffian stakes, and dam of the 2001 Remsen Stakes winner Saarland;

Gold Fever (Forty Niner—Lead Kindly Light, by Majestic Light), winner of the NYRA Mile and three other stakes and sire of the millionaire Gold Mover early in his stud career.

Champions Bred

Ogden Mills Phipps

Rhythm
1989 champion two-year-old colt

Inside Information
1995 champion older female

Storm Song
1996 champion two-year-old filly
(bred in partnership with W.S. Farish)

Cynthia Phipps

Christmas Past
1982 champion three-year-old filly

Stakes Winners Bred by the Ogden Phipps Family

Ogden Phipps

White Cockade, b.g. 1933, Diavolo—White Favor, by *Sir Gallahad III
The Admiral, br.c. 1946, War Admiral—Big Hurry, by Black Toney
Busanda, blk.f. 1947, War Admiral—Businesslike, by Blue Larkspur
Striking, b.f. 1947, War Admiral—Baby League, by Bubbling Over
Auditing, br.c. 1948, Count Fleet—Businesslike, by Blue Larkspur
Ancestor (stp.), b.c. 1949, Challedon—Bloodroot, by Blue Larkspur
Great Captain, dk.br.c. 1949, War Admiral—Big Hurry, by Black Toney
One Throw, ch.c. 1949, Some Chance—White Favor, by *Sir Gallahad III
Bassanio, b.c. 1950, Bimelech—Portia, by *Rhodes Scholar
Flirtatious, b.f. 1950, Menow—Flitabout, by Challedon
Torch of War, b.c. 1950, Bimelech—Bellicose, by *Boswell
Landscaping (stp.), br.g. 1951, Bimelech—Puccoon, by Bull Lea
Searching, b.f. 1952, War Admiral—Big Hurry, by Black Toney
Affable (stp.), br.g. 1953, *Ambiorix—Graciously, by *Sir Gallahad III
Dining Alone, br.c. 1953, Eight Thirty—Quarantaine, by *Isolater
Point of Order, b.c. 1953, Menow—Turmeric, by Bimelech
Bureaucracy, br.c. 1954, Polynesian—Busanda, by War Admiral
Harmonizing, ch.g. 1954, Counterpoint—Baby League, by Bubbling Over
Clear Road, ch.f. 1957, Hasty Road—Tiny Request, by Requested
Sarcastic, b.f. 1957, *Ambiorix—Caustic, by Menow
Funloving, b.f. 1958, Tom Fool—Flitabout, by Challedon

My Flag

Hitting Away, b.c. 1958, *Ambiorix—Striking, by War Admiral
Shavetail, ch.c. 1958, Jet Pilot—Bushleaguer, by War Admiral
The Sport (stp.), dk.b.g. 1958, *Ambiorix—Flirtatious, by Menow
Broadway, b.f. 1959, Hasty Road—Flitabout, by Challedon
Comic, b.c. 1959, Tom Fool—His Duchess, by *Blenheim II
Subtle, b.c. 1959, *Princequillo—Punctilious, by Better Self
Fashion Verdict, b.f. 1960, *Court Martial—So Chic, by *Nasrullah
Royal Ascot, b.c. 1960, *Princequillo—Glamour, by *Nasrullah
The Ibex, ch.c. 1960, Hill Prince—Resourceful, by Shut Out
Bupers, dk.b.c. 1961, Double Jay—Busanda, by War Admiral
Jove (stp.), ch.g. 1962, Bold Ruler—Turmeric, by Bimelech
Open Hearing, b.f. 1962, *Court Martial—Oak Cluster, by *Nasrullah

Staunchness, b.c. 1962, Bold Ruler—Tiny Request, by Requested
Buckpasser, b.c. 1963, Tom Fool—Busanda, by War Admiral
Destro, ch.f. 1963, *Ribot—Ingenuity, by My Request
Poker, b.c. 1963, Round Table—Glamour, by *Nasrullah
Great White Way, b.c. 1964, *Court Martial—Broadway, by Hasty Road
Jaunty (stp.), b.g. 1965, *Ambiorix—Glamour, by *Nasrullah
Lucretia Bori, b.f. 1965, Bold Ruler—*Arietta II, by *Tudor Minstrel
Queen of the Stage, b.f. 1965, Bold Ruler—Broadway, by Hasty Road
Vitriolic, b.c. 1965, Bold Ruler—Sarcastic, by *Ambiorix
Beau Brummel, b.c. 1966, Round Table—So Chic, by *Nasrullah
King of the Castle, b.c. 1966, Bold Ruler—Ambulance, by *Ambiorix
Reviewer, b.c. 1966, Bold Ruler—Broadway, by Hasty Road
Pass the Drink, b.c. 1967, Swaps—St. Bernard, by Hill Prince
The Pruner, dk.b/br.c. 1967, *Herbager—Punctilious, by Better Self
Boucher, ch.c. 1969, *Ribot—Glamour, by *Nasrullah (SW in England/Ireland)
Numbered Account, b.f. 1969, Buckpasser—Intriguing, by Swaps
Outdoors, b.c. 1969, *Herbager—Open Hearing, by *Court Martial
Will Hays, b.c. 1969, Bold Ruler—Broadway, by Hasty Road
My Great Aunt, b.f. 1970, Bold Ruler—Aunt Edith II, by Primera (SW in France)
Plastic Surgeon, b.c. 1970, Dr. Fager—Fashion Verdict, by *Court Martial
Snobishness, b.f. 1970, *Forli—So Social, by Tim Tam (SW in France)
Operating, ch.f. 1971, Dr. Fager—Cutting, by Bold Ruler
Ward McAllister, b.c. 1971, Bold Ruler—So Social, by Tim Tam
Bubbling, ch.f. 1972, Stage Door Johnny—Sparkling, by Bold Ruler
Landscaper, ch.c. 1972, *Herbager—My Boss Lady, by Bold Ruler
Our Hero, dk.b/br.c. 1972, Bold Ruler—Dorine, by Aristophanes
Cunning Trick, b.c. 1973, Buckpasser—Intriguing, by Swaps
Effervescing, ch.c. 1973, *Le Fabuleux—Sparkling, by Bold Ruler
Critical Cousin, ch.f. 1974, Reviewer—Aunt Edith II, by Primera
Fabulous Time, b.g. 1974, *Le Fabuleux—Her Prerogative, by Buckpasser
Hasty Reply, b.c. 1974, Pronto—So Social, by Tim Tam (SW in France)
How Curious, b.c. 1974, Buckpasser—Intriguing, by Swaps
Pas de Deux, ch.c. 1974, Nijinsky II—So Chic, by *Nasrullah
Land of Eire, ch.g. 1975, *Herbager—Irish Jay, by Double Jay
The Liberal Member, ch.c. 1975, Bold Reason—Lady Pitt, by Sword Dancer
Blitey, b.f. 1976, Riva Ridge—Lady Pitt, by Sword Dancer
King of Mardi Gras, dk.b/br.c. 1976, Tentam—Carnival Queen, by *Amerigo
Le Notre, b.c. 1976, *Herbager—Open Hearing, by *Court Martial
Private Account, b.c. 1976, Damascus—Numbered Account, by Buckpasser
Relaxing, b.f. 1976, Buckpasser—Marking Time, by To Market
Comptroller, b.c. 1977, Reviewer—Bank of England, by Buckpasser
Banner Gala, dk.b/br.f. 1978, Hoist the Flag—So Social, by Tim Tam
Sartorialy Perfect, b.c. 1978, Gallant Romeo—Fashion Verdict, by *Court Martial
Dance Number, b.f. 1979, Northern Dancer—Numbered Account, by Buckpasser
Russian Roubles, b.c. 1980, Nijinsky II—Squander, by Buckpasser
Personal Flag, dk.b/br.c. 1983, Private Account—Grecian Banner, by Hoist the Flag
Soar to the Stars, b.c. 1983, Danzig—Flitalong, by *Herbager
Social Business, b.f. 1983, Private Account—So Social, by Tim Tam

Cadillacing, b.f. 1984, Alydar—Relaxing, by Buckpasser
Dancing All Night, ch.f. 1984, Nijinsky II—Blitey, by Riva Ridge
Personal Ensign, b.f. 1984, Private Account—Grecian Banner, by Hoist the Flag
Polish Navy, b.c. 1984, Danzig—Navsup, by Tatan
Stacked Pack, b.c. 1984, Majestic Light—Con Game, by Buckpasser
Dancing Spree, ch.c. 1985, Nijinsky II—Blitey, by Riva Ridge
Seeking the Gold, b.c. 1985, Mr. Prospector—Con Game, by Buckpasser
Easy Goer, ch.c. 1986, Alydar—Relaxing, by Buckpasser
Fantastic Find, dk.b/br.f. 1986, Mr. Prospector—Blitey, by Riva Ridge
Fast Play, dk.b/br.c. 1986, Seattle Slew—Con Game, by Buckpasser
Confidential Talk, b.c. 1987, Damascus—Confidentiality, by Lyphard
Easy Now, dk.b/br.f. 1989, Danzig—Relaxing, by Buckpasser
Miner's Mark, b.c. 1990, Mr. Prospector—Personal Ensign, by Private Account
Strolling Along, dk.b/br.c. 1990, Danzig—Cadillacing, by Alydar
Heavenly Prize, dk.b/br.f. 1991, Seeking the Gold—Oh What a Dance, by Nijinsky II
My Flag, ch.f. 1993, Easy Goer—Personal Ensign, by Private Account

Ogden Phipps and Wheatley Stable
Glamour, b.f. 1953, *Nasrullah—Striking, by War Admiral
Mako (stp.), dk.b.g. 1960, *Tulyar—Puccoon, by Bull Lea
Impressive, b.c. 1963, *Court Martial—High Voltage, by *Ambiorix
Marking Time, ch.f. 1963, To Market—Allemande, by Counterpoint
My Boss Lady, ch.f. 1963, Bold Ruler—Striking, by War Admiral
Great Power, dk.b/br.c. 1964, Bold Ruler—High Voltage, by *Ambiorix
Big Advance, dk.b/br.f. 1966, Bold Ruler—Stepping Stone, by *Princequillo
King Emperor, b.c. 1966, Bold Ruler—Irish Jay, by Double Jay
Army Court, ch.f. 1967, *Court Martial—Explorer, by *Nasrullah (SW in Fr)
Daring Young Man, dk.b/br.c. 1969, Bold Lad—Batter Up, by Tom Fool

Ogden Phipps and Stuart S. Janney III
Mesabi Maiden, b.f. 1993, by Cox's Ridge—Steel Maiden, by Damascus

Mrs. Ogden Phipps
Politico, b.c. 1967, Right Royal—Tendentious, by Tenerani (SW in England)
Straight and True, dk.b/br.g. 1970, Never Bend—Polly Girl, by *Prince Bio
Top Command, ch.c. 1971, Bold Ruler—Polly Girl, by *Prince Bio
Oh So Choosy, ch.f. 1978, Top Command—Snobishness, by *Forli (SW in England)
Forever Command, ch.f. 1982, Top Command—Pretty Special, by Riverman
My Big Boy, dk.b/br.g. 1983, Our Hero—Pretty Special, by Riverman
Stated, b.f. 1983, Hawaii (SAf)—Fashoda, by Restless Wind

Ogden Mills Phipps
Time Tested, b.c. 1962, Better Self—Past Eight, by Eight Thirty
Out of the Park, b.c. 1966, Hitting Away—Medici, by Bold Ruler
Celestial Lights, b.f. 1971, Bold Ruler—Irradiate, by *Ribot
In Hot Pursuit, b.f. 1971, Bold Ruler—Lady Be Good, by Better Self
Primed, b.c. 1971, Pronto—Bold Princess, by Bold Ruler (SW in Ireland)
Shady Character, ch.c. 1971, Graustark—Eastern Princess, by *Nasrullah

Unknown Heiress, b.f. 1971, Bagdad—Medici, by Bold Ruler
Alpine Lass, b.f. 1972, Bold Ruler—Castle Forbes, by *Tulyar
Intrepid Hero, b.c. 1972, *Forli—Bold Princess, by Bold Ruler
Majestic Light, b.c. 1973, Majestic Prince—Irradiate, by *Ribot
Popular Hero, b.g. 1973, Nijinsky II—Brave Lady, by *Herbager
Fluorescent Light, dk.b.c. 1974, *Herbager—Irradiate, by *Ribot
Resolver, b.f. 1974, Reviewer—Lovely Morning, by Swaps
Squander, b.f. 1974, Buckpasser—Discipline, by *Princequillo
Quick Turnover, b.c. 1975, Buckpasser—Bold Consort, by Bold Ruler
Flitalong, br.f. 1976, *Herbager—Pleasant Flight, by Bold Ruler
At Ease, dk.b.c. 1977, Hoist the Flag—The Bride, by Bold Ruler (SW in Argentina)
Highest Regard, b.f. 1977, Gallant Romeo—Bold Example, by Bold Lad
Posse, ch.c. 1977, *Forli—In Hot Pursuit, by Bold Ruler (SW in England)
Quick as Lightning, b.f. 1977, Buckpasser—Clear Ceiling, by Bold Ruler (SW in England)
Good Economics, b.f. 1978, Damascus—Pennygown, by *Herbager
Hail Emperor, dk.b.c. 1978, Graustark—Queen Empress, by Bold Ruler
Heavenly Match, b.f. 1978, Gallant Romeo—The Bride, by Bold Ruler
On a Cloud, b.g. 1978, Val de l'Orne (Fr)—Pleasant Flight, by Bold Ruler
Chess Move, dk.b.f. 1979, Avatar—Queen's Gambit, by Bold Ruler
Play for Love, ch.c. 1979, Jacinto—Winning Trick, by Damascus
Stratospheric, b.f. 1979, Majestic Light—Clear Ceiling, by Bold Ruler (SW in England)
Stuttering, b.f. 1979, Ack Ack—Call Me Madam, by Bold Ruler
Tantalizing, b.c. 1979, Tom Rolfe—Lady Love, by Dr. Fager
Waitlist, ch.c. 1979, Avatar—Renounce, by Buckpasser
Erin Bright, b.c. 1980, Majestic Light—Irish Manor, by Bold Ruler
Infinite, b.f. 1980, Majestic Light—Clear Ceiling, by Bold Ruler
Hot Rodder, b.g. 1981, *Forli—In Hot Pursuit, by Bold Ruler (SW in Germany)
Time for a Change, ch.c. 1981, Damascus—Resolver, by Reviewer
Duty Dance, b.f. 1982, Nijinsky II—Discipline, by *Princequillo
Freedom's Choice, ch.c. 1982, *Forli—Full of Hope, by Bold Ruler (SW in Italy)
Lay Down, b.g. 1984, Spectacular Bid—Impish, by Majestic Prince
Mining, ch.c. 1984, Mr. Prospector—I Pass, by Buckpasser
Awe Inspiring, b.c. 1986, Slew o' Gold—Highest Regard, by Gallant Romeo
Late as Usual, b.c. 1986, *Forli—In Hot Pursuit, by Bold Ruler
Personal Business, b.f. 1986, Private Account—Heavenly Match, by Gallant Romeo
Adjudicating, dk.b.c. 1987, Danzig—Resolver, by Reviewer
Rhythm, b.c. 1987, Mr. Prospector—Dance Number, by Northern Dancer
Get Lucky, b.f. 1988, Mr. Prospector—Dance Number, by Northern Dancer
Tax Collection, dk.b/br.c. 1988, Private Account—Resolver, by Reviewer
All Gone, b.c. 1990, Fappiano—Squander, by Buckpasser
Dispute, b.f. 1990, Danzig—Resolver, by Reviewer
Educated Risk, b.f. 1990, Mr. Prospector—Pure Profit, by Key to the Mint
Iron Gavel, ch.g. 1990, Time for a Change—Sealed Bid, by Mr. Prospector
Living Vicariously, dk.b/br.c. 1990, Time for a Change—Extravagant Woman, by Alydar
Private Light, b.f. 1990, Private Account—Illuminating, by Majestic Light
Inside Information, b.f. 1991, Private Account—Pure Profit, by Key to the Mint

Party Manners, b.c. 1991, Private Account—Duty Dance, by Nijinsky II

Polish Treaty, b.f. 1991, Danzig—Infinite, by Majestic Light

Recognizable, dk.b/br.f. 1991, Seattle Slew—Highest Regard, by Gallant Romeo

Serious Spender, dk.b/br.c. 1991, Seattle Slew—Squander, by Buckpasser

In Conference, b.f. 1992, Dayjur—Personal Business, by Private Account

Accelerator, dk.b/br.c. 1994, A.P. Indy—Get Lucky, by Mr. Prospector

Diamond, ch.c. 1995, Mr. Prospector—Pure Profit, by Key to the Mint

Shake the Dice, ch.c. 1998, Boundary—Option Contract, by Forty Niner

Ogden Mills Phipps and William S. Farish

Storm Song, b.f. 1994, Summer Squall—Hum Along, by Fappiano

Cynthia Phipps

Christmas Bonus, dk.b/br.f. 1978, Key to the Mint—Sugar Plum Time, by Bold Ruler

Christmas Past, gr.f. 1979, *Grey Dawn II—Yule Log, by Bold Ruler

Lead Kindly Light, b.f. 1983, Majestic Light—Arabian Dancer, by Damascus

On Retainer, b.c. 1983, Damascus—Ten Cents a Dance, by Buckpasser

For Kicks, ch.f. 1985, Topsider—Ten Cents a Dance, by Buckpasser

Out of Place, ch.c. 1987, Cox's Ridge—Arabian Dancer, by Damascus

Versailles Treaty, b.f. 1988, Danzig—Ten Cents a Dance, by Buckpasser

Gold Fever, ch.c. 1993, Forty Niner—Lead Kindly Light, by Majestic Light

Emanating, dk.b/br.f. 1996, Cox's Ridge—Lead Kindly Light, by Majestic Light

Jackpot, ch.c. 1998, Seeking the Gold—Frolic, by Cox's Ridge

Saarland, b.c. 1999, Unbridled—Versailles Treaty, by Danzig

Phipps Stable

Furlough, b.f. 1994, Easy Goer—Blitey, by Riva Ridge

Country Hideaway, b.f. 1996, Seeking the Gold—Our Country Place, by Pleasant Colony

Oh What a Windfall, b.f. 1996, Seeking the Gold—Oh What a Dance, by Nijinsky II

Cat Cay, b.f. 1997, Pleasant Colony—Cadillacing, by Alydar

Finder's Fee, dk.b/br.f. 1997, Storm Cat—Fantastic Find, by Mr. Prospector

Traditionally, ch.c. 1997, Mr. Prospector—Personal Ensign, by Private Account

Pure Prize, ch.c. 1998, Storm Cat—Heavenly Prize, by Seeking the Gold

Storm Flag Flying, dk.b/br.f. 2000, Storm Cat—My Flag, by Easy Goer

Sources: *The Blood-Horse* Archives and The Jockey Club

13

Louis E. Wolfson

Most distinguished ladies and gentlemen of the Turf over the years have embraced the philosophy that breeding one's own good horses trumps the pleasures of having purchased them. After Louis E. Wolfson won the 1978 Kentucky Derby with his homebred Affirmed, he gave his own eloquent slant on this leit motif:

"I bought Raise a Native as a yearling at Saratoga. I bought Roman Brother as a two-year-old in Florida; he was Horse of the Year in 1965, the best horse I ever raced until this one. It is a great feeling to own a champion, but it is not the same thing as racing one you have bred. I spent twenty years of my life to get this one. I brought him into this world. I bred and raced his sire. I raised him at my farm. The feeling you get when a horse you have bred and raised has a good chance to win the Kentucky Derby — well, it is something that you just cannot go out and buy.

"So I said (to pre-race offers), 'Don't talk to me about this horse. He's not for sale at any price.'"

Even a sporting fellow has to have some sort of fall-back position in his economic life to speak with such high-minded finality. After all, the specific offer Wolfson was addressing had been for $8 million, and that was a quarter-century ago. In Wolfson's case the economic wherewithal to deal in the upper strata of sportsmanship, in an already expensive arena, might have come as an alternative to what he had hoped as a young man to achieve — play professional football and then coach. In 1931 Yale had a player named Albie Booth, whose fame was celebrated in legend and verse. University of Georgia had a six-foot-two scrapper named Lou Wolfson. When Wolfson tackled Booth, it was the flesh-and-blood version of the Yaleman, not the legend, and there seemed nothing poetic about the shoulder injury that ensued. Wolfson's football dreams were dashed.

Born on January 28, 1912, Wolfson had been a versatile athlete at Andrew Jackson High School in Jacksonville, Florida, before heading to Georgia. After the injury he stayed in college one more year and, despite the shoulder, was sound enough to letter in basketball and wrestling. He then went back to Jacksonville and joined the scrap-metal business of his father, a Russian emigrant.

How Wolfson would have done as a pro football player no one can know. How he did as a businessman was of Hall of Fame material. Starting with a loan of $10,000, he launched Florida Pipe and Supply Company with his father and brother. He parlayed profits from one corporation into another, became a force in Florida politics, and wound up controlling the industrial complex known as Merritt-Chapman & Scott. He presently was in position to control $153 million worth of shares in Montgomery Ward and, even though he failed to win a proxy fight, was a winner of sorts when the board chairman resigned and was succeeded by a man acceptable to Wolfson.

By his middle forties Louis E. Wolfson, former footballer, was in position to seek some sporting leisure in a big way. He was already a fan of Thoroughbred racing, betting five hundred or one thousand dollars a race when he visited the track, and in 1958 he decided to get into the game as an owner. (He later quipped that once he had his own stable, he reduced his betting level to no more than one hundred and two hundred dollars, betting solely on his own horses, and so, in a manner of speaking, "saved a lot of money buying my first horse.")

Wolfson purchased his first horses from a friend, Miami restaurant owner Charlie Block, who suggested the name of Harbor View for Wolfson's stable, Wolfson then living on Biscayne Bay. Wolfson later sent Block and the latter's trainer, A.G. Robertson, to Keeneland with $125,000 to spend on yearlings. They spent $114,000 of it and came home with four horses, one of whom was Francis S., a *Royal Charger colt purchased from Leslie Combs II's Spendthrift Farm. Francis S. won the Wood Memorial and three other stakes and earned $215,914.

Robertson served as Wolfson's trainer, too, but only briefly. When the new owner asked fellow Florida horseman Fred Hooper for a recommendation, Hooper sent him to Ivan Parke, but Parke then recommended that Wolfson invite his brother, Burley Parke, to come out of retirement. Burley Parke had trained major winners such as Occupy, Occupation, Free For All, and Blue Delight and had retired after guiding champion *Noor to his famous four victories over Citation in 1950.

A decade later Burley Parke took on the rapidly growing Harbor View Stable, and he trained seventeen stakes winners, including champions Raise a Native and Roman Brother and Champagne Stakes winner Roving Minstrel, as well as Francis S. Harbor View ranked third among owners in earnings in 1960, the third year of its existence, and was second

Patrice and Louis Wolfson

INGER DRYSDALE

in both 1964 and 1965.

Wolfson had a good time naming horses for famous people and family members: Francis S. for Frank Sinatra; Omarbrad for General Omar Bradley; Garwol, Marwol, and Stevward for sons Gary, Marty, and Steve. He also named a filly Royal Patrice, in honor of the daughter of famed trainer Hirsch Jacobs. (The widower Wolfson later married Patrice Jacobs.) Of the above named, all but Omarbrad became stakes horses.

Wolfson was unfulfilled by his success in buying good horses and recognized the lure of breeding one's own. He bought a farm in the burgeoning horse country of Ocala, Florida, affixed the name Harbor View to the land as well as to the enterprise, although there was no harbor to view in the area, and began acquiring broodmares. He had made a masterstroke as a businessman, owner, and prospective breeder in the purchase of Raise a Native in 1962. A chestnut son of the great gray horse Native Dancer, Raise a Native was bred by Mr. and Mrs. Cortright Wetherill and was out of the Case Ace mare Raise You. Raise a Native was sold at the 1961 Keeneland fall mixed sale for $22,000. In the days before sales of well-bred weanlings were common, this figure stood as an all-time record! The buyer was Mrs. E.H. Augustus of Virginia, who resold him the following summer at Saratoga, where Wolfson bought the colt for $39,000.

Raise a Native was precocious as well as brilliant. He won all his four races with dash and finality, leading the veteran trainer Hirsch Jacobs to declare him the best two-year-old he had ever seen. Raise a Native then bowed a tendon on the eve of the Sapling Stakes and was retired with earnings of $45,955, barely over his purchase price but of scant importance to his escalated value. Wolfson sent him to stand at Combs' Spendthrift Farm. He stood him for several years before syndicating him for $2,625,000.

He became one of the great sires and sires of sires of the twentieth century. His seventy-eight stakes winners (9 percent) included champions such as Majestic Prince. Three of Raise a Native's sons became leading sires: Exclusive Native, Alydar, and Mr. Prospector.

In the rapid acquisition of a broodmare band, Wolfson found a pearl in the mare Exclusive. Bred by Alfred Vanderbilt and sold to Major Albert Warner before Wolfson bought her, Exclusive was by Shut Out—Good Example, by Pilate. She was a moderate runner but produced five stakes winners, four of them for Wolfson. Among the first crop of foals sired by Raise a Native was a chestnut out of Exclusive. This one Wolfson named for the colt's own family, to wit, Exclusive Native. Although he was outranked by Vitriolic, What a Pleasure, and other two-year-olds of 1967, Exclusive Native was a high-class racehorse, winning the Sanford Stakes at two and the Arlington Classic at three. Exclusive Native won four of thirteen races, earned $169,013, and joined his sire at Spendthrift.

In 1968 Wolfson's business career hit a crevice for the first time. He was convicted on what the writer Jim Bishop described as a "technical violation" of Securities and Exchange Commission policy, accused of selling unregistered stock. Wolfson fought the verdict, maintaining his business enemies had targeted him. "I feel sorry for Wolfson because I think he is a great American," Bishop wrote. "I am not his apologist; he needs justice, not mercy."

Wolfson suffered the indignity of serving nine months in prison, being released early in 1970. To be protective of the reputation of the sport he had grown to love, he swiftly had largely dissolved his broodmare band and young horses. A draft of 102 was sold at Saratoga for $989,700. Two of his sons at that time were running a separate operation, Happy Valley Farm, and they purchased a number of the horses. Two years later Happy Valley had a reduction sale, and Wolfson purchased twenty-four mares.

Somewhat ironically, it was right after this storm-tossed time that Harbor View reached a pinnacle, leading the national breeders' list in 1970 with $1,515,861 and again in 1971 with $1,739,214. Much of the foundation stock of that success had been sold.

North America's leading breeder was busy gathering a second broodmare band that at times numbered more than one hundred. Early the next year, 1972, Wolfson made a pivotal acquisition at the Keeneland January sale. (This was a key year personally, too, for in December 1972, Wolfson wed Patrice Jacobs.) Wolfson bought the Crafty Admiral mare Won't Tell You from F. Eugene Dixon Jr. at that Keeneland sale. Won't Tell You brought $18,000, an extremely modest price for a mare in foal to Raise a Native, who by then was renowned as the sire of classic winner Majestic Prince.

Won't Tell You had won five races in twenty-three starts and earned $21,210. She was from the producing Volcanic mare Scarlet Ribbon, whose winning dam Native Valor, was by *Mahmoud. The next dam, Native Gal, had produced the champion Royal Native and the high-class stakes winner Billings.

Won't Tell You's foals were moderate enough that not long after she had produced Affirmed, Wolfson resold her, for $5,500 at the Keeneland fall mixed

TONY LEONARD

Exclusive Native

sale. She had been sent to Exclusive Native in part, Wolfson said, because "my wife, Patrice, noticed *Sir Gallahad III back behind Crafty Admiral on the mare's side, and so we doubled up on that because it was back behind Exclusive Native, too."

This modest form of inbreeding actually was to *Teddy, the sire of *Sir Gallahad III. *Teddy was the sire of Case Ace, broodmare sire of Raise a Native, and on the bottom half of the pedigree, his son *Sir Gallahad III appeared as the sire of Crafty Admiral's sire, Fighting Fox.

As to the choice of Exclusive Native instead of Raise a Native as a match for Won't Tell You, Wolfson further explained to Kent Hollingsworth for *The Blood-Horse* in 1978 that, "I think Exclusive Native is the better sire. I have bred to them both, of course, for many years, and I have the general impression of those I have had in training that a greater percentage of the Exclusive Natives are sound and that more of them are capable of going a distance than have our horses by Raise a Native."

While this might have been written off as a man letting his admiration for a homebred over a purchase color his thinking, within two years he had on his hands a two-time leading sire in Exclusive Native! Moreover, two years after Exclusive Native became the sire of the eleventh (and through 2004 the last) Triple Crown winner, he became the sire of the second filly (Genuine Risk) to win the Kentucky Derby. No other stallion in history has that combination — a Triple Crown winner and a filly Derby win-

ner — to his credit. Exclusive Native eventually sired sixty-six stakes winners (13 percent), and he is the broodmare sire of ninety-seven stakes winners.

The Exclusive Native—Won't Tell You colt named Affirmed resembled his distant relative Royal Native, both being chestnuts with a graceful stream of white down the face. Affirmed was a light, bright chestnut, not especially masculine in appearance, but he had a heart of steel and a reservoir of class that, combined with dashing speed and classic stamina, made him a racehorse of engulfing style and efficiency.

Early in his two-year-old season he met the Raise a Native colt Alydar who would be his rival for the next two years. Alydar was more powerful, more charismatic, and for a time even Wolfson was captured by the Calumet Farm colt

"I thought Alydar was the best two-year-old I had seen since Raise a Native," he said. "Then when Affirmed beat him by a half-length in the Hopeful and came right back to beat him in the Futurity, I thought he was the best." Wolfson went so far as to ask his trainer, Laz Barrera, to try to buy Alydar, or an interest in him, to race in the pink-and-black Harbor View colors. "See, I didn't think these two good colts should keep racing against each other. I was satisfied by then that Affirmed was the better, and I would run them in different races."

Ironically, then, a tower of sportsmanship would have inadvertently denied American Thoroughbred racing its greatest sustained rivalry had Calumet been willing to deal. As matters developed, Affirmed

Exclusive Native, 1965	Raise a Native, 1961	Native Dancer, 1950	Polynesian (Unbreakable—Black Polly)
			Geisha (Discovery—Miyako)
		Raise You, 1946	Case Ace (**Teddy**—Sweetheart)
			Lady Glory (American Flag—Beloved)
	Exclusive, 1953	Shut Out, 1939	Equipoise (Pennant—Swinging)
			Goose Egg (*Chicle—Oval)
		Good Example, 1944	Pilate (Friar Rock—*Herodias)
			Parade Girl (Display—Panoply)
AFFIRMED	Crafty Admiral, 1948	Fighting Fox, 1935	***Sir Gallahad III** (**Teddy**—Plucky Liege)
			Marguerite (Celt—*Fairy Ray)
		Admiral's Lady, 1942	War Admiral (Man o' War—Brushup)
			Boola Brook (*Bull Dog—Brookdale)
Won't Tell You, 1962	Scarlet Ribbon, 1957	Volcanic, 1945	*Ambrose Light (Pharos—*La Roseraie)
			Hot Supper (Gallant Fox—Big Dinner)
		Native Valor, 1948	*Mahmoud (*Blenheim II—Mah Mahal)
			Native Gal (***Sir Gallahad III**—Native Wit)

and Alydar carried on their spirited battles through the remainder of their two-year-old season and through the most hotly contested of Triple Crowns. Affirmed had the upper hand, although Alydar did beat him in the Champagne in their next start after the Futurity. Affirmed then won the Laurel Futurity to clinch the two-year-old championship.

In 1978 both came into the Kentucky Derby with unblemished three-year-old records. Affirmed won that one handily, was tested by Alydar to win the Preakness by only a neck, and then faced his persistent tormentor at the mile and a half of the Belmont Stakes. The rivalry had been engaging from the spring of their two-year-old days. Now, it became sublime. They battled through the second half of the Belmont, and Alydar forged to the front in the stretch. The teenage jockey champion Steve Cauthen switched his whip and slapped Affirmed on the left side for the first time. Affirmed probably needed no prodding to produce his best, but perhaps the element of surprise was helpful. At any rate he surged back, held at bay once more the exceptional racer to his right, and won the eleventh Triple Crown by a head on courage as well as class.

The only later meeting was a messy affair, in which Affirmed finished first but was disqualified for having crowded Alydar on the far turn. Thus, the score stood at ten meetings, seven wins for Affirmed, three officially for Alydar. Incredibly, five of Affirmed's victories had come by a half-length or less.

In the autumn of 1978, Affirmed faced the previous year's Triple Crown winner, Seattle Slew, in racing's first meeting of two winners of the classic sweep. A great older horse is anticipated to beat a great three-year-old, and such was the case. The following year it was Affirmed's turn, and he defeated the great three-year-old Spectacular Bid in the mile and a half Jockey Club Gold Cup back at Belmont. That was the final race of his career and closed a smashing handicap campaign of seven consecutive wins after two early-season defeats. He had been Horse of the Year at three despite Seattle Slew, and he repeated in that honor at four.

Wolfson, the sportsman and traditionalist, had said in the aftermath of the Triple Crown that he would not regard Affirmed as a "great" horse overall until he had proven himself as a weight carrier. That he did at four, his seven triumphs including the Californian under 130 pounds and the Hollywood

Gold Cup under 132. Affirmed had become racing's all-time leading earner with $2,393,818, having won twenty-two of twenty-nine races. Harbor View, having won the statistical battles years before, was accorded additional quality distinctions, being voted the Eclipse Award for breeder in 1978 and for owner in both 1978 and 1979.

Affirmed was an exceptional sire, although Alydar fulfilled predictions by becoming a better one. Affirmed endured a strange pattern of location, standing at two famous farms whose financial fortunes bedeviled them. Thus, he went from Spendthrift to Calumet Farm, before Wolfson and key adviser John Williams delivered him as a revered stallion to the Bell family's Jonabell Farm. Affirmed was at stud at Jonabell for a decade. He died at twenty-six in 2001 and is buried on the property, now owned by Sheikh Maktoum and known as Darley at Jonabell. A previously commissioned statue of Affirmed will stand outside a new stallion complex on the property. Such a tribute from a respectful farm owner who had no prior connection to the horse emphasizes the status of Affirmed in the international Thoroughbred community.

Affirmed sired eighty-five stakes winners (10 percent), one of whom was Harbor View's sterling two-time turf female champion Flawlessly. Others by Affirmed included Canadian Triple Crown winner Peteski, the stylish internationalist Zoman, Irish classic winner Trusted Partner, and the enduring sprinter-miler Affirmed Success. Affirmed is also the broodmare sire of ninety-six stakes winners, including Breeders' Cup Classic and Dubai World Cup winner Pleasantly Perfect.

In 1978, the year Affirmed won the Triple Crown and was Horse of the Year for the first of two seasons, Wolfson raced another champion, the two-year-old filly It's in the Air. This filly, who tied with Candy Eclair for the division title, was in the first wave of success for the Raise a Native stallion Mr. Prospector. It's in the Air was bred by Wolfson's sons' Happy Valley Farm and was from a daughter of Queen Nasra, she a *Nasrullah mare who had been purchased by the father and produced the good stakes winner Native Royalty. The bloodline was retrieved by the sons through the Francis S. mare A Wind Is Rising, who was the dam of It's in the Air. After Happy Valley sold It's in the Air for $25,000 to Jerry Frankel, the filly was promising enough that the senior Wolfson bought her for $300,000. After earning an Eclipse Award in the Harbor View silks, It's in the Air came back the next year to upset Calumet's champion Davona Dale in the Alabama. The Harbor View filly also won the Vanity and Ruffian among a four-year campaign total of nine major stakes and had a career mark of sixteen wins in forty-three starts and earnings of $892,339. (She foaled three stakes winners and is the second dam of Gainsborough Farm's French classic winner Musical Chimes.)

Wolfson bred ten stakes winners by Raise a Native, twelve by Exclusive Native, and twelve by Affirmed, most of these individually, but also including various partnerships. Thus, the shadows from his early purchase of Raise a Native have been lengthy and lovely. Of his other early purchases that proved successful on the racetrack, Roman Brother was a gelding and Roving Minstrel died young, while Francis S. was only a moderate success at stud.

A Triple Crown winner is the obvious pinnacle for a breeder (although Belair Stud and Calumet Farm had two each). In the quarter-century since Affirmed, Wolfson has continued with consistent success. He was instrumental in helping his wife's family maintain a few of their best mares even before their wedding. Patrice Jacobs Wolfson's father, Hirsch Jacobs, had died in 1970, and he left the wish that the horses be sold. "He thought the stable was too big for me to manage," his widow, Ethel D. Jacobs, said. The family balanced sentiment against economics and decided not to divest itself totally of its Thoroughbred holdings. "It was through the support and aid of Lou Wolfson that our current stable was made possible," Mrs. Jacobs said in the 1980s. "I owe him a great deal of gratitude."

"I think he was afraid of the natural hazards of the business," Wolfson said of Jacobs's plan for his family. "He didn't want Mrs. Jacobs to have any real wor-

ries or financial problems. I had a deep feeling and love for the family and simply went along with their feelings about continuing the stable."

One step involved Wolfson buying out longtime Jacobs partner Isidor Bieber's interest in certain mares. The combination of old Bieber-Jacobs Stable bloodlines and Harbor View bloodlines has been instrumental in the ongoing success. An early example was Desiree, who was by Wolfson's Raise a Native and out of the Bieber-Jacobs' champion Straight Deal. (Straight Deal was by Hail to Reason, the spectacular 1960 juvenile colt champion who raced in the colors of Patrice Jacobs. In a complex series of family connections, Miss Jacobs admitted to some dis-

Desiree

comfort with the declaration by her own father that Wolfson's Raise a Native was the best two-year-old he had ever seen. One assumes she did not know at the time that she would eventually be the wife of Raise a Native's owner.) Desiree, a foal of 1973, won the Santa Barbara Handicap in 1977. To the cover of Affirmed's tormentor Seattle Slew, Desiree foaled Adored. Bred by Harbor View and Mrs. Jacobs, Adored won seven stakes, including the Santa Margarita and Delaware handicaps, and earned nearly $900,000.

Desiree was but one of three stakes winners who

were bred by the Harbor View-Jacobs combine and foaled from Straight Deal. To the cover of Affirmed's sire, Exclusive Native, Straight Deal foaled Belonging, winner of the Typecast Stakes and the dam of Lane's End Farm's successful Danzig sire Belong to Me. The other of the Straight Deal stakes winners was Reminiscing, by Never Bend. Reminiscing, a foal of 1974, won the Sequoia, La Habra Stakes, and La Potranca Stakes, and then added to the sequence by foaling the stakes winners Commemorate, Persevered, and Premiership. Yet another foal from Straight Deal was So Endearing, whose son Qualify (by Danzig) won the Del Mar Futurity and finished second in the Breeders' Cup Juvenile Colts in 1986.

The first Breeders' Cup, in 1984, had a considerable Wolfson-Jacobs' influence. Commemorate (by Exclusive Native), identified above as a stakes winner from Straight Deal's daughter Reminiscing, raced for Windfields Farm. In the inaugural Breeders' Cup Sprint, Commemorate was second to Eillo. That same day, a Harbor View homebred, Outstandingly, was named the winner of the first Breeders' Cup Juvenile Fillies via disqualification,

BARBARA D. LIVINGSTON

Caress

and she also was the champion two-year-old filly of that season. Outstandingly was by Exclusive Native—La Mesa, by Round Table. In addition to the $1 million Breeders' Cup race for her division, Outstandingly won the $500,000 Hollywood Starlet Stakes. She retired with a record of ten wins in twenty-eight starts and earnings of $1,412,206.

Outstandingly's dam, La Mesa, was from Finance, a daughter of *Nasrullah and the grand mare Busanda (dam of Buckpasser). This was a segment of the *La Troienne female family nurtured by the Phipps family. Mrs. Wolfson's father, Hirsch Jacobs, had turned to this bloodline years earlier in the acquisitions of Searching and of Straight Deal's family, and the Wolfsons looked to the same well in 1979 in purchasing La Mesa privately.

Finance was bred by Ogden Phipps and then acquired by William K. Taylor, manager of the Hancock family's Claiborne Farm, where the Phipps horses are boarded. Finance was sold again, to Roger S. Braugh, who bred La Mesa and sold her for $92,000 to Normal Silberman at a mixed sale at Hollywood Park. The Wolfsons bought La Mesa early

			Polynesian (Unbreakable—Black Polly)
		Native Dancer, 1950	Geisha (Discovery—Miyako)
	Raise a Native, 1961		Case Ace (*Teddy—Sweetheart)
		Raise You, 1946	Lady Glory (American Flag—Beloved)
Exclusive Native, 1965			Equipoise (Pennant—Swinging)
		Shut Out, 1939	Goose Egg (*Chicle—Oval)
	Exclusive, 1953		Pilate (Friar Rock—*Herodias)
		Good Example, 1944	Parade Girl (Display—Panoply)
OUTSTANDINGLY			Prince Rose (Rose Prince—Indolence)
		*Princequillo, 1940	*Cosquilla (*Papyrus—Quick Thought)
	Round Table, 1954		Sir Cosmo (The Boss—Ayn Hali)
		*Knight's Daughter, 1941	Feola (Friar Marcus—Aloe)
La Mesa, 1970			Nearco (Pharos—Nogara)
		*Nasrullah, 1940	Mumtaz Begum (*Blenheim II—Mumtaz Mahal)
	Finance, 1955		War Admiral (Man o' War—Brushup)
		Busanda, 1947	Businesslike (Blue Larkspur—*La Troienne)

in 1979. La Mesa foaled the stakes winners Lovelier, in addition to Outstandingly, and also was the dam of the Affirmed filly La Affirmed. La Affirmed produced the Harbor View homebred Caress, a Storm Cat filly who won three grade III stakes among a total of seven stakes wins and earned $666,076. Caress in turn foaled Sky Mesa, whom Wolfson sold as a yearling in 2001 and who won the Hopeful and Breeders' Futurity for John Oxley. Sky Mesa represented the success Wolfson has had in going back to the auction market, but now as a seller instead of a buyer. Sky Mesa, in the second crop by the Claiborne Farm stallion Pulpit, brought $750,000 from Oxley as a Keeneland September yearling.

The most spectacular sale that adviser Williams developed for Wolfson had come in 1984. Williams, who then was managing Spendthrift, where the Harbor View mares were boarded, sent into the Keeneland July sale a Seattle Slew—Desiree colt that brought $6.5 million from the Darley Stud Management group of Sheikh Mohammed. In 2001 Eaton Sales sold for Wolfson a Storm Cat—La Affirmed colt for $5.5 million at the Keeneland September auction.

Champions Bred
Athenian Idol
1973 champion steeplechaser
Affirmed
1977 champion two-year-old colt
1978 champion three-year-old colt
1978 Horse of the Year
1979 champion handicap male
1979 Horse of the Year
It's in the Air
1978 champion two-year-old filly
Outstandingly
1984 chamption two-year-old filly
Flawlessly
1992 champion turf female
1993 champion turf female

Harbor View's homebred champion Flawlessly was an Affirmed filly from La Confidence, a daughter of the English Triple Crown winner and international sire Nijinsky II. La Confidence represented another of the top female families. Her dam, La Dame Du Lac, was a Round Table filly from Cosmah, she a Broodmare of the Year and dam of Halo and Tosmah. La Confidence was bred by Celler Stables and was purchased by Wolfson as a 1981 Saratoga sale yearling for $400,000.

Flawlessly was a promising stakes winner on the East Coast at two for trainer Dick Dutrow and then was sent to the masterful trainer Charlie Whittingham in California. She prevailed as the champion female on grass in both 1992 and 1993, following Miesque as only the second to have won the Eclipse Award twice in that division since its establishment in 1979. In addition to being a model of consistency, Flawlessly was the mistress of repetition: she won both the grade I Matriarch Stakes and Ramona Handicap three times each among a total of nine grade I stakes victories. Flawlessly, inducted into racing's Hall of Fame in 2004, had a career record of sixteen wins in twenty-eight starts and earnings of $2,572,536.

		Raise a Native, 1961	**Native Dancer** (Polynesian—Geisha)
			Raise You (Case Ace—Lady Glory)
	Exclusive Native, 1965	Exclusive, 1953	Shut Out (Equipoise—Goose Egg)
			Good Example (Pilate—Parade Girl)
Affirmed, 1975		Crafty Admiral, 1948	Fighting Fox (*Sir Gallahad III—Marguerite)
			Admiral's Lady (War Admiral—Boola Brook)
	Won't Tell You, 1962	Scarlet Ribbon, 1957	Volcanic (*Ambrose Light—Hot Supper)
			Native Valor (*Mahmoud—Native Gal)
FLAWLESSLY		Northern Dancer, 1961	Nearctic (Nearco—*Lady Angela)
			Natalma (**Native Dancer—Almahmoud**)
	Nijinsky II, 1967	Flaming Page, 1959	Bull Page (Bull Lea—Our Page)
			Flaring Top (Menow—Flaming Top)
La Confidence, 1980		Round Table, 1954	*Princequillo (Prince Rose—*Cosquilla)
			*Knight's Daughter (Sir Cosmo—Feola)
	La Dame du Lac, 1973	Cosmah, 1953	Cosmic Bomb (*Pharamond II—Banish Fear)
			Almahmoud (*Mahmoud—Arbitrator)

Wolfson had sold his Florida farm in the late 1970s and boarded mares in several states. After the demise of the Combs family's Spendthrift Farm, Wolfson has boarded the majority of his mares at Williams' Ballindaggan Farm and now at Mr. and Mrs. Williams' new farm, Elmwood, outside Versailles, Kentucky. Primarily on his own but also involving family and outside partnerships, Louis Wolfson so far has bred a total of ninety-five stakes winners.

He had harbored an ambition to breed his own, rather than buy, and that was an ambition abundantly fulfilled.

Stakes Winners Bred by Harbor View Farm

Morry E., b.g. 1959, Crafty Admiral—Tuonine, by Reaping Reward

Master Dennis, ch.g. 1960, *Alcibiades II—Tuonine, by Reaping Reward

Irvkup, ch.g. 1961, *Alcibiades Ii—Exclusive, by Shut Out

Gary G., b.g. 1963, Sailor—*Avenida II, by Souverain

Mellow Marsh, ch.f. 1963, *Seaneen—Exclusive, by Shut Out

Abifaith, ch.f. 1964, *Seaneen—Sherry Jen, by Sun Again

Duke Cannon (stp.), ch.c. 1964, *Alcibiades II—*Polly Toogood, by Darius

Georgia Joe, b.c. 1964, *Wolfram—Joan's Cathy, by *Alcibiades II

American Native, ch.c. 1965, Raise a Native—Sherry Jen, by Sun Again

Bo Rama, b.g. 1965, Quiz Star—*Mosul II, by Mossborough

Exclusive Native, ch.c. 1965, Raise a Native—Exclusive, by Shut Out

Kernel Marty (stp.), b.g. 1965, Kerne—Fascinatin Fanny, by Citation

Sparkling Native, ch.c. 1965, Raise a Native—Sparkle, by Count Fleet

Creeque Alley, dk.b/br.f. 1966, *Alcibiades II—Dancing Lark, by Native Dancer

End of Time, ch.c. 1966, Francis S.—Truckle Time, by *Fair Truckle

Happy Intellectual, dk.b/br.g. 1966, *Wolfram—Bright n'Gay, by Citation

Lion Sleeps, b.g. 1966, Johnasark—What a Bird, by Papa Redbird

Doc Kope, dk.b/br.g. 1967, Stevward—Jet Wave, by Ace Destroyer

Exclusive Dancer, ro.f. 1967, Native Dancer—Exclusive, by Shut Out

Native Heritage, ch.c. 1967, Raise a Native—Tim's Princess, by Tim Tam

Native Royalty, b.c. 1967, Raise a Native—Queen Nasra, by *Nasrullah

Oh Fudge, ch.g. 1967, *Alcibiades II—Fascinatin Fanny, by Citation

Sign of the Times, ch.f. 1967, Francis S.—Jolly Princess, by *Princequillo

Sun Lover, ro.f. 1967, Nasomo—Sunshine Gino, by Sun Again

Athenian Idol, ch.g. 1968, *Alcibiades II—Dottys Dream, by Francis S.

Sonny Says Quick, dk.b/br.f. 1968, Stevward—Dancing Lark, by Native Dancer

Swift Pursuit, b.c. 1968, *Alcibiades II—Swift Lass, by Johnasark

Life Cycle, b.c. 1969, *Wolfram—Katira, by Correspondent

Specialamente, b.g. 1969, *Alcibiades II—Rather Special, by Third Brother

Faithful Girl, ch.f. 1970, Exclusive Native—Charlo, by Francis S.

Swifty Gal, b.f. 1970, Nimble Nate—Extra Guest, by Cravat

Green Gambados, b.c. 1971, Swaps—Cargreen, by *Turn-to

Princely Native, ch.c. 1971, Raise a Native—Charlo, by Francis S.

Raisela, b.f. 1971, Raise a Native—La Verde, by Yatasto

Due Diligence, b.c. 1972, Stevward—Rather Special, by Third Brother

Featherfoot, b.c. 1972, Bold Native—Magic Garden, by Francis S.

Miss Florida, ch.g. 1972, Bold Native—Hey Dolly, by Francis S.

With Dignity, ch.c. 1972, Bold Native—Swinging Melody, by Francis S.

Life's Hope, b.c. 1973, Exclusive Native—Tim Marie, by Tim Tam

Root Cause, ch.g. 1973, Exclusive Native—Any Port, by Sailor

Affiliate, ch.c. 1974, Unconscious—Swinging Doll, by Raise a Native

All Arranged, b.c. 1974, Stevward—Miss Francie, by Francis S.

Little Happiness, b.f. 1974, Raise a Native—Tim Marie, by Tim Tam

Affirmed, ch.c. 1975, Exclusive Native—Won't Tell You, by Crafty Admiral

Breezing On, dk.b/br.c. 1976, Stevward—Shalimar Gardens, by Raise a Native

Eloquent, ch.f. 1976, Exclusive Native—Mona Pucheau, by Prince Taj

Love You Dear, ch.f. 1976, Bold Native—Won't Tell You, by Crafty Admiral

A Little Affection, ch.f. 1977, King Emperor—Chicken Little, by Olympia

Gratification, ch.c. 1977, Princely Native—Bird Island, by Tumiga

Hail to the Queen, b.f. 1977, Native Royalty—Instinctively, by Francis S.

Andy's Wish, ch.c. 1978, Transalantic—Hey There (Fr), by American Native

For Safekeeping, ch.f. 1978, Stevward—Happy Huldah, by Happy Nasrullah

Imperial Lass, ch.f. 1978, Princely Native—Gusellie, by One-Eyed King

The Captain, ch.c. 1979, Stevward—Lonesome Highway, by Francis S.

Dominating Dooley, ch.c. 1980, Teddy's Courage—Frajan, by Francis S.

Treasured One, ch.c. 1981, Exclusive Native—Monteen, by Hail to Reason

Affirmance, ch.f. 1982, Affirmed—Politician, by Buckpasser

Entitled To, ch.c. 1982, Golden Act—Ragtime Girl, by Francis S.

Happy Bid, b.c. 1982, Spectacular Bid—Little Happiness, by Raise a Native

Outstandingly, b.f. 1982, Exclusive Native—La Mesa, by Round Table

Solidified, dk.b/br.c. 1982, Tell—Swingin Axe, by The Axe II

Festivity, gr.f. 1983, Spectacular Bid—Dancing On, by Dancer's Image

Lovelier, dk.b/br.f. 1984, Affirmed—La Mesa, by Round Table

Only Companion, b.c. 1984, Exclusive Era—Companionship, by Princely Native

Perfecting, dk.b/br.c. 1985, Affirmed—Cornish Colleen, by Cornish Prince

Stedes Wonder, b.f. 1985, Native Royalty—Bestowed, by Stevward

Endow, b.c. 1986, Flying Paster—Kindheartedness, by Exclusive Native

Love and Affection, b.f. 1986, Exclusive Era—A Little Affection, by King Emperor

Reaffirming, ch.c. 1986, Affirmed—Crafty Alice, by Crafty Admiral

Zoman, ch.c. 1987, Affirmed—A Little Affection, by King Emperor (SW in Ireland/France/U.S.)

Flawlessly, b.f. 1988, Affirmed—La Confidence, by Nijinsky II

Elegant Buck, gr.c. 1990, Silver Buck—Elegantly, by Danzig
Caress, dk.b/br.f. 1991, Storm Cat—La Affirmed, by Affirmed
Lady Affirmed, ro.f. 1991, Affirmed—Festivity, by Spectacular Bid
Perfect, b.c. 1992, Affirmed—La Confidence, by Nijinsky II
Fortitude, b.c. 1993, Cure the Blues—Outlasting, by Seattle Slew
Renewed, dk.b/br.c. 1994, Lost Code—Nifty, by Roberto
Sky Mesa, b.c. 2000, Pulpit—Caress, by Storm Cat

Harbor View Farm and Ethel D. Jacobs
Desiree, b.f. 1973, Raise a Native—Straight Deal, by Hail to Reason
Reminiscing, b.f. 1974, Never Bend—Straight Deal, by Hail to Reason
Belonging Deal, b.f. 1979, Exclusive Native—Straight Deal, by Hail to Reason
Adored, b.f. 1980, Seattle Slew—Desiree, by Raise a Native
Premiership, b.c. 1980, Exclusive Native—Reminiscing, by Never Bend

Harbor View Farm and Spendthrift Farm
Exclusive One, ch.c. 1979, Exclusive Native—La Jalouse, by Nijinsky II
Lover Boy Leslie, ch.c. 1980, Raise a Native—Clairvoyance, by Round Table
My Darling One, ch.f. 1981, Exclusive Native—Princess Marshua, by Prince John

Princess Claire, ch.f. 1981, Princely Native—Clairvoyance, by Round Table

Harbor View Farm, Dr. Walter Jacobs, Dr. Robert Jaffe
Dice Passer, b.f. 1984, Affirmed—Pass the Dice, by Buckpasser

Harbor View Farm and Manganaro Stables
Lil's Memory, ch.f. 1992, Affirmed—Giboulee Era, by Giboulee

Mr. & Mrs. Louis E. Wolfson and Ethel D. Jacobs
Commemorate, b.c. 1981, Exclusive Native—Reminiscing, by Never Bend
Persevered, b.c. 1984, Affirmed—Reminiscing, by Never Bend
Qualify, b.c. 1984, Danzig—So Endearing, by Raise a Native
Notoriety, b.g. 1993, Affirmed—Endearingly, by Lyphard
Compassionate, b.f. 1995, Housebuster—Adored, by Seattle Slew

Mr. & Mrs. Louis E. Wolfson, Ethel D. Jacobs, and Seahorse Farm
Mais Oui, b.f. 1987, Lyphard—Affirmatively, by Affirmed (SW in France)

Source: *The Blood-Horse* Archives

CHAPTER

14

W.T. Young

Quality always appealed to W.T. Young. When he and his wife were a young couple, he recalled, they would tend to go some time between purchases of furniture, for example, but save up enough to buy a distinctive antique when they did buy something. Young's combination of the three Ds of his business life — discipline, determination, and discernment — eventually made it possible to acquire much of what he wanted, without a great deal of waiting.

Born in Lexington, Kentucky, on February 15, 1918, William T. Young graduated from the University of Kentucky with high distinction in mechanical engineering in 1939. At the time of his death at eighty-five on January 12, 2004, he had become regarded, and revered, as one of the giants of Kentucky life, of American business, and of the sport of the Thoroughbred. As a silvery-white-haired senior citizen, he presented all the gentility, generosity, and graceful accent anyone might ascribe to the image of a Southern gentleman of the best sort.

As a young businessman, returning from military service as a major, he showed the heart of an investment gambler, as well. Young established Big Top peanut butter and sold it to Procter & Gamble, which decided to call it Jif. He also was a chairman of Royal Crown Cola, and his other enterprises have included the Humana health care group and his own warehousing firm in Lexington.

Financial success was accompanied by a sense of civic duty, not the least of which was contributing of $5 million for the University of Kentucky to establish a library that bears his name and is a resource not just for one campus but for the entire state. A collector of art, Young also helped preserve for Kentucky the unusual loveliness and thought-provoking history embedded in the old Shaker colony known as Pleasant Hill, and he crossed town from his alma mater to serve as chairman of historic Transylvania University in Lexington.

Obliquely reminiscent of his days of his antique-buying discipline, Young did not enter Thoroughbred racing until he could do it in style. Even his acquisition and development of the immaculate Overbrook Farm had an undercurrent of the shrewd investor, however, for the property is in the path of Lexington's expansion, which all but guarantees its significance as a financial resource for whichever generation of his descendants might choose otherwise for its use. Its future as a farm seems certain for the present as, following his death in 2004, the family has continued the operation. Young's son, William Jr., and daughter Lucy have key roles, and Mr. Young's grandson, Chris Young, has taken on the job of racing manager.

W.T. Young would have been the last to suggest that luck is not a major part of success in Thoroughbred breeding, but he brought business principles to the endeavor. He conceived early that racing was the key to maximizing the possibility of multi-fold expansion of a horse's value. A yearling with good relatives and fine conformation might sell

215

well but thereafter is out of the seller's hands and management. A yearling retained as a racehorse might fail to achieve anything, but one really good stallion with a stylish pedigree would be worth multiples of what he had been worth as a sale yearling.

With astounding alacrity, W.T. Young came upon a horse to justify, magnify, and exemplify this philosophy. Storm Cat was his name, and the route to his greatness was fraught with examples of how the old adage of breeding the best to the best can fill or empty, your pockets according to the whims of chance.

Young was forthcoming in his praise of luck in the scenarios that placed Storm Cat atop the list of fashionable, and profit-producing, sires and fortuitously placed him in the Overbrook Stud barn. In the late 1970s Young had developed the lovely property of Overbrook and was ready to stock the place with horses. As will happen, bloodstock agents of various stripe were eager to help. Young turned to his friend John A. Bell III, master of Jonabell Farm and a Lexington citizen of stature known for integrity in all his business dealings. Veterinarian Bob Copelan and venerable adviser Vic Heerman also were consulted, and lifetime friend and business/philanthropy associate Alex Campbell also was on hand for initial guidance. None of these are the sort to hustle a fellow into an expensive package. Surely, there was comfort in this knowledge, but to go first class meant paying big numbers to somebody for the germination of a broodmare band.

Enter Dr. Bill Lockridge, who has on occasion said to the author that he understands that his compulsion to go not just first-class but luxury class has had its downside. In the late 1970s Lockridge's project of the moment was Ashford Stud, which he was developing grandly with Robert Hefner of Oklahoma. Around that same time, Lockridge promoted to Young a package of three high-class mares: Three Troikas, winner of the Prix de l' Arc de Triomphe; Terlingua, a spectacular racing daughter of the great Secretariat; and Cinegita, another stakes winner by Secretariat.

If there is one formula in breeding that gives the best chance of success, it is to deal in quality; these mares promised just that. It is not surprising to report, however, that the best of these mares as a racehorse, Three Troikas, turned out to be the least successful as a breeding animal. Lady Luck was not,

W.T. Young

ANNE M. EBERHARDT

after all, named for her predictable nature. Terlingua and Cinegita, however, were the stuff of dividends beyond logical expectations.

An additional cog in the scenario headed toward ultimate success was Storm Bird. As a son of the enormously successful and fashionable stallion Northern Dancer, Storm Bird had been a $1 million yearling bought by the high-rolling international team headed by Robert Sangster. He won all his five races at two in 1980 and was top-ranked on both the English and Irish juvenile handicaps. The following year a bizarre turn was handed this tale, when a disgruntled ex-employee broke into the stable and hacked at the colt's mane and tail. Despite his trauma Storm Bird was not taken out of training, but it was announced in the grand style of the day that an American group had purchased him on the basis value of $30 million. Storm Bird raced once more, failed to win, and was retired to Ashford. There his mates the following year included Terlingua (Secretariat—Crimson Saint, by Crimson Satan). An early star in the launching of D. Wayne Lukas as a Hall of Fame Thoroughbred trainer, Terlingua had defeated colts in the Hollywood Juvenile Championship and won six other stakes. (Her dam, Crimson Saint, also foaled the stakes winners Pancho Villa, Alydariel, and Royal Academy, the latter a Breeders' Cup mile winner and a major international stallion.)

"It really has to be considered pure luck," Young emphasized to *The Blood-Horse* in addressing the existence and career of Storm Cat. "First of all, Bill Lockridge is responsible for the mating that produced Storm Cat. He talked me into buying Terlingua. Ashford owned half, and I owned half. I later bought them out when they got into financial trouble."

The lucky butterfly's flitting was not complete: "Storm Cat (the Storm Bird—Terlingua colt) was entered in the 1984 Keeneland summer yearling sale, but Keeneland asked us to take him out and sell him instead in September because he tested positive (for equine viral arteritis). I couldn't understand that decision, so I decided to race him."

Young turned Storm Cat over to trainer Jonathan Sheppard, and the Overbrook youngster came within a nostril of being named 1985 champion two-year-old colt. He had finished second in the World Appeal Stakes and won the grade I Young America Stakes before being run down in the final stride by Tasso in the Breeders' Cup Juvenile. The gentleman known as "Mr. Young" did not induce many images of embarrassment, but on that occasion he learned a lesson never to be forgotten when he took the optimistic word of those around him that he had won the photo and started down to the winner's circle. Tasso's nose had prevailed, but so had Lady Luck in her taste for irony. Young later said that he had been offered $8 million for Storm Cat, but, "because he lost, his value was less, and I kept him."

So, the relative newcomer-sportsman, whose belief that raising one's own stallion prospects was the way to go, had undergone three nudges into following his own idea. Storm Cat entered stud at Overbrook with a record of four wins in eight starts and earnings of $570,610. Knock $70,610 off that lifetime racing record, and you have the fee for a single breeding that Storm Cat would command almost twenty years later.

As of mid-2004, Storm Cat had already passed the one hundred-stakes-winner mark, siring 119 (11 percent). He had led the North American sire list twice

Storm Cat

ANNE M. EBERHARDT

SKIP DICKSTEIN

Tabasco Cat

project, and the care, attention, and settling down turned the classy chestnut into a classic winner. Tabasco Cat won the 1994 Preakness and then underscored the growing indication of Storm Cat's ability to sire stayers when he added the mile and a half Belmont Stakes. (Ordinarily, such a campaign would secure a championship, but Holy Bull's form outside the Triple Crown was so gaudy that he was champion three-year-old and Horse of the Year.)

Tabasco Cat also induced a footnote in Turf history when the makers of Tabasco sauce showed their corporate image to be ironically bland and challenged any future use of the name for horses. Tabasco Cat won eight of eighteen races and earned $2,347,671. He returned to stand at Overbrook and was sold to Japan in 2000 as the market's interest waned, perhaps prematurely; he left behind developing stakes winners such as grade I winners Habibti and Snow Ridge.

Cat Thief, by Storm Cat, illustrated other purchases into major families that have marked Overbrook's growth to the level of having one hundred mares. (Young always emphasized he was not a horseman, but he led his team of horsemen and advisers and made many final decisions. Ric Waldman, brought on board to appraise the bloodstock at Campbell's suggestion, later moved his office to Overbrook and has a key and continuing advisory role.) Lukas purchased Train Robbery for Young for $600,000 at the 1988 Keeneland summer sale. She was by Alydar and out of Track Robbery, 1982's champion older filly or mare. Train Robbery won the grade III Monmouth Park Breeders' Cup Handicap and Honeybee Stakes, plus the non-graded Remington Park Oaks and Rolling Meadows Stakes, and she placed in several grade I races.

Cat Thief earned nearly $4 million, or an average of about $1 million for every victory. His own spotty win-loss record did not seem to measure his class

(1999-2000), ranked second once in England/Ireland (2000), and led the juvenile sire list here six times. Storm Cat's best among Overbrook's own included Preakness-Belmont winner Tabasco Cat and Breeders' Cup Classic winner Cat Thief. Other breeders and/or buyers benefited from Storm Cat with juvenile filly champion Storm Flag Flying; the "Iron Horse" of European group I races, Giant's Causeway; and such other grade/group I winners as Sharp Cat, Raging Fever, November Snow, Sardula, Desert Stormer, Black Minnaloushe, Aldiza, Hennessy, High Yield, Aljabr, and Hold That Tiger. Many of Storm Cat's best runners have been fillies, and, in the expected sequence, he is a developing presence as a broodmare sire, with forty stakes winners in that context.

Tabasco Cat (Storm Cat—Barbicue Sauce, by Sauce Boat), raced by Young in partnership with David G. Reynolds, owns a place in history that combines tragedy and glory. He was the horse who broke loose one morning in the barn of trainer Lukas and lumbered into his son and assistant, Jeff. The young man's life hung in the balance for some time before he recovered. The senior Lukas, noting a tendency in the barn to resent the colt, took him on as a special accurately. For example, Cat Thief won only two of thirteen races at three, but they were the grade I Breeders' Cup Classic and Swaps Stakes. He was third in the Kentucky Derby. At four he failed to win in ten starts but placed in five graded races. Cat Thief had won two of seven as a juvenile, taking the Lane's End Breeders' Futurity at Keeneland, and was third in the Breeders' Cup Juvenile. This came to a career record of four wins in thirty starts, placings in fourteen stakes, and earnings of $3,951,012. Cat Thief now stands at Overbrook.

Further illustrating the impact of Storm Cat on Overbrook's success, Young bred thirty-two stakes winners by the stallion Inevitably, some raced for other owners. Despite his philosophy of keeping homebreds, practicality dictated that some yearlings be marketed. Young maintained a lucrative sales arm, and those he bred and sold include the international stallion Hennessy, plus Sardula and November Snow. Hennessy (Storm Cat—Island Kitty, by Hawaii), winner of the grade I Hopeful and two grade II races at two, made an immediate mark at stud with the Breeders' Cup Juvenile winner and international champion Johannesburg. (In the same crop as Hennessy was the Overbrook-bred Honour and Glory, a son of Relaunch. Honour and Glory was a major stakes winner and then led the juvenile sire list of 2000, one year before Hennessy led that list. Since Storm Cat had led in 1998-99 and 2002, this meant five years of leadership for horses bred by Young.)

Another son illustrating Storm Cat's rising mark as a sire of sires was Harlan. Young raced Harlan in partnership with Arthur Hancock III. The horse went to stand at Hancock's Stone Farm, where he died young, but not before siring the grade I winner and classics-placed Menifee as well as grade I winner Harlan's Holiday.

The Blood-Horse Stallion Register for 2004 listed ninety sons of Storm Cat at stud, those not mentioned above including Storm Boot, Stormy Atlantic, Tale of the Cat, and Belmont Stakes runner-up Vision and Verse.

The success of Storm Cat as a stallion and sire of stallions was the upside of fortune for W.T. Young. As is true of all breeders, he also saw the other side. One of the most dazzling stallion prospects he ever raced — in truth, probably seen as an even better prospect than Storm Cat — was Grand Canyon (Fappiano—Champagne Ginny, by L'Enjoleur). Grand Canyon was bred by Lin-Drake Farm of Florida and purchased by Lukas for Young for $825,000 at the 1988 Keeneland summer sale. Although outvoted for juvenile colt honors because he was second to Rhythm in the Breeders' Cup Juvenile of 1989, Grand Canyon created a reputation as a major classic contender with his dazzling scores in the Kentucky Jockey Club Stakes and Hollywood Futurity late that year. He had won the Norfolk Stakes earlier. Grand Canyon, however, succumbed to laminitis the next year. Another major loss was Union City. A high-class prospect by Private Account, Union City suffered a fatal injury in the Preakness Stakes.

STORM CAT				
Storm Bird, 1978	Northern Dancer, 1961	Nearctic, 1954	**Nearco** (Pharos—Nogara)
			*Lady Angela (Hyperion—Sister Sarah)
		Natalma, 1957	Native Dancer (Polynesian—Geisha)
			Almahmoud (*Mahmoud—Arbitrator)
	South Ocean, 1967	New Providence, 1956	Bull Page (Bull Lea—Our Page)
			*Fair Colleen (Preciptic—*Fairvale)
		Shining Sun, 1962	Chop Chop (Flares—*Sceptical)
			Solar Display (Sun Again—Dark Display)
Terlingua, 1976	Secretariat, 1970	Bold Ruler, 1954	*Nasrullah (**Nearco**—Mumtaz Begum)
			Miss Disco (Discovery—Outdone)
		Somethingroyal, 1952	*Princequillo (Prince Rose—*Cosquilla)
			Imperatrice (Caruso—Cinquepace)
	Crimson Saint, 1969	Crimson Satan, 1959	Spy Song (Balladier—Mata Hari)
			*Papila (Requiebro—Papalona)
		Bolero Rose, 1958	Bolero (Eight Thirty—Stepwisely)
			First Rose (Menow—Rare Bloom)

Golden Attraction

BARBARA D. LIVINGSTON

Besides Storm Cat, another of the most spectacular of Overbrook's winners traced to that purchase of Three Troikas, Terlingua, and Cinegita. This was Flanders, whose dam, Starlet Storm, was from the last-named. In breeding Starlet Storm, Overbrook also turned to Storm Cat's sire, Storm Bird. Racing only at three, Starlet Storm was undefeated in two starts and was retained for breeding.

A breeder who set out to breed and stand his own stallions, and who was initially successful at it, might be tempted to rely primarily on his own breeding shed. Young did not fall into that temptation, and he reached out to a number of the best sires on other farms. Starlet Storm was bred first to Alydar at Calumet Farm and the next year to Seeking the Gold at Claiborne Farm. The Alydar foal was moderate; the Seeking the Gold foal, a champion.

Seeking the Gold was a young, unproven stallion at the time, but was a very high-class racehorse by the great stallion Mr. Prospector and had the richness of a Phipps family behind him. Flanders, in his second crop, was turned over to Lukas. Although a West Coast-oriented trainer through most of his career, Lukas had adapted to the time-honored pattern of bringing out well-bred juveniles at Saratoga. The chestnut Flanders appeared there on August 10, 1994, and won a maiden race by seven and a half lengths. Nineteen days later she was a division leader, dominating the grade I Spinaway to win by

nearly five lengths. She was a front-runner, but when Lukas moved her up to a mile, she had no trouble winning the Matron by three and a quarter lengths; she had trouble in hanging on to the victory, though, for a technical medication finding resulted in a disputed disqualification. Her status as an officially unbeaten racehorse was thus denied.

The Frizette Stakes at a mile and a sixteenth resulted in one of the larger margins ever recorded for a grade I race. Flanders needed the work and was ridden out by jockey Pat Day long after the race was secure, and she rambled home twenty-one lengths in front. Then came the Breeders' Cup Juvenile Fillies, one of those epics that could stand alone as a syllabus on the Thoroughbred. Flanders was hooked this time, and she was hooked by a filly we were later to recognize as something special in her own right — Hall of Famer Serena's Song. They were battling after a half-mile, and they were battling turning for home. They were battling at the furlong poll, and they were battling at the wire. Flanders prevailed by a head at 2-5, but she was vanned off lame and did not race again. Her record officially stands at four wins in five starts and earnings of $805,000. She was voted the champion juvenile filly of 1994. She quickly became a champion who foaled a champion. Her 1997 Seattle Slew filly, Overbrook's Surfside, bridged the century turnover as the champion three-year-old-filly of 2000.

At two Surfside had been close to championship consideration by winning the grade I Starlet and Frizette. Three dashing scores in Santa Anita stakes in the winter of her three-year-old season, including two grade Is, encouraged Lukas to try her against colts in the Santa Anita Derby, but Surfside finished fifth. She was not seen under colors again until autumn, when she finished second in the Raven Run Stakes and in Spain's Breeders' Cup Distaff. A four-length victory over males in the grade I Clark Handicap at a mile and an eighth completed a strangely configured campaign for a three-year-old filly and earned her the Eclipse Award for the division. Surfside was retired to the breeding shed after only two unsuccessful starts at four and had a career record of eight wins in fifteen starts and earnings of $1,852,987.

Young had also continued to look for racing prospects from outside. The day Flanders won the Breeders' Cup Juvenile Fillies, Young also was co-owner of the Juvenile Colt winner and juvenile male champion, Timber Country. He owned that Lukas yearling purchase in partnership with Graham Beck and the owners of Serena's Song, Robert and Beverly Lewis. Timber Country came back to win the Preakness the following year.

The year after Flanders, Overbrook again raced the homebred juvenile filly champion, although this one did not win on Breeders' Cup Day. Golden Attraction was by the Claiborne stallion Mr. Prospector and was out of a filly who had delivered Young one of a Kentuckian's most sentimental victories. The dam was Seaside Attraction, whom Young had purchased for $1,050,000 as a weanling in the Warner L. Jones Jr. dispersal at Keeneland in 1987. The filly was bred in partnership by Jones, owner of Hermitage Farm and board chairman of Churchill Downs, with Three Chimneys Farm owner Robert N. Clay and his father, Albert Clay. Seaside Attraction was by Triple Crown winner and leading sire Seattle Slew and out of Kamar, by Key to the Mint. Kamar was Broodmare of the Year in 1990, the year Seaside Attraction, one of her four stakes winners, won the venerable Kentucky Oaks for Young.

Seaside Attraction's Mr. Prospector filly Golden Attraction did not deal in such gaudy margins as Flanders, but she won many of the same key races. Lukas started this one on the West Coast, where she won her debut at Hollywood Park, but she worked her way East, winning the Debutante at Churchill Downs before hitting Saratoga. There she won the Schuylerville and Spinaway, with a road trip down to Monmouth Park, where she was second in the Sorority. Golden Attraction beat Cara Rafaela by a neck in the Matron and My Flag by three parts of a length in the Frizette, but on the muddy track of Breeders' Cup Day at Belmont, both of those beat her. The Overbrook filly's superb record of six wins in eight starts prevailed in voting, however, and she was the champion juvenile filly of 1995. She won two of

three at three and was retired with eight wins in eleven starts and earnings of $911,508. Seaside Attraction also foaled the Florida Derby winner Cape Town and San Miguel Stakes winner Cape Canaveral for Overbrook.

A fourth champion two-year-old in three years (counting Timber Country) was homebred Boston Harbor. He, too, was by an outside stallion, Three Chimney's Breeders' Cup Juvenile winner Capote. Boston Harbor's dam, Harbor Springs, by the Northern Dancer stallion Vice Regent, was a $500,000 purchase as a Keeneland summer yearling in 1990. Harbor Springs won the Wishing Well Stakes and six other races and earned $123,038. She represented a family of upward mobility. There were a few nice stakes horses produced in the third and fourth generations of Boston Harbor's pedigree, and then the second dam produced Groovy, who broke through the class barrier to become an Eclipse Award-winning sprinter in the latter 1980s.

Boston Harbor, the first foal from Harbor Springs, made his debut for Lukas in late May of the

Grindstone

BARBARA D. LIVINGSTON

Churchill Downs spring-summer meeting of 1996, winning a maiden race by five lengths and the Bashford Manor Stakes by four. At Saratoga he was second to Kelly Kip in the Sanford, and he then was shipped back to Kentucky to win the Ellis Park Juvenile by six lengths and the Kentucky Cup Juvenile by seven. At Keeneland he won the Breeders' Futurity by a half-length. The sweep of the four Kentucky stakes victories qualified him for a $1 million bonus provided via the Kentucky Thoroughbred Association and Kentucky Thoroughbred Owners and Breeders Association. This was the second handy million picked up by Young via this short-lived program. His Mountain Cat had swept the designated four stakes to win the bonus in 1992. (Prior to establishment of the Kentucky Cup Day, the designated Turfway Park race in the series at that time was the Alysheba Stakes.)

The Breeders' Cup was held at Toronto's Woodbine in 1996, and Boston Harbor led throughout, holding off Acceptable to win by a neck. He had won six of seven races and prevailed as Eclipse Award winner in the juvenile male division. He finished fourth in his only start at three before unsoundness problems sent him to stud at Overbrook. Boston Harbor won six of eight and earned $1,934,605.

Young, a member of The Jockey Club (as is his daughter, Lucy Hamilton), was voted the Eclipse Award for breeders in 1994, and his other honors include selection as Thoroughbred Club of America honor guest at its annual Testimonial Dinner in 1996.

Young is the breeder to date of 106 stakes winners.

In 1996 the victory that resides as the seed and flower of horsemen's dreams came to W.T. Young. He won the Kentucky Derby with a homebred. Again he emphasized the importance of luck. Young's good friend and business associate Carl Pollard had approached him to help with a fund-raising auction to benefit the Kentucky Derby Museum, of which Pollard was chairman. Young bid $30,000 on a donated season to 1990 Kentucky Derby winner Unbridled. At the time, it was seen as a generous bid, although Unbridled's later record, both before and after his early death, indicated he might have been destined to greatness as a sire.

The Unbridled season was matched to Buzz My Bell, a Drone mare who had won the Spinaway and Adirondack stakes in 1983 and had been purchased as a broodmare by Overbrook for $2,050,000 at the 1985 Keeneland November sale. Named Grindstone, the Unbridled—Buzz My Bell colt won early for Lukas and Young at two, then at three prepped for the classics with a victory in the Louisiana Derby and a second in the Arkansas Derby. On May 4, 1996, Grindstone was part of a Kentucky Derby entry with the Overbrook stablemate Editor's Note — a yearling purchase destined to win the Belmont Stakes five weeks later. Grindstone turned for home fourth, with Cavonnier seemingly home free, but in a grinding, slashing stretch rally jockey Jerry Bailey brought the Overbrook colt relentlessly to the leader's throat.

Since Storm Cat's harrowing loss in the Breeders' Cup Juvenile eleven years before, Young had schooled himself not to approach any winner's circle that might not welcome him. He and daughter Lucy and lifelong friend Campbell fretted and fidgeted in Young's front-row box above the finish line of Churchill Downs. The minutes passed slowly, the minutes passed torturously, and then the telltale lights went up.

W.T. Young, Kentuckian, had won the Derby. The winner's circle awaited.

Champions Bred

Flanders
1994 champion two-year-old filly

Golden Attraction
1995 champion two-year-old filly

Boston Harbor
1996 champion two-year-old colt

Surfside
2000 champion three-year-old filly

Stakes Winners Bred by W.T. Young

Northern Eternity, b.f. 1983, Northern Dancer—Hopespringseternal, by Buckpasser (SW in England)

Storm Cat, dk.b/br.c. 1983, Storm Bird—Terlingua, by Secretariat

Storm Star, dk.b/br.f. 1983, Storm Bird—Cinegita, by Secretariat (SW in England)

Chapel of Dreams, ch.f. 1984, Northern Dancer—Terlingua, by Secretariat

Fabrina, ch.f. 1984, Storm Bird—Equanimity, by Sir Ivor

Quirinetta, ch.f. 1984, Ardross—Tanapa (Fr), by Luthier (SW in South Africa)

Luthier's Launch, ro.f. 1986, Relaunch—Tanapa (Fr), by Luthier

Shy Tom, ch.c. 1986, Blushing Groom (Fr)—Island Kitty, by Hawaii (SAf)

Captain Starbuck, b.g. 1987, Cure the Blues—Storm Star, by Storm Bird

Carson City, ch.c. 1987, Mr. Prospector—Blushing Promise, by Blushing Groom (Fr)

Patches, dk.b/br.f. 1987, Majestic Light—Miss Betty, by Buckpasser

Rouse the Louse, ch.c. 1987, Irish River (Fr)—Nuit d'Amour, by Restless Native

Cuddles, b.f. 1988, Mr. Prospector—Stellarette, by Tentam

Dodge, dk.b/br.c. 1988, Mr. Prospector—Storm Star, by Storm Bird

Fancy Ribbons, ch.f. 1988, Blushing Groom (Fr)—My Yellow Bird, by Raise a Native

Nine Carat, b.c. 1988, Slew o' Gold—Fortune's Folly, by Graustark (SW in France)

Northern Park, b.c. 1988, Northern Dancer—Mrs Penny, by Great Nephew (SW in France)

Nucleus, b.g. 1988, Nureyev—Nellie Forbes, by Secretariat

Blacksburg, b.g. 1989, Seattle Slew—Devil's Sister, by Alleged

Border Cat, b.g. 1989, Storm Cat—Muriesk, by Nashua

Hickman Creek, b.c. 1989, Seattle Slew—Miss Betty, by Buckpasser

Natural Nine, dk.b/br.c. 1989, Mogambo—Next Fall, by Alleged

Western Territory, ch.g. 1989, Storm Cat—Buzz My Bell, by Drone

Accommodating, dk.b/br.f. 1990, Akarad—Siala (Fr), by Sharpman (SW in France)

Bonus Award, b.c. 1990, Pancho Villa—Water Girl (Fr), by Faraway Son (SW in U.S./Dominican Republic)

Flirting Dancer, b.c. 1990, Shareef Dancer—Mill Loft (Ire), by Mill Reed (SW in Hong Kong)

Fort Chaffee, ch.c. 1990, Mr. Prospector—Till Eternity, by Nijinsky II

Future Storm, ch.c. 1990, Storm Cat—Sea Sands, by Sea-Bird (Fr) (SW in Italy/U.S.)

Mountain Cat, dk.b/br.c. 1990, Storm Cat—Always Mint, by Key to the Mint

Raggedy Edge, b.c. 1990, Procida—Next Fall, by Alleged

Aucilla, b.f. 1991, Relaunch—My Cherie Amour, by Sham

Barodet, b.c. 1991, Afleet—Endurable Heights, by Graustark (SW in France)

Capias, b.c. 1991, Alleged—Smooth Bore, by His Majesty (SW in England)

Cat Attack, dk.b/br.f. 1991, Storm Cat—Harp Strings (Fr), by Luthier

Syourinomegami, dk.b/br.f. 1991, Mr. Prospector—Cinegita, by Secretariat (SW in Japan)

Three Angels, dk.b/br.f. 1991, Halo—Three Troikas (Fr), by Lyphard (SW in France)

Truth Or Die, b.g. 1991, Proud Truth—Baffling Ballerina, by Northern

CHERYL MANISTA

Flanders and Surfside, her 1997 Seattle Slew filly

Dancer (SW in France)

Boone's Mill, ch.c. 1992, Carson City—Shy Miss, by Secretariat

Country Cat, b.f. 1992, Storm Cat—La Affirmed, by Affirmed

Deputy Bodman, b.c. 1992, Deputy Minister—Buzz My Bell, by Drone

Flanders, ch.f. 1992, Seeking the Gold—Starlet Storm, by Storm Bird

Ocean Cat, dk.b/br.f. 1992, Storm Cat—Pacific Princess, by Damascus

Red Carnival, b.f. 1992, Mr. Prospector—Seaside Attraction, by Seattle Slew (SW in England)

Cat Affair, ch.f. 1993, Storm Cat—Tesio's Love, by Tom Rolfe

Golden Attraction, b.f. 1993, Mr. Prospector—Seaside Attraction, by Seattle Slew

Grindstone, dk.b/br.c. 1993, Unbridled—Buzz My Bell, by Drone

Hennessy, ch.c. 1993, Storm Cat—Island Kitty, by Hawaii (SAf)

Hishi Natalie, b.f. 1993, Seattle Slew—Devil's Sister, by Alleged (SW in Japan)

Honour and Glory, b.c. 1993, Relaunch—Fair to All, by Al Nasr (Fr)

Reallyaroan, dk.b/br.f. 1993, Imperial Falcon—Tiz a Looker, by Mr. Prospector

Boston Harbor, b.c. 1994, Capote—Harbor Springs, by Vice Regent

Lady Carson, b.f. 1994, Carson City—Heathers Surprise, by Best Turn

Leestown, dk.b/br.c. 1994, Seattle Slew—Bright Candles, by El Gran Senor

Pearl City, ch.f. 1994, Carson City—Island Kitty, by Hawaii (SAf)

Queen of Money, ch.f. 1994, Corporate Report—Chandelier, by Majestic Light

Ticket Counter, ch.f. 1994, Deposit Ticket—Peppermint Lane, by Alydar

Cape Town, b.c. 1995, Seeking the Gold—Seaside Attraction, by Seattle Slew

Grand Slam, dk.b/br.c. 1995, Gone West—Bright Candles, by El Gran Senor

Relinquish, b.f. 1995, Rahy—Sun Blush, by Ogygian

Western City, gr/ro.c. 1995, Carson City—Fire Alarm, by Secretariat

Cape Canaveral, dk.b/br.c. 1996, Mr. Prospector—Seaside Attraction, by Seattle Slew

Cat Thief, ch.c. 1996, Storm Cat—Train Robbery, by Alydar

China Storm, b.f. 1996, Storm Cat—China Bell, by Seattle Slew

Hidden City, dk.b/br.c. 1996, Carson City—Pacific Hideaway, by Seattle Slew

Imperfect World, ch.f. 1996, Carson City—Mais Oui, by Lyphard (SW in U.S./France)

Lake William, dk.b/br.c. 1996, Salt Lake—Sol de Terre, by Mr. Prospector

Mountain Range, dk.b/br.g. 1996, Mountain Cat—Proper Form, by Deputy Minister

Tactical Cat, gr/ro.c. 1996, Storm Cat—Terre Haute, by Caro (Ire)

Apollo Cat, dk.b/br.f. 1997, Storm Cat—Hopespringsforever, by Mr. Prospector

Katz Me If You Can, dk.b/br.f. 1997, Storm Cat—Cuddles, by Mr. Prospector

Lady Showtime, b.f. 1997, Pembroke—San Isabel, by Alysheba

Penny's Gold, b.f. 1997, Kingmambo—Penny's Valentine, by Storm Cat

Surfside, b.f. 1997, Seattle Slew—Flanders, by Seeking the Gold

City Fair, ch.f. 1998, Carson City—Encorevous, by Storm Bird

Gold Trader, dk.b/br.c. 1998, Storm Cat—Golden Attraction, by Mr. Prospector

Boston Common, b.g. 1999, Boston Harbor—Especially, by Mr. Prospector

Four Corners, b.g. 1999, Salt Lake—Kettle Ridge, by Mr. Prospector

Gold Dollar, b.c. 1999, Seattle Slew—Miss Prospector, by Crafty Prospector

Jump Start, dk.b/br.c. 1999, A.P. Indy—Steady Cat, by Storm Cat

Steaming Home, b.f. 1999, Salt Lake—County Fair, by Mr. Prospector (SW in Ireland)

Storm Commander, b.c. 1999, Storm Cat—Picco Bello, by Seeking the Gold

Boston Park, dk.b/br.c. 2000, Boston Harbor—Maple Creek, by Forty Niner

Buffythecenterfold, b.f. 2000, Capote—Augusta Springs, by Nijinsky II

Country Romance, dk.b/br.f. 2000, Saint Ballado—Cuddles, by Mr. Prospector

My Boston Gal, b.f. 2000, Boston Harbor—Western League, by Forty Niner

Foolishly, dk.b/br.f. 2001, Broad Brush—City Band, by Carson City

Quick Action, b.c. 2001, Carson City—Indian Sunset, by Storm Bird

Shy Lil, b.f. 2001, Lil's Lad—Cheyenne City, by Carson City

W.T. Young (Overbrook Farm) in various partnerships

Tom Register, dk.b/br.c. 1978, Groton—Matushka, by Tom Fool (SW in Panama)

Harlan, dk.b/br.c. 1989, Storm Cat—Country Romance, by Halo

Joy Baby, dk.b/br.f. 1989, Storm Cat—Babe's Joy, by King of the Sea

November Snow, b.f. 1989, Storm Cat—Princess Alydar, by Alydar

Senate Appointee, dk.b/br.f. 1989, Storm Cat—Kermis, by Graustark

Timber Cat, dk.b/br.g. 1989, Storm Cat—Sweet Valentine, by Honey Jay

Mistle Cat, ro.c. 1990, Storm Cat—Mistle Toe, by Maribeau (SW in France/Italy/England)

O. P. Cat, dk.b/br.f. 1990, Storm Cat—Princess S., by Assagai Jr.

Tempest Dancer, b.f. 1990, Storm Cat—Honor an Offer, by Hoist the Flag

Delineator, b.c. 1991, Storm Cat—Mountain Climber, by Grey Dawn II (Fr)

Sardula, b.f. 1991, Storm Cat—Honor an Offer, by Hoist the Flag

Stellar Cat, dk.b/br.f. 1991, Storm Cat—Sweet Valentine, by Honey Jay

Tabasco Cat, ch.c. 1991, Storm Cat—Barbicue Sauce, by Sauce Boat

Cat Appeal, dk.b/br.f. 1992, Storm Cat—Amyark, by Caro (Ire)

Memories of Ronny, dk.b/br.f. 1992, Storm Cat—Pass Away, by Swing Pass

Aristis, dk.b/br.c. 1993, Storm Cat—Freesia, by Cox's Ridge (SW in South Africa)

Magic Storm, dk.b/br.f. 1999, Storm Cat—Foppy Dancer, by Fappiano

Sources: *The Blood-Horse* Archives and The Jockey Club

John C. Mabee

Six out of seven consecutive years during the 1990s, the leading Thoroughbred breeder in North America hailed from a Midwestern state not consistently regarded as racehorse country. John and Betty Mabee, of Iowan origins, led the list in 1992 and 1998, and fellow Iowa native Allen Paulson took top honors from 1994 to 1997.

The Mabees came from little Seymour, Iowa, where John was born on August 21, 1921. He liked to race his mare named Bird against other kids' horses, but he was beset by pneumonia and bronchitis as a youngster and would visit the public library looking for information on locations with climates better suited to his health. He settled on San Diego, California, but went there only after having married his childhood sweetheart, Betty Murphy.

He took work in San Diego as a journeyman carpenter, then, having saved two thousand dollars, bought a little grocery story in 1944. This set Mabee on a career path that grew into the Big Bear chain of supermarkets, and by 1957 he had not only the urge but the means to get into horse racing. The couple attended the Del Mar yearling sales with a budget of $15,000 in mind and stayed right on target, bringing home one yearling for nine thousand dollars and two for three thousand dollars.

It was a decade later, in 1968, that the Mabees' first stakes winner was bred, but soon the modest aspect of their operation would be a thing of the past. In the early 1970s the Mabees purchased property near Ramona that became their Golden Eagle Ranch. The

more than two-dozen grocery stores eventually were sold, in 1994 and 1995, but Mabee retained the land on which many of the stores were located and was one of the largest landholders in San Diego County. He also developed Golden Eagle Insurance, but ran afoul of a regulatory issue that many felt was a politically motivated effort to take over the company and was forced to divest himself of any interest in the insurance market.

The setback did not abort the momentum Mabee had created in the horse business, and by the time of his death at eighty in 2002, he and his wife had bred 182 stakes winners.

As Mabee's involvement in breeding developed, he also took more of a leadership role in the local racetrack, Del Mar. From its beginnings in 1937, Del Mar had always been long on charm but in some ways was a country cousin to the big city Santa Anita Park and Hollywood Park, up the coast in Los Angeles. Mabee became president and chairman of Del Mar Thoroughbred Club and headed its rise in business and purses and its eventual overhaul into a huge, modern plant that still managed to retain its casual resort charm. The San Diego *Union-Tribune* described this effort as an "$80-million project in the early 1990s that resulted in a new grandstand" and credited Mabee's long run at the racetrack as resulting in "an era of unprecedented growth in attendance, betting handle, and national recognition."

"He was the father of modern Del Mar," affirmed Del Mar's general manager and succeeding president, Joseph W. Harper. "He was the power behind all of us."

The Mabees' commitment to their adopted home included generous support of various local charities, especially those benefiting children. Mabee also owned 20 percent interest in the National Football League's San Diego Chargers for a time and attempted to bring a National Basketball Association franchise to the city after it lost the Rockets to Houston. The *Union-Tribune* quoted mayor Dick Murphy describing Mabee after his death as "a genuinely decent person (whose) generosity and friendship will be missed."

Through 1979 Golden Eagle had bred a total of fourteen stakes winners. Thereafter, the momentum was dramatic, as the broodmare band grew to some two hundred, of which about one-third would be boarded at prime farms in Kentucky. The reality of the business of the Turf has been that it does not work economically to keep a really top-class stallion in California. Mabee responded by creating a division in Kentucky, where he could utilize some of the best

CANDACE RUSHING

John and Betty Mabee

sires in the world. At the same time he held fast to give his fellow California breeders major benefits from his own success. Standing at Golden Eagle are three high-class sons of Seattle Slew: the well-proven General Meeting and Avenue of Flags and the promising young Event of the Year.

General Meeting is an excellent example of Mabee's concept that good breeding stock needs to have good pedigree top and bottom behind it. In addition to being by the Triple Crown winner and major sire Seattle Slew, General Meeting is from an Alydar mare, who is in turn from the Nijinsky II mare Summertime Promise, a stakes winner bred by Paul Mellon of Rokeby Farm. The Mabees' homebred General Meeting was a nice West Coast stakes winner without reaching a level that Kentucky breeders clamored immediately for him too loudly to resist. He won the Volante Handicap and Bradbury Stakes and placed in the Hollywood Futurity and Hollywood Prevue, establishing a career mark of four wins in seventeen starts and earnings of $441,125. Through mid-2004, he had sired twenty-four stakes winners,

Seattle Slew, 1974	Bold Reasoning, 1968	Boldnesian, 1963	Bold Ruler (*Nasrullah—Miss Disco)
			Alanesian (**Polynesian**—Alablue)
		Reason to Earn, 1963	Hail to Reason (*Turn-to—Nothirdchance)
			Sailing Home (Wait a Bit—Marching Home)
	My Charmer, 1969	Poker, 1963	Round Table (*Princequillo—*Knight's Daughter)
			Glamour (*Nasrullah—Striking)
		Fair Charmer, 1959	Jet Action (Jet Pilot—Busher)
			Myrtle Charm (Alsab—Crepe Myrtle)
GENERAL MEETING	Alydar, 1975	Raise a Native, 1961	Native Dancer (**Polynesian**—Geisha)
			Raise You (Case Ace—Lady Glory)
Alydar's Promise, 1983		Sweet Tooth, 1965	On-and-On (*Nasrullah—Two Lea)
			Plum Cake (Ponder—Real Delight)
	Summertime Promise, 1972	Nijinsky II, 1967	Northern Dancer (Nearctic—Natalma)
			Flaming Page (Bull Page—Flaring Top)
		Prides Promise, 1966	Crozier (*My Babu—Miss Olympia)
			Hillbrook (I Will—Johann)

which at 6 percent is not distinguished, but they include two outstanding runners in General Challenge and Excellent Meeting, as well as a third grade I winner in Magical Allure.

Big, long General Challenge was an exciting gelding who won four of the key races on the West Coast: the Santa Anita Handicap, Del Mar's Pacific Classic, the Santa Anita Derby, and Charles H. Strub Stakes. He had earned $2.8 million by that time; having been set aside to recover from a series of problems, he made a brief, unsuccessful comeback in 2003.

When he contested the Kentucky Derby in 1999, General Challenge was part of a stunning showcase of Golden Eagle's prominence. He and the homebred filly Excellent Meeting raced for the Mabees, while the Del Mar Futurity winner Worldly Manner — a colt they had bred and sold to Sheikh Mohammed bin Rashid al Maktoum of Dubai for a reported $5 million — was also in the race. General Challenge was whacked mightily in the first turn and finished eleventh, while the filly Excellent Meeting ran an excellent race to be fifth, and Worldly Manner loomed boldly in the upper stretch before tiring to seventh.

Excellent Meeting, who won three grade I filly stakes in California, also had finished second, beaten a half-length by champion Silverbulletday, in the 1998 Breeders' Cup Juvenile Fillies. She earned $1.4 million.

Avenue of Flags, another of the Golden Eagle golden Seattle Slew triumvirate, ran only three times. He won twice and was second to Deposit Ticket in the 1990 Hollywood Juvenile Championship. He is out of the stakes winner Beautiful Glass, by the Buckpasser stallion Pass the Glass. Avenue of Flags is inbred 4x4 to Bold Ruler. Standing at Golden Eagle, he has sired nineteen stakes winners (5 percent), but they include the grade I Oak Leaf winner Notable Career, La Canada Stakes winners Fleet Lady and Belle's Flag, and Fall Highweight Handicap winner Victor Avenue.

"When I was learning the business, I ended up with several fillies that I had bought to race without much thought about breeding," Mabee told *The Blood-Horse* in a 1994 interview. "Today, when I buy broodmares, I'm looking for mares by broodmare sires which have good records, such as Secretariat, Nijinsky II, and Northern Dancer. Then the mares

Excellent Meeting

227

SHIGEKI KIKKAWA

Best Pal

Good pedigree top and bottom was also important to him in sire selection: "I don't believe in breeding to sires who have been produced by nothing mares, even if the sires have been excellent race horses themselves. You just don't find many good stallions from poor female families. I want to see the pedigrees there on both the top and bottom lines of the stallions, too."

Mabee parried numerous offers from pedigree advisors to mate his mares for him and enjoyed the challenge of designing many of the matings himself. "I don't claim to be Wayne Lukas (record-breaking Hall of Fame trainer) in picking out a good equine athlete," he said, "but I do think I've had a chance over the years to learn from experience. I take pride in reaching my own decisions about horses, and I would like to feel I can pick out a better horse than a lot of these consultants."

Along the way Mabee said he developed generalizations that included a leeriness of South American mares as broodmare prospects as well as mares that were more than thirteen years old.

"He never used a calculator or computer," mar-

have to have a look about them. Their physical appearance has to suggest that they would be able to produce athletic foals.

"I definitely want a decent pedigree for a mare. I am not going to buy an individual with a weak bottom line, even if she is by a good broodmare sire. The mare doesn't necessarily have to have a super race record, but her pedigree has to be strong on both the top and bottom. I wouldn't buy a mare who is by some unknown stallion and has a very weak bottom line, even if she is a stakes winner."

	Habitat, 1966	Sir Gaylord, 1959	*Turn-to (*Royal Charger—*Source Sucree)
			Somethingroyal (*Princequillo—Imperatrice)
		Little Hut, 1952	Occupy (*Bull Dog—Miss Bunting)
*Habitony, 1974			Fancy Racket (*Wrack—Ultimate Fancy)
	Courteous Lady, 1960	*Gallant Man, 1954	*Migoli (Bois Roussel—*Mah Iran)
			*Majideh (*Mahmoud—Qurrat-al-Ain)
		Fleurlea, 1951	Bull Lea (*Bull Dog—Rose Leaves)
			Blue Lily (Blue Larkspur—Ship Ablaze
BEST PAL	King Pellinore, 1972	Round Table, 1954	*Princequillo (Prince Rose—*Cosquilla)
			*Knight's Daughter (Sir Cosmo—Feola)
		Thong, 1964	Nantallah (*Nasrullah—Shimmer)
Ubetshedid, 1980			*Rough Shod II (Gold Bridge—Dalmary)
	Ubetido, 1975	Sir Wiggle, 1967	Sadair (*Petare—Blue Missy)
			*Wiggle II (Rego—*Sweet Nymph)
		Udontmeanit, 1965	Gray Phantom (*Ambiorix—Grey Flight)
			Lady Bourbon (Bolingbroke—Bourbon Girl)

veled Mrs. Mabee. "He had a built-in computer in his head that retained things far beyond normal, going back on all the bloodlines. When you gamble like he did, you'd better know where you're putting your chips. And he knew."

The recognition of Mabee's breeding program was demonstrated in one of the most testing of crucibles, the Keeneland July sale, where he topped the 2000 venue with a Mr. Prospector—Molly Girl filly purchased by Satish Sanan for $3.6 million.

The signature, beloved horse among Mabee's many outstanding runners was Best Pal. This was a homebred by *Habitony, a horse the late racing official Jimmy Kilroe once told us he thought would be the next "important" horse on the California circuit. *Habitony was both success and disappointment. He had an internationally appealing pedigree, combining the miler quality of his European-raced sire Habitat with the stamina of his American-raced broodmare sire, Belmont Stakes winner *Gallant Man. *Habitony's dam, Courteous Lady, was a stakes-winning daughter of the Bull Lea mare Fleurlea. *Habitony won the Sunny Slope and Norfolk stakes at two in 1976 and the Santa Anita Derby in 1977.

Mabee recalled having bought *Habitony with a partner as a yearling in Ireland and then selling him at two, but he wound up standing the horse at stud. *Habitony sired twenty-seven stakes winners (6 percent) of which Best Pal was the best, and another was the durable sprinter Richter Scale.

Best Pal was a gelding from Ubetshedid, whose sire, King Pellinore, was a high-class stakes winner in California. Mabee must have waived more than one of his criteria in the decision to utilize Ubetshedid. She was unplaced in her only start. While her dam, Ubetido, placed in good California stakes, she was by the undistinguished Sir Wiggle and out of an unraced daughter of Gray Phantom. Furthermore, the next dam, Lady Bourbon, was also unraced.

Vindicating whatever Mabee saw in this family, Best Pal became the all-time leading California-bred in earnings, with $5,668,245 (since surpassed by

Jeanne Jones

Tiznow) and one of the most popular. Golden Eagle has not bred an American classic winner or champion, but Best Pal came close. He was second to Strike the Gold in the 1991 Kentucky Derby and won a portfolio of the most revered races in California, as well as the Oaklawn Handicap at Hot Springs, Arkansas. He gave the Mabees victories in time-honored events such as the Santa Anita Handicap, Hollywood Gold Cup, Swaps Stakes, and Charles H. Strub Stakes. Perhaps his victory in a new race, however, was even more gratifying. With the inaugural Pacific Classic in 1991, Mabee realized the ambition to have a $1-million event at his pet track, Del Mar. Best Pal defeated older horses Twilight Agenda and Unbridled to win that first running.

Event of the Year

"It brought tears to my eyes," said Gary Jones, one of several horsemen who trained Best Pal and who had him at the time. "And after the race, instead of going to the directors' room, Mr. Mabee was back at the barn drinking beer with my crew. You'll never see people like that again."

Best Pal joined the likes of Native Diver as a Cal-bred gelding whose origins and longevity ingrained him in the hearts of California fans, while also gaining national recognition. He won eighteen of forty-seven races from two through eight. Best Pal was retired to the ranch but died young and was buried near the Golden Eagle office. Few breeders ever had a pal that good.

Perhaps coming even closer to an Eclipse Award than Best Pal was the sterling Golden Eagle home-bred Jeanne Jones. A daughter of the great racehorse and Kentucky sire Nijinsky II, Jeanne Jones was another foaled from Avenue of Flags' dam, Beautiful Glass. Jeanne Jones was the second choice going into the Breeders' Cup Juvenile Fillies at Hollywood Park in 1987 and had a six-length lead in the stretch, only to be caught on the wire by eventual two-year-old filly champion Epitome. Jeanne Jones later was third in the Hollywood Starlet Stakes. A couple of inches' difference in the Breeders' Cup, and then stopping for the year, almost certainly would have made her champion. Jeanne Jones came back at three to win the grade I Fantasy at Oaklawn Park.

The seven millionaire earners bred by the Mabees were Best Pal, General Challenge, Dramatic Gold, Nostalgia's Star, Excellent Meeting, Early Pioneer, and Event of the Year. Also among their most important horses were Souvenir Copy, Diablo, River Special, Individualist, Big Pal, Worldly Ways, and Fantastic Look.

An unusual aspect of the Mabee breeding program has been a recent influence far away, in Germany, where the Mabee mare Spirit of Eagles has produced two group I winners for the Gestut Faehrhof: Silvano and Sabiango.

Statistically, Mr. and Mrs. Mabee achieved singular status as breeders for 1992, with record one-year earnings of $7,026,627, and again led breeders for 1998, with $8,225,102. They were voted the Eclipse Award for excellence as breeders three times — in 1991, 1997, and 1998 — and won multiple awards as owners and breeders nationally from the Thoroughbred Owners and Breeders Association.

Mabee battled cancer, came back, suffered several strokes, came back again, and then succumbed, eventually, at eighty, in 2002. Mrs. Mabee, who survives along with their son Larry, is carrying on Golden Eagle.

Stakes Winners Bred by John C. Mabee

Quite a Day, b.g. 1968, Prince Khaled—Princess Imbros, by Imbros

Plenty of Style, b.c. 1970, Hillsdale—Miss Carousel, by Nashville

Bird of Courage, b.g. 1973, Great Career—Fancy Jet, by Nearctic

Happy Bettor, b.f. 1973, Crazy Kid—Kissing Ring, by Prince Khaled

Beausejour Boy, gr.c. 1975, Verbatim—Kahoolawe, by Warfare

True Lady, b.f. 1975, *Le Fabuleux—Dancerina, by Never Bend

Beau's Eagle, ch.c. 1976, Golden Eagle II—Beaufield, by Maribeau

Fantastic Girl, b.f. 1976, Riva Ridge—Lady of Elegance, by Tom Fool

Sis C., b.f. 1976, Nijinsky II—Fond Hope II, by Sir Ivor

Splendid Girl, b.f. 1976, Golden Eagle II—Coccinea, by Jaipur

Laura's Jet, ch.f. 1978, Wajima—Fancy Jet, by Nearctic

Another Bid, gr.c. 1979, Little Current—Sweet Melissa, by Crazy Kid

Beautiful Glass, b.f. 1979, Pass the Glass—Beautiful Spirit, by Bold Bidder

Devine Look, b.f. 1979, Wajima—Lady of Elegance, by Tom Fool

Current Hope, ro.c. 1980, Little Current—Kahoolawe, by Warfare

Write Off, b.c. 1980, Wajima—Rare Lady, by Never Bend

Akabir, b.c. 1982, Riverman—Lypatia (Fr), by Lyphard

Amazing Courage, b.c. 1982, Splendid Courage—Sainte Jeanne, by Battle Joined

Nostalgia's Star, dk.b/br.c. 1982, Nostaglia—Aunt Carol, by Big Spruce

Water Into Wine, ro.c. 1982, State Dinner—Ubetido, by Sir Wiggle

Aloma's Tobin, dk.b/br.c. 1983, J. O. Tobin—Aloma, by Native Charger

Beau's Leader, ch.c. 1983, Beau's Eagle—Lady Maxwell, by Crowned Prince

Bravo Fox, b.c. 1983, Northern Baby—Fantastic Girl, by Riva Ridge (SW in Ireland)

Darby Fair, dk.b/br.c. 1983, Darby Creek Road—Fair Advantage, by Raja Baba

Lady Pastor, dk.b/br.f. 1983, Flying Paster—Youthful Lady, by Youth

Thrill Show, b.c. 1983, Northern Baby—Splendid Girl, by Golden Eagle II

Truly Nureyev, ch.c. 1983, Nureyev—True Lady, by *Le Fabuleux

Become a Lord, b.c. 1984, Lord Avie—Become a Star, by First Balcony

David's Bird, b.c. 1984, Storm Bird—Splendid Girl, by Golden Eagle II

Double the Charm, ch.f. 1984, Nodouble—Album, by Never Bend

Exotic Eagle, b.c. 1984, Beau's Eagle—Exotic Visage, by Tom Rolfe

Fairly Affirmed, b.c. 1984, Affirmed—Fair Advantage, by Raja Baba

Half a Year, ch.c. 1984, Riverman—Six Months Long, by Northern Dancer (SW in England)

Mount Laguna, ch.g. 1984, Huguenot—Laura's Star, by Key to the Kingdom

Young Flyer, dk.b/br.f. 1984, Flying Paster—Youthful Lady, by Youth

Art College, b.f. 1985, Hostage—Melissa's Choice, by Little Current

Art's Angel, b.c. 1985, Czaravich—Laura's Star, by Key to the Kingdom

Basic Rate, ch.c. 1985, Valdez—Foolish Girl, by Foolish Pleasure

Book Collector, b.f. 1985, Irish River—Lypatia (Fr), by Lyphard

Floral Magic, dk.b/br.f. 1985, Affirmed—Rare Lady, by Never Bend

Jeanne Jones, b.f. 1985, Nijinsky II—Beautiful Glass, by Pass the Glass

Vive, ch.f. 1985, Nureyev—Viva Regina, by His Majesty

Winter Rates, dk.b/br.f. 1985, Bates Motel—Heremkemmese, by *Forli

Advocate Training, b.c. 1986, Desert Wine—Desire, by Graustark

Annual Date, dk.b/br.c. 1986, Nostalgia—Dancing Role, by Droll Role

Beautiful Melody, ch.f. 1986, Alydar—Beautiful Spirit, by Bold Bidder

Fantastic Look, ch.f. 1986, Green Dancer—Fantastic Girl, by Riva Ridge

Noble and Nice, ch.f. 1986, Valdez—Somebody Noble, by *Vaguely Noble

Pirate Army, dk.b/br.c. 1986, Roberto—Wac, by Lt. Stevens (SW in England)

Splendid Career, ch.g. 1986, Northern Baby—Sir Ivor's Sorrow, by Sir Ivor

Annual Reunion, dk.b/br.f. 1987, Cresta Rider—Love for Life, by *Forli

Bel's Starlet, ro.f. 1987, Bel Bolide—Vigor's Star, by Vigors

Devine Force, dk.b/br.g. 1987, Habitony—Rare Lady, by Never Bend

Diablo, dk.b/br.c. 1987, Devil's Bag—Avilion, by Cornish Prince

Good Courage, b.g. 1987, Bates Motel—Goodside, by Topsider

Great Event, dk.b/br.c. 1987, Al Nasr (Fr)—Melissa's Diamond, by Chieftain

Individualist, b.g. 1987, Bel Bolide—Become a Star, by First Balcony

Lady Lavina, gr.f. 1987, Habitony—Rare Gal, by Caro (Ire)

Lovely Habit, b.f. 1987, Habitony—Big Spirit, by Big Spruce

Oh Sweet Thing, dk.b/br.f. 1987, Hostage—Quiet Quality, by Chieftain

Splendid Dream, b.f. 1987, Bel Bolide—Truly Splendid, by Naskra

Bel's Starlet

Best Pal, b.g. 1988, Habitony—Ubetshedid, by King Pellinore

Fantastic Ways, b.f. 1988, Secretariat—Fantastic Girl, by Riva Ridge

General Meeting, b.c. 1988, Seattle Slew—Alydar's Promise, by Alydar

Lively Music, dk.b/br.f. 1988, Habitony—Lady Pastor, by Flying Paster

Lycius, ch.c. 1988, Mr. Prospector—Lypatia (Fr), by Lyphard (SW in England)

Man From Eldorado, b.c. 1988, Mr. Prospector—Promising Girl, by Youth (SW in England/U.S.)

Maraakiz, ch.c. 1988, Roberto—River of Stars, by Riverman (SW in Turkey)

Paperback Habit, b.f. 1988, Habitony—Ubetido, by Sir Wiggle

Recent Arrival, dk.b/br.g. 1988, Lemhi Gold—Incredible Idea, by Youth

Spending Bucks, b.g. 1988, Spend a Buck—Sis C., by Nijinsky II

True Brave (stp.), b.c. 1988, Dancing Brave—True Lady, by *Le Fabuleux (SW in France)

Big Pal, b.g. 1989, Beau's Eagle—Big Spirit, by Big Spruce

Big Sur, b.c. 1989, Alydar—Laday, by Lyphard

Burnished Bronze, b.g. 1989, Seattle Slew—Splendid Pride, by Exclusive Native

Darling Dame, b.f. 1989, Lyphard—Darling Lady, by Alleged

Gala Dinner, b.f. 1989, Habitony—Quiet Quality, by Chieftain

Great Decision, b.g. 1989, Habitony—Lexingstar, by Raise a Native

Majestic Velvet, ch.f. 1989, Bel Bolide—Lady Cup, by Raise a Cup

My True Lady, b.f. 1989, Seattle Slew—Lady for Two, by Storm Bird

Old Top Hat, ch.g. 1989, Beau's Eagle—Private Hatter, by Private Account

Ready Effort, dk.b/br.g. 1989, Habitony—Foolish Girl, by Foolish Pleasure

Total Tempo, b.g. 1989, Habitony—Desire, by Graustark

Art of Living, ch.c. 1990, Alydar—Nervous Pillow, by Nervous Energy

Boating Pleasure, ch.g. 1990, Flying Paster—Committee Boat, by Secretariat

Devil Diamond, b.c. 1990, Devil's Bag—Melissa's Diamond, by Chieftain

Evening Highlight, dk.b/br.f. 1990, Al Nasr (Fr)—Candelight Service, by Blushing Groom (Fr)

Finest City, b.f. 1990, Lyphard—True Lady, by *Le Fabuleux

Souvenir Copy

Friendly Bells, ch.f. 1990, Bel Bolide—Wonderful Friend, by Czaravich

Likeable Style, b.f. 1990, Nijinsky II—Personable Lady, by No Robbery

Noble Year, ch.g. 1990, Half a Year—Somebody Noble, by *Vaguely Noble

Pracer, gr.f. 1990, Lyphard—Shindy, by Roberto (SW in Italy)

River Special, ch.c. 1990, Riverman—Nijinska Street, by Nijinsky II

Spiritual Star, b.f. 1990, Soviet Star—Excellent Spirit, by Damascus

Al's River Cat, b.c. 1991, Riverman—Comical Cat, by Exceller

Annual Dance, b.f. 1991, Nostalgia—Dancing Role, by Droll Role

Dance With Grace, b.f. 1991, Mr. Prospector—Dancing Tribute, by Nureyev

Dancing Habit, b.g. 1991, Habitony—This Is My Dance, by Marshua's Dancer

Dramatic Gold, b.g. 1991, Slew o' Gold—American Drama, by Danzig

Garden of Roses, ch.f. 1991, Half a Year—Laura's Bouquet, by Beau's Eagle

Ladies Cruise, b.f. 1991, Fappiano—Youthful Lady, by Youth

Majestic Style, b.c. 1991, Nureyev—Fantastic Girl, by Riva Ridge

Night Letter, b.g. 1991, Shahrastani—Ballena Ridge, by Riva Ridge

Queen of the River, ch.f. 1991, Riverman—Nijinska Street, by Nijinsky II

River Flyer, dk.b/br.c. 1991, Riverman—Young Flyer, by Flying Paster

Winning Pact, b.c. 1991, Alydar—Six Months Long, by Northern Dancer

Yearly Tour, ch.f. 1991, Half a Year—Victorian Village, by L'Emigrant

Awesome Thought, ch.c. 1992, Flying Paster—Incredible Idea, by Youth

Beau Jingles, b.c. 1992, Riverman—Glamorous Beau, by Assert

Dance Treat, ch.f. 1992, Nureyev—Office Wife, by Secretariat (SW in France)

Day Jewels, b.f. 1992, Slew o' Gold—Laday, by Lyphard

Feathered Friend, ch.g. 1992, Beau's Eagle—Friendly One, by Bel Bolide

Funny Tale, dk.b/br.g. 1992, Habitony—Comical Story, by Assert (Ire)

Save One for Me, dk.b/br.g. 1992, Habitony—This Is My Dance, by Marshua's Dancer

Shake Hand, b.f. 1992, Mr. Prospector—Dancing Tribute, by Nureyev (SW in Japan)

Tereshkova, b.f. 1992, Mr. Prospector—Lypatia (Fr), by Lyphard (SW in France/UAE)

Advancing Star, b.f. 1993, Soviet Star—Fair Advantage, by Raja Baba

Agenda for the Day, dk.b/br.g. 1993, Avenue of Flags—Private Meeting, by Golden Eagle II

Air Bag, b.g. 1993, Devil's Bag—Lady Hardwick, by Nijinsky II

Avenue of Gold, b.f. 1993, Avenue of Flags—Golden Garden, by Golden Eagle II

Avenue Shopper, dk.b/br.f. 1993, Avenue of Flags—Exciting Gal, by Master Willie (GB)

Beaming Year, ch.f. 1993, Half a Year—Apalachee's Star, by Apalachee

Cat's Career, b.c. 1993, Mr. Prospector—Comical Cat, by Exceller

Choose One, ch.f. 1993, Half a Year—Melissa's Choice, by Little Current

Cold Blood, b.g. 1993, Capote—Laura's Star, by Key to the Kingdom

Dance Sequence, ch.f. 1993, Mr. Prospector—Dancing Tribute, by Nureyev (SW in England)

General Idea, dk.b/br.f. 1993, General Meeting—Incredible Idea, by Youth

Sasha's Pal, dk.b/br.f. 1993, Habitony—Goodside, by Topsider

Singing Year, ch.f. 1993, Half a Year—Candlelight Song, by Seattle Song

Young and Daring, b.f. 1993, Woodman—Youthful Lady, by Youth

Best Star, dk.b/br.c. 1994, Seattle Slew—Center Court Star, by Secretariat

Classy Prospector, ch.c. 1994, Mr. Prospector—Fantastic Look, by Green Dancer

Crowning Meeting, b.g. 1994, General Meeting—Fitted Crown, by Chief's Crown

Eternal Vigilance, ch.g. 1994, Half a Year—Homecoming Crown, by Chief's Crown

Fleet Lady, dk.b/br.f. 1994, Avenue of Flags—Dear Mimi, by Roberto

Idealistic Cause, dk.b/br.f. 1994, Habitony—Special Idea, by Cresta Rider

Precious Peace, b.f. 1994, Habitony—Incandescent, by Halo

Princely Price, b.g. 1994, General Meeting—Lace Dancer, by Irish River (Fr)

Reading Habit, b.f. 1994, Half a Year—Paperback Habit, by Habitony

Really Happy, dk.b/br.f. 1994, Avenue of Flags—Dear Laura, by Youth

Winning Habits, dk.b/br.g. 1994, Habitony—Golden Friendship, by Golden Eagle II

Worldly Ways, ch.c. 1994, Generous (Ire)—Fantastic Ways, by Secretariat (Bred in England)

Career Collection, b.f. 1995, General Meeting—River of Stars, by Riverman

Dance the Avenue, dk.b/br.g. 1995, Avenue of Flags—Dancing Role, by Droll Role

Early Pioneer, ch.g. 1995, Rahy—Golden Darling, by Slew o' Gold

Event of the Year, dk.b/br.c. 1995, Seattle Slew—Classic Event, by Mr. Prospector

Fiscal Year, dk.b/br.f. 1995, Half a Year—Fiscal Gold, by Slew o' Gold

Full Moon Madness, ch.g. 1995, Half a Year—Soft Charm, by Secretariat

Magical Allure, b.f. 1995, General Meeting—Rare Lady, by Never Bend

Meiner Love, dk.b/br.c. 1995, Seeking the Gold—Heart of Joy, by Lypheor (GB) (SW in Japan)

Moonlight Meeting, ch.g. 1995, General Meeting—Moonlight Elegance, by Nijinsky II

Mother's Meeting, b.f. 1995, General Meeting—Mother Bear, by Flying Paster

Post a Note, b.g. 1995, Avenue of Flags—Utterly Elegant, by Key to the Mint

Souvenir Copy, dk.b/br.c. 1995, Mr. Prospector—Dancing Tribute, by Nureyev

Absolute Harmony, ch.c. 1996, Seeking the Gold—Fantastic Look, by Green Dancer

Avenue of Style, dk.b/br.g. 1996, Avenue of Flags—Dancing Style, by Danzig

Daring General, dk.b/br.g. 1996, General Meeting—Flying Belle, by Flying Paster

Doneraile Court, dk.b/br.c. 1996, Seattle Slew—Sophisticated Girl, by Stop the Music

Dream of Gifts, ch.f. 1996, Half a Year—Charming Dreamer, by Miswaki

Excellent Meeting, b.f. 1996, General Meeting—Fitted Crown, by Chief's Crown

General Challenge, ch.g. 1996, General Meeting—Excellent Lady, by Smarten

Magic of Sunrise, b.f. 1996, Woodman—Country Cruise, by Riverman

Sunday Stroll, b.g. 1996, General Meeting—Bel Darling, by Bel Bolide

Time to Meet, b.f. 1996, General Meeting—Above all Time, by Private Account

Worldy Manner, dk.b/br.c. 1996, Riverman—Lady Pastor, by Flying Paster

Ancient Traveler, b.g. 1997, Rahy—Long Time Ago, by Conquistador Cielo

Annual Rainfall, dk.b/br.f. 1997, Thunder Gulch—Annual Reunion, by Cresta Rider

Crown of Crimson, b.f. 1997, Seattle Slew—Fitted Crown, by Chief's Crown

Designed for Luck, ch.g. 1997, Rahy—Fantastic Look, by Green Dancer

Favorite Funtime, b.f. 1997, Seeking the Gold—Promising Girl, by Youth

Worldly Ways

Minor Details, dk.b/br.f. 1997, General Meeting—Sparkling Star, by Lyphard

New Advantage, dk.b/br.c. 1997, Dayjur—Somebody Noble, by *Vaguely Noble

Notable Career, dk.b/br.f. 1998, Avenue of Flags—Excellent Lady, by Smarten

Orange Glow, ch.f. 1998, River Special—Mock Orange, by Alydar

Shining Nuggets, dk.b/br.g. 1998, Half Term—Storybook Fair, by Avenue of Flags

Song of the Moment, dk.b/br.f. 1998, River Special—Memorable Moment, by Secretariat

Victory Ride, b.f. 1998, Seeking the Gold—Young Flyer, by Flying Paster

Wild and Wise, dk.b/br.g. 1998, Avenue of Flags—Terrys Wild Again, by Wild Again

Funny Meeting, dk.b/br.c. 1999, General Meeting—Aerial Spirit, by Nureyev

Martha's Music, b.f. 1999, Sultry Song—Danzig Key, by Danzig

Chief Planner, dk.b/br.c. 2000, General Meeting—Flag of Freedom, by Fappiano

Long Term Wish, b.f. 2000, Souvenir Copy—Warmest Wishes (GB), by Nashwan

Yearly Report, b.f. 2001, General Meeting—Fiscal Year, by Half a Year

Souvenir Gift, b.f. 2002, Souvenir Copy—Alleged Gift, by Alleged

Mr. and Mrs. John C. Mabee

Western Hemisphere, ch. f, 2001, General Meeeting—Excellent Lady, by Smarten

Mrs. John C. Mabee

Sterling Cat, gr/ro f. 2002, Event of the Year—Stormy Regina, by Storm Cat

Sources: *The Blood-Horse* Archives and The Jockey Club

CHAPTER

16

Allen Paulson

Spotting opportunity and making it work in his favor were major ingredients in Allen Paulson's business career, and they carried over into his success as a Thoroughbred owner and breeder. In less than two decades, he established, expanded, and contracted Brookside Farm and his broodmare band; bred stakes winners at an astounding rate — 113 of them; and developed some of the best horses of the 1990s and beyond.

The Breeders' Cup, inaugurated in 1984, quickly soared to the top rung of ambition in Thoroughbred racing internationally, and Paulson forged a better record in the new, glamorous event than any other participant. Horses he, his wife, or estate owned solely or in partnership have won nine Breeders' Cup races, and Paulson bred six of them — both figures topping those lists. Paulson was North America's leading breeder in earnings four times, annually from 1994 to 1997, bred and raced four Eclipse Award winners and campaigned four others he had purchased, and won Eclipse Awards both as breeder and owner. In addition to his run as leading breeder, he also topped the owners' list twice, in 1995 and 1997. Most notably, Paulson was identified as the sporting breeder and owner of one great horse, Cigar, and as co-owner of one brilliant comet, Arazi.

All this success in his avocation followed a personal success story in aviation that ranks in the best traditions of Horatio Alger. Lest this statement smack of hyperbole, we hasten to defend it by pointing out that Paulson was, in fact, awarded the Horatio Alger

Association's Award for Distinguished Americans. His other honors not entailing a statuette of a horse included the American Academy of Achievement's Golden Plate Award, the National Aeronautics Association Harmon Trophy, the prestigious Wright Brothers Trophy, and five honorary degrees from universities.

None of it had come easily.

Paulson was born on April 22, 1922, near Clinton, Iowa, and before he was a teenager he had gone to work to help a family financially crippled by the Depression. His father moved to California, where the young son worked on a dairy. The mechanics of milking cows had none of the fascination of the mechanics of airplanes, and Paulson would wash planes for local stunt pilots in return for rides.

At nineteen Paulson went to work for Trans World Airlines at thirty cents an hour. Acumen, hard work, and imagination were his tools, and he worked his way up to foreman and then to flight engineer. His hands-on experience helped turn an observation into concrete results, to wit, development of a valve that could lubricate aircraft cylinders more effectively. This development was not merely handed over to his employer. Paulson secured the patent and sold the invention to airlines all over the world, including TWA.

That was the first of five patents he was awarded. He also became a crack flyer, and he and his crew set thirty-five flight records in two around-the-world journeys on Gulfstream jets. As founder and owner of the company, Paulson knew the equipment from the ground up. He also became the first distributor for Learjet. Eventually, his company would have annual sales in excess of $30 million.

By the 1980s Gulfstream had been developed into Gulfstream Aerospace Technologies, and in 1985 Chrysler Corporation purchased the company. Its annual gross earnings were more than $1 billion. Paulson repurchased control in 1990.

Along the way Paulson dabbled in racing with inexpensive horses, and then in the early 1980s he decided to re-enter the sport on a major level. About a decade later he also indulged another personal passion, golf, and, typically, did it in a businesslike way, purchasing not a membership in a golf club but a whole golf club, Del Mar Country Club in California. He maintained a home and business headquarters in Savannah, Georgia, for years but spent much of his later years in California with his wife, Madeleine, and her daughter, Dominique. He also developed a western division of his Kentucky-based Brookside Farm, near Bonsal, California. (Such was the environment of an aviator's business affairs that his two revered poodles, Frosty and Lucky, leapt about gleefully at the sound not of a doorbell or car door slam, but the vibrato hum of the helicopter Paulson flew to and from work.)

Allen Paulson with Cigar

Paulson burst onto the racing scene from California. In 1983 he paid a record $800,000, at the California Thoroughbred Breeders Association's juvenile sale, to buy the Super Concorde colt Cardell, whose three-year-old brother, Croeso, had recently won the Florida Derby at 85-1. Paulson had the benefit of good advice, having hooked up with the widely experienced racetrack veterinarian Jack Robbins. The aviation magnate also bought six others for $1,335,000. At the Keeneland summer yearling sale that year, Paulson was overshadowed by the Dubai versus English/Irish duel that saw the record for a single yearling pushed above $10 million. Still, he was the leading American buyer, spending $5,435,000 on ten yearlings, and he equaled the

world record for a filly by paying $2.5 million for a Northern Dancer yearling out of aging Epsom Oaks winner *Valoris II. At Saratoga the next month he bought five more for $1.62 million, again ranking as the most conspicuous among domestic spenders. Paulson made the trip highly profitable by taking orders from Dubai for three executive jets, including a spiffy new model priced at about $14 million.

In the spring of 1984, word came of Paulson's purchase of 425 acres of Kentucky land from Victor Green, the soft-spoken West Virginia businessman who would etch his name into racing history as co-breeder (with John R. Gaines of Gainesway Farm) of the brilliant European champion Dancing Brave. After selling some then buying more acreage,

Paulson expanded Brookside to 1,600 acres. Brookside's main division was near Versailles and was a sprawling property on which were built handsome barns with the steeply pitched roofs and cupolas of Bluegrass tradition. Veteran Kentucky horseman Ted Carr managed the farm, and for some years Emmanuel de Seroux was a key adviser. (Years later, de Seroux's wife, Laura, a former assistant to Hall of Famer Charlie Whittingham, trained Paulson-bred Azeri to a Breeders' Cup Distaff victory and Horse of the Year honors for 2002.)

In acquiring the farm, Paulson made it clear his ultimate goal was to breed his own horses and be only a "spot buyer" at yearling sales, as he put it. Paulson had a business plan, and strong opinions. He stressed "real good conformation and real good breeding," he told *The Blood-Horse* at Keeneland in the summer of 1984. "I also take into account what the sire of the yearling stands for. When I buy one for close to the stud fee, I feel like I have a real good deal. Sometimes I buy a filly that I don't think will be a racehorse. I bought a Northern Dancer filly for cheaper ($335,000) than I could get a breeding to the sire. She should be a good broodmare."

While market breeders live in hopes of major new buyers appearing at their sale barns, newcomers who are big spenders also tend to evoke behind-the-back comment. Small talk was not a strong suit of Allen Paulson's, nor was glib horse chatter, his status as an animal lover notwithstanding. This, coupled with the visibility of his large investments in the farm

and in breeding stock, created a wait-and-see mentality. Would this be one of those big spenders who are taken for a ride and then get out of the business?

Even impressive immediate results did not assuage all doubts. Estrapade won Paulson's first Eclipse Award, as champion turf female in 1986; she had cost him $4.5 million from the Bluegrass Farm consignment to the Keeneland November sale as a horse in training. The expensive Northern Dancer—*Valoris II filly carried the name Savannah Dancer to victories in the Del Mar Oaks and three other stakes, earning $360,600. Cardell was a stakes winner, albeit a moderate one. Then the 1988 foal crop bred by Paulson included sixteen stakes winners, but none rose high enough on the radar screen of major racing that many noticed. The impression that he was not getting rewards commensurate with his investment somehow lingered.

If Paulson even was aware of what jaded pros thought of him, that insight did not induce him to forsake his own judgment. While some had doubts, he happily set about populating the ranks of major winners with names duplicating those of aviation checkpoints around his "other" world. And, as matters transpired, it did not take him long to become the nation's leading breeder.

In the acquisition of Strawberry Road (Aust) as a stallion for Brookside, Paulson used a virtual reverse template for popularity in an American-based sire prospect. Pedigree? Exotic, but not in a fashionable way. Strawberry Road was by Whiskey Road, a son of

			Northern Dancer, 1961	Nearctic (**Nearco**—*Lady Angela)
				Natalma (Native Dancer—Almahmoud)
		Nijinsky II, 1967	Flaming Page, 1959	Bull Page (Bull Lea—Our Page)
				Flaring Top (Menow—Flaming Top)
Whiskey Road, 1972			Sailor, 1952	Eight Thirty (Pilate—Dinner Time)
		Bowl of Flowers, 1958		Flota (Jack High—Armada)
			Flower Bowl, 1952	*Alibhai (**Hyperion**—Teresina)
STRAWBERRY ROAD				Flower Bed (*Beau Pere—*Boudoir II)
		Rich Gift, 1959	Princely Gift, 1951	*Nasrullah (**Nearco**—Mumtaz Begum)
				Blue Gem (Blue Peter—Sparkle)
Giftisa, 1974			Riccal, 1953	Abernant (Owen Tudor—Rustom Mahal)
				Congo (Bellacose—Kong)
		Wahkeena, 1963	Red Jester, 1948	Red Mars (**Hyperion**—Red Garter)
				Climax (Nightmarch—Rejoice)
			Royal Souci, 1952	Regal Diamond (Treasure Hunt—Relative)
				Carefree (Harum Scarum—Hine Maka)

Nijinsky II, but one cast off to stand in the Antipodes, where Strawberry Road was bred; the dam was by Rich Gift, whoever that was. Racing style? Specialized in longer distances, and on grass at that, rather than the sprinter/miler speed on dirt that sends American horses to stud amid great expectations.

While other breeders would not flock to Strawberry Road as a commercial star in the making, what Paulson saw was "one of the most super horses ever. He won at six furlongs and at a mile and a half, and he was a champion on two continents. He had good conformation, a lot of stamina, and I think he passes that off." Strawberry Road won major races in Australia and Europe and finished second in the 1985 Breeders' Cup Turf. He won twenty-one of fifty starts, earning $1,655,678. Paulson bred eighteen stakes winners by Strawberry Road. Among the best of those, all raced by Paulson, are the following:

• Fraise (out of the purchased established major winner Zalataia, by Dictus), winner of the 1992 Breeders' Cup Turf (in the colors of Madeleine Paulson) over champion Sky Classic;

• Dinard (out of Daring Bidder, by Bold Bidder), winner of the 1991 Santa Anita Derby over Best Pal;

• Fowda (out of Al Balessa, by Rare Performer), winner of the 1992 Spinster Stakes over champions Paseana and Meadow Star;

• Ajina (out of Winglet, by Alydar), winner of the 1997 Breeders' Cup Distaff and that year's champion three-year-old filly;

Dahlia's Dreamer

• Escena (out of Claxton's Slew, by Seattle Slew), winner of the 1998 Breeders' Cup Distaff and that year's champion older female.

Strawberry Road died in 1995 at age sixteen from complications of a bacterial infection. Thus, other breeders were not fully able to capitalize on having seen his quality demonstrated to them. (One who believed early on was Virginia Kraft Payson, who bred the Strawberry Road filly Strawberry Reason, a foal of 1992. From Strawberry Reason, Mrs. Payson bred 2002 Breeders' Cup Juvenile winner and champion two-year-old colt Vindication.)

Jade Hunter was another key stallion acquisition for Brookside's stallion roster. This one fit the U.S. model more comfortably, for he is a son of the great and greatly popular Mr. Prospector. Jade Hunter did not hit his peak until four, however, when he won the 1988 Gulfstream Park and Donn handicaps in Paulson's silks. He retired with six wins in fourteen starts and earnings of $407,260. Jade Hunter has sired thirty-nine stakes winners (5 percent) through mid-2004, and, again, Paulson bred most of his top echelon of runners. Jade Hunter (out of the Pharly mare Jadana, an Irish-bred) reached new heights in 2002 as the sire of the Horse of the Year. The thirteen stakes winners Paulson bred by Jade Hunter include the following:

• Azeri (out of Australian-bred Zodiac Miss, by Ahonoora), winner of the Breeders' Cup Distaff and champion older female as well as Horse of the Year for Paulson's family trust in 2002, then repeat winner as champion distaffer again in 2003;

• Yagli (out of Nijinsky's Best, by Nijinsky II), winner of the grade I Manhattan, United Nations, Gulfstream Park Breeders' Cup Handicap, etc., and earner of $1,702,121;

• Stuka (out of Caerleon's Success, by Caerleon), winner of the 1994 Santa Anita Handicap;

• Diazo, (out of Cruella, by Tyrant), winner of the grade I Pegasus Handicap and Strub Stakes in 1993 and 1994, respectively.

Another stallion Paulson bought into during the horse's racing

E.M. JESSEE

Theatrical holding off Trempolino in the 1987 Breeders' Cup Turf

career was the Irish-bred Theatrical. Although, he, too, had the stigmas of being late to develop and being a grass specialist, Theatrical demonstrated enough racing class to create an aura of at least some fashion internationally. He capped a season of six grade I wins by holding off Prix de l'Arc de Triomphe winner Trempolino to win the Breeders' Cup Turf in 1987. He was the champion male on grass that year, and Paulson groused a bit that Ferdinand got Horse of the Year off fewer victories because Ferdinand's were scored on dirt. Theatrical's other major triumphs included the Turf Classic, Man o' War, Bowling Green, Sword Dancer, and Hialeah Turf Cup, and he had a career record in Europe and North America of ten wins in twenty-two starts and earnings of $2,940,036.

Paulson bought controlling interest in Theatrical and sent the son of Nureyev—Tree of Knowledge, by Sassafras, to Brookside. By mid-2004 Theatrical had sired some sixty-seven stakes winners (8 percent). Other breeders had great success with him, and Paulson held his own in use of the horse. The more than two-dozen stakes winners he bred by Theatrical include the following:

• Zagreb (out of Sophonisbe, by Wollow), winner of the 1996 Irish Derby and champion three-year-old in Ireland;

• Madeleine's Dream (out of L'Attrayante, by Tyrant), winner of the 1993 French One Thousand Guineas;

• Geri (out of Garimpeiro, by Mr. Prospector), winner of the grade I Oaklawn Handicap in 1996, etc., and earner of $1,707,980;

• Dahlia's Dreamer (out of Dahlia, by *Vaguely Noble), winner of the grade I Flower Bowl Handicap in 1994;

• Startac (out of Tenga, by Mr. Prospector), winner of the grade I Secretariat Stakes in 2001;

• Pharma (out of Committed, by Hagley), winner of the grade I Santa Ana Handicap in 1996.

Since Paulson was a golf devotee, the breeding of one other particular grade I winner by Theatrical must have been particularly pleasing. This was Broadway Flyer, bred in partnership with Stonereath Farm and professional golf champion Gary Player. Broadway Flyer (out of Serena, by Jan Ekels) won the Sword Dancer Invitational Handicap in 1996.

Another stakes winner sired by Theatrical and bred by Paulson was Mendocino, a modest stakes winner in France who emerged as the sire of Starine, winner of the 2002 Breeders' Cup Filly and Mare Turf.

It might be seen as somewhat typical of the perversity of the Turf that two of Paulson's potentially best sire prospects were Blushing John, who was largely disappointing, and Cigar, who was sterile! Blushing John (Blushing Groom—La Griffe, by Prince John)

Eliza

ANNE M. EBERHARDT

Eliza won her juvenile championship by getting one and one-sixteenth miles at that early age. Eliza was a 1990 foal from Daring Bidder, whose sire, Bold Bidder, was a champion son of Bold Ruler and sire of two Kentucky Derby winners in Cannonade and Spectacular Bid. George Swift bred Daring Bidder in Virginia and Paulson purchased her for $125,000 at the Fasig-Tipton Kentucky summer yearling sale in 1983, the year of Paulson's initial high visibility in the auction arena. She never raced but also foaled major winner Dinard.

Eliza won the Arlington-Washington Lassie and Alcibiades Stakes before her one-length triumph over Educated Risk in the Breeders' Cup Juvenile Fillies clinched her Eclipse Award. She had a career record of five wins in twelve starts and earnings of $1,095,316. (Paulson family banter had it that it was for Eliza that Madeleine traded back Cigar to her husband.)

Among discernible patterns in Paulson's Thoroughbred breeding history was the contribution of Seattle Slew mares. One, Solar Slew, produced his masterpiece, Cigar. Another, Claxton's Slew, foaled the champion Escena and other stakes winners Humbel and Showlady. The Seattle Slew mare Royal Strait Flush foaled three Brookside-bred stakes winners, including Tenga, dam of Startac. Another daughter of Triple Crown winner Seattle Slew was Savannah Slew, dam of stakes winners Admiralty and Astra.

Paulson extended Brookside's connection to Seattle Slew when he bought that champion's dam, My Charmer. My Charmer was eighteen when Paulson paid $2.6 million to buy her from the dispersal of Hermitage Farm owner Warner L. Jones Jr. (and partners) at Keeneland in 1987. She was in foal to Lyphard.

My Charmer was not successful for Paulson, but he got better dividends from another distinguished old mare he purchased. This was the international champion Dahlia, who was eighteen when Paulson bought her for $1.1 million at the 1988 Nelson Bunker Hunt dispersal at Keeneland. Dahlia already had foaled major winners Dahar, Rivlia, Delegant, and Wajd, and for Paulson she foaled the Flower Bowl Handicap winner Dahlia's Dreamer (by Theatrical) and Jersey Derby and Lexington Stakes winner Llandaff (by Lyphard). Dahlia became an honored pensioner at Brookside in 1996 and

was bred by Franklin Groves and acquired to race in Paulson's silks. Blushing John won the French Two Thousand Guineas and two other French stakes before being returned to North America. Blushing John earned his Eclipse Award as 1989 leading older male by winning the Pimlico Special, Hollywood Gold Cup, and several other good races and finishing third behind three-year-olds Sunday Silence and Easy Goer in the Breeders' Cup Classic.

The pedigrees noted above bespeak a wide mixture of bloodlines, the products of a breeder who liked to deal in large numbers and whose outlook was not seriously narrowed by any particular theory or dogma. While Strawberry Road and Theatrical brought a concentration of stamina, outside stallion seasons and broodmare acquisitions did not uniformly stress that quality. Committed, for example, was a crack sprinter before being acquired by Paulson to foal for him the stakes winners Pharma, Committed Dancer, and Hap.

Mt. Livermore was seen as a sire of speed (sprint champion Housebuster), but Paulson's homebred retained that status after Texan Gerald Ford bought the property and renamed it Diamond A Farms. Dahlia was thirty-one when she died in 2001 and was buried on the property.

Even though the dam of Seattle Slew proved less than successful for Paulson, Seattle Slew mares were quite influential in the Paulson breeding program. Escena's dam, Claxton's Slew, had been bred by Norman Hicks in Virginia and purchased by Paulson for $450,000 at Saratoga in 1985. She became a stakes winner in Ireland. Claxton's Slew was out of the Raise a Native mare Nutmeg Native, whose second dam was Sequence, from the vaunted Spendthrift Farm family going back to Myrtlewood. Seattle Slew also traced to Myrtlewood, his fifth dam.

In 1996, at three, Escena won the Fantasy Stakes and was beaten a neck by Pike Place Dancer in the Kentucky Oaks and by division champion Yanks Music in the Mother Goose. She got on a roll in 1998 at five, winning four consecutive stakes, including the Apple Blossom and Vanity handicaps for trainer Bill Mott. After a nose defeat by Aldiza in the Go for Wand Handicap, she faded to sixth in the Personal Ensign Handicap and thus was second choice to three-year-old Banshee Breeze for the Breeders' Cup Distaff. Escena took the lead early and stood off the favorite's torrid challenge to win by a nose. She was named champion of her division and had a career

record of eleven wins in twenty-nine starts and earnings of $2,962,639.

Escena's hammer price was $3.25 million when she was knocked down to Reynolds Bell Jr., agent, at the 1999 Keeneland January horses of all ages sale. The following year, her dam, Claxton's Slew, brought a winning bid of $3.3 million from Coolmore at the same sale.

A year before she won the Breeders' Cup Distaff, Escena had finished behind the favorite, Sharp Cat, in the same event. For the Paulson household, however, this caused little disappointment, for Sharp Cat was beating Escena not for first, but for second, and well in command on the front was Ajina, another homebred by Strawberry Road.

Ajina's dam, homebred Winglet, won the 1991 Princess Stakes and placed in the 1992 La Canada Stakes. She was out of Highest Trump, a champion filly raced in England and Ireland. Paulson purchased Highest Trump, by Bold Bidder, for $1.05 million from the 1988 January dispersal of Nelson Bunker Hunt's horses. Hunt had paid $2.5 million for her four years earlier. Highest Trump also became the second dam of English champion Bahri, sire of 2001 Prix de l'Arc de Triomphe winner Sakhee. Highest Trump's family also traces to Myrtlewood.

Ajina signaled exceptional class for trainer Mott when she won the Demoiselle Stakes by seven lengths late in her 1996 juvenile campaign. The following year she won the Mother Goose and historic Coaching Club American Oaks before running second in the Alabama and (to older distaff champion Hidden Lake) in the Beldame. In the Breeders' Cup Distaff, Ajina lurked near the pace and powered ahead in the stretch to win by two lengths to clinch her Eclipse Award for three-year-old filly. She had a career record of seven wins in seventeen starts and earnings of $1,327,915. Her dam, Winglet, in foal to Storm Cat, topped the Brookside consignment to the 1999 Keeneland November breeding stock sale at $4.6 million when bought by the Coolmore organization.

Ajina

BARBARA D. LIVINGSTON

The Paulsons with Arazi

Kentucky Derby. Ralph Wilson, owner of the Buffalo Bills of the National Football League and a longtime horseman, bred Arazi. The colt was a chestnut son of the top sire Blushing Groom and out of the vaunted family of Native Partner. Arazi's dam, Danseur Fabuleux, was by Northern Dancer, who had done for small and stocky what Arazi seemed about to do for small and dainty.

In one of those happenings that must have made Paulson chuckle about his own way of going, vis-à-vis the tutoring that other owners received from their own agents, Arazi had been sniffed at by many an agent for unsightly forelegs. The weanling had been bypassed by many and was bought only by one — Paulson — for $350,000.

Of the Eclipse Award winners Paulson raced but had not bred, Theatrical might be said to be the more accomplished than Blushing John, Estrapade, and Arazi. In terms of magnetism, however, none of the others rivaled Arazi as he rocketed home in the 1991 Breeders' Cup Juvenile. By the time the event took place on that autumn afternoon at Churchill Downs, Paulson already had a major victory on the day. Opening Verse, among the horses trained for a time by Dick Lundy, had won the Breeders' Cup Mile earlier on the program at 26-1. Bred by Jacques Wimpfheimer and Associates, Opening Verse (The Minstrel—Shy Dawn, by *Grey Dawn II) had been a nice stakes winner in England and after his return to this country had won the Oaklawn Handicap on dirt and the Early Times Turf Classic on grass among six stakes wins prior to the Breeders' Cup. In the Mile he rallied from fifth to win from Val des Bois in a finish in which the odds on the first four horses ranged in price from 16-1 to 41-1. In a tweak of irony, the camera caught the first two favorites, In Excess and Tight Spot, in a dead heat — for ninth.

Later that day Arazi's victory in the Breeders' Cup Juvenile was so spectacular as to install him instantly in the minds of racing fans as a super horse in the making and clear-cut choice for the following year's

Arazi was among the prospects Paulson sent to France, to trainer François Boutin. In 1988, three years before Arazi's starring year, Paulson had experimented with bringing a promising juvenile back to America in the autumn for the Breeders' Cup and had run third with Tagel, behind Is It True and Easy Goer. After Arazi had blazed through six consecutive victories in France to secure ranking as the best two-year-old there, Paulson bade Boutin to Churchill Downs. There, Arazi was favored although being asked to overcome the outside post position in a field of fourteen and his first start on dirt. He circled the field, and as announcer Tom Durkin was gearing up to describe a battle with Bertrando, he was left to gasp "and Arazi runs right past him!" The little colt galloped home by five lengths.

Arazi had arthroscopic surgery to remove chips in both knees, and when he came back at three he failed in the Kentucky Derby. The ability to elicit high emotion in humankind's expectation for a race-horse is one of the charms of the Turf, but the downside is the irrational response to disappointment. Arazi's later victories did little to retrieve his hyped status. Always the businessman, Paulson had sold an interest in Arazi to Sheikh Mohammed, who was no stranger to either triumph or travail on the racetrack.

Few owners find themselves with one horse that creates a hubbub comparable to that engendered by Arazi. Amazingly, Paulson saw it again within five years. This time, the kleig lights were on Cigar. Although the name was derived from an aviation checkpoint west of Tampa in the Gulf of Mexico, it resonated in another context to the public. Cigar, like Seabiscuit in another era, tripped nimbly across the image line from relatively privileged origins, to be seen as something of an upstart.

Cigar resulted from another of Paulson's stallion acquisitions. This one had ironies cloaked in contradiction. As a flashy and high-class scion of the Northern Dancer sire line, Palace Music had to be rated at least a solid prospect. Instead, he was a statistical failure, but, then again, he did sire Cigar! The muses of genetics long ago decided that a stallion does not have be a great sire in order to be the sire of a great horse.

Palace Music was by one of Northern Dancer's three Epsom Derby winners, The Minstrel (sire of Opening Verse), and was out of Come My Prince, by Prince John. He was bred by historic Mereworth Farm of Kentucky and was sold as a Keeneland yearling for $130,000 to Nelson Bunker Hunt. Paulson bought an interest in Palace Music as a racehorse during a career that included victory over champion Pebbles in the Champion Stakes in England and a grade I victory in this country in the mile and one-eighth John Henry Handicap. Palace Music was beaten a head by Last Tycoon in the 1986 Breeders' Cup Mile. A year earlier he had appeared to finish second, though handily beaten, to Cozzene in the Mile but was disqualified to ninth for lugging in and bothering another runner. In 1987 Palace Music entered stud at Hunt's Bluegrass Farm, then stood the Southern Hemisphere season that year in New Zealand. He went to Brookside for a couple of North American springs before going to Australia permanently. It was at Brookside that he sired Cigar.

The Seattle Slew mare who foaled Cigar was Solar Slew, who was attractive enough to Paulson that he had bought her for $510,000 when she was two and consigned to the 1984 California Thoroughbred Breeders sale. Solar Slew was the top-priced juvenile filly sold at auction that year.

Solar Slew was bred by Carl J. Maggio and the international dealers, the Murty Brothers (Dwayne and Wayne) of Murty Farm of Kentucky. The Murtys long had had connections in South America, as well as Europe, and it had been Wayne Murty who brought to Ogden Phipps the opportunity to buy the South American mare *Dorine, destined to become the second dam of Phipps' unbeaten champion Personal Ensign.

Beneath the glamour of her sire Seattle Slew, Solar Slew had a hardy background. She was out of Gold Sun, winner of fourteen races and a group I winner in Argentina. Gold Sun produced Jungle Gold (by Master Willie), a stakes winner who was third in the Irish One Thousand Guineas. Solar Slew's broodmare sire was Solazo, one of the hardy, but not high-fashion horses bred by King Ranch (Chapter 3). Solazo (Beau Max—*Solar System II, by Hyperion) won two stakes at six and was a half brother to Selima Stakes winner La Fuerza and to two other stakes winners.

Solar Slew failed to win in seven tries for Paulson and was retired. Her first foal was unraced, but her second foal, Mulca, who was by Raised Socially, was a standout in the limited sphere of Puerto Rico. Solar Slew's fourth foal was Cigar, who was foaled at the Pons family's Country Life Farm in Maryland, where the mare had been sent to be bred to Corridor Key.

Having had unsuccessful foals by upscale stallions such as *Forli and then Strawberry Road, the non-winner with the high price tag clearly had dropped in the pecking order of Brookside mares. Paulson sold Solar Slew privately for about $20,000 to Haras La Madrugada, an Argentine outfit that clearly would appreciate her family. Later, Cigar's exploits sent Paulson back to buy her again, and he paid $150,000 for her as a thirteen-year-old mare. In two purchases she had cost a combined $660,000 and in one sale had grossed about $20,000. This is hardly the sort of cost-reward ratio that made Paulson a success, but in Solar Slew's case he was glad to do the math and get her back. As matters transpired, Cigar was her one vehicle of glory, but he was all anyone might ask.

Cigar broke his maiden on dirt, but his pedigree seemed to make him a likely grass prospect. He floundered in that context, however, for young trainer Alex Hassinger, then was sent east to future Hall of Famer Bill Mott and returned to dirt. It was late in the colt's four-year-old season in 1994 when things turned around. An eight-length victory in an allowance mile encouraged a huge step up to the

BARBARA D. LIVINGSTON

Madeleine Paulson, Allen Paulson, Jerry Bailey, and Bill Mott

grade I NYRA Mile. Cigar won by seven lengths (the race later was renamed the Cigar Mile in his honor).

The following winter, momentum continued with an allowance victory and then a win in the Donn Handicap marred by the breakdown that prompted the retirement of defending Horse of the Year Holy Bull. The baton had been passed, but clumsily. Cigar then streaked on, taking the Gulfstream Park Handicap, Oaklawn Handicap, and Pimlico Special, all by daylight. Next came a four-length victory in the Massachusetts Handicap, a cross-country trip to win the Hollywood Gold Cup, and then a return east to win both the Woodward and Jockey Club Gold Cup. It was a campaign reminiscent of earlier eras of rugged older campaigners, encompassing various tracks, miles of travel, and different conditions and distances.

Cigar arrived at the 1995 Breeders' Cup on the brink of a perfect season of ten wins in ten starts. It harked back to the ten-for-ten season of Tom Fool in 1953 and the nine-for-nine of Spectacular Bid in 1980. On a dark, rainy day at Belmont Park, Cigar rallied to the lead with jockey Jerry Bailey one more time and raced through the stretch to win by two and a half lengths to seal a year's perfection.

Despite his status as Horse of the Year and with twelve consecutive victories, Paulson chose not to cash in yet on Cigar's stud value. He would race his champion, an entire male, at six. By then Cigar had captured the public's imagination well beyond the confines of racing. Madeleine Paulson, vivacious and comely and with a flair for public relations, catered to the public with taste. Cigar was transported into Manhattan on a van festooned with a Cigar banner, and he and Bailey made a dignified ceremonial appearance at a horse show in packed Madison Square Garden.

To Mott fell the pressure of bringing back a six-year-old off three and a half months, with a dozen straight victories. The streak was extended, first by a second Donn Handicap. The Maktoum family of Dubai had created the richest race in the world in the $4-million Dubai World Cup. The travel and lack of precedent were no deterrents to Paulson's sporting nature. The man who had flown around the world in record time was happy for his horse to fly halfway around, or so. In the desert night Cigar and Bailey again forged to the front at the proper moment and stood off another American traveler, Soul of the Matter. L'Carriere completed an American-bred and -owned sweep.

Back home Cigar added another Massachusetts Handicap and then won a specially arranged, but legitimate, race at Arlington Park called the Citation Challenge. The name commemorated the fact that Cigar was poised to equal the longest major winning streak in the modern era, the sixteen in a row won by the great Citation in 1948–50. This Cigar managed with aplomb.

One thing modern horses, even of Cigar's stardom level, are rarely asked is to carry high weights. Cigar's imposts finally came close to the echelons of the best old timers' experience, as he carried 128 in the Donn and 130 in the Mass Cap and the Arlington race.

The chosen opportunity to win his seventeenth consecutive race and break the shared mark came in the environs of Paulson's golf club, Del Mar. The Pacific Classic was where the streak ended, however. Siphon set a torrid pace and Bailey felt he had to chase it, and Cigar — even Cigar — could not keep it up, being run down by Dare and Go in the stretch. Bailey was criticized for chasing a pace of :45 4/5, 1:09 1/5, and 1:33 3/5 in a one and a quarter-mile race. Siphon was not cheap speed, however. He was a high-class racehorse, a grade I winner at the distance. Imagine the outcry that would have been heaped upon him if Bailey had let a sixteen-race winning streak end with a timid ride in which he allowed a front-runner to steal away to a long easy lead and never be challenged!

Back east Cigar righted things by winning the Woodward by four lengths but lost in the Jockey Club Gold Cup and Breeders' Cup, convincing many that he had lost just a whit of his sharpness as his sixth year neared conclusion. Nevertheless, he repeated as Horse of the Year. Although the glamour of the Triple Crown had not been a part of his career, Cigar was probably as famous and beloved a racehorse as any since Secretariat and Ruffian.

He went to stud with

appropriate fanfare and an all-time earnings record of $9,999,815. His career mark was nineteen wins in thirty-three starts, but once he got good, he had won seventeen of twenty. Cigar's public was soon stunned to learn of his sterility, but while a business setback, the development had the positive outcome of having him permanently on display, as a revered dignitary at the Kentucky Horse Park. Cigar was elected to the Hall of Fame in 2002, the first year he was eligible.

Despite Cigar's retirement, Paulson again was the leading breeder and leading owner the next year. He and Madeleine had already begun to reduce the size of the breeding and racing operation. A reduction consignment to Keeneland in January 1995 had sold 104 horses for $2,010,600, an average of $19,333.

At the 2000 Keeneland January sale, Claxton's Slew, as noted above, brought $3.3 million, and Cigar's eighteen-year-old dam, Solar Slew, was purchased by Lakland Farm for $2.8 million. The Paulsons sold sixteen mares at that sale for $9,195,000, as part of a reduction.

By then Paulson had developed cancer, and therein launched into one battle he could not win. He died at seventy-eight in the summer of 2000. That November

Azeri

245

the Allen Paulson Living Trust sold fifty horses at the Keeneland breeding stock sale. Solar Slew sold again for $1.3 million to Brookdale, agent, and the Strawberry Road mare Adel sold for $1.1 million to Diamond A Farm. The dispersal grossed $12,791,000, to average $255,820 — a tribute to the success and quality the one-time big-spending new guy had achieved and maintained.

Azeri's Epilogue

After Paulson's death, horses began running in the name of the Allen Paulson Living Trust. Madeleine Paulson was not involved in the operation of the stable and developed her own plans for continued involvement in racing and for the Del Mar Country Club.

One of Paulson's sons, Michael, became active in management, but there were conflicts within the management of the trust that for a time seemed to give rise to the spectacle of a Horse of the Year being auctioned a few months after the Eclipse Awards were handed out. This eventuality was avoided by negotiation, and Azeri in 2003 continued to race under Michael Paulson's management.

Azeri represented another illustration of Allen Paulson's worldly view and the lack of constraints on his thinking by market fashion. Having had wonderful success with the Australian Strawberry Road, he was not frightened away by Australian blood. He was not breeding primarily to sell, after all, and the good and sturdy racing and breeding stock from that country was a positive; any perceived fashion negative was not of overriding importance.

Azeri's dam, Zodiac Miss (Aust), was bred by the

Champions Bred

Mulca
1991 Puerto Rican champion three-year-old filly

Eliza
1992 champion two-year-old filly

Cigar
1995 champion handicap male
1995 Horse of the Year
1996 champion handicap male
1996 Horse of the Year

Ajina
1997 champion three-year-old filly

Escena
1998 champion handicap female

Mi Amigo Guelo
2000 Puerto Rican champion two-year-old colt

Azeri
2002 champion handicap female
2002 Horse of the Year
2003 champion handicap female

In Europe:

Nixon
1992 French champion two-year-old filly

Zagreb
1996 Irish champion three-year-old colt

Theoretically
2000 Irish champion three-year-old filly

---

wonderfully exotic sounding Tasman Breeding Venture No. 1. She was by the good, but not great, English sire Ahonoora and out of unraced Capricornia, by the Northern Dancer stallion Try My Best. Zodiac Miss won two stakes sprinting in Australia before Paulson bought her. She raced a couple of times at Santa Anita, placing once, before retirement. Azeri, by Jade Hunter, was her third foal. Azeri drew a top bid of $110,000 when put through the 1999 Keeneland September yearling sale. Heatherway, agent, was listed as buyer, but she wound up racing for the trust.

Through 2002 Azeri had a Cigar-like record, but she had not gone through a desultory early chapter prior to hitting her stride as he had. Unraced at two, she had run eleven times at three and four, winning ten. At four she won eight of nine, of which seven were in succession. That streak included the Santa Margarita, Apple Blossom, Milady, Vanity — all grade I races — and the grade II Clement L. Hirsch and Lady's Secret, and concluded with the Breeders' Cup Distaff. Her dominating performance in the Distaff at Arlington International swept her to the first Horse of the Year title for a female since Lady's Secret in 1986.

Paulson thus joined the ranks of breeders of more than one Horse of the Year, which include Mrs. Henry Carnegie Phipps (Seabiscuit and Bold Ruler), Meadow Stud (Hill Prince and Secretariat), Paul Mellon (Fort Marcy and Arts and Letters), William S. Farish (co-breeder of A.P. Indy, Charismatic, and Mineshaft), and, of course, Calumet Farm (Citation, Whirlaway, Twilight Tear, etc.).

Stakes Winners Bred by Allen Paulson

River Mike, dk.b/br.c. 1986, Icecapade—Aspern (Fr), by Riverman (SW in Germany)

Royal Danzig, b.c. 1986, Danzig—Royal Strait Flush, by Seattle Slew

Committed Dancer, b.c. 1987, Nijinsky II—Committed, by Hagley (SW in Ireland)

Linli, dk.b/br.f. 1987, Flying Paster—Spring Loose, by *Noholme II

Ozal, ch.c. 1987, Lyphard—L'Attrayante (Fr), by Tyrant (SW in France)

Power Bidder, b.f. 1987, Lines of Power—Daring Bidder, by Bold Bidder

Berry Road, b.f. 1988, Strawberry Road (Aust)—Positioned, by Cannonade

Cudas, dk.b/br.c. 1988, Seattle Song—Aspern (Fr), by Riverman (SW in France/U.S.)

Dhaka, gr.f. 1988, Icecapade—Cloudy Day Sunny, by Buckaroo

Dinard, dk.b/br.c. 1988, Strawberry Road (Aust)—Daring Bidder, by Bold Bidder

Dorky, b.f. 1988, Flying Paster—Going Raja, by Well Decorated

Fowda, ch.f. 1988, Strawberry Road (Aust)—Al Balessa, by Rare Performer

Fraise, b.c. 1988, Strawberry Road (Aust)—Zalataia (Fr), by Dictus

Ganges, gr.c. 1988, Riverman—Paloma Blanca, by Blushing Groom (Fr) (SW in France/U.S.)

Loach, b.c. 1988, Lines of Power—Scarlet Rain, by Rainy Lake

Mulca, dk.b/br.f. 1988, Raised Socially—Solar Slew, by Seattle Slew (SW in Puerto Rico)

Richard's Lass, ch.f. 1988, Allen's Prospect—My Account, by Private Account

Sha Tha, dk.b/br.f. 1988, Mr. Prospector—Savannah Dancer, by Northern Dancer (SW in France/U.S.)

Tivli, ch.f. 1988, Mt. Livermore—Bold Boston, by Bold Forbes

Varney, gr.g. 1988, Dahar—Spectacular Lady, by Spectacular Bid

Winglet, b.f. 1988, Alydar—Highest Trump, by Bold Bidder

Xray, b.c. 1988, Allen's Prospect—Galafest, by Cornish Prince

Brier Creek, dk.b/br.c. 1989, Blushing Groom (Fr)—Savannah Dancer, by Northern Dancer (SW in England)

Cleone, dk.b/br.c. 1989, Theatrical (Ire)—Sauna (Aust), by Loosen Up

Dahlia's Dreamer, ch.f. 1989, Theatrical (Ire)—Dahlia, by *Vaguely Noble

Mendocino, dk.b/br.c. 1989, Theatrical (Ire)—Brorita, by Caro (Ire) (SW in France)

Misako Togo, b.f. 1989, Theatrical (Ire)—Eastland, by Exceller (SW in Ireland)

Miss Lenora, b.f. 1989, Theatrical (Ire)—L'Attrayante (Fr), by Tyrant

Sing For Free, b.f. 1989, Theatrical (Ire)—Dumdedumdedum, by Star de Naskra

Tropico, gr.c. 1989, Herat—Paloma Blanca, by Blushing Groom (Fr)

Winnetka, ch.f. 1989, Theatrical (Ire)—Zalataia (Fr), by Dictus

Bucharest, ch.g. 1990, Strawberry Road (Aust)—Trudie Domino, by King of Macedon

Cigar, b.c. 1990, Palace Music—Solar Slew, by Seattle Slew

Corby, b.c. 1990, Dahar—La Escala, by Brave Shot

Diazo, ch.c. 1990, Jade Hunter—Cruella, by Tyrant

Eliza, b.f. 1990, Mt. Livermore—Daring Bidder, by Bold Bidder

Llandaff, ch.c. 1990, Lyphard—Dahlia, by *Vaguely Noble

Madeleine's Dream, dk.b/br.f. 1990, Theatrical (Ire)—L'Attrayante (Fr), by Tyrant (SW in France)

Siebe, b.c. 1990, Dahar—Cloudy Day Sunny, by Buckaroo

Stuka, ch.c. 1990, Jade Hunter—Caerleon's Success (Ire), by Caerleon

Tenga, dk.b/br.f. 1990, Mr. Prospector—Royal Strait Flush, by Seattle Slew (SW in France)

Vinista, ch.f. 1990, Jade Hunter—Sky Ninski, by Nijinsky II

Wende, b.f. 1990, Herat—Riverton, by Riverman

Wixon, b.f. 1990, Fioravanti—Forli's Fair, by Thatch (Bred in France; SW in France)

Apolda, gr.f. 1991, Theatrical (Ire)—Paloma Blanca, by Blushing Groom (Fr)

Ayanka, ro.f. 1991, Jade Hunter—Al's Charm, by Al Hattab

Dayev, b.g. 1991, Strawberry Road (Aust)—Saratoga Moon, by Saratoga Six

Duda, dk.b/br.f. 1991, Theatrical (Ire)—Noble Times (NZ), by Drums of Time

Hode, b.g. 1991, Dahar—Roxanne's Capade, by Icecapade

Jade Flush, ch.f. 1991, Jade Hunter—Royal Strait Flush, by Seattle Slew

Pharma, b.f. 1991, Theatrical (Ire)—Committed, by Hagley

Puzar, b.c. 1991, Herat—Cloudy Day Sunny, by Buckaroo

Sudana, ch.f. 1991, Dahar—Miss Carlotita (Arg), by Masqued Dancer

Yappy, gr.c. 1991, Herat—Griddle (Arg), by Off Shore Gamble

Yokohama, gr.c. 1991, Theatrical (Ire)—Griddle (Arg), by Off Shore Gamble

Admiralty, dk.b/br.c. 1992, Strawberry Road (Aust)—Savannah Slew, by Seattle Slew

Assurance (stp.), b.g. 1992, Herat—Cerlynne (Aust), by Cerreto (Ire)

Dowty, ch.c. 1992, Irish River (Fr)—Miss Carlotita (Arg), by Masqued Dancer

Geri, ch.c. 1992, Theatrical (Ire)—Garimpeiro, by Mr. Prospector

Humbel, b.c. 1992, Theatrical (Ire)—Claxton's Slew, by Seattle Slew (SW in Ireland)

Rice, ch.g. 1992, Blushing John—Estrapade, by *Vaguely Noble

Born Twice, b.f. 1993, Opening Verse—Slew Boyera, by Seattle Slew

Clure, ch.c. 1993, Theatrical (Ire)—Garimpeiro, by Mr. Prospector

Crimson Road, ch.f. 1993, Strawberry Road (Aust)—Bejat, by Mr. Prospector

Escena, b.f. 1993, Strawberry Road (Aust)—Claxton's Slew, by Seattle Slew

Gordi, ch.c. 1993, Theatrical (Ire)—Royal Alydar, by Alydar (SW in England/Ireland)

Lenta, ch.f. 1993, Theatrical (Ire)—Gold Nickle, by Plugged Nickle

Penne, ch.f. 1993, Jade Hunter—Pero Yo Se (Arg), by Proud Arion

Tanja, b.f. 1993, Allen's Prospect—Fleet Road, by Magesterial

Yagli, ch.c. 1993, Jade Hunter—Nijinsky's Best, by Nijinsky II

Yokama, ch.f. 1993, Irish River (Fr)—Dicken's Miss, by *Vaguely Noble

Zagreb, b.c. 1993, Theatrical (Ire)—Sophonisbe, by Wollow (SW in Ireland)

Ajina, dk.b/br.f. 1994, Strawberry Road (Aust)—Winglet, by Alydar

Chitka, ch.f. 1994, Jade Hunter—Royal Herat, by Herat

Garbu, b.c. 1994, Strawberry Road (Aust)—Sauna (Aust), by Loosen Up

Glok, b.g. 1994, Theatrical (Ire)—Gozo Baba, by Raja Baba

Moonlight Paradise, b.f. 1994, Irish River (Fr)—Ottomwa, by Strawberry Road (Aust) (SW in England)

My Sugar Magnolia, b.f. 1994, Opening Verse—Conquistadoria, by Conquistador Cielo

Oturi, b.g. 1994, Strawberry Road (Aust)—Pearlie Gold, by Lot o' Gold (SW in Canada)

Road to Seattle, dk.b/br.c. 1994, Strawberry Road (Aust)—Vands, by Seattle Song

Trample, ch.c. 1994, Trempolino—Spectacular Native, by Spectacular Bid

Tuzia, b.f. 1994, Blushing John—Fleet Road, by Magesterial

Zede, dk.b/br.c. 1994, Strawberry Road (Aust)—Index's, by Storm Bird

Adel, dk.b/br.f. 1995, Strawberry Road (Aust)—Prankstress, by Foolish Pleasure

Cali, ch.f. 1995, Strawberry Road (Aust)—Lyphard's Starlite, by Lyphard

Hustler, b.c. 1995, Strawberry Road (Aust)—Over All, by Mr. Prospector

Jewels Togo, b.g. 1995, Jade Hunter—Misako Togo, by Theatrical (Ire)

Astra, b.f. 1996, Theatrical (Ire)—Savannah Slew, by Seattle Slew

Glick, b.c. 1996, Theatrical (Ire)—Bejat, by Mr. Prospector

Hap, ch.c. 1996, Theatrical (Ire)—Committed, by Hagley

Jarf, b.g. 1996, Mt. Livermore—Whiskey's Gift (Aust), by Whiskey Road

Mort, ch.g. 1996, Jade Hunter—Cruella, by Tyrant

Rize, dk.b/br.g. 1996, Theatrical (Ire)—Campagnarde (Arg), by Oak Dancer (GB)

Del Mar Show, b.c. 1997, Theatrical (Ire)—Prankstress, by Foolish Pleasure

Mary Kies, ch.f. 1997, Opening Verse—Flying Girl (Fr), by Rusticaro

Rob's Spirit, b.c. 1997, Theatrical (Ire)—Winglet, by Alydar

Serial Bride, ch.g. 1997, Stuka—Risky Bride, by Blushing Groom (Fr)

Theoretically, dk.b/br.f. 1997, Theatrical (Ire)—Aspern (Fr), by Riverman (SW in Ireland)

Walkslikeaduck, dk.b/br.c. 1997, Blushing John—Nabla, by Theatrical (Ire)

Angel Gift, b.f. 1998, Allen's Prospect—Whiskey's Gift (Aust), by Whiskey Road

Azeri, ch.f. 1998, Jade Hunter—Zodiac Miss (Aust), by Ahonoora (GB)

Jade's Ace, b.f. 1998, Jade Hunter—Dahar's Best, by Dahar

Megantic, b.c. 1998, Theatrical (Ire)—Jade Flush, by Jade Hunter

Mi Amigo Guelo, dk.b/br.c. 1998, Dehere—Izana, by Theatrical (Ire) (SW in Puerto Rico)

Premeditation, ch.c. 1999, Afternoon Deelites—Berga, by Jade Hunter

Presidio Heights, dk.b/br.g. 1998, Allen's Prospect—Kalfo, by Dahar

Startac, b.c. 1998, Theatrical (Ire)—Tenga, by Mr. Prospector

Stylish, b.f. 1998, Thunder Gulch—Miss Lenora, by Theatrical (Ire)

Sarah Jade, ch.f. 1999, Jade Hunter—Scotch and Dry (Aust), by Whiskey Road

Showlady, b.f. 1999, Theatrical (Ire)—Claxton's Slew, by Seattle Slew

Allen Paulson and Nelson Bunker Hunt

Prudent Offer, ch.c. 1984, Providential—Domitia, by Royal Ski

Allen Paulson, Stonereath Farms, and Gary Player

Broadway Flyer, b.c. 1991, Theatrical (Ire)—Serena (SAf), by Jan Ekels (SW in England/U.S.)

Madeleine Paulson

Major Bowen, b.g. 1994, Blushing John—Raise a Herat, by Herat

Dominique's Joy, b.f. 1995, Strawberry Road (Aust)—Madeleine's Joy, by Theatrical (Ire)

Laura's Lucky Boy, b.c. 2001, Theatrical (Ire)—Corridora Slew (Arg), by Corridor Key

Rock Hard Ten, dk.b/br.c. 2001, Kris S.—Tersa, by Mr. Prospector

Madeleine Paulson, W.S. Farish, and Skara Glen Stables

Broadway Gold, dk.b/br.f. 2002, Seeking the Gold—Miss Doolittle, by Storm Cat

Sources: *The Blood-Horse* Archives and The Jockey Club

CHAPTER

17

Nelson Bunker Hunt

Nelson Bunker Hunt has never easily fit the mold of a monied horse breeder. He is a jovial teetotaler, a born-to-wealth Texan who once played big in commodities while known to travel coach class, an innovator who owned the first shuttle stallion, a patron of the grandstand who has won Europe's gaudiest prizes.

Nevertheless, there are a lot of familiar qualities there: staunch family man, dedicated horse lover, and ardent sportsman, with that combination of realist and dreamer without which Thoroughbred racing might have floundered years ago.

For the criteria of this book, as laid out in the introduction, Hunt seems almost an anti-hero. While the focus here is meant to be on those who bred principally for the North American racing scene, or market, Hunt carved out much of his success by sending young stock to excel in Europe.

The horses, though, kept coming back to vaunt him.

Homebreds Dahlia, Youth, Trillion, and Estrapade returned from Europe to earn championship honors in North America, while horses produced by Hunt's Bluegrass Farm made him the leading breeder here in 1986 and 1987.

During the 1930s Hunt accompanied his father, tycoon H.L. Hunt, to Arlington Downs in his native Texas, before legislation scuttled the sport in the Lone Star State for fifty years or so. Then, in prep school, Hunt drew as a roommate one Edward Stephenson. Over the next decade they stayed in touch sufficiently that when Stephenson called one day to suggest they go to a horse sale in Kentucky, Hunt went along with the idea.

Hunt learned a lesson that the Doc Bonds, George Swinebroads, Laddie Dances, and Humphrey and John Finneys of the world of gavel and chant have tried to keep under their hats, to wit: "If you don't plan to buy, don't go to the sale." The school chums bought six horses that did not amount to much.

About a year after he had divested himself of his interest in those horses, Hunt fell into a conversation with the trainer Frank Christmas, who prevailed upon the young entrepreneur to try the claiming game. Years later Hunt still supported that route as "the best way for a new man to get into the business, better than buying yearlings. With claimers you learn about the business, and you get into action sooner."

Still, class was its own siren song. In 1956 Stephenson (who owned a farm in Virginia) and Hunt stepped into the big time by paying a reported $110,000 for the French horse *Master Boing within hours of his winning the Washington, D.C., International at Laurel. If a fellow wanted to be in the claiming game, *Master Boing was the ticket. He failed to sire a stakes winner.

Hunt went his way in a professional sense, being involved in the top levels of the oil and silver markets, and had vast expanses of cattle ranches. He was at parties with presidents and was photographed presenting gifts to world leaders.

Hunt also chose to deal with distinguished fellows in his sideline pursuit of the Thoroughbred. Tom Smith (with a retro fame in 2003 for having trained Seabiscuit) was recruited to handle the horses, while pedigree advice was supplied by Joseph A. Estes, former editor of *The Blood-Horse* and a tireless and eloquent disciple of class in the dam. In later years the similarly eloquent and worldly Abram S. Hewitt came into the fold, and it was on his testimony that we personally gained a perspective about Hunt that, even when he owned six hundred or so horses, he was knowledgeable about each of them. Over the years other key retained or employed horsemen included Larry Richardson, Bob Bricken, Dan Midkiff III, and Bill Taylor.

Hunt at first had followed the lead of Stephenson in buying a farm in Virginia but soon decided that since Kentucky was where the top stallions were, that is where his horses should be. He sold the farm and started sending his mares to Henry White's Plum Lane Farm in Kentucky. Later he transferred his mares to the Hancock family's Claiborne Farm and, still later, established his own Bluegrass Farm. Bluegrass grew into several massive divisions of acreages in several Central Kentucky counties.

"I am a kind of high-number operator," Hunt said of the scope of his operation as it stood in the 1970s. "If you only had, say, five mares, you might never raise a good horse, even if you did things right and had good stock. With as many mares as I have, I figure I can come up with at least one good horse every few years. You can figure there will be a total of maybe 60,000 Thoroughbreds born each year in the United States, Europe, Argentina, New Zealand, and Australia. Out of that 60,000 there will be, what, twelve top ones? That's part of the appeal, seeing if you can beat the odds, because in numbers the odds are against you."

He also maintained a ranch, largely for training, in his native Texas, which, like Robert Kleberg Jr. of King Ranch before him, he believed provided an environment and soil type that benefited a potential racehorse: "There is a lot of sand, and it seems to help their feet. You almost never see a contracted frog there."

Hunt expanded to New Zealand and was a pioneer at standing horses in both the Northern Hemisphere and Southern Hemisphere in the 1970s.

"I was in New Zealand on other business, and I

Nelson Bunker Hunt

MILT TOBY

thought it was the greatest pastureland I had ever seen," he once told us for an article in *The Blood-Horse*. "I mentioned to Jim Shannon that I might be interested in buying a farm, and he took me around to show me a place he knew was available. It had been a dairy farm. I said, 'Well, if I go back home I might change my mind, so why don't I just buy it now?' " He had spent almost an entire morning on the project.

Shannon, a combination journalist, bloodstock agent, and hands-on horseman, had earned a place in Australian Turf history when he and his wife traveled the ocean with Sailor's Guide en route to his victory in the 1958 Washington, D.C., International. He later was hired by Hunt as a sort of worldwide general manager for what grew into a huge international Thoroughbred operation.

Hunt's Dahlia twice won the King George VI and Queen Elizabeth Diamond Stakes at Ascot. In the photos of the occasion when Shannon represented Hunt in accepting the trophy from Queen Elizabeth II, his expression seems to reveal the wonderment of "how in the world did I come to be here?" Hunt told us that the year he personally accepted the trophy from the Queen, he later was on a public transfer bus at John F. Kennedy airport, lugging the trophy home by hand, when he was asked "Are those real diamonds?" and realized he should be careful in how visibly he brandished such an object. (Low-fare travel and inexpensive restaurants and accommodations probably were exaggerated in descriptions of Hunt during his heyday. Then again, we once personally saw him board an international flight carrying a tattered white cardboard box — presumably full of business papers — held together by string, and join us back in the coach section.)

Hunt bred a couple of minor stakes winners at the end of the 1950s, and then in the next decade he bred the good mare Amerigo Lady. *Amerigo was a fierce but fine son of Nearco who stood at Stephenson's Kilmaurs Stud Farm and, in a promising career ended by an early death, sired champion Fort Marcy. His daughter Amerigo Lady was out of an Australian mare, *Lady Sybil, by Count Rendered. Amerigo Lady had had some success racing for Hunt before his tax people advised that it would be helpful to sell

her, so he passed her along to Fort Marcy's breeder-owner, Paul Mellon, who either had better advisers or was in a different sort of tax year. Amerigo Lady won eight stakes from coast to coast among thirteen wins in fifty-four starts and earned $416,465.

A breakthrough for Hunt as a breeder and also as an owner came in 1964, with the foaling of the filly he named *Gazala II. The filly was conceived in the United States but foaled in France. She was by Dark Star (son of Australian *Royal Gem II), the 1953 Kentucky Derby winner who was a good sort of sire but off the top fashion screen. She was out of *Belle Angevine, by L'Amiral, from a solid French family. At three in 1967, *Gazala II won both the filly classics in France, the Poule d'Essai des Pouliches (Guineas) and Prix de Diane (Oaks). A fellow who would buy a French-bred stakes winner or a New Zealand farm precipitously, who would embrace Australian breed-

*Vaguely Noble

ing for the American scene, and who had won classics with a combination of Australian-French bloodlines, could already be said to be adventurous, or impulsive. At the least, he was not limited to conventional wisdom about what was fashionable. In the purchase of *Pamplona II, Hunt went farther down a path less taken.

While Argentina was recognized as a high-class source of Thoroughbreds, largely through years of

Dahlia

importations from England, other South American countries were seen internationally as down the scale. Peru was obscure, but on an early 1960s trip to South America, where he knew mares could be bought more cheaply than in the United States or Europe, he learned of a mare who in 1959-60 had become the first filly to win the Peruvian Triple Crown. He asked to see her and wound up buying her. The mare's name was *Pamplona II, and she was a product of one of those combinations of lesser representatives of classic English bloodlines that one might expect. Her sire, Postin, was an outstanding sire in Peru and was by Hyperion's half brother Hunter's Moon. (It was another horse of this name, *Hunter's Moon IV, who sired the great American steeplechaser Neji.) *Pamplona II's third dam was Irish Oaks winner Conversation Piece.

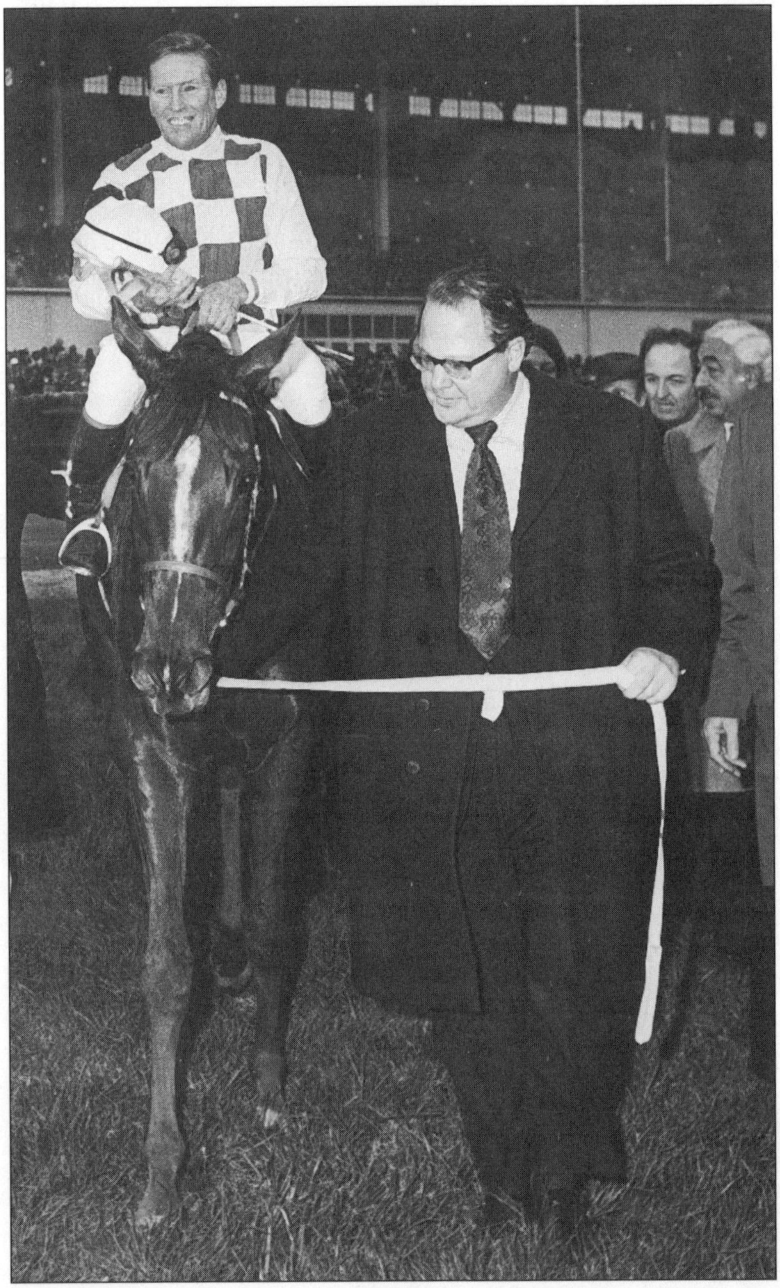

While fashion mavens might have wondered what he was thinking, Hunt brought *Pamplona II home, and she proceeded to produce two European-raced classic winners among four stakes winners. The Estes doctrine of class in the dam, more or less exclusive of pedigree worries, was certainly working for Hunt.

A key opportunity arrived in 1967, when a colt named *Vaguely Noble came up for sale in England. *Vaguely Noble was a handsome number from the breeding establishment of Major Lionel B. Holliday. Holliday died late in 1965, and his son, Brook Holliday, was still dealing with death duties two years later. When *Vaguely Noble won the Observer Gold Cup in such fashion as to insure a high price on the open market, Brook Holliday decided it was prudent to sell.

*Vaguely Noble was by Sir Winston Churchill's good horse Vienna and out of Noble Lassie,

252

by the great Nearco. (In a testimony to the ecumenical nature of the Turf and the world of sport, Vienna, owned by an Allied hero of World War II, stood at stud first in England, then in France, and eventually was sent to Japan.) The female family traced to Brulette (fourth dam), winner of the Epsom Oaks and Goodwood Cup, with mares by Big Game and Hyperion in between.

Such are the patterns of the sale market that top-class racehorses seldom come up for auction during their active careers. The 37,500 guineas paid for Flying Fox in 1900 was still carried in publications as the record for a horse in training when two-year-old *Vaguely Noble entered the Newmarket sale ring in December of 1967. Bidding opened at more than twice that figure, 80,000, and he was knocked down a few minutes later for a world-record 136,000 guineas ($342,720). The flashy California bloodstock agent Albert Yank placed the winning bid on behalf of Dr. Robert A. Franklyn.

Yank outbid Hunt's representative but wasted little time in suggesting that Dr. Franklyn might welcome a partner. Hunt was wary — "Owning a horse with another man can be a problem, like two men having the same girlfriend." Franklyn was not exactly a proven commodity on the Turf. He had never owned a racehorse before. Adhering to the moral principle that humans are put on Earth to help one another, Franklyn had crafted the specialty of surgically raising the self-esteem of ladies who felt their natural bosoms were only vaguely notable.

Perhaps offended by attitudes encountered in Europe, Franklyn decreed that his wife was as much a lady as anyone else and so would thence be known as "Lady Wilma Franklyn." Plus, there was talk of running *Vaguely Noble in gold-tinted aluminum plates.

Still, the colt was an incredibly attractive prospect. In the end, the depth of the horse's glamour overcame the shallowness of the good doctor's glitz, and Hunt bought in. *Vaguely Noble was not eligible for the Epsom Derby, so the selected trainer, Etienne Pollet, prepared him for two major French tests. After victories in lesser races, he finished third the Grand Prix de Saint-Cloud. If this suggested he had been overrated, he put that impression right in the fall, when the classy Epsom Derby winner, Sir Ivor, came calling. *Vaguely Noble smote Sir Ivor in the climactic Prix de l'Arc de Triomphe.

John R. Gaines would establish at his Kentucky farm, Gainesway, one of the world's greatest concentrations of international sire power. *Vaguely Noble was syndicated on an evaluation of $5 million in 1969 and became part of a Gainesway line-up that was to include Blushing Groom, Lyphard, and Riverman. Hunt over time bought some of Franklyn's shares and achieved majority interest in *Vaguely Noble.

*Vaguely Noble became an internationally acclaimed stallion, who sired seventy stakes winners (9 percent). He was a good fit for Hunt in the sense that Europe was a favored venue for a key part of the

*Vaguely Noble, 1965	Vienna, 1957	Aureole, 1950	**Hyperion (Gainsborough—Selene)** / Angelola (Donatello II—Feola)
		Turkish Blood, 1944	Turkhan (*Bahram—Theresina) / Rusk (Manna—Baby Polly)
	Noble Lassie, 1956	Nearco, 1935	Pharos (Phalaris—Scapa Flow) / Nogara (Havresac—Catnip)
		Belle Sauvage, 1949	Big Game (*Bahram—Myrobella) / Tropical Sun (**Hyperion**—Brulette)
DAHLIA	Honeys Alibi, 1952	*Alibhai, 1938	**Hyperion (Gainsborough—Selene)** / Teresina (Tracery—Blue Tit)
Charming Alibi, 1963		Honeymoon, 1943	*Beau Pere (Son-in-Law—Cinna) / Panoramic (Chance Shot—Dustwhirl)
	*Adorada, 1947	*Hierocles, 1939	Abjer (Asterus—Zariba) / Loika (Gay Crusader—Coeur a Coeur)
		Gilded Wave, 1938	Gallant Fox (*Sir Gallahad III—Marguerite) / *Ondulation (*Sweeper—Frizette)

Texan's career. As a resident of Dallas, Hunt was a long way from any American racetrack at that time, so trips to see his horses in, say, New York or California, involved considerable travel, anyway. Also, he had frequent business dealings in Europe and since "horses stand training better in Europe," he said, there was "the humanitarian aspect of just hating to see the horses get hurt."

*Vaguely Noble sired European classic winners from ruggedly bred American mares, from the Peruvian Triple Crown winner, from classic fillies themselves. Hunt held each of these cards, and he cashed in on all three.

The best horse Hunt bred was in *Vaguely Noble's first crop. She was Dahlia.

A lengthy but elegant chestnut, Dahlia was from a dam who was an arch example of the importance of racing class. Charming Alibi was by Honeys Alibi, who sired only seven stakes winners, and there was not much fashionable about her dam, *Adorada II, by *Hierocles. Still, Charming Alibi was an effective race mare, winning four stakes among sixteen wins in seventy-one starts to earn $110,483. Henry White bought her at Hunt's behest from C.F. Parker, for $80,000.

*Gazala II had been trained for Hunt by Jack

Cunnington Jr., but the brigade of *Vaguely Nobles would be in the hands of one Maurice Zilber. An Egyptian with a flamboyance that was opposite to the low-key public personality of Hunt, Zilber would make younger Turf writers of today think Bob Baffert, for example, was just "sort of" colorful and that D. Wayne Lukas and Bobby Frankel were lacking in self-confidence. Zilber's bravado needed outstanding horses to avoid his seeming comical, and Hunt saved him from himself. For his part, Hunt seemed bemused by Zilber and his posturing.

The swoosh of a Dahlia drive to victory in the Hunt silks of dark green-and-light green blocks was breathtaking, and she gave the world plenty of chances to revel in the moment. The cocky and dedicated writer Tony Morris was moved to imply a challenge of sorts in *The Bloodstock Breeders' Review* with regards to the first (as a three-year-old) of Dahlia's victories in the King George VI and Queen Elizabeth Stakes, in 1973: "This was no defeat for the males. It was humiliation. They dictated the pace that destroyed them. They broke under the ferocious pressure. They were devoured by the filly who bided her time in the rear until that explosion of killer pace made her Ascot performance one of the greatest ever seen—anywhere. Secretariat fans may write me c/o *The Bloodstock Breeders' Review*."

Dahlia's earlier European victories had included the Irish Oaks, and late in the year she came back to America to defeat older males again in the Washington, D.C., International. The next year brought victories in the Grand Prix de Saint-Cloud and Benson and Hedges Gold Cup, plus another King George. In her annual invasion in the autumn, Dahlia won the Man o' War and Canadian International before being upset seeking a second Washington, D.C., International. The Turf category of Eclipse Awards was not split into male and female divisions until 1979, but Dahlia won it anyway, in 1974.

At five Dahlia's temperament was working against her, and she

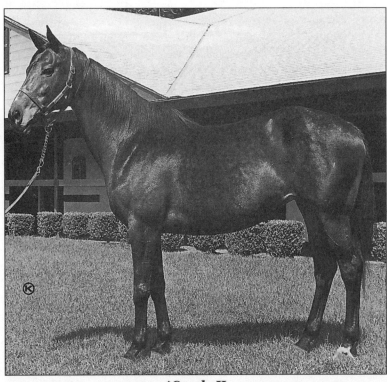

Gazala II

THE BLOOD-HORSE

Exceller with his connections

won only one important race, a second Benson and Hedges. Her annual trip to America netted no victories, although Hunt's Nobiliary won the D.C., International with Dahlia in arrears.

A frustrating aspect of Dahlia's career was that she could never defeat her contemporary female foe, Allez France. Hunt wanted her to beat Allez France off the course and in the record books, getting to the top in earnings. Thus, he decided to keep Dahlia in training at six in 1976. Remaining in America with Charlie Whittingham, she did little in most of her thirteen starts but jumped up to win an allowance race and then score in the Hollywood Invitational. She had a record of fifteen wins in forty-eight starts and earnings of $1,543,139 to surpass Allez France and all other distaffers to that time.

Dahlia's career earnings barely surpassed the auction price of her yearling half brother of that year. "If you have as many horses as I do," Hunt said, "you have to have some program of selling, but I really am not carried away with selling yearlings."

The market appeal of a colt in the first crop of Triple Crown winner Secretariat and from the dam of Dahlia made one Hunt-bred a natural for the auction ring in the spirit of balancing costs and income.

Keeneland added a digit space to its price board, and the colt brought $1,500,000, more than twice the previous yearling mark of $715,000. Purchased by a Canadian partnership, he was named Canadian Bound. He never raced.

The sales became an essential part of Hunt's Bluegrass Farm strategy. He topped the Keeneland July sale in 1976 with an average of $366,111 and again in 1978 with 21 averaging $251,476. He continued to say that he did not relish selling yearlings, but it was good business strategy.

During the time Dahlia was the most famed of Hunt's horses, he had turned over to Zilber some other compelling homebreds. The classic winner *Gazala II foaled a *Vaguely Noble colt named Mississipian, who in 1973 won the key French juvenile race, the Grand Criterium.

In 1975 the aforementioned Nobiliary showed sufficient brilliance in winning the Prix Saint-Alary in late May to embolden Hunt and Zilber to send her to Epsom. This adventure was not for the Oaks, however, but for the Derby on the first Wednesday in June. In more than two centuries only six fillies have won the Epsom Derby and none have won since Fifinella in 1916. Nobiliary acquitted herself, well, nobly, finishing second while beaten three lengths by the very good Derby winner Grundy. She also won the Prix de la Grotte that year, in addition to the Washington, D.C., International.

Nobiliary was another by *Vaguely Noble and was out of the good race mare Goofed, whom Hunt had purchased after she produced the good little racehorse and exceptional sire Lyphard. From Goofed, Hunt also bred the Sailor colt Barcas, winner of the Bowling Green Handicap.

The next year, 1976, the disappointment of Dahlia's having only a single stakes victory was virtually lost in an amazing run of success for the Hunt stable. Again, the emphasis was abroad, but as in the case of Dahlia and Nobiliary, ended with an American racing chapter. That year Hunt won the Epsom Derby and French Derby with homebreds and the Grand Prix de Paris and classic Prix Royal-Oak with a purchased colt.

The Epsom Derby winner was Empery, who was by *Vaguely Noble and out of that Peruvian Triple Crown winner, *Pamplona II. That mare had long since proven the wisdom of her purchase, for Empery was her fourth stakes winner. Moreover, she had already achieved distinction as the dam of a classic winner when her Pampered Miss won the 1970 Poule d'Essai des Pouliches (French One Thousand Guineas) for Hunt.

Empery won only one stakes, but it was the Epsom Derby, and the great rider Lester Piggott had accept-

ed the mount. Four days later Hunt and Zilber won the French version of the Epsom Derby, the Prix du Jockey-Club. This victory came with Youth, another son of *Gazala II and thus a half brother to Mississipian. Youth was sired by the Claiborne Farm stallion Ack Ack, Horse of the Year in 1971.

Whereas Empery had one day of glory, Youth had several and suggested the possibility that he was a truly great horse. After his dashing three-length triumph in the French Derby, he was unplaced in the King George but rebounded to be third behind Ivanjica in the Prix de l'Arc de Triomphe.

Zilber loved to come to North America, and Hunt was all for the big fall turf races. At Woodbine, Youth won the Canadian International by four lengths, and at Laurel he dazzled in the Washington, D.C., International, winning by ten lengths. The Arc winner was third.

Oddly, bad advice had prompted Hunt's decision to keep Empery and Youth rather than sell them as yearlings. The Keeneland summer sale had closed so Hunt was planning to send them to the Saratoga August sale of 1974. Then a tax adviser told him that it was not a good time to sell, so he did not. As Hunt told us a few years later, the tax advice, as it turned out, "had been wrong."

Youth and Empery both were retired and put into a combined syndication based on a dual evaluation of $12 million. In a pattern that harked back to Calumet Farm's Bull Lea, *Vaguely Noble's sons were consistently failures at stud, and Empery fit that pattern. Youth was an indifferent sire, as well, save for one bright spot in the form of 1983 Epsom Derby winner Teenoso.

In 1976, when Hunt was winning classics with homebreds, he also had Exceller in the stable. The same age as Youth and Empery, Exceller had been purchased on the spur of the moment at Keeneland in 1974. The colt was by *Vaguely Noble and out of the good stakes winner Too Bald (by Bald Eagle). He was bred by the widow of major Turfman Charles W. Engelhard. One of Hunt's advisers, Ted

Youth

Curtin, spotted the colt earlier and, although they were at the sale looking for fillies, suggested that Hunt jump in when the horse was languishing in the ring. Hunt, majority owner of *Vaguely Noble and breeder of twenty-one stakes winners by the horse, picked up another *Vaguely Noble colt for $25,000.

After winning the Grand Prix de Paris and Prix Royal-Oak at three in the shadow of Empery and Youth, Exceller was a major campaigner for several more years. At four he won the Coronation Cup in England, the Grand Prix de Saint-Cloud in France, and the Canadian International as Hunt and Zilber continued their autumn invasions. At five and six Exceller campaigned in this country, under Whittingham. Hunt had hoped Dahlia would prove herself on dirt and she did not, but Exceller did. He won several major stakes on both surfaces in California and then scored one of the most memorable of victories when he rallied from twenty-two lengths back to edge Seattle Slew in the Jockey Club Gold Cup. Exceller won fifteen of thirty-three starts and earned $1,674,587.

(Hunt was a blameless character in two bizarre situations regarding outstanding horses. Years after failing at stud in this country, Exceller wound up in Scandinavia where his latest owner would not maintain him, and he was killed. Hunt gladly would have rescued him had he any way of knowing the peril. Earlier, a good filly in Hunt's European stable, Carnauba, had been kidnapped and was saved at the eleventh hour from going to the killers.)

The same year Exceller won the Jockey Club Gold Cup, a Hunt homebred was ranked as one of the best three-year-old fillies in North America. Amazer was a *Vaguely Noble filly from yet another good racemare acquisition, Sale Day, a To Market filly who won the Spinster Stakes and Falls City Handicap. Amazer won a group III race in France and then was returned to this country to win the Yellow Ribbon Stakes, a race emerging as one of the best for distaffers on grass.

Hunt continued on his way, always looking for prospects to buy. "As far as I can tell," he said, "the big, successful breeding operations have begun to fail at the point they stopped buying." He could not resist *Hurry Harriet II after she defeated Allez France in England's Champion Stakes, although she was by the obscure sire Yrrah Jr. *Hurry Harriet II foaled the San Juan Capistrano winner Load the Cannons, by Cannonade.

The Turf was ever fascinating.

"There is a sort of four-way compensation," Hunt said. "One is the business side, another is sport, another the gamble of it, and the last is athletics. After all, the Thoroughbred racehorse is the greatest

Empery

athlete of them all. They have been bred for some 250 years for only one thing, and that is not true in other sports." On a different level Hunt also admitted that one of the appeals of being deeply involved in the sport was that "nearly everybody I meet in racing is interesting," a phenomenon that he could not assign universally to the conference tables of the business world.

Among the more unusual championships ever earned in this country was that of the Hunt-bred mare Trillion. She was the champion who did not win a race! A 1974 foal by leading sire Hail to Reason, Trillion represented another acquisition, being from Margarethen, a stakes winner herself and

Trillion

Champion Stakes. She became the first filly ever to win the Irish Two Thousand Guineas.

The singular mare Dahlia had also been a remarkable producer. For Hunt she produced the millionaires Dahar and Rivlia and the major winner Delegant. Dahar (by Lyphard) won seven of twenty-nine races including the San Juan Capistrano and five other stakes in this country and France and earned $1,207,286. Rivlia (by Riverman) won the Hollywood Invitational and two other grade I races in California and earned $1,005,041. Delegant (by *Grey Dawn II) also won the San Juan Capistrano and earned $624,941.

Numbers, which had been part of Hunt's chosen approach, eventually became overwhelming: Business numbers probed by the government; numbers of dollars or years in fines, or probation, or purgatory — whatever you call it; bankruptcy numbers for the family's Placid Oil Company; numbers, numbers, numbers. The numbers nightmare stemmed from the collapse of the silver market, in which Hunt and his brother William Herbert had invested heavily. The Hunts later were found guilty of attempting to corner the market and heavy fines imposed by a federal judge forced Nelson Bunker Hunt to file for bankruptcy.

In 1987, a year before his bankruptcy, Hunt announced, "because of business commitments in the oil and gas industry, I reluctantly have made the decision to disperse my entire Thoroughbred holdings." The numbers of the game, as he played it, would turn out to be impressive that year. For the second year in succession, he was the leading breeder in North American earnings, with more than $5 million each season. In foal crops of 1982–85, which would feed the racing ranks of the two years of leadership, he bred thirty-five stakes winners. It had taken more than 2,400 starts in each year (1986–87) and 281 and then 324 wins to achieve that status, though. He was also the leading breeder in number of wins in both years.

In January of 1988, Keeneland sold 580 horses for Hunt, the largest dispersal sale in history. They

a pearl from the classic family of Marguerite (dam of Gallant Fox, etc.). Trillion was a major winner in Europe, and in 1979 she contested a sequence of four of the best races against males on grass in this country. She was second in all four, beaten a neck by Golden Act in the Canadian International, a neck by Bowl Game in the Turf Classic, nearly two lengths by Balzac in the Oak Tree Invitational, and three-parts of a length by Bowl Game in the Washington, D.C., International. There was no qualm about her being voted the Eclipse Award in the first year distaffers had their own turf category.

Especially appealing about Trillion was that she was bred by Hunt and raced in partnership with his old school pal, Edward Stephenson. (One can only doubt that, even in their most innocent days, either suggested to the other that, "This game's so easy you can have a champion without even winning a race!")

Bred to Riverman, Trillion in 1982 foaled a mare of similar ruggedness in Triptych. Bred by Hunt and Stephenson, Triptych was consigned by agent Vic Heerman to the 1983 Keeneland summer yearling sale, where she was purchased for Alan Clore for $2.15 million. She became the all-time leading distaff earner in Europe, with $2,706,175. Triptych won nine group I races, and won several major races against males, including the Coronation Cup and grossed $46,911,800, to average $27,500. Among the six individuals that brought $1 million or more each was beloved Dahlia, not in foal at that time and eighteen but worth $1.1 million in the eyes of another rather star-struck horseman, Allen Paulson. Dahlia's Northern Dancer colt brought $1.3 million. Named Wajd and sent to Europe, he won three French stakes and earned $229,342 as the fourth stakes winner bred by Hunt from Dahlia. Paulson bred the stakes winners Dahlia's Dreamer and Llandaff from Dahlia, raising her total to six. The Hall of Famer Dahlia lived to the age of thirty-one and was buried on Diamond A Farm, which had been Paulson's Brookside Farm at the time he acquired her.

Paulson had been a major customer before. In 1985 he had purchased the Hunt-bred Estrapade for $4.5 million at Keeneland, where she was entered in the November mixed sale as a horse in training (a la *Vaguely Noble, her sire). Estrapade was out of Klepto, who also foaled the Hunt-bred Isopach, a Reviewer colt who was a major winner in Italy. Estrapade had won the Yellow Ribbon and three other California stakes for Hunt, after a European campaign of three stakes victories. For Paulson she won the Arlington Million the next year and was voted the champion turf mare.

Champions Bred

Dahlia
1974 champion turf female

Youth
1976 champion turf male

Trillion
1979 champion turf female

Estrapade
1986 champion turf female

In Europe:

Gazala II
1967 French champion three-year-old filly

Dahlia
1973 Irish champion three-year-old
1973 English champion three-year-old filly
1973 English Horse of the Year
1974 English champion older female
1974 English Horse of the Year
1975 English champion older female

Mississipian
1973 French champion two-year-old colt

Youth
1976 French champion three-year-old colt

Super Concorde
1977 French champion two-year-old colt

Trillion
1978 French champion older female
1979 French champion older female

Friendswood
1982 Italian champion two-year-old filly

Antiqua
1987 German champion two-year-old colt

Nelson Bunker Hunt & Edward L. Stephenson

Triptych
1984 French champion two-year-old filly
1986 English champion older female
1986 French champion older female
1987 English champion older female
1987 French champion older female
1987 Irish champion older female
1988 French champion older female

Among earlier Hunt horses that were sold and had beneficial ramifications for other breeders was the stakes-winning mare Cinegita, who for W.T. Young became the second dam of champion Flanders and third dam of champion Surfside.

Nelson Bunker Hunt bred 150 stakes winners, of which Estrapade was the fourth champion in this country.

After years of absence from the game, Hunt is back, early in the twenty-first century, as an active, and adventurous, owner. He is now in his late seventies. The passion for racing is intact, and so is the smile.

Stakes Winners Bred by Nelson Bunker Hunt

Kings Creek (stp.), dk.b/br.g. 1959, *King's Evidence—*Snow Line ll, by Ujiji

Ramshorn Creek (stp.), dk.b/br.g. 1959, *King's Evidence—*Sovereign Case, by *Royal Charger

Walden, b.c. 1963, Mr. Music—*Ma Princesse, by Prince Bio

Amerigo Lady, b.f. 1964, *Amerigo—*Lady Sybil, by Count Rendered

Cocktail Music, blk.f. 1964, Mr. Music—*Lucena, by Fascinador

Deauville, b.f. 1964, Ridan—*Ma Princesse, by Prince Bio

*Gazala ll, br.f. 1964, Dark Star—*Belle Angevine, by L'Amiral (Bred in France; SW in France)

Spire, b.f. 1964, Carry Back—*Torrecila, by Sideral

Sports Event, b.f. 1964, T. V. Lark—*Pamplona ll, by Postin

Anglo Peruvian, b.f. 1965, *My Babu—*Pamplona ll, by Postin

David Boy, dk.b/br.g. 1965, *Amerigo—*Wiggle ll, by Rego

Derby Day Boy, ch.c. 1966, *My Babu—Roman Ronda, by The Rhymer

Gris Vitesse, gr.f. 1966, *Amerigo—*Matchiche ll, by Mat de Cocagne (SW in France)

Le Courtillet, b.c. 1967, L'Epinay—Chimere Fabuleuse, by Coaraze (SW in France)

Mincemeat, ch.g. 1967, *Gustav—*La Nene, by Aristophanes

Pampered Miss, ch.f. 1967, Sadair—*Pamplona ll, by Postin (SW in France)

Sir Wiggle, dk.b/br.c. 1967, Sadair—*Wiggle ll, by Rego

Snow Man, ch.g. 1967, *Gustav—*Lucena, by Fascinador

Acclimatization, b.c. 1968, Clem—*Lola Montes lll, by Birikil

Clems Match, gr.c. 1968, Clem—*Matchiche ll, by Mat de Cocagne (SW in France)

Crimson Clem, ch.c. 1968, Clem—*La Nene, by Aristophanes

Lucky Traveler, b.f. 1968, Derring-Do—*Homeward Bound ll, by Alycidon

Falaise, dk.b/br.c. 1969, *Pretendre—*Festiva, by Espace Vital (SW in Ireland)

My Friend Paul, b.c. 1969, *Pretendre—Darling Adelle, by Polynesian (SW in France/Australia)

South Island, b.f. 1969, Ridan—Lily Pons, by Licencioso (SW in France)

Dahlia, ch.f. 1970, *Vaguely Noble—Charming Alibi, by Honeys Alibi (SW in U.S./Europe)

Barcas, ch.c. 1971, Sailor—Goofed, by *Court Martial

Brigand, b.c. 1971, *Noholme ll—*Lady Sybil, by Count Rendered

Busiris, b.c. 1971, Ridan—Lily Pons, by Licencioso (SW in France)

Justimus, b.c. 1971, *Indian Chief ll—*Farmer's Daughter ll, by Agricola (SW in Ireland)

Mississipian, b.c. 1971, *Vaguely Noble—*Gazala ll, by Dark Star (SW in France)

*Adieu ll, b.f. 1972, Tompion—Armoricana, by Bel Baraka (Bred in France; SW in Canada)

Amphioxus, ch.g. 1972, Clem—For Smitty, by Hillsdale (SW in Belgium)

Canvasser, ch.c. 1972, T. V. Lark—Flying Legs, by Palestine

Margravine, dk.b/br.f. 1972, Hail to Reason—Margarethen, by *Tulyar (SW in France)

Nobiliary, ch.f. 1972, *Vaguely Noble—Goofed, by *Court Martial

Principium, ch.g. 1972, Majestic Prince—Las Canas, by Academico

Creation, ch.f. 1973, *Noholme ll—*New Fashion ll, by Sideral

Diagramatic, dk.b/br.c. 1973, Sir Wiggle—Miss Suzy, by Agasajo (SW in France)

Empery, b.c. 1973, *Vaguely Noble—*Pamplona ll, by Postin (SW in England)

Palatable, b.c. 1973, Tom Rolfe—Ameri Belle, by *Amerigo (SW in England)

Spiranthes, b.f. 1973, *Vaguely Noble—Shenow, by Eternal Bim (SW in England)

Youth, b.c. 1973, Ack Ack—*Gazala ll, by Dark Star (SW in U.S./France)

Billion, ch.c. 1974, Restless Wind—*Festiva, by Espace Vital (SW in England)

Nice Balance, b.c. 1974, High Echelon—*Wiggle ll, by Rego (SW in England)

Sporting Yankee, b.c. 1974, *Vaguely Noble—Sale Day, by To Market (SW in England)

Trillion, b.f. 1974, Hail to Reason—Margarethen, by *Tulyar (SW in France)

Amazer, ch.f. 1975, *Vaguely Noble—Sale Day, by To Market (SW in U.S./France)

Celebrated, b.c. 1975, Native Charger—Faneuil Hall, by Bolinas Boy (SW in Belgium)

Claude M., ch.g. 1975, Piko—Vally Holme, by *Noholme ll

Flaunter, gr.f. 1975, High Echelon—*Fast Ride ll, by Sicambre

Piko's Honey, ch.f. 1975, Piko—Holme At Once, by *Noholme ll

Super Concorde, dk.b/br.c. 1975, Bold Reasoning—Prime Abord, by Primera (SW in France)

Trillionaire, b.f. 1975, *Vaguely Noble—Amerigo's Fancy, by *Amerigo (SW in England)

Valour, b.c. 1975, *Vaguely Noble—Louisador, by Indian Hemp (SW in Germany/France)

Anthurium, dk.b/br.c. 1976, Never Bend—Wanika, by Sadair (SW in France)

Brown Gravy, dk.b/br.f. 1976, Bold Reason—*Wiggle ll, by Rego

Fast, ch.c. 1976, Bold Bidder—Tipping Time, by Commanding ll

Most Noble, b.c. 1976, *Vaguely Noble—Castle Flower, by *Noholme ll

Paleocene, b.f. 1976, *Vaguely Noble—Louisador, by Indian Hemp (SW in France)

Phosphurian, b.g. 1976, Ace of Aces—*Fast Ride ll, by Sicambre (SW in Ireland)

Quadrupler, b.f. 1976, *Vaguely Noble—Alota Calories, by Candy Spots (SW in Italy)

Apalachee Chief, b.c. 1977, Apalachee—Dumptys Lady, by Dumpty Humpty

Captain General, b.c. 1977, *Vaguely Noble—Charming Alibi, by Honeys Alibi

Cinegita, b.f. 1977, Secretariat—Wanika, by Sadair

Cream, ch.g. 1977, Prove Out—Fiddling Jimmer, by Jimmer

Globe, dk.b/br.c. 1977, Secretariat—Hippodamia, by Hail to Reason

Gonzales, b.c. 1977, *Vaguely Noble—*Gazala ll, by Dark Star (SW in Ireland)

Isopach, b.c. 1977, Reviewer—Klepto, by No Robbery (SW in Italy)

Rapanello, ch.c. 1977, His Majesty—Louisador, by Indian Hemp (SW in Italy)

Rightly Noble, b.c. 1977, Noble Decree—Aryshire Lass, by Slamruler

Solar Current, ch.c. 1977, Little Current—Amerigo's Fancy, by *Amerigo

Texas Gem, gr.c. 1977, *Canonero ll—Gray Mirage, by Bold Bidder

Three Leaders, b.f. 1977, Noble Decree—Edies Double, by Nodouble

Wealth, b.f. 1977, Ace of Aces—Queen Minasseh, by Greek Page

Master Ace, b.c. 1978, Ace of Aces—Tipping Time, by Commanding II
Silky Baby, b.c. 1978, What a Pleasure—*Gazala II, by Dark Star (SW in France)
T. V. Barb, ch.f. 1978, Empery—Barbsie, by T. V. Lark
Adept, b.f. 1979, Grey Dawn II—Magnabid, by Bold Bidder
Buchanette, b.f. 1979, Youth—Duke's Little Gal, by Duke of Dublin (SW in U.S./France)
Discovered, b.c. 1979, Royal Ski—Bedazzler, by Dr. Knighton
Friendswood, b.f. 1979, *Vaguely Noble—Summer Point, by Summer Tan (SW in Italy)
Hainesville, dk.b/br.c. 1979, Mississippian—For Smitty, by Hillsdale
Jacksboro, b.c. 1979, Ace of Aces—House of Cards, by Promised Land
Rockwall, b.c. 1979, Cannonade—Fiddling Jimmer, by Jimmer
Vidor, dk.b/br.f. 1979, *Vaguely Noble—Prestissimo, by Bold Reasoning (SW in France)
Best of Both, dk.b/br.c. 1980, J. O. Tobin—Gazala II, by Dark Star
Carocake, gr.f. 1980, Caro (Ire)—Cake, by Hail to Reason (SW in France)
Cutter Sark, ch.c. 1980, Sauce Boat—Mrs. Full Charge, by Barbizon
Estrapade, ch.f. 1980, *Vaguely Noble—Klepto, by No Robbery (SW in U.S./France)
Load the Cannons, b.c. 1980, Cannonade—*Hurry Harriet II, by Yrrah Jr. (SW in U.S./France)
Barbery, b.c. 1981, Empery—Barbsie, by T. V. Lark
Champion Pilot, b.c. 1981, Exceller—Jet to Market, by Farm to Market
Dahar, b.c. 1981, Lyphard—Dahlia, by *Vaguely Noble (SW in U.S./France)

Eastland, b.f. 1981, Exceller—Crimson Lass, by Crimson Satan (SW in U.S./France)
Enduring, ch.c. 1981, Run Dusty Run—Ole Honey, by Bold and Brave
How Ya Doon, b.f. 1981, Matsadoon—Indialucie, by Sir Ivor
Satch, ch.f. 1981, Dust Commander—Saturnina, by Ballydonnell
Amorphous, gr.c. 1982, Sir Wiggle—Waterbuck, by The Axe II
Armilary, ch.c. 1982, Decies II—Gold Drain, by *Forli (SW in Ireland)
Capable (stp.), b.c. 1982, *Vaguely Noble—Prime Abord, by Primera (SW in Ireland)
Lock's Dream, b.f. 1982, Youth—Trillionaire, by *Vaguely Noble
Rivlia, b.c. 1982, Riverman—Dahlia, by *Vaguely Noble (SW in U.S./France)
Stater, b.c. 1982, Proud Birdie—Steak Diane, by White Gloves II
Victoriate, ro.c. 1982, Vigors—Coin Silver, by Buckpasser
Barger, b.f. 1983, Riverman—Trillion, by Hail to Reason (SW in France)
Crafty Bo, dk.b/br.c. 1983, Decies II—Bovina, by Crafty Khale
Imprimatur (NZ), ch.c. 1983, Imposing (Aust)—Calera (NZ), by Zamazaan (Fr) (Bred in New Zealand; SW in Australia)
Just a Knockout, b.c. 1983, Unconcscious—Ozona, by Verbatim
Kraemer, b.f. 1983, Lyphard—Rich and Riotous, by Empery
Mandeville, b.c. 1983, Explodent—Honorary, by Hail to Reason
Nimoy, dk.b/br.c. 1983, Liloy (Fr)—Turn in Space, by Turn to Reason
Nobiluomo, dk.b/br.c. 1983, *Vaguely Noble—Golferette, by Mr. Randy (SW in Italy)
Reloy, b.f. 1983, Liloy (Fr)—Rescousse (Fr), by Emerson (SW in U.S./France)

Estrapade

Rosedale, b.c. 1983, *Vaguely Noble—Ivory (Fr), by Riverman (SW in U.S./Italy)

Swink, b.c. 1983, Liloy (Fr)—Swiss, by *Vaguely Noble (SW in U.S./France)

Vacherie, ch.f. 1983, Vigors—Bastonera (Arg), by Con Brio II

Bestofbothworlds, dk.b/br.f. 1984, Globe—Intensive, by Sir Wiggle

Conquer, b.c. 1984, Conquistador Cielo—Snow Plow, by Royal Ski

Delegant, gr.c. 1984, Grey Dawn II (Fr)—Dahlia, by *Vaguely Noble

French Impulse, b.c. 1984, Perrault—Matidia, by Tom Rolfe

Heartleaf, dk.b/br.f. 1984, Lyphard's Wish (Fr)—Pretty Leader, by Mr. Leader

Icetrain, gr.c. 1984, Icecapade—Travois, by Navajo

Motley, b.c. 1984, Best Turn—Tipping Time, by Commanding II (SW in U.S./France)

Nyssia, ch.c. 1984, Dust Commander—Our Dancing Girl, by Solo Landing

Stargrass, b.f. 1984, Plugged Nickle—McAllister Miss, by Ward McAllister

Sweettuc, gr.f. 1984, Spectacular Bid—Sweet Simone (Fr), by Green Dancer

Talinum, ch.c. 1984, Alydar—Water Lily (Fr), by Riverman

Yucca, gr.c. 1984, Vigors—Bold Brat, by Boldnesian

Antiqua, dk.b/br.c. 1985, Lypheor (GB)—Paradise (Fr), by Brigadier Gerard (SW in Germany/U.S.)

Commander Time, b.f. 1985, Dust Commander—Tempus Fugit, by Salvo

Dance Renee, dk.b/br.f. 1985, Green Dancer—Hurry Renee, by Villamor (SW in Ireland)

Delighter, b.f. 1985, Lypheor (GB)—Amazer, by *Vaguely Noble (SW in U.S./France)

If Liloy, b.f. 1985, Liloy (Fr)—If You Prefer, by Olympiad King

Intensive Command, ch.c. 1985, Dust Commander—Intensive, by Sir Wiggle

Lively One, dk.b/br.c. 1985, Halo—Swinging Lizzie, by The Axe II

Pattern Step, b.f. 1985, Nureyev—Tipping Time, by Commanding II

Queen's Master, ch.c. 1985, Master Derby—Queen's Family, by New Prospect

Rose of Clare, b.f. 1985, Isopach—Cry Havoc, by Ramirez

Seattle Sangue, b.f. 1985, Seattle Slew—Sangue (Ire), by Lyphard

Tricky Johnny, b.c. 1985, Clever Trick—Johnnyette, by Stage Door Johnny

Ask Me a Question, b.c. 1986, Liloy (Fr)—Brent's Star, by Brent's Prince

Cajetano (stp.), b.c. 1986, Run the Gantlet—Intensive, by Sir Wiggle (SW in France/Switzerland)

Dancing Lindsay, b.f. 1986, Liloy (Fr)—Bold Brat, by Boldnesian

Declare Your Wish (stp.), dk.b/br.g. 1986, Lyphard's Wish (Fr)—Declared, by Tom Rolfe

Lord Nelson, b.g. 1986, Big Spruce—Fast Tipper, by Exceller (SW in Puerto Rico/Panama)

Lunelle, b.f. 1986, Exceller—Dia (Fr), by Tyrant

Presidential, ch.c. 1986, Vice Regent—Sister Sal, by Delta Judge

River Warden, b.c. 1986, Riverman—Sweet Simone (Fr), by Green Dancer (SW in U.S./France)

Smiling Spruce, b.c. 1986, Spruce Needles—Smiling Sun, by Smiling Jack

Tarage, b.f. 1986, Vice Regent—Timotara, by Secretariat

Tomahawk Lancer, dk.b/br.g. 1986, Providential (Ire)—Purdah Star, by Bold Forbes

Behaving Dancer, ch.f. 1987, Green Dancer—Behaving Lady, by Ambehaving

Centerfold Miss, b.f. 1987, Explodent—Thasos, by *Vaguely Noble

Explosive Ele, b.f. 1987, Explodent—Shroud, by *Vaguely Noble

Lord Sreva, b.g. 1987, *Vaguely Noble—Lyphard's Lady, by Lyphard

Palace Chill, ch.f. 1987, The Minstrel—Castle Royale, by Slady Castle

Prince Syn, dk.b/br.c. 1987, Sham—Hempens Syn, by Symmetric (SW in Japan)

Wajd, ch.f. 1987, Northern Dancer—Dahlia, by *Vaguely Noble (SW in France)

Cannon Green, b.g. 1988, Cannonade—Dusan, by Green Dancer

Nelson Bunker Hunt and John Gaines

Oilfield, dk.b/br.c. 1973, Hail to Reason—Ole Liz, by Double Jay

Beaconaire, b.f. 1974, *Vaguely Noble—Ole Liz, by Double Jay (SW in France)

Gain, ch.c. 1976, Mississipian—Miss Ribot, by Sir Ribot (SW in France)

Au Point, b.c. 1980, Lyphard—Quillo Queen, by *Princequillo

Krotz, b.f. 1983, Exceller—Miss Ribot, by Sir Ribot (SW in U.S./Germany)

Nelson Bunker Hunt and Edward L. Stephenson

Triptych, dk.b/br.f. 1982, Riverman—Trillion, by Hail to Reason (SW in Europe)

Nelson Bunker Hunt and Allen Paulson

Prudent Offer, ch.c. 1984, Providential (Ire)—Domitia, by Royal Ski

Source: *The Blood-Horse* Archives

William S. Farish

William S. Farish made his first major headlines in racing not as a breeder but as the owner of Bee Bee Bee, upset winner over Riva Ridge in the 1972 Preakness. In the ensuing three decades and counting, Farish has probably held more reins than anyone else in racing who is not a jockey. He has bred more than 185 stakes winners, including three Horses of the Year among six Eclipse Award champions, three Belmont Stakes winners, two Preakness winners, and a Kentucky Derby winner. Farish has developed Lane's End Farm into one of the world's top breeding, selling, and stallion operations; has topped both the Keeneland July and September sales; and was involved in the sale of the most expensive yearling in history. In leadership roles on the Turf, Farish has been chairman of Churchill Downs and the American Horse Council, vice chairman of The Jockey Club, and a major supporter of equine research via the Grayson-Jockey Club Research Foundation and Maxwell H. Gluck Center. He has recruited major owners as partners in many of his stakes winners and has hosted presidents and Queen Elizabeth II. Oh, yes, one of the horses he bred in partnership who did not win a championship, a classic, or even a stakes, became one of the world's most influential stallions: Danzig.

Outside the Turf, Farish has served on the boards of multiple corporations; helped run a political campaign for eventual President George Bush (the elder); chaired the Ephraim McDowell Cancer Foundation and the Houston Chapter of the National Urban League; and served on the boards of Rice University, Baylor College of Medicine, and Transylvania University. President George W. Bush appointed him as the U.S. ambassador to Great Britain at the beginning of his presidential term in 2001. This is a glamorous sounding but demanding post in which Farish's more public moments included representing his country when Her Majesty directed that "The Star-Spangled Banner" be played at Buckingham Palace in the aftermath of September 11, 2001. Farish served as ambassador until he resigned in the early summer of 2004.

Soft-spoken, quick to smile, Farish has achieved all the above without any sense of bravado. He learned to give a forceful, eloquent speech on weighty world issues, but he never lost the knack for friendly horse talk. (There was plenty of precedent in the English post for combining international affairs with Thoroughbred racing. John Hay Whitney of Greentree Stud was a distinguished ambassador to England during the Eisenhower Administration. Farish added a bit of sporting distinction to his ambassadorial role during serious times by winning the historic Epsom Oaks with homebred Casual Look in 2003.)

Farish had no particular connection to Kentucky originally. He was born in Houston on March 17, 1939. He grew up in Texas, where his grandfather, William S. Farish Sr. — a University of Mississippi law graduate — pioneered this and that in the oil industry and emerged as president of Humble Oil by the early 1920s. In the 1930s, as chairman of Jersey Standard, Farish moved his family to New York City. As would his grandson, the original William S. Farish found his industrial prominence drawing him into government, and he served on the petroleum committee of the Council for National Defense. His son, William S. Farish Jr., was killed in an airplane accident during World War II.

The next generation, the Ambassador Farish of the early twenty-first century, dropped the III from his name some years ago. He is commonly known as "Will." His son Bill became key to the management of Lane's End while his father was in England. The younger Farish earlier had a stint in the White House as a special assistant to the first President Bush and also spent some years in finance before moving to the farm.

Will Farish attended the University of Virginia. He was a stockbroker for a time and later became president of Fluorex, an international mining and exploration operation, and also of Navarro Exploration Company. He was a founding director of the New York holding company Eurus and of Capital National Bank of Houston.

Racing was also in his family. His grandfather Farish had raced horses, and then the widow, Libbie Rice Farish, raced in partnership with daughter Martha Gerry as Lazy F Ranch. Lazy F's greatest distinction was as breeder and owner of Forego, three-time Horse of the Year in the 1970s. Mrs. Gerry, who continues Lazy F, presently chairs the executive committee of the National Museum of Racing.

Will Farish's first venture into owning Thoroughbreds was in partnership with his father-in-law, the late Bayard Sharp, whose earlier horses had included Hannibal and Troilus. (The high-class racehorse and successful Lane's End stallion Dixieland Band raced for the Sharps.) Farish also fell into his family's interest in polo and thereby became associated with Del Carroll and the latter's patron, Michael Phipps. These friendships helped hone Farish's ability as a hands-on horseman — on the ground and in the saddle — and led to professional connections as well. Carroll trained Bee Bee Bee and other good horses for Farish before the trainer's death when he was apparently thrown from the horse Sportin' Life.

Farish's first stakes winner raced in his own name was Kaskaskia, a Catullus colt purchased for $9,700

BILL STRAUS

William Farish (right) with Queen Elizabeth II

as a two-year-old. Kaskaskia won the 1967 Youthful and Juvenile stakes. In later years such victories would not rank a horse high in Farish's history, but at the time it garnered full honors, i.e., a Richard Stone Reeves portrait. Bee Bee Bee, Farish's upstart Preakness winner, was part of a package purchased from William S. Miller of Chicago. At that time, Farish was concentrating some of his growing numbers of bloodstock in Maryland to take advantage of one of the pioneering state breeders benefit programs. It was not long, however, until he found Kentucky a logical locus. He joined with veteran horseman Warner L. Jones Jr. in some partnerships, the most spectacular of which eventually would include ownership of My Charmer, dam of 1977 Triple Crown winner Seattle Slew.

Farish owned 1,200-acre Huisache Ranch back home in the Houston area, but in the late 1970s he purchased some land outside Versailles, Kentucky. This was the beginning of Lane's End, which today sprawls over some two thousand acres and has a management arrangement for more nearby land owned by the Niarchos family.

Any numerically large stallion operation has a revolving-door reality as some make it big and most do not. Among the seventeen advertised stallions who have earned some permanency at Lane's End are A.P. Indy, Dixieland Band, Belong to Me,

Kingmambo, and Gulch. The young brigade includes Came Home, Lemon Drop Kid, Stephen Got Even, and Mineshaft. Illustrating the harsh realities of bloodstock breeding, among those to have been shipped elsewhere in recent years were Kentucky Derby winners Alysheba, Spend a Buck, and Sea Hero; Personal Ensign's grade I-winning son Miner's Mark; and the grass champion Manila. The Turf is a fascinating game, not an easy one.

As Farish wrote in the foreword to the book *Dynasties*: "Breeders are not dealing in certainties, but percentages. That's why I look for horses that are correct, had at least grade I speed at a mile, and have good bone. I think that is the high-percentage approach."

Some of the Farish horses are still sent to Texas. Occasionally, the Texas stallions are pasture-bred, and Farish has observed some of the difference in "what we ask of a stallion and what he wants to do." For example, it has been observed on the Texas spread that a stallion in a field with a group of mares, when sensing in his natural instincts "that four

mares need to be covered, he will accomplish that all in one night."

A.P. Indy, his half brother Summer Squall, and Lemon Drop Kid all descend from one of Farish's most important purchases, the Missy Baba bloodline. Farish bought into the family of Michael Phipps' great foundation mare Missy Baba on several fronts. In addition to bolstering his own growing broodmare band in an important way, he also acquired the stallion Raja Baba, a son of Missy Baba by Bold Ruler. Raja Baba became the North American leading sire in 1980. (A Raja Baba season used for the mare Nalees Flying Flag resulted in the filly Sacahuista, bred in partnership by Farish with G. Watts Humphrey Jr. Sacahuista won the 1987 Breeders' Cup Distaff and was named champion three-year-old filly that year.)

Missy Baba's dam was the 1941 Irish Oaks winner *Uvira II, who was bred and raced by the Aga Khan and then had a series of buyers before winding up being inherited by Michael Phipps (her ninth owner) from his father, John S. Phipps. *Uvira II foaled

Weekend Surprise

Missy Baba for Michael Phipps, and although this *My Babu filly was only a maiden winner she was retained for breeding. She eventually foaled six stakes winners, four of which were bred by Phipps before his death in 1973. Raja Baba, Farish's first major stallion, was among them.

Missy Baba's brood also included Gay Missile, a Sir Gaylord filly that continued the family's success. Gay Missile's foals included the stakes-winning Buckpasser filly Lassie Dear, who was purchased for $76,000 as a yearling. She raced in the name of one of Farish's key partners, W.S. Kilroy, a Houston resident also with connections to the oil business. Lassie Dear was retired in 1978 with five wins in twenty-six starts, and Farish acquired an interest by swapping with Kilroy half ownership of another mare, Bold Bikini. Bold Bikini then foaled the 1985 Irish Derby winner Law Society, bred by Farish and Kilroy. Lassie Dear produced four stakes winners, including the Secretariat filly Weekend Surprise, who raced in Farish's green-and-gold silks.

Weekend Surprise won the 1982 Schuylerville and Golden Rod stakes at two and overall had seven wins in thirty-one starts and earnings of $402,892. "I told Mr. Kilroy I thought Weekend Surprise might be the best mare we'd ever had, because of the talent she had, the pedigree, the whole works," Farish would say later. "There were a number of reasons why I liked the mating of Secretariat and Lassie Dear, and I liked it physically, a fact that we strongly emphasize." Secretariat was stoutly built, the mare delicate

and feminine. Farish added that Weekend Surprise "was not a typical Secretariat. She's quite feminine, very, very attractive, but no resemblance to him. To see her progeny is interesting, because there is a strong female-side pull coming through this whole family. I guess it has to go back to Missy Baba or *Uvira II." Farish further explained that he believes "the sire and dam are equally important, and the fact that some sires stamp their foals does change the importance of the mare. A mare does not have to put her stamp on her foals in order for her to have an influence on their abilities."

Weekend Surprise became Broodmare of the Year for 1992. By that time two of her sons had won classic races and one was a Horse of the Year. Bred to the Northern Dancer stallion Storm Bird in 1986, she produced the racy little Summer Squall, who was knocked down to Cot Campbell's Dogwood Stable for $300,000 at the 1988 Keeneland July yearling sale. Trained by Neil Howard, who primarily is Farish's private trainer but has accepted a few other horses, Summer Squall was unbeaten at two when he won the Hopeful and three other stakes. The following year he won the Preakness after being second to Unbridled in the Kentucky Derby. (Unbridled was by the Lane's End stallion Fappiano, so it was an "emotion neutral" Derby for Farish, he said.) As one of many shareholders in Summer Squall, Farish was involved in winning the Maryland classic a second time. Summer Squall battled a tendency to bleed and had quarter cracks throughout his three years of

Seattle Slew, 1974	Bold Reasoning, 1968	Boldnesian, 1963	**Bold Ruler** (*Nasrullah—Miss Disco)
			Alanesian (Polynesian—Alablue)
		Reason to Earn, 1963	Hail to Reason (*Turn-to—Nothirdchance)
			Sailing Home (Wait a Bit—Marching Home)
	My Charmer, 1969	Poker, 1963	Round Table (*Princequillo—*Knight's Daughter)
			Glamour (*Nasrullah—Striking)
		Fair Charmer, 1959	Jet Action (Jet Pilot—Busher)
			Myrtle Charm (Alsab—Crepe Myrtle)
A.P. INDY	Secretariat, 1970	**Bold Ruler**, 1954	*Nasrullah (Nearco—Mumtaz Begum)
			Miss Disco (Discovery—Outdone)
Weekend Surprise, 1980		**Somethingroyal**, 1952	*Princequillo (Prince Rose—*Cosquilla)
			Imperatrice (Caruso—Cinquepace)
	Lassie Dear, 1974	Buckpasser, 1963	Tom Fool (Menow—Gaga)
			Busanda (War Admiral—Businesslike)
		Gay Missile, 1967	Sir Gaylord (*Turn-to—Somethingroyal)
			Missy Baba (*My Babu—*Uvira II)

campaigning but still won thirteen of twenty races and earned $1,844,282.

He was returned to Lane's End to stud. Despite some fertility problems, he proved a significant stallion and sired two champions bred in partnership by Farish. Summer Squall's thirty-two stakes winners (9 percent) through mid-2004 include the champions Charismatic and Storm Song. The stallion was pensioned in summer 2004.

Personal connections were involved in both partnerships. Kelly Farish, wife of Farish's son, Bill, is a stepdaughter of Ogden Mills (Dinny) Phipps. Phipps is chairman of The Jockey Club, of which Ambassador Farish remains as long-time vice chairman. Although the Phipps family retains its historic association with the Hancock family's Claiborne Farm insofar as where its mares are boarded and key stallion prospects stand, Farish and Phipps also have some partnership activity. They bought the mare Minstress, by The Minstrel, from the 1985 Newstead Farm dispersal for $1.3 million and bred from her the Fappiano filly Hum Along. Bred to Summer

Squall, Hum Along produced a filly who was purchased from Phipps and Farish by the avid Summer Squall loyalist Campbell of Dogwood Stable. Named Storm Song, this filly won the 1996 Breeders' Cup Juvenile Fillies and Frizette Stakes and was voted clear-cut champion of her division.

Charismatic was bred by crossing Summer Squall with the Drone mare Bali Babe, who went back to Alfred Vanderbilt's mare Good Example (also dam of the outstanding producer Exclusive). Bali Babe was owned by the family of Dr. Ben Roach and his son, Tom, at their historic Parrish Hill Farm in Midway, Kentucky. Farish is a patient of Dr. Roach's practice. Farish only rarely enters into foal-sharing agreements, but this was an attractive combination, he said, because the big, sturdy mare "is the type that mates perfectly with Summer Squall."

A dashing chestnut, the Summer Squall—Bali Babe colt was purchased privately for $200,000 as a weanling for Robert and Beverly Lewis. Turned over to trainer D. Wayne Lukas, Charismatic emerged suddenly in the spring of 1999 to win the Kentucky

Charismatic winning the 1999 Kentucky Derby

Lemon Drop Kid

the star. Lemon Drop Kid also represents Farish's putting into practice his philosophy of looking for stallion prospects that show first-class speed at a mile. The sire in question, Kingmambo, had miler speed, and much more, going for him. A son of superb sire Mr. Prospector, Kingmambo was foaled from the two-time Breeders' Cup Mile winner Miesque (by Nureyev). Bred and raced by the Niarchos family, Kingmambo is one of two classic winners foaled from Miesque, whose daughter East of the Moon won the mile Poule d'Essai des Pouliches (French One Thousand Guineas) and mile and a half Prix de Diane (French Oaks). Kingmambo won the Poule d'Essai des Poulains (French Two Thousand Guineas) at a mile, and he also won two other group I races at a mile, the Prix du Moulin de Longchamp and Royal Ascot's St. James's Palace Stakes. Those victories, and a career mark of five wins in thirteen starts to earn $734,804, were on the record for all to see. Farish had specific observations to give him additional reason to like Kingmambo.

"One thing about Kingmambo that fascinated me was that his three grade I mile victories were run in extremely heavy going," Farish recalled, "and he would win with a flourish, with a whole lot left. I believe horses winning at a mile over the deep courses of Europe, uphill and downhill, actually show that they have enough stamina for a mile and an eighth or a mile and a quarter over here. To take this thought further, I believe a top miler in Europe is more likely to get classic horses here than that same horse would in England. I think the softer going creates greater demand for stamina."

Lemon Drop Kid excelled in the only remaining important stamina test at one and a half miles on

Derby and Preakness. He was pulled up with a career-ending leg fracture at the end of the Belmont Stakes, failing in his Triple Crown bid. The aptly named Charismatic was named three-year-old champion and Horse of the Year and stood a few years at Lane's End before being sold to Japan.

The Belmont winner in Charismatic's year was a kinsman, also bred by Farish. This was Lemon Drop Kid, bred in partnership with Kilroy. It marked the first time one breeder had accounted for a combined Triple Crown with more than one horse in the same season since 1880. (The pioneering breeder A.J. Alexander bred Derby winner Fonso and Preakness-Belmont winner Grenada that year, although the three races were not grouped and recognized as the Triple Crown until a half-century later.)

Lemon Drop Kid, who was sold to Jeanne Vance for $200,000 as a Keeneland September yearling, represented another dividend from the acquisition of Missy Baba's granddaughter Lassie Dear. His dam, Charming Lassie, by Triple Crown winner Seattle Slew—Lassie Dear, won her only start. She is the dam of four stakes winners, of which Lemon Drop Kid is dirt in this country when he won the 1999 Belmont. That summer he added the one and a quarter-mile Travers. The following year Lemon Drop Kid seemed to have a bead on Horse of the Year as he swept the Suburban, Brooklyn, Whitney, and Woodward. His demanding race in the Woodward might have drained him, for he faded late in the year. He did earn older male championship honors, and he went to stud at Lane's End with ten victories in twenty-four starts and earnings of $3,245,370.

Weekend Surprise followed her Preakness-winning son Summer Squall with a Belmont-winning son in A.P. Indy. At the time Summer Squall was in his classic spring, Farish noted that one of "the most interesting" of the colts in his upcoming Keeneland consignment was the half brother. Just as Weekend Surprise had not been a typical Secretariat physically, the yearling from Weekend Surprise was "a unique Seattle Slew; he is not typical. He's a bit more refined and racy." The colt was inbred 4x3 to Secretariat's sire, Bold Ruler.

The combination of looks and star power in the pedigree made that colt the highest-priced yearling of 1990. He was purchased for $2.9 million by British Bloodstock Association Ireland for Tomonori Tsurumaki and given the name A.P. Indy. He was trained by Neil Drysdale. After removal of an undescended testicle freed his action, A.P. Indy blossomed into championship form. Despite the colt's nagging injuries, Drysdale had him ready to win the Belmont Stakes and Breeders' Cup Classic at three in 1992,

when he was Horse of the Year. Weekend Surprise technically was the seventh mare to produce two winners of any of the races now known as the Triple Crown; she was without precedent in that achievement during the era in which the three races have been recognized as such an important and singular target of the Turf. A.P. Indy also won the Santa Anita Derby, Hollywood Futurity, and two other stakes in a career of eight wins from eleven starts to earn back slightly more than his purchase price, $2,979,815. He was elected to the National Museum of Racing's Hall of Fame in 2000.

In addition to A.P. Indy and Lemon Drop Kid, Farish was co-breeder of another Belmont Stakes winner, Bet Twice, who won the race in 1987 in the colors of prominent sportsman Bob Levy and his family. Bet Twice was by the Nijinsky II stallion Sportin' Life out of Golden Dust, by Dusty Canyon, and was bred in partnership with E.J. Hudson Sr., a Texan and business association who was following in his son's footsteps. E.J. Hudson Jr. had been a co-breeder of Sportin' Life with Farish.

Farish bought back into A.P. Indy and secured the colt's return to Lane's End as a stallion. The horse has been an immediate success, through mid 2004 showing sixty-nine stakes winners (10 percent), including Tempera, Aptitude, Golden Missile, Tomisue's Delight, Pulpit, Old Trieste, Stephen Got Even, and A P Valentine. Two of the best fillies Farish has raced in recent years, Runup the Colors and Secret Status, are A.P. Indy fillies.

	Raise a Native, 1961	Native Dancer, 1950	Polynesian (Unbreakable—Black Polly)
			Geisha (Discovery—Miyako)
		Raise You, 1946	Case Ace (*Teddy—Sweetheart)
Mr. Prospector, 1970			Lady Glory (American Flag—Lady Comfey)
	Gold Digger, 1962	Nashua, 1952	*Nasrullah (Nearco—Mumtaz Begum)
			Segula (Johnstown—*Sekhmet)
		Sequence, 1946	Count Fleet (Reigh Count—Quickly)
			Miss Dogwood (*Bull Dog—Myrtlewood)
PROSPECTORS DELITE	Hoist the Flag, 1968	Tom Rolfe, 1962	*Ribot (Tenerani—Romanella)
			Pocahontas (Roman—How)
		Wavy Navy, 1954	War Admiral (Man o' War—Brushup)
			Triomphe (Tourbillon—Melibee)
Up the Flagpole, 1978	The Garden Club, 1966	*Herbager, 1956	Vandale (Plassy—Vanille)
			Flagette (Escamillo— Fidgette)
		Fashion Verdict, 1960	*Court Martial (Fair Trial—Instantaneous)
			So Chic (*Nasrullah—Striking)

Runup the Colors descends from another key acquisition. As early as 1974, Farish bred the good-class stakes winner Nostalgia from an *Herbager mare, The Garden Club, who was acquired from the Phipps family's broodmare band. The Garden Club was from Fashion Verdict, one of the caravan of major producers tracing to *La Troienne. In addition to Nostalgia, Farish bred the Hoist the Flag filly Up the Flagpole from The Garden Club. Up the Flagpole won the Delaware Oaks and then added her own chapter to the continuing excellence of the family by foaling no fewer than seven stakes winners.

The A.P. Indy filly Runup the Colors was one of them, and in 1997 she won the historic Alabama for Farish as well as duplicating her dam's Delaware Oaks triumph. Another of Up the Flagpole's brood was the Mr. Prospector filly Prospectors Delite, who won the Acorn and Ashland for Farish and in turn foaled an A.P. Indy filly, Stephen Hilbert's Tomisue's Delight, who won the Ruffian and earned more than $1 million.

Tomisue's Delight was only the beginning for Prospectors Delite, who rapidly lived up to her female family's tendency to produce stakes winners in multiples. The stakes winners Delta Music, Monashee Mountain, and Rock Slide came next, and then in 1999 Prospectors Delite foaled Mineshaft.

Bred by Farish in another partnership with one of his associates, attorney James Elkins Jr., the colt would become another star for A.P. Indy.

Mineshaft was first sent to Europe, where he fared moderately, winning three of nine races at three, but no major ones. Returned to race on dirt in the United States, he blossomed for trainer Howard and won seven times at four in 2003. All seven were stakes, and they included some of the most important races on the calendar, such as the grade I Jockey Club Gold Cup, Pimlico Special, Suburban Handicap, and Woodward Stakes. By year's end Mineshaft was suggesting true greatness, but a leg injury prompted his withdrawal from Breeders' Cup Classic consideration and his retirement to Lane's End. Nevertheless, he was voted Horse of the Year and older male division champion. Mineshaft had won ten of eighteen races and earned $2,283,402.

Following A.P. Indy and Charismatic, Mineshaft became the third Horse of the Year of whom Farish was a breeder. No individual had been the breeder of more than

Mineshaft

ANNE M. EBERHARDT

two Horses of the Year since Warren Wright Sr.'s Calumet Farm turned out five — Whirlaway, Twilight Tear, Armed, Citation. and Coaltown — in the 1940s.

In 2000 Farish won the Kentucky Oaks, Mother Goose Stakes, and Florida Oaks with an A.P. Indy home-bred, the filly Secret Status. He had also bred her dam, the Alydar filly Private Status, who won the 1994 Bourbonette Stakes. Private Status was out of the *Con Brio II mare Miss Eva. *Con Brio II was by the great *Ribot, who died in 1972 just as Farish emerged as a lead-ing horseman.

It is possible, although unlikely, that A.P. Indy might eventually surpass Danzig and become the best stallion and sire of sires Farish ever bred. It is certain, however, that A.P. Indy will forever stand more than $10 million below the highest-priced yearling Farish ever sold. If Federal Reserve chief Alan Greenspan had been tracking the Thoroughbred market of the 1980s instead of the stock market of the 1990s, he

RICK SAMUELS

Kentucky Oaks winner Secret Status and William Farish

would probably have found the phrase "irrational exuberance" inadequate to describe what he was see-ing. The Keeneland July yearling sale, which had averaged in the fifty thousands of dollars in the early 1970s, hit the half-million dollar mark for its average in 1983. The next year it reached $544,681. Since the price of potential racing stock is supposed to be linked to earning potential, this represented a disas-sociation from economic fundamentals. The avail-able purse money's increase in that decade was not tenfold, but only about threefold.

In 1985 — the year before a rather serious market decline set in — Farish and his friends Warner L.

Jones Jr. and W.S. Kilroy took to the market a horse whose credentials probably matched those of any other yearling ever offered. The handsome colt was by the great sire and English Triple Crown winner Nijinsky II and out of the dam of American Triple Crown winner Seattle Slew, already an emerging force as a stallion and the sire of 1984 Derby-Belmont winner Swale. Moreover, the dam, My Charmer, had also produced another classic winner in Lomond, the 1983 English Two Thousand Guineas victor. The leading buyer on the interna-tional scene during the 1970s was a group whose most visible figure was Robert Sangster of England,

ANNE M. EBERHARDT

Sacahuista

who had purchased Lomond privately as a weanling. By the 1980s the Maktoums, the ruling family of the benign, oil-rich Emirate of Dubai, had challenged that standing. Also, trainer D. Wayne Lukas had become one of the major players at the sales, representing various wealthy businessmen from the West Coast. When the Nijinsky II—My Charmer colt was led into the ring, the determination of two of these teams was at full throb. "Luck came into it because at least two major syndicates wanted him," partner Jones said. "One was D. Wayne Lukas' syndicate and the other was Robert Sangster's group."

A bidding duel resulted that eventually saw the English team prevail at $13.1 million, nearly $3 million more than the previous record. Jones, the garrulous horseman who bred 1953 Kentucky Derby winner Dark Star, had also consigned an earlier record colt, a Bold Ruler yearling named One Bold Bid. That colt's all-time record high of $170,000 only two decades earlier suddenly seemed to represent a totally different game.

Named Seattle Dancer, the $13.1-million yearling barely made an interest payment on his purchase price, although he did win two group II races in Ireland, the Gallinule and the Derrinstown Stud Derby Trial and was second in the group I Grand Prix de Paris. He earned $152,413 and later sired 1996 Kentucky Oaks winner Pike Place Dancer. My Charmer

retained her value to the extent that when Jones held a major dispersal for health reasons at Keeneland in 1987, she brought $2.6 million from Allen Paulson although the mare was eighteen at the time.

Seattle Dancer and Lomond were among twenty stakes winners Farish bred in association with Jones, owner of Hermitage Farm outside Louisville. There was an international eventuality to their successes, for those they bred in Kentucky together included the French Oaks winner Northern Trick, French champion Rousillon (sire of Melbourne Cup winner Vintage Crop), and Irish Two Thousand Guineas runner-up Procida, in addition to Lomond and Seattle Dancer. Farish succeeded Jones as chairman of Churchill Downs and thus accepted a responsibility toward the older man's life-long devotion to the Kentucky Derby.

As mentioned above, Will Farish co-bred the remarkable stallion Danzig. Pas de Nom, the dam of Danzig, was a daughter of the high-class racehorse but indifferent sire, Admiral's Voyage. The filly's second dam was the English Oaks winner Steady Aim, but Pas de Nom brought only $4,700 as a yearling. Racing for Beverly Roma, Pas de Nom won the 1970 Inferno Stakes in Canada and later was purchased by Farish and Marshall Jenney, owner of Derry Meeting Farm in Pennsylvania. For Farish and Jenney, Pas de Nom won the Virginia Belle Stakes and divisions of the Jasmine and Seashore stakes in 1971. She was a good filly with considerable speed and had a career record of nine wins from forty-two races and earnings of $121,741. From eleven named foals, she failed to produce a stakes winner and became immortal in the process.

Her second foal, Danzig, bred in partnership by Farish and Jenney, was purchased by Henryk de Kwiatkowski for $310,000 in 1978 at Saratoga, where Jenney traditionally sold his yearlings. Danzig showed brilliance for trainer Woody Stephens but could not be kept sound very long in any given sequence. He won all his three races with style and flair but had to be retired before being tested in stakes company. He earned $32,400. Seth Hancock of Claiborne Farm syndicated him for $80,000 per share, a case in which arguably "irrational exuberance" generated one of life's all-time bargains. Danzig was pensioned at the end of the 2004 breeding season, but he had long become one of the greatest sires in the incredible pantheon of Northern Dancer. Although never having stood dual-hemisphere duty as have several of his contemporaries in the highest echelons, Danzig has sired 181 stakes winners (18 percent) and led the North American sire list three times in succession, 1991–93. In so doing, he became the first stallion since Bold Ruler to lead the list more than twice consecutively.

Danzig's best among many outstanding runners include Dance Smartly, Dayjur, Chief's Crown, Versailles Treaty, Polish Precedent, and Danehill. The last named is a bellwether among the many important sons of Danzig as stallions. Having alternated between hemispheres for parts of his career, Danehill raced ahead of his sire in stakes winners, with 255 stakes winners. Of other Danzig connections in the current sire lines, his son Chief's Crown sired an Epsom Derby winner in Erhaab, while a son of Chief's Crown, Grand Lodge, sired the English Derby, Irish Derby, and Prix de l'Arc de Triomphe winner Sinndar.

Six horses bred by Farish and in partnerships have won nine Eclipse Awards. Farish also was voted an Eclipse Award for breeder in 1992 and in 1999.

Champions Bred

William S. Farish & G. Watts Humphrey

Sacahuista
1987 champion three-year-old filly

William S. Farish & W.S. Kilroy

A.P. Indy
1992 champion three-year-old colt
1992 Horse of the Year

Lemon Drop Kid
2000 champion three-year-old colt

William S. Farish & O.M. Phipps

Storm Song
1996 champion two-year-old filly

William S. Farish & Parrish Hill Farm

Charismatic
1999 champion three-year-old colt
1999 Horse of the Year

W.S. Farish, J. Elkins, & T. Webber Jr.

Mineshaft
2003 champion handicap male
2003 Horse of the Year

In Puerto Rico:

William S. Farish

Egregio
1981 champion two-year-old colt

William S. Farish & J.W. Backer

Forbidden Queen
2003 champion two-year-old filly

In Europe:

William S. Farish

Flagbird
1995 European champion older mare
1995 English champion older mare
1995 Irish champion older mare
1995 Italian champion older mare

William S. Farish & W.S. Kilroy

Law Society
1984 Irish champion two-year-old colt
1985 Irish champion three-year-old colt

Wolfhound
1993 English champion older horse
1983 European champion older horse

William S. Farish & Warner L. Jones Jr.

Northern Trick
1984 French champion three-year-old filly
1984 French Horse of the Year

Rousillon
1985 French champion miler

William S. Farish & E.J. Hudson Sr.

Or Acier
1989 Italian champion colt

William S. Farish & G. Watts Humphrey

Misil
1991 Italian champion miler
1992 Italian champion older horse

Stakes Winners Bred by William S. Farish

Bold Gun, ch.c. 1972, Victriolic—Frances Flower, by Derrick

Winter Fox, b.g. 1972, Maribeau—Paula, by Nizami II

Bear the Palm, ch.g. 1974, *Hawaii—Neat'n Sweet, by Attention Mark

Nostalgia, ch.c. 1974, Silent Screen—The Garden Club, by *Herbager

Bee Golden Girl, ch.f. 1976, Better Bee—Golden Beach, by Djeddah

Clever Trick, dk.b/br.c. 1976, Icecapade—Kankakee Miss, by Better Bee

Strike Your Colors, b.c. 1976, Hoist the Flag—Bold Bikini, by Boldnesian

Lines of Power, b.c. 1977, Raise a Native—Exotic Garden, by Bold Ruler

Loveshine, dk.b/br.f. 1977, Gallant Romeo—Kankakee Miss, by Better Bee

Seed the Cloud, gr.f. 1978, Al Hattab—Right as Rain, by Rasper II

Truly Bound, b.f. 1978, In Reality—Natashka, by Dedicate

Up the Flagpole, dk.b/br.f. 1978, Hoist the Flag—The Garden Club, by *Herbager

Blushing Cathy, ch.f. 1979, Blushing Groom (Fr)—The Garden Club, by *Herbager

Clever Shot, ch.c. 1979, Torsion—Kankakee Miss, by Better Bee

Dance Country, b.f. 1979, Big Spruce—Grand Char, by Crimson Satan (SW in Argentina)

Egregio, b.g. 1979, Singh—Summer Sky, by *Sky High II (SW in Puerto Rico)

Hot Ski, b.f. 1979, Royal Ski—Hot Road, by Road House

Garthorn, dk.b/br.c. 1980, Believe It—Garden Verse, by Cyane

Permissible Tender, gr.f. 1980, Al Hattab—Oh So Bold, by Better Bee

Quay Ya Ho, ch.c. 1980, Master Derby—Native Hostess, by Raise a Native

Star of the Road, b.c. 1980, Ramirez—Hot Gravey, by Javelin

Belle of Jove, ro.f. 1981, Northern Jove—Mrs. Herman, by Map Maker

Land of Oz, b.c. 1981, Fluorescent Light—Miss Maite, by Swoon's Son

Nitap, dk.b/br.c. 1981, Amerrico—Winter Frolic, by Admiral Vee

Tomalu, dk.b/br.f. 1981, Accipiter—In Triumph, by Hoist the Flag

Charleston Rag (Ire), ch.f. 1982, General Assembly—Music Ville (Ire), by Charlottesville

Rapide Pied, ch.f. 1982, Raise a Native—Nancy Jr., by Tim Tam (SW in France)

Weekend Delight, dk.b/br.f. 1982, Clever Trick—Udder Delight, by Raja Baba

Danzidea, b.c. 1984, Danzig—Merry Thought, by Haveago

Fold the Flag, ch.f. 1984, Raja Baba—Up the Flagpole, by Hoist the Flag

Hide the Kaz, dk.b/br.f. 1985, Kid Kas—Hidden Bonus, by Sauce Boat

Allied Flag, b.c. 1986, Danzig—Up the Flagpole, by Hoist the Flag

Brush Aside, b.c. 1986, Alleged—Top Twig, by High Perch (SW in England)

Ebros, b.c. 1986, Mr. Prospector—Scuff, by Forli (Arg)

Skatingonthinice, gr.f. 1986, Icecapade—Rain Shower, by Bold Forbes

A Touch of Kas, dk.b/br.c. 1987, Kid Kas—No Llama, by Troon Road

Cautious Dee, b.f. 1987, With Caution—Hillybob, by Personality

Down the Flag, ch.c. 1987, Stalwart—Hunt's Lark, by Knightly Dawn (SW in England)

Fashion Delight, b.f. 1987, Fappiano—Charleston Rag (Ire), by General Assembly

Thakib, dk.b.c. 1987, Sovereign Dancer—Eternal Queen, by Fleet Nasrullah (SW in England)

An Enemy of My Own, b.f. 1988, Enemy Number One—Unexpected Arrival, by Sauce Boat

Long View, b.f. 1988, Damascus—Up the Flagpole, by Hoist the Flag

Snowy Owl, dk.b/br.c. 1988, Storm Bird—Nafees, by Raja Baba (SW in England)

Withallprobability, b.f. 1988, Mr. Prospector—Sulemeif, by Northern Dancer

Camilla Blu, b.f. 1989, Storm Bird—Nafees, by Raja Baba

Hadnot, b.g. 1989, With Caution—Hillybob, by Personality

Line In The Sand, ch.c. 1989, Mr. Prospector—Really Lucky, by Northern Dancer

Position of Power, b.g. 1989, Lines of Power—Noranc, by Northern Prospect

Prospectors Delite, ch.f. 1989, Mr. Prospector—Up the Flagpole, by Hoist the Flag

Quiet Enjoyment, b.c. 1989, Ogygian—Broas, by Wig Out

Misspitch, b.f. 1990, Waquoit—Duo Disco, by Spring Double

Dove Hunt, b.c. 1991, Danzig—Hunt's Lark, by Knightly Dawn

Exotic Moves, ch.f. 1991, Miswaki—Syrian Dancer, by Damascus

Flagbird, dk.b.f. 1991, Nureyev—Up the Flagpole, by Hoist the Flag (SW in France/Ireland/Italy)

Makhraj, dk.b/br.c. 1991, Gulch—Sunny Smile, by Boldnesian (SW in Sweden)

Private Status, ch.f. 1991, Alydar—Miss Eva (Arg), by Con Brio

Kadrmas, ch.g. 1992, Gulch—Sulemeif, by Northern Dancer

Top Account, dk.b/br.c. 1992, Private Account—Up the Flagpole, by Hoist the Flag

Special Request, b.f. 1993, Sovereign Dancer—Special Power, by Lines of Power

Make Haste, b.f. 1994, Cure the Blues—Pure Speed, by Gulch

Runup the Colors, dk.b/br.f. 1994, A.P. Indy—Up the Flagpole, by Hoist the Flag

Digitalize, dk.b/br.f. 1995, Dayjur—Dancer's Candy, by Noble Dancer (GB)

Quick Lap, b.f. 1995, Nicholas—Pure Speed, by Gulch

Swear by Dixie, dk.b/br.c. 1995, Dixieland Band—Under Oath, by Deputed Testamony

Bounce Back (stp), ch.g. 1996, Trempolino—Lattaquie (Fr), by Fast Topaze (SW in France)

Queen's Word, b.f. 1996, Kingmambo—Under Oath, by Deputed Testamony

Shabby Chic, b.f. 1996, Red Ransom—Style Setter, by Manila (SW in France)

Wind Tunnel, b.f. 1996, Summer Squall—Tivli, by Mt. Livermore

Burning Roma, b.c. 1998, Rubiano—While Rome Burns, by Overskate

Deputy Strike, dk.b/br.g. 1998, Smart Strike—Political Process, by Deputy Minister

Overview, b.c. 1998, Kingmambo—Long View, by Damascus

Quick Tip, dk.b/br.f. 1998, Unaccounted For—Mystic Moves, by Green Dancer

Music Club, b.f. 1999, Dixieland Band—Long View, by Damascus (SW in France)

Casual Look, b.f. 2000, Red Ransom—Style Setter, by Manila (SW in England)

Midway Road, b.c. 2000, Jade Hunter—Fleet Road, by Magesterial

Nine Pines, b.f. 2000, Pine Bluff—Ninth Trestle, by Forty Niner

Strike Rate, b.f. 2000, Smart Strike—Harda Arda, by Nureyev

Shadow Cast, ch.f. 2001, Smart Strike—Daily Special, by Dayjur

Shanghied, b.c. 2001, A.P. Indy—Fleet Road, by Magesterial

Expect Will, b.c. 2002, Valid Expectations—Amazing Trace, by Triple Sec

274

W.S. Farish and Warner L. Jones Jr./Hermitage Farm

Hoist Emy's Flag, b.f. 1979, Hoist the Flag—Natural Sound, by Olden Times

Jordy's Baba, b.f. 1979, Raja Baba—Celeberty Dancer, by Northern Dancer

Permeability, gr.f. 1979, Olden Times—Silk 'n' Set, by Determine

Sweet Slew, b.f. 1980, Seattle Slew—Trick Chick, by Prince John

Heartlight, b.f. 1981, Majestic Light—Hinterland, by *Tulyar

Northern Trick, ch.f. 1981, Northern Dancer—Trick Chick, by Prince John (SW in France)

Procida, dk.b/br.c. 1981, Mr. Prospector—With Distinction, by Distinctive

Rousillon, dk.b/br.c. 1981, Riverman—Belle Dorine, by Marshua's Dancer (SW in France/England)

Smoken Tobin, b.c. 1982, J. O. Tobin—Fire Tail, by Prince John

El Corazon, b.c. 1983, Raja Baba—Dan's Dream, by Your Host

Trick Question, b.c. 1985, Lyphard—Trick Chick, by Prince John (SW in France)

Whow, dk.b/br.f. 1985, Spectacular Bid—Hooplah, by Hillary

Shades of Peace, b.g. 1986, Hero's Honor—Compassionately, by Hail to Reason (SW in France)

Bright Candles, ch.f. 1987, El Gran Senor—Christmas Bonus, by Key to the Mint

Colour Chart, b.f. 1987, Mr. Prospector—Rainbow Connection, by Halo (SW in France)

East Royalty, b.g. 1987, Far Out East—Princess Ribot, by *Ribot

Victory Piper, ch.c. 1987, Nijinsky II—Arisen, by Mr. Prospector (SW in Ireland)

W.S. Farish, Warner L. Jones Jr., W.S. Kilroy

Lomond, b.c. 1980, Northern Dancer—My Charmer, by Poker (SW in England/Ireland)

Argosy, b.c. 1981, Affirmed—My Charmer, by Poker (SW in Ireland)

Seattle Dancer, b.c. 1984, Nijinsky II—My Charmer, by Poker (SW in Ireland)

W.S. Farish and W.S. Kilroy

Cerada Ridge, ch.f. 1977, Riva Ridge—Bold Bikini, by Boldnesian

Weekend Surprise, b.f. 1980, Secretariat—Lassie Dear, by Buckpasser

Easy Step, ch.f. 1982, Overskate—Sahsie, by Forli (Arg)

Law Society, dk.b/br.c. 1982, Alleged—Bold Bikini, by Boldnesian (SW in Ireland/England)

Spectacular Spy, gr.c. 1982, Spectacular Bid—Lassie Dear, by Buckpasser

Alfarazdq, ch.c. 1983, Exclusive Native—Bold Bikini, by Boldnesian (SW in Austria)

Chuck n Luck, ch.c. 1983, Stalwart—Tie a Bow, by Dance Spell

Legal Bid, ch.c. 1984, Spectacular Bid—Bold Bikini, by Boldnesian (SW in England)

Connie's Gift, b.f. 1986, Nijinsky II—Connie Knows, by Buckpasser

Consent, b.g. 1986, Clever Trick—So Cozy, by Lyphard

Bite the Bullet, b.c. 1987, Spectacular Bid—Lassie's Lady, by Alydar

Summer Squall, b.c. 1987, Storm Bird—Weekend Surprise, by Secretariat

A.P. Indy, dk.b/br.c. 1989, Seattle Slew—Weekend Surprise, by Secretariat

Bien Bien, ch.c. 1989, Manila—Stark Winter, by Graustark

Shuailaan, ch.c. 1989, Roberto—Lassie's Lady, by Alydar (SW in England)

Wolfhound, ch.c. 1989, Nureyev—Lassie Dear, by Buckpasser (SW in England/France)

Grand Jewel, b.c. 1990, Java Gold—Flying Rumor, by Alydar

Special Alert, b.f. 1990, Gulch—So Cozy, by Lyphard

Braari, b.f. 1991, Gulch—So Cozy, by Lyphard (SW in England)

Foxhound, b.c. 1991, Danzig—Lassie Dear, by Buckpasser (SW in France)

Sovereign Sage, b.g. 1992, Sovereign Dancer—Charming Lassie, by Seattle Slew

Special Moves, ch.f. 1992, Forty Niner—So Cozy, by Lyphard

Buying Rain, dk.b/br.g. 1993, Gulch—Selling Sunshine, by Danzig

Gold Token, b.c. 1993, Mr. Prospector—Connie's Gift, by Nijinsky II

Brulay, b.f. 1995, Rubiano—Charming Lassie, by Seattle Slew

Lemon Drop Kid, b.c. 1996, Kingmambo—Charming Lassie, by Seattle Slew

Stephen Got Even, b.c. 1996, A.P. Indy—Immerse, by Cox's Ridge

Garcia Marquez, ch.g. 1997, Gulch—Weekend Storm, by Storm Bird (SW in Hong Kong)

Welcome Surprise, dk.b/br.f. 1997, Seeking the Gold—Weekend Surprise, by Secretariat

Statue of Liberty, dk.b/br.c. 2000, Storm Cat—Charming Lassie, by Seattle Slew (SW in England)

William S. Farish and E. J. Hudson Sr.

Really Royal, b.f. 1978, In Reality—Royal Suspicion, by Bagdad

Alleged Jr., b.g. 1980, Alleged—Nancy Jr., by Tim Tam

Oakbrook Lady, b.f. 1981, Exceller—Savy, by Tom Rolfe

Bet Twice, b.c. 1984, Sportin' Life—Golden Dust, by Dusty Canyon

Or Acier, b.c. 1986, Lypheor (GB)—Secorissa, by Secretariat (SW in France, Italy)

William S. Farish, E. J. Hudson Sr., and E. J. Hudson Jr.

Fair Judgment, b.c. 1984, Alleged—Mystical Mood, by Roberto (SW in England, Ireland, U.S.)

Dixieland Dream, dk.b/br.f. 1986, Dixieland Band—Par Three, by Alleged

Art Sebal, b.c. 1988, Arctic Tern—Par Three, by Alleged (SW in Italy)

Freewheel, ch.f. 1989, Arctic Tern—Dinner Surprise, by Lyphard

Everhope, b.f. 1993, Danzig—Battle Creek Girl, by His Majesty

William S. Farish and E. J. Hudson

Private Man, b.c. 1988, Private Account—Lyphard's Princess, by Lyphard

Spanish Parade, dk.b/br.f. 1988, Roberto—Nijit, by Nijinsky II

Spendaccione, b.c. 1989, Spend a Buck—Donna Wichita (Ger), by Windwurf (SW in Italy)

Linda Coqueta, gr/ro.f. 1994, Rubiano—Mystical Mood, by Roberto

Parade Queen, b.f. 1994, A.P. Indy—Spanish Parade, by Roberto

William S. Farish and E. J. Hudson Jr. Irrevocable Trust

Parade Leader, b.c. 1997, Kingmambo—Battle Creek Girl, by His Majesty

Line Rider, b.c. 1999, Danzig—Freewheel, by Arctic Tern

William S. Farish and Hudson Interests Ltd.

Parade Ground, b.c. 1995, Kingmambo—Battle Creek Girl, by His Majesty

Storm Alert, dk.b/br.f. 1995, Storm Cat—Speed Dialer, by Phone Trick

William S. Farish, E.J. Hudson Jr., et al.

Sportin' Life, b.c. 1978, Nijinsky II—Homespun, by Round Table

Folk Art, b.f. 1982, Nijinsky II—Homespun, by Round Table
Mashaallah, b.c. 1988, Nijinsky II—Homespun, by Round Table
Signal Tap, b.c. 1991, Fappiano—South Sea Dancer, by Northern Dancer

William S. Farish and W. Temple Webber Jr.
Dixieland Gambler, ch.g. 1992, Dixieland Brass—Rain Gauge, by Sauce Boat
Grand Charmer, b.c. 1992, Lord Avie—Regal Feeling, by Clever Trick
Political Whit, dk.b/br.c. 1993, Lines of Power—Political Parody, by Doonesbury
Power of Prayer, dk.b/br.g. 1993, Lines of Power—Dewanaplay, by Dewan
Mocha Express, dk.b/br.c. 1994, Java Gold—Gayla, by Lord Gaylord
Ring My Bell, dk.b/br.c. 1995, Idabel—Regal Feeling, by Clever Trick
Globalize, ch.c. 1997, Summer Squall—Sugar Hill Chick, by Fit to Fight
Golden Hurricane, ch.f. 1998, Gold Fever—Hurricane Alert, by Storm Cat

William S. Farish, W. Temple Webber Jr., and James Elkins
Tomisue's Delight, ch.f. 1994, A.P. Indy—Prospectors Delite, by Mr. Prospector
Delta Music, dk.b/br.f. 1995, Dixieland Band—Prospectors Delite, by Mr. Prospector
Monashee Mountain, b.c. 1997, Danzig—Prospectors Delite, by Mr. Prospector (SW in Ireland)
Postponed, b.c. 1997, Summer Squall—Bridal Tea, by Gulch
Secret Status, ch.f. 1997, A.P. Indy—Private Status, by Alydar
Rock Slide, b.c. 1998, A.P. Indy—Prospectors Delite, by Mr. Prospector
Alumni Hall, ch.c. 1999, A.P. Indy—Private Status, by Alydar
Mineshaft, dk.b/br.c. 1999, A.P. Indy—Prospectors Delite, by Mr. Prospector

William S. Farish and Bayard Sharp
Amber Pudding, b.f. 1968, Ambernash—Cream Pie, by Boston Doge
Meringue Pie, b.f. 1978, Silent Screen—Bavarian Cream, by Nashua
Monsagem, b.c. 1986, Nureyev—Meringue Pie, by Silent Screen (SW in England)
Pie in Your Eye, ch.g. 1989, Spend a Buck—Meringue Pie, by Silent Screen
Pie's Lil Brother, dk.b/br.g. 1999, Roar—Meringue Pie, by Silent Screen

William S. Farish and G. Watts Humphrey Jr.
Sacahuista, b.f. 1984, Raja Baba—Nalees Flying Flag, by Hoist the Flag
Misil, ro.c. 1988, Miswaki—April Edge, by The Axe II (SW in Italy)
Gold Streamer, b.f. 1996, Forty Niner—Banner Dancer, by Danzig

William S. Farish and Hannibal Horse Co.
So Cozy, ch.f. 1980, Lyphard—Special Warmth, by Lucky Mike
Dr. Schwartzman, b.c. 1981, Fluorescent Light—Stark Winter, by Graustark
Fantasy Lover, b.f. 1983, Raja Baba—Stark Winter, by Graustark
Special Power, b.f. 1983, Lines of Power—Special Warmth, by Lucky Mike
Integra, b.f. 1984, Torsion—Clever Miss, by Kaskaskia

William S. Farish in various other partnerships
Silent Dignity, gr.c. 1976, Damascus—Croquet, by *Court Martial
El Baba, b.c. 1979, Raja Baba—Hail to El, by Hail to All
Raft, dk.b/br.c. 1981, Nodouble—Gangster of Love, by Round Table (SW in France)
Bedside Promise, ch.c. 1982, Honest Pleasure—Enchanted Native, by Native Charger
Hot Debate, b.c. 1982, Overskate—Hot Rumor, by Swift Ruler
Lil Saucy, b.c. 1984, Sauce Boat—Table Vice, by Round Table
Fappavalley, b.c. 1985, Fappiano—Seven Valleys, by Road At Sea
Slewfoot Seven, b.g. 1988, Lines of Power—Table Vice, by Round Table
Retsel, b.c. 1989, Sportin' Life—In One Piece, by Borzoi
Naughty Notions, dk.b/br.f. 1991, Relaunch—Taras Charmer, by Majestic Light
Clever Tish, b.f. 1994, Clever Trick—Spend a Dream, by Spend a Buck
Storm Song, b.f. 1994, Summer Squall—Hum Along, by Fappiano
Swiss Yodeler, ch.c. 1994, Eastern Echo—Drapeau, by Raja Baba
Jazz Club, b.c. 1995, Dixieland Band—Hidden Garden, by Mr. Prospector
Charismatic, ch.c. 1996, Summer Squall—Bali Babe, by Drone
Darling My Darling, b.f. 1997, Deputy Minister—Roamin Rachel, by Mining
Hurricane Carter, dk.b/br.g. 1997, Summer Squall—Ivy Coast, by Fappiano
Silver Streaker, b.f. 1998, Smart Strike—Silver Trainor, by Silver Hawk (SW in Puerto Rico)
Asong for Billy, b.g. 1999, Unbridled's Song—Adarling, by Alleged
Gulch Approval, dk.b/br.g. 2000, Gulch—Classic Approval,. by With Approval
Tigress Bythetail, gr/ro.f. 2000, Unaccounted For—Empress Sirrima, by Waquoit
Forbidden Queen, dk.b/br.f. 2001, Halory Hunter—Last Portrait, by Sunny Clime (SW in Puerto Rico)
Gradepoint, dk.b/br.c. 2001, A.P. Indy—Class Kris, by Kris S.
Broadway Gold, dk.b/br.f. 2002, Seeking the Gold—Miss Doolittle, by Storm Cat

Sources: *The Blood-Horse* Archives and The Jockey Club

CHAPTER

19

Seth W. Hancock

In 1972 the loss of arguably the strongest individual in international breeding created questions about the future for the world-renowned Claiborne Farm. His younger son, in his early 20s, stepped forward boldly and quickly assuaged doubts that Claiborne could carry on much as it had.

Arthur B. "Bull" Hancock Jr. died that September 1972 after a short illness. He had taken over operation of Claiborne Farm from his father, A.B. Hancock Sr., and had built upon its sturdy legacy to make Claiborne the most influential Thoroughbred operation in the world. Hancock was a breeder, an adviser to the many patrons who boarded mares at his farm, and a stallion man with the savvy and clout to put together a deal to stand virtually any young horse he wanted. Hancock's will created roles for both of his sons, Arthur III and Seth, and created an advisory role for three friends and associates, i.e., Ogden Phipps, William Haggin Perry, and Charles Nuckols. Hancock also directed that, for financial security reasons, the farm return to the practice of selling yearlings, a practice his own father had followed but that he personally had abandoned.

A ship needs a captain, not two, and before long the older son, Arthur III, struck out on his own. He established and expanded nearby Stone Farm and within a decade achieved something neither his father nor grandfather had managed, the racing of a homebred Kentucky Derby winner.

The younger son, Seth Hancock, born on July 22, 1949, was only twenty-three when his father died. He had attended the University of the South and then transferred to the University of Kentucky, where he earned a degree in agriculture. The plan had been for him to spend extended times working in the various divisions of Claiborne. This long-term education was abrogated, and Seth was soon in charge.

If the racing world speculated much about Seth Hancock's credentials, he made two major moves early that established his savvy to a high degree of satisfaction. At the dispersal sale of his father's stock, Seth reached in on his own account and purchased a Damascus colt for $90,000. By the end of the year, he had made a major move in keeping with the history of Claiborne when he arranged the syndication of budding superhero Secretariat for more than $6 million, a record at the time.

"Secretariat was a total stroke of luck. It was obviously more because of my father's association with Mr. (C.T.) Chenery than my association with his daughter, Penny," Hancock told the author some years later in an interview for *Spur* magazine. "But, it happened, and I think people thought 'This guy knows what he's doing.'"

This thought was reinforced and expanded to another area early in 1974, when that Damascus colt, Judger, won the Florida Derby and Blue Grass Stakes in Hancock's own colors. (When Secretariat won the Triple Crown of 1973, his runner-up in the first two races of the series was Sham, a Pretense colt Sigmund Sommer had purchased from the Bull Hancock dispersal.)

Although buoyed in his own self-confidence by such dramatic early success, Hancock was struggling with one aspect of his legacy, the selling of yearlings. Like his father, Hancock preferred racing to selling. He has said repeatedly that "I don't think I'm a very good salesman," and he chafed under the yoke of the unintended consignor.

The fact is, however, that Claiborne had some spectacular sales under Seth's low-key style of promotion. The summer after his father's death, Claiborne sold a Bold Ruler colt, Wajima, for a record at the time, $600,000. The colt went on to be the champion three-year-old of 1975 for East-West Stable. In the same crop Claiborne sold the Sir Ivor filly Ivanjica for $180,000, and at four she won the Prix de l'Arc de Triomphe for the Wertheimer family.

Although Claiborne's stock-in-trade over many years was the acquisition and standing of world-class stallions, the Hancocks were not stubborn about recognizing the merits of another man's horse. Among outside shares Claiborne had was one in Northern Dancer, the little Canadian-bred who forced the industry to re-evaluate small horses as runners and breeding stock. One of his most diminutive sons was the 1977 colt from Special, which was in Seth Hancock's Keeneland consignment in the summer of 1978. Countering his size, the colt had style, and Nureyev brought $1.3 million, then the second-highest price ever paid for a yearling. He was purchased by Stavros Niarchos. Natural salesman or not, a guy does not sell a Keeneland sale topper without knowing what he is doing.

Nureyev's dam, Special, was by *Forli, an Argentine champion whom Bull Hancock had imported during the 1960s. Special had failed to win in her only start but was out of Thong, a stakes-placed full sister to the starring triumvirate of Moccasin, Ridan, and Lt. Stevens. While Seth Hancock would shrink from any competitive comparison to his forebears, in Nureyev he might well have bred a better stallion than either his father or grandfather ever bred (although Bull could make something of a case for Round Table). Nureyev was a high-class racehorse and his stallion career ranked with the best of the Northern Dancers. He has sired 137 stakes winners (17 percent), the most distinguished of which included Prix de l'Arc de Triomphe winner Peintre Celebre, Eclipse Award winner Theatrical, two-time Breeders' Cup Mile winner

Seth Hancock

Miesque, additional Breeders' Cup Mile winner Spinning World, plus Zilzal, Soviet Star, and Sonic Lady.

The influence Seth Hancock has had on international breeding also included another flow through Nureyev's dam, Special. In 1974 Hancock bred Special to Bold Reason; as matters would turn out, Bold Reason was not a particularly successful sire, but was a high-class racehorse and an interesting resource for the Thoroughbred gene pool. Bold Reason was sired by the sire of sires Hail to Reason and was out of the *Djeddah mare Lalun. He was thus a half brother to Never Bend, who became an important influence internationally through his great son Mill Reef. The Bold Reason—Special filly foaled in 1975 did not bring much glamour to the sale ring, although by then the family (pre-Nureyev) was already seen to have produced Special's English/Irish champion brother Thatch, plus King Pellinore. Presumably hurt in the market by the fact that she was by Bold Reason, she brought only $40,000 from the British Bloodstock Agency group representing Robert Sangster.

The filly was named Fairy Bridge and was sent abroad, where she won her only two races and was retired. Fairy Bridge's second foal was by Northern Dancer and was none other than Sadler's Wells, a classic winner who has become an amazing stallion,

278

a sire of 239 stakes winners through mid-2004, dominant in English/Irish statistics for more than the last decade. Moreover, Fairy Bridge foaled Sadler's Wells' full brother Fairy King, sire of Arc winner Helissio and Epsom Derby winner Oath.

Special was a key example of Seth Hancock's belief in strong female families.

"Generally, I believe in the old saying that the family is stronger than the individual," he wrote in the foreword to the 1999 book on broodmares, *Matriarchs* (Eclipse Press). This sentiment echoed his father's philosophies and his own father's before him.

"Connected to that line of reasoning is the fact that the racing record alone is not always a true measure. When you attempt to judge the breeding potential of a filly, I believe it is important to have as much information as possible. A filly might have run only three times and not gotten much done and yet still have had ability. An extreme example for Claiborne has been the mare Special. She only raced once and was unplaced, but I believe my father learned she was a bleeder. A lot of things can come up that keep a horse from showing its real potential."

He once amplified this for *The Blood-Horse*: "With broodmares, I think pedigree is definitely the most important. I would place conformation second and race record third. I would rather have a mare that had a good pedigree than a mare that was a really good racemare that didn't have any pedigree."

In addition to Nureyev and Fairy Bridge, Special also foaled two major stakes winners retained to race for Claiborne in Number and Bound, both by the homestanding sire Nijinsky II. Number won three graded stakes, including the Hempstead and Firenze handicaps, and earned $301,793. She is the dam of the stakes winners Jade Robbery, Numerous, and Chequer. Bound won the Churchill Downs Budweiser Breeders' Cup Handicap and is the dam of stakes winner Limit.

The first champion officially bred on Seth Hancock's watch was Revidere. As a foal of 1973, Revidere was a holdover result of a mating designed by his father and William Haggin Perry (see chapter on A.B. Hancock Jr., Vol. I), who was the family's partner in some thirty-five crops of homebreds. Revidere was by the Bold Ruler stallion Reviewer, renowned as the sire of the great Ruffian. The dam was Quillesian, by the leading Claiborne sire *Princequillo. Whereas most of the mares brought

into the Claiborne-Perry partnership originated with the Hancocks, Perry contributed Revidere's family, for Quillesian was out of one of Perry's early separate purchases, the top-class filly Alanesian.

Revidere tended to look like she was sore and did not race at two. She was shopped for $35,000 but had no buyer. Although bred in the name of Claiborne, Revidere raced as the sole property of Perry. With David Whiteley masterfully bringing her along, Revidere emerged as the champion three-year-old filly of 1976, winning the Coaching Club American Oaks, Ruffian Stakes, Monmouth Oaks, and two other stakes. She won eight of eleven races and earned $330,019.

As was true of his father and grandfather, the total impact of Seth Hancock is not a matter of record. How many major horses he has had influence in breeding is not reflected in printouts or *Stud Book* certificates. One bit of advice that is well documented, though, utilized another representative of the Alanesian family and produced the singular horse named Seattle Slew.

In 1963 Alanesian foaled the Bold Ruler colt Boldnesian, who won the Santa Anita Derby in a short career of four wins in five starts. Boldnesian's best runners included the Jersey Derby winner Bold Reasoning, who raced for the Kosgrove Stable of Charles Hargrove and William Kosnick and entered stud at Claiborne in the spring after Bull Hancock's death.

Ben Castleman, a Kentucky racing commissioner, small-scale horse breeder, and tavern owner, called Seth and "told him what mares I had, asked him to check over their pedigrees, and see what stallions he had open that I could breed to, that I could afford to breed to. He called me and said I could breed My Charmer to Bold Reasoning. He was just going to stud. My Charmer was the third mare ever taken to Bold Reasoning. She caught on the first cover, and there came Seattle Slew. Seth Hancock selected that mating."

Castleman sold Seattle Slew for $17,500 to Dr. Jim and Sally Hill and Mickey and Karen Taylor at the Fasig-Tipton Kentucky summer sale of 1975. This launched an improbable journey whose ramifications on the breed continue in the early twenty-first century. Slew was unbeaten through the Triple Crown and became an exceptional stallion.

Hancock was neither the breeder nor owner of

Seattle Slew, but he developed his own way of bene-fiting from the admixture of genetics whereby he helped orchestrate the creation of Seattle Slew. The most dramatic avenue was Swale, who carried the Claiborne colors to a longed-for Kentucky Derby win before a stunning and inexplicable death meant he could not carry on the line at stud.

This side of a Walter Brennan movie, it is hard to conjure up a tale of old Kentucky that combines sen-timent, horsemanship, and bittersweet success more poignantly than the true story of Swale.

Swale was a foal of 1981. By then Seth Hancock was well established as president of Claiborne, and he had decided, as his father had done nearly thirty years before, to alter his strategy from being a mar-ket breeder to working out a way to keep Claiborne-breds to race. He and Perry formed a partnership known as Raceland that invited in some special asso-ciates as well as maintaining interests in the horses for themselves.

"All the mares included were owned by Claiborne

and Mr. Perry," Seth explained. "So, technically, Claiborne sells a quarter interest (in the mares' off-spring) to Dell (one of Hancock's sisters), Ed Cox (a client), and myself, and Mr. Perry sells one-quarter interest to Peter Brant (also a client)." The Claiborne entity included Hancock, his sisters Dell and Clay, and their mother, Waddell Hancock.

Mrs. Hancock recalled the birth of Swale's dam, Tuerta, as an unhappy occasion. Bull Hancock held a Kentuckian's reverence for the Kentucky Derby and had a burning ambition to win it. He had bred and raced a lot of good horses but only started two in the Derby, finishing unplaced with Dunce and third with Dike. At the time, only one filly, Regret, had won the Derby and that had happened back in 1915, so it was just taken as a matter of fact that if a man were going to have a Derby horse it would have to be a colt.

"When Swale's dam, Tuerta, was foaled," Mrs. Hancock recalled some years later, "the man in the barn had to tell him, 'Mr. Hancock, it's a filly.' Well,

Swale

280

that made Bull furious, and then the man had to say, 'Mr. Hancock, I hate to tell you, but she's only got one eye.' Bull went into one of his rages and kicked a bucket down the barn."

Well, one-eyed Tuerta was good enough to win the Blue Hen Stakes at two. She thus was the last stakes winner that Bull Hancock saw win in the Claiborne colors, and this encouraged the family to select her to hold out of the dispersal. (A total of thirty-five Claiborne horses in training and yearlings sold at Belmont Park in the fall of 1972, brought $2,580,000 to average $73,714.)

Tuerta raced on after two, also winning the Long Island Handicap and Chrysanthemum Handicap. She had nine wins in twenty-five starts and earnings of $125,912. Her first four foals included four winners, of which Illuminate and Sight were stakes-placed.

Tuerta represented one of the lengthy tapestries that involved generations of the Hancock men and four generations of mares. Her third dam, *Highway Code, was an English daughter of the great sire Hyperion and was out of Book Law, winner over colts in England's classic St. Leger. A foal of 1939, *Highway Code was imported in 1950, by which time Bull Hancock was running the farm for his ailing father. *Highway Code, winner of a small stakes, was bred to the treasured Claiborne acquisition *Nasrullah in his first season at the farm and in 1952 foaled the filly Courtesy.

Courtesy was a winner who finished second in the Ashland Stakes. She then foaled three stakes winners, Knightly Manner, Dignitas, and Respected. Courtesy's other foals included the Double Jay mare Continue, who was a modest winner and then produced five stakes winners. Tuerta was one of these, and the others were Perpetual, List, Yamanin, and File.

The breeding of *Forli to Continue produced Tuerta, who was inbred 3x4 to Hyperion since *Forli was by the Hyperion stallion Aristophanes. In 1980 Seth Hancock called upon a resource of his own making and bred Tuerta to Seattle Slew. The resulting foal was a husky, dark colt who picked up the nickname of Foghorn for his tendency to snore while lying peacefully in the field.

In the previous crop the Claiborne-Perry mare Alluvial, by Buckpasser, had foaled a Seattle Slew colt who raced for the Seattle Slew owners and other partners under a foal-swap arrangement. This was Slew o' Gold, who was bred by Claiborne. Racing for the Taylors, Hills, et al., Slew o' Gold was the champion three-year-old of 1983 and champion older male of 1984. He won two runnings of both the Jockey Club Gold Cup and Woodward Stakes, plus the Marlboro Cup, Whitney Stakes, and Wood Memorial. Slew o' Gold won twelve of twenty-one starts and earned $3,533,534. He was a moderate sire, with twenty-nine stakes winners (5 percent).

Seattle Slew, then, though an outside stallion, was enmeshed in the ongoing march of Claiborne by the time his son Swale came along. Taken into the Raceland partnership, Swale raced in the Claiborne silks and was trained by Woody Stephens, the Kentucky native who had become something of a folk hero of the Turf. At two Swale was overshadowed in Stephens' barn by Mr. and Mrs. James P. Mills' Devil's Bag, the dynamic juvenile champion. Still, Swale got in his moments, winning the historic Futurity Stakes at Belmont and the Breeders' Futurity at Keeneland, as well as the Saratoga Special and Young America Stakes.

Devil's Bag did not extend his domination at three and was not in the Kentucky Derby. By Derby Day, Swale had won the Hutcheson and Florida Derby but had a loss in the mud at Keeneland in his last prep. Stephens was in poor health, and the day-to-day handling of the colt was turned over to Mike Griffin, but Woody remained in the picture, literally and figuratively. With Laffit Pincay Jr. aboard, Swale raced prominently from the start, took over on the far turn, and rolled home by three and a quarter lengths. An ambition long cherished for Claiborne had been achieved. Woody rallied to meet the owners and the colt in the Derby winner's circle, where the veteran trainer had foregathered with Cannonade ten years earlier.

Seth Hancock sensed that two weeks between races was not Swale's game, but, well, the Triple Crown is the Triple Crown. Swale made little impression on his field in the Preakness Stakes, but with three weeks between races he dominated the Belmont Stakes — another first for the Claiborne colors. Coastal (Majestic Prince—Alluvial, by Buckpasser) was another who had been bred in the Claiborne name but raced as Perry's sole property when he won the Belmont in 1979.

Eight days after the Belmont Stakes, while being

bathed at Stephens' barn at Belmont Park, Swale collapsed without warning. The flame did not even flicker before dying.

The impact of the sudden death of a star was emotional to the Hancocks, but it also was a major setback in a business sense. While Secretariat had been a successful sire, by 1984 it was no longer presumed he would come up with a series of major horses. The sire line of Bold Ruler, who had been so dominant, was waning in many branches, and Seattle Slew was emerging as its most powerful weapon. A Derby-Belmont winner by Seattle Slew would have been a significant addition to any stallion barn.

Seth Hancock harbors a sense of the romance of racing, and of following in the footsteps of his ancestors, but he has developed his own sense of realities, too.

"The death of Swale, I just can't understand still," he said years later. "But maybe he wasn't going to be a good stallion anyway. I think with those type of things, if you dwell on them, they'll tear you all to pieces, so you need to try to put it in a light that you can accept, and go on."

Swale had won nine of fourteen races and earned $1,583,660. He was voted the champion three-year-old of 1984. He helped Seth Hancock ascend to a distinction his forebears had achieved, that as leading breeder of the year. The Claiborne-breds earned a total of $5,554,012 in North America in 1984.

Seth Hancock would make good use of the Seattle Slew blood some years later, and he certainly made good additional use of the female family that had produced Swale. File, one of the stakes winners from Tuerta's dam Continue, was a daughter of Tom Rolfe. Tom Rolfe had been bred by Raymond Guest via the great European champion *Ribot. The Preakness winner and three-year-old champion of 1965, Tom Rolfe had been recruited to Claiborne's stallion roster. File won a division of the Cinderella Stakes among five wins in twenty-two starts and earned $73,774.

In 1985 File foaled Forty Niner, a Claiborne homebred who again illustrated that the third generation of Hancocks was on a par with the predecessors. Forty Niner was by Mr. Prospector, who had become one of Seth Hancock's stallion acquisitions that ranked with the best of those brought to Claiborne by his father and grandfather.

Claiborne had long been the home of a sequence of great sires — *Sir Gallahad III, *Blenheim II, *Princequillo, Double Jay, *Nasrullah, Bold Ruler, Round Table, Damascus, Sir Ivor, Nijinsky II. The acquisition of Mr. Prospector, arguably has been the best coup during Seth Hancock's tenure at Claiborne. Mr. Prospector (Raise a Native—Gold Digger, by Nashua) had topped the Keeneland summer sale of 1971 when Spendthrift Farm sold him for $220,000 to A.I. Savin. The colt flashed brilliant speed, but injuries restricted his career to a pair of sprint stakes, some sensational clockings, and an aura of prodigious potential unfulfilled. Savin had wanted him as a stallion prospect in the first place

Mr. Prospector, 1970	Raise a Native, 1961	Native Dancer, 1950	Polynesian (Unbreakable—Black Polly)
			Geisha (Discovery—Miyako)
		Raise You, 1946	Case Ace (*Teddy—Sweetheart)
			Lady Glory (American Flag—Beloved)
	Gold Digger, 1962	Nashua, 1952	*Nasrullah (Nearco—*Mumtaz Begum)
			Segula (Johnstown—*Sekhmet)
		Sequence, 1946	Count Fleet (Reigh Count—Quickly)
			Miss Dogwood (*Bull Dog—Myrtlewood)
FORTY NINER	Tom Rolfe, 1962	*Ribot, 1952	Tenerani (Bellini—Tofanella)
			Romanella (El Greco—Barbara Burrini)
		Pocahontas, 1955	Roman (*Sir Gallahad III—*Buckup)
File, 1976			How (*Princequillo—*The Squaw II)
	Continue, 1958	Double Jay, 1944	Balladier (Black Toney—Blue Warbler)
			Broomshot (Whisk Broom II—Centre Shot)
		Courtesy, 1952	*Nasrullah (Nearco—*Mumtaz Begum)
			*Highway Code (Hyperion—Book Law)

and duly sent him to stud at his own Aisco Farm in Florida.

Mr. Prospector got the champion filly It's in the Air in his first crop. Soon his runners created a status for the young sire that made the idea of keeping him in Florida less and less practical. The young horseman Peter Brant suggested to Seth that Mr. Prospector might be worthy of Claiborne, and soon the deal was made. Mr. Prospector moved to Kentucky for the 1981 breeding season.

Seth Hancock was a youngster in the days

Forty Niner beats Seeking the Gold

when the Claiborne stallion Bold Ruler led the sire list seven times in succession and a total of eight times. In Mr. Prospector — brought to him by a contemporary of his own rather than his father's — Seth had a stallion that he eventually felt need take a backseat to no other horse:

"Bold Ruler was a wonderful sire...but I don't think he was any better sire than Danzig (and) Mr. Prospector. Mr. Prospector's sons are blockbusters, and his daughters are wonderful producers. If there has been a better sire in our lifetime, or in history, than him, I'd like to know who it was."

Seth has resisted the modern trends toward book sizes of well over one hundred, surmising that no stallion with such a book size can have a uniform quality to his mares; he just does not like to breed a stallion to a mare that does not rate. When honored by the Thoroughbred Club of America in 2000, he spoke of having increased book sizes to between eighty and eighty-five mares and 130 covers, and noted that it made the stallions "mentally tired. If our stallions are experiencing this, then what about stallions that have 150 mares and 200 covers?" Hancock also has resisted the short-term business temptation to hike stud fees to the maximum the market will bear.

Mr. Prospector was one that soared to the top under Hancock's management. The great stallion, who died at twenty-nine in 1999, is the sire of 180 stakes winners (15 percent), and broodmare sire of 274 others to date. He led the American sire list of 1987 and 1988; led the broodmare sire list annually from 1997 to 2003; and topped the juvenile sire list in 1979 and 1987.

Mr. Prospector's legacy is enormous. Among his best racehorses were Conquistador Cielo, Forty Niner, Gulch, Queena, Gold Beauty, Seeking the Gold, Fappiano, Rhythm, Gone West, Golden Attraction, and Fusaichi Pegasus. His sons at stud include Fappiano, Gulch, Seeking the Gold, Kingmambo, Woodman, and Miswaki. The sequence continues. For example:

Fappiano sired Kentucky Derby winner Unbridled, in turn the sire of Derby winner Grindstone as well as Halfbridled, Empire Maker, Unbridled's Song, Banshee Breeze, Red Bullet, and Anees. (Grindstone has since sired 2004 Belmont Stakes winner Birdstone.) Another son of Fappiano, Quiet American, sired Kentucky Derby-Preakness winner Real Quiet;

Seeking the Gold is the sire of international star Dubai Millennium and American champions Heavenly Prize and Flanders;

Gulch is the sire of Derby-Belmont winner Thunder Gulch, in turn the sire of Preakness-Belmont winner and Horse of the Year Point Given; Kingmambo is the sire of Belmont winner and champion Lemon Drop Kid;

Woodman is the sire of Preakness-Belmont winner Hansel, Preakness winner Timber Country, and European champions Bosra Sham and Hector Protector;

Gone West is the sire of Belmont winner Commendable, European champion Zafonic, Breeders' Cup Mile double winner Da Hoss, and mile record holder Elusive Quality (in turn the sire of classic winner Smarty Jones).

Clearly, Mr. Prospector ranks with the elite of Claiborne's proud history of stallions.

To return to the theme linking Derby winner Swale's family with Mr. Prospector, it will be recalled that the half siblings of Swale's dam Tuerta included the stakes winner File. In 1984 Hancock bred File to Mr. Prospector, and the resulting foal was a dashing, darkish chestnut who was given the prospecting name of Forty Niner. At two Forty Niner came to hand well enough for Woody Stephens — recovered, but in the twilight of his grand career — that he added a second Futurity for Seth, winning by three lengths, then won the Champagne by four and a half. Kentuckians through and through, Hancock and

Stephens raced Forty Niner next in the Breeders' Futurity at Keeneland, and Forty Niner just scraped home by a nose over a horse named Hey Pat. Earlier, Forty Niner had won the Sanford Stakes at Saratoga. Forty Niner did not venture west for the Breeders' Cup Juvenile Colts of 1987 but won the Eclipse Award for the two-year-old colt division.

At three Forty Niner missed by a neck catching Winning Colors in the Kentucky Derby, but his campaign was highlighted by two sterling duels with Seeking the Gold. Also by Mr. Prospector, Seeking the Gold was bred at Claiborne by the Phipps family, and he got to within a nose of Forty Niner in both the Haskell Invitational and the Travers. Claiborne had never won the historic Travers, and that was one of the victories that implanted Forty Niner on the first rung of horses in the sentiments of one Seth W. Hancock.

Forty Niner ran a brilliant race when he got within a neck of the older champion Alysheba in the Woodward Stakes, then won the NYRA Mile before a fourth-place finish in the Breeders' Cup Classic. He won eleven of nineteen races and earned $2,726,000. In the stud at Claiborne, he was not the immediate success that some might have expected, although he then became very hot, and he joined the farm's ranks of prominent sires and ranked second in earnings in 1996. Despite the stallion's early promise, Hancock accepted the offer of $10 million to sell him to Japan. Hancock later dealt successfully for Mr. Prospector's grandson Unbridled, who died after only a few years at Claiborne.

Horses left behind by Forty Niner included the fellow Travers winner Coronado's Quest, who was bred by Stuart S. Janney III, foaled at Claiborne, and returned there to stud. (Coronado's Quest later was also sold to Japan.) Other key offspring of Forty Niner to date include Belmont Stakes winner Editor's

BARABARA D. LIVINGSTON

Lure

Note, the leading juvenile sire Distorted Humor (sire of Derby-Preakness winner Funny Cide), plus Roar, Sunday Break, Twining, Gold Fever, and End Sweep.

Such are the dynamics of the stallion market from time to time that a decade after setting a record in syndicating Secretariat for about $6 million, Hancock found himself dealing with a horse valued at six times as much. The worldly Henryk de Kwiatkowski had come onto the scene, and Woody Stephens was training for him. They purchased a Mr. Prospector colt at Saratoga, and under the name of Conquistador Cielo, this fellow won the Metropolitan Handicap and Belmont Stakes in spectacular form within one week in 1982.

The Florida-bred Consquistador Cielo (Mr. Prospector—K D Princess, by Bold Commander) was syndicated by Seth on de Kwiatkowski's behalf for a startling $36 million! This broke the record of $30 million, which had been the valuation of Storm Bird. He in turn had beaten Seth's record of $20 million to syndicate Spectacular Bid in 1980. Conquistador Cielo, Horse of the Year in 1982, was a good, reliable stallion without hitting the top ranks. He got sixty-nine stakes winners (7 percent), including the Jockey Club Gold Cup winner Wagon Limit and the Hollywood Gold Cup winner Marquetry. The latter has become something of a speed influence as the sire of two Breeders' Cup Sprint winners, Artax and Squirtle Squirt.

Spectacular Bid, winner of the Kentucky Derby and Preakness in 1979 and unbeaten at four, was another Horse of the Year Seth Hancock secured to stand at Claiborne. He eventually left for Milfer Farm in New York. The son of Bold Bidder—Spectacular, by Promised Land was the sire of forty-four stakes winners (6 percent). It is an indication of how high the historic bar at Claiborne Farm remains that such stallions in the context of their surroundings might be perceived as disappointing.

One who was definitely of the other stripe — successful beyond any reasonable expectation — was Danzig. Again, this was a yearling purchased by de Kwiatkowski and trained by Stephens. He was bred by William S. Farish and Marshall Jenny's Derry Meeting Farm and was by Northern Dancer—Pas de Nom, by Admiral's Voyage. Indicating the caprices of nature, Admiral's Voyage had been a very good racehorse but a failure as a sire.

Danzig flashed brilliant speed, but Stephens could not keep him sound long enough to get into stakes company. Danzig won all his three races and earned only $32,400, but commanded a price of $2,880,000 when syndicated for $80,000 for each of thirty-six shares. It was a price based on potential and pedigree. Hancock's father had stood the non-stakes winner Nantallah with success, but by and large the history of Claiborne's stallion barn has not dealt very often in horses that never won a stakes races. Danzig was undeterred by that lack of distinction and presently joined the ten previous Claiborne Farm stallions to have led the general list at least once. (Forty Nine became the twelfth leading sire for Claiborne in 1996, as ranked by *Daily Racing Form* although *The Blood-Horse*'s more inclusive international figures ranked Cozzene the leader that year.) Danzig followed Bold Ruler as the second of that distinguished group to lead as many as three years in succession. He topped the list in 1991-93, according to *The Blood-Horse*.

Danzig, pensioned from study duty in July 2004, was the sire of an exceptional number of stakes winners and an exceptional percentage. His 182 account for 17 percent of his foals, and they range from the champion sprinter Dayjur, to the Canadian Triple Crown winner Dance Smartly, to the Belmont Stakes winner Danzig Connection, to the crack turf miler Lure. Danzig's son Danehill was an international phenomenon at stud. Having stood in both hemispheres, he became a force in both, with a total of 256 stakes winners as of mid-2004. Danzig's champion son Chief's Crown sired one Epsom Derby winner in Erhaab. In turn, Chief's Crown's champion son Grand Lodge sired the Derby and Arc de Triomphe winner Sinndar.

Hancock made good use of the best stallions with Claiborne-owned mares. He bred a dozen stakes winners by Mr. Prospector and has bred ten by Danzig for Claiborne. (Danzig was retired at 27 in the summer of 2004). The best of his Danzigs was Lure. A handsome colt foaled in 1989, Lure was from Endear, a high-class Alydar filly who won the grade I Hempstead Handicap and two other stakes for Claiborne. Endear died young, after producing only two foals.

Lure first made a mark on dirt, dead-heating with Devil His Due in the Gotham Stakes at three in 1992. He was trained by Shug McGaughey, who also is the trainer for the Phipps family, long-time Claiborne clients. After some disappointments with Lure,

McGaughey put him on grass, and the Danzig colt ended the year with a victory over a top-class international field in the Breeders' Cup Mile. The following year his campaign again was climaxed by a score in the Breeders' Cup Mile, following victories in four other stakes. Still in action at five, Lure stretched out to win the Caesars International at Atlantic City at a mile and three-sixteenths and also won the Elkhorn and Bernard Baruch. He was retired with fourteen wins in twenty-five starts and earnings of $2,515,289. Lure stunned Claiborne by turning up virtually sterile and was purchased via payoff of a fertility insurance policy. He made some progress and stood for a time at Ashford Stud but later was returned to Claiborne as an honored pensioner. Lure is the sire of Beverly D. Stakes winner England's Legend and the champion Irish two-year-old Orpen.

Danzig is already the broodmare sire of some 116 stakes winners. Clearly he is one of the most important links in the amazing genetic chain of his sire Northern Dancer.

Seth Hancock inherited the management of Northern Dancer's best racing son, Nijinsky II, who had been syndicated in 1970. Nijinsky II, a large, lengthy bay unlike most of the Northern Dancers physically, had won the English Triple Crown. He broadened the appeal of Northern Dancer blood internationally, and Hancock again bred not only a good racehorse but also a highly important stallion in the Nijinsky II horse Caerleon. This colt represented another case of a strong female family, albeit one that Claiborne had owned for only two generations. Regal Gleam, the champion two-year-old filly of 1966, had been purchased from Bieber-Jacobs Stable. Her Round Table filly Foreseer won three races for Claiborne and was stakes-placed, and then produced Caerleon and three other stakes winners. The $654,080-earner Yonder was a Claiborne homebred foaled from Foreseer's daughter Far.

Caerleon was sold to the Sangster team and won the French Derby and England's Benson & Hedges Gold Cup in 1983. In the stud he got 101 stakes winners, including the 1991 Epsom Derby and Irish Derby winner Generous.

In the name of his own separate Cherry Valley Farm and Perry's Gamely Corporation, Hancock also bred the Nijinsky II colt Shadeed, winner of England's Two Thousand Guineas and himself a major stallion. Shadeed's dam, Continual, by Damascus, represents the same female line that produced Swale and Forty Niner.

In summary, as the breeder of Nureyev and Caerleon as well as the breeder of the dam of Sadler's Wells, the soft-spoken, stay-at-home Kentuckian Hancock has exerted profound influence on the recent history of European racing.

Another important offspring of Nijinsky II for Claiborne was the big filly State, who raced for Dell Hancock. State was by Nijinsky II and out of Monarchy, a stakes-winning full sister to the exalted Claiborne-bred champion Round Table. State was only a modest winner, but her family was certainly one of those that looked likely to be stronger than the individual. In 1980 State foaled a filly by Honest Pleasure; the latter was a brilliant racehorse but had an offbeat female family and generally was not a success at stud. State's daughter was named Narrate and won

Pulpit

ANNE M. EBERHARDT

the Falls City Handicap and Princess Doreen Stakes for Claiborne.

The naming of this line illustrated the long-standing Hancock tendency to use one-word names, many with only one syllable. With Narrate, the naming pattern switched from one line of thinking to another: Monarchy's daughter State was now used not as a political entity, but as the spoken word. Then, when Narrate foaled a Mr. Prospector filly in 1989, the nuance of naming went theological. The filly was named Preach, and she had her say, winning the grade I Frizette Stakes as well as the Bourbonette Stakes. Preach won four of fifteen races and earned $304,656.

In addition to his respect for Seattle Slew's outright qualities, Hancock also saw the stallion's bloodlines as a helpful outcross to the inevitable combines of Mr. Prospector and Northern Dancer blood that marks his and other top-class operation's broodmare bands. He sent Preach to Seattle Slew's son A.P. Indy, the 1992 Horse of the Year who stands at Lane's End Farm. A.P. Indy is out of the high-class Secretariat mare Weekend Surprise, and the next dam was by Buckpasser, a Phipps champion who stood at Claiborne. So, while there was plenty that was familiar in the pedigree of A.P. Indy, it still provided the outcross.

In the final year of his life, Bull Hancock still tended to quote his own father in explaining his own approach and ideas. Likewise, on the subject of inbreeding in general, Seth wrote that "some of Claiborne's best families have come from mares my father bought in England, and we still try to buy mares over there that might be outcrosses to all the Northern Dancer and Mr. Prospector…When I see a horse inbred 3x3 to Raise a Native, I kind of shudder. But I also think it makes a difference which individual horses are involved." (A very recent example of the pattern of buying mares in England is Maid for Walking, whom Claiborne purchased in England and who is the dam of stakes winner Patrol and 2004 graded stakes winner Stroll.)

After Lure's first victory in the Breeders' Cup Mile, Hancock again referred — perhaps deferred — to his father: "My father had a saying that a good bull is half of a herd, and a bad bull is all of it. I think what he meant was that if you have a good sire, then you have half a chance, and then you must have a good broodmare to have a full chance. If you have a bad sire, you've got no chance."

Named Pulpit, the A.P. Indy—Preach colt was spectacular early in the winter of 1997. Indeed, Hancock's mother, the experienced and wise lady of many a season of racing, was somewhat miffed by what she saw as the overhype the media gave the lightly experienced colt. Pulpit won the Fountain of Youth, as the media had suspected, but then in the Florida Derby he proved mortal, as Mrs. Hancock had known any horse might. Pulpit then rolled home impressively, however, in the Blue Grass Stakes and gave high hopes of another Derby triumph. He injured himself in the classic, however, finishing fourth, and was retired with much of his potential still seemingly untapped. Pulpit went to stud at Claiborne with four wins in six starts and earnings of $728,200. He has made an immediate mark at stud. His first crop included the Dubai Classic and Super Derby winner Essence of Dubai, and Pulpit's second crop brought out the top prospect Sky Mesa, winner of the Hopeful and Breeders' Futurity at two, as well as the aforementioned Stroll. The next crop included Wood Memorial winner Tapit.

The senior A.B. Hancock essentially ran Claiborne Farm for about forty years; Bull Hancock ran it for twenty-five years. Seth Hancock has now held the reins for thirty years and counting, and he has maintained quality and high status amid many changes in the market and the international Thoroughbred scene. Key individuals, such as yearling man John Sosby (now retired) and stallion/broodmare manager Gus Koch, have had the lasting relationship with the family and the farm that marked the previous generations' management. Hancock strolls through the fields to make sure he is familiar with each of the perhaps three hundred mares that might be on the place from time to time. He studies results of clients' mares in which he has no input in matings, just as he studies those where he does. He feels an obligation to make sure his occasional absences to attend a race, sale, etc., are brief.

The Phipps family, first associated with Hancock's grandfather, has continued to board mares and welcome Hancock's input. While that relationship is key to Claiborne, Hancock is his own man. They are not joined at the hip in such matters as stallion acquisition. Dinny Phipps explained that there is no automatic connection or seal of approval before Hancock selects and negotiates for a new stallion prospect. "Sometimes we are consulted, and sometimes we support the horse, and sometimes we don't," said

Phipps. The impact of the connection is obvious and positive, though, as the Phippses have returned to Claiborne such successful young stallions as Bold Ruler, Buckpasser, Reviewer, Majestic Light, and, more recently, Seeking the Gold.

Hancock doesn't put himself in the limelight as a matter of choice, but he is a strong leader in the industry. The Breeders' Cup might very well have died aborning had he not thrown the weight of Claiborne Farm behind it by signing up his stallions and paying the appropriate fees so their offspring one day could be eligible to compete in Breeders' Cup races. Hancock is a member of The Jockey Club, as are his brother Arthur and his sister Dell. Seth was honored by the Thoroughbred Club of America as its testimonial dinner guest in 2000, thus receiving an honor conferred in earlier years on both his father and grandfather.

While the relationship with the Phippses has been a mainstay, any business such as Claiborne has the need to reach out to additional clients from era to era. William Haggin Perry, who had continued his close connection to the Hancocks, passed away in 1993. Five years later, in the Keeneland January sale, his widow and Claiborne jointly dispersed thirty-four horses for $21,205,000, averaging $623,676.

In recent years Claiborne has purchased, bred, and raced major winners in partnership with Adele Dilschnneider, who is a close friend of Seth's other sister, Clay. These winners include Roar, Mighty, and Arch.

Hancock has reached out to purchase in the yearling market, and Mighty and Arch illustrate his understanding that fresh blood is always important.

Champions Bred

Seth W. Hancock

Revidere
1976 champion three-year-old filly

Slew o' Gold
1983 champion three-year-old colt
1984 champion handicap male

Swale
1984 champion three-year-old colt

Forty Niner
1987 champion two-year-old colt

In Europe:

Fairy Bridge
1977 Irish champion two-year-old filly

Belted Earl
1982 Irish champion sprinter
1982 Irish champion older horse

Caerleon
1983 French champion three-year-old colt

Cherry Valley & Edward A. Cox Jr.

Bakharoff
1985 English champion two-year-old colt

In United Arab Emirates:

Cherry Valley & Edward A. Cox Jr.

Emperor Jones
1995 champion older horse

In North America:

Arthur Hancock & Leone J. Peters

Risen Star
1988 champion three-year-old colt

Another illustrative recent acquisition is Horse Chestnut, who was a champion in exotic South Africa and combines familiar with lesser-known bloodlines. His sire, Fort Wood, could hardly be more fashionably bred, being by Sadler's Wells and one of nine stakes winners from the remarkable mare Fall Aspen. Fort Wood won the Grand Prix de Paris. The dam of Horse Chestnut is London Wall, whose sire, Col Pickering, was by Wilwyn, winner of the first Washington, D.C., International but not a mainstay of American/European pedigrees. (Horse Chestnut's pedigree presents enticing name possibilities, Col Pickering being a son of the mare Julie Andrews, who was a daughter of Eliza Doolittle.)

The recent Claiborne stallion acquisition Monarchos, winner of the 2001 Kentucky Derby for John Oxley, is an outcross for Mr. Prospector blood. Typically, there is some Claiborne history to Monarchos' pedigree, for his sire, Maria's Mon, is a grandson of the longtime Claiborne/Phipps stallion Majestic Light.

Swale, Forty Niner, Lure, Caerelon, and Nureyev, are part of an aggregate of 168 stakes winners (worldwide) bred so far by Seth Hancock as Claiborne and with various partnerships. This figure includes a head start the young Hancock got when he and former Claiborne farm manager W.K. Taylor bred the stakes winners Decidedly D and Princess Doubleday in 1969 and 1970.

Brother Arthur

In the meantime Arthur Hancock III has forged a pleasing success of his own, so there is no sense of one son succeeding and the other not. Arthur is most often associated with Sunday Silence. Although he did not

ANNE M. EBERHARDT

Arthur Hancock III

terms of the Kentucky Derby when Gato Del Sol, whom he bred and raced in partnership with Leone J. Peters, won the roses in Arthur's colors in 1982. In 1989 Sunday Silence gave him a second Derby in his silks as the son of Halo upset Easy Goer, a magnificent champion Seth had raised for Ogden Phipps.

Arthur and Peters also bred and sold the 1988 Preakness-Belmont winner and three-year-old colt champion Risen Star in partnership, and Arthur bred and sold the 2000 Kentucky Derby winner Fusaichi Pegasus in partnership with Robert McNair's Stonerside Farm. Other major winners Arthur Hancock III has bred include the Epsom Derby runner-up Hawaiian Sound and the Futurity, Hopeful, and Flamingo Stakes winner Tap Shoes. Arthur B. Hancock III has bred a total of sixty-five stakes winners.

Two horses bred by Arthur, Strodes Creek and Menifee, finished second in the Derby, and Menifee's major triumphs in the colors of Hancock and James Stone included another revered Kentucky target, the Blue Grass Stakes.

So, three decades after the death of A.B. Hancock Jr., there can be only be one conclusion: "Bull's boys have done him proud."

breed this champion, he raised him and raced him in partnership. Arthur had gotten a jump on his family in

Stakes Winners Bred by Seth Hancock and Arthur Hancock III

Claiborne Farm (Seth Hancock)

An Act, dk.b/br.c. 1973, Pretense—Durga, by Tatan
Fabled Monarch, b.c. 1973, *Le Fabuleux—Monarch,
  by *Princequillo
Loop, dk.b/br.f. 1973, Round Table—Lea Moon, by *Nasrullah
Nantequos, b.c. 1973, Tom Rolfe—Moccasin, by Nantallah
Over to You, ch.c. 1973, Buckpasser—Rose Bower, by *Princequillo
Revidere, ch.f. 1973, Reviewer—Quillesian, by *Princequillo
Sir Vincent, ch.c. 1973, Sir Ivor—Vanita, by Masked Light
  (SW in Italy)
Sir Wimborne, b.c. 1973, Sir Ivor—Cap and Bells, by Tom Fool
  (SW in Ireland/England)
Brahms, b.c. 1974, Round Table—Moccasin, by Nantallah
  (SW in Ireland)
Din, ro.f. 1974, Drone—Durga, by Tatan
Flag Officer, b.c. 1974, Hoist the Flag—Batteur, by Bold Ruler
Kulak, b.g. 1974, *Herbager—Respected, by Round Table
Lady Capulet, ro.f. 1974, Sir Ivor—Cap and Bells, by Tom Fool
Tiller, ch.g. 1974, *Herbager—Chappaquiddick, by Relic
True Colors, b.c. 1974, Hoist the Flag—Rose Bower, by *Princequillo
Caption, b.f. 1975, Riva Ridge—Title, by Bold Ruler
Ethnarch, dk.b/br.c. 1975, Buckpasser—Reveille II, by Star Kingdom
Palmistry, dk.b/br.f. 1975, *Forli—Foreseer, by Round Table
Peat Moss, b.g. 1975, *Herbager—Moss, by Round Table
Bagfull, ch.c. 1976, Bagdad—Aesthetic, by Mr. Trouble
Coastal, ch.c. 1976, Majestic Prince—Alluvial, by Buckpasser
File, ch.f. 1976, Tom Rolfe—Continue, by Double Jay
Poppycock, b.g. 1976, Dewan—Shinnecock, by Tom Fool
Romeo Lima (stp.), b.g. 1976, Buckpasser—Respected, by Round Table
Ballare, b.f. 1977, Nijinsky II—Morgaise, by Round Table
Espadrille, b.f. 1977, Damascus—Thong, by Nantallah
Belted Earl, ch.c. 1978, Damascus—Moccasin, by Nantallah
  (SW in Ireland)
Edge, ch.f. 1978, Damascus—Ponte Vecchio, by Round Table
Midnight Mine, b.c. 1978, Riva Ridge—Desert Love, by *Amerigo
Park's Policy, b.c. 1978, Tell—Aware, by Buckpasser
Shimmy, b.f. 1978, Nijinsky II—Amalesian, by *Ambiorix
Syria, dk.b/br.f. 1978, Damascus—Santiago Lassie, by Vertex
Doodle, ch.f. 1979, Dewan—Actual, by Round Table
Lords, b.c. 1979, Hoist the Flag—Princessnesian, by *Princequillo
  (SW in Ireland)
Lucence, ch.c. 1979, Majestic Light—Veroushka, by Nijinsky II
Maniches, b.f. 1979, Val de l'Orne (Fr)—Shahadish, by Northern
  Dancer
Number, b.f. 1979, Nijinsky II—Special, by *Forli
Scuff, ch.f. 1979, *Forli—Moccasin, by Nantallah
Bon Gout, b.f. 1980, Dewan—Tastefully, by Hail to Reason
Caerleon, b.c. 1980, Nijinsky II—Foreseer, by Round Table (SW in
  England/Ireland/France)
Descent, ch.f. 1980, Avatar—Alyne Que, by Raise a Native
Narrate, dk.b/br.f. 1980, Honest Pleasure—State, by Nijinsky II
Proof, ch.c. 1980, Believe It—Face the Facts, by *Court Martial
Routine, ch.f. 1980, Northern Dancer—Chappaquiddick, by Relic
Slew o' Gold, b.c. 1980, Seattle Slew—Alluvial, by Buckpasser
Top Competitor, b.c. 1980, Sir Ivor—Shinnecock, by Tom Fool
Allusion, b.f. 1981, Mr. Prospector—Touch, by *Herbager
Flippers, ch.f. 1981, Coastal—Moccasin, by Nantallah
Swale, dk.b/br.c. 1981, Seattle Slew—Tuerta, by *Forli

Vision, b.c. 1981, Nijinsky II—Foreseer, by Round Table
Encolure, b.c. 1982, Riva Ridge—Jabot, by Bold Ruler
Double Feint, gr.c. 1983, Spectacular Bid—State, by Nijinsky II
Savings, dk.b/br.c. 1983, Buckfinder—Reveille (Aust), by Star
  Kingdom
Swear, ch.c. 1983, Believe It—Suave, by Majestic Prince
Claim, b.c. 1985, Mr. Prospector—Santiago Lassie, by Vertex
Forty Niner, ch.c. 1985, Mr. Prospector—File, by Tom Rolfe
Dibs, ro.f. 1986, Spectacular Bid—State, by Nijinsky II
Gild, ch.f. 1986, Mr. Prospector—Veroushka, by Nijinsky II
Asia, dk.b/br.g. 1987, Danzig—Syria, by Damascus
Search, dk.b/br.g. 1987, Mr. Prospector—Loan, by Buckfinder
Discover, b.c. 1988, Cox's Ridge—Find, by Mr. Prospector
Bag, ch.c. 1989, Devil's Bag—Allusion, by Mr. Prospector
Hitch, b.f. 1989, Cox's Ridge—Knot, by Majestic Light
Preach, b.f. 1989, Mr. Prospector—Narrate, by Honest Pleasure
Region, ch.g. 1989, Devil's Bag—State, by Nijinsky II
Anchor, b.c. 1990, Polish Navy—Syria, by Damascus
Swank, ch.c. 1991, Topsider—Suave, by Majestic Prince
Thread, b.f. 1991, Topsider—Knot, by Majestic Light
Announce, ch.c. 1992, Forty Niner—State, by Nijinsky II
Haint, dk.b/br.c. 1994, Devil's Bag—Realm, by Mr. Prospector
Pulpit, b.c. 1994, A.P. Indy—Preach, by Mr. Prospector
Basic Trainee, b.g. 1995, Majestic Light—Acquire, by Spectacular Bid
Conserve, b.g. 1996, Boundary—Slew and Easy, by Slew o' Gold
Trip, ch.f. 1997, Lord At War (Arg)—Tour, by Forty Niner
Watch, ch.f. 1997, Lord At War (Arg)—Watch the Time, by Timeless
  Moment
Joke, b.f. 1998, Phone Trick—Tour, by Forty Niner
Poker Brad, b.g. 1998, Go for Gin—Gild, by Mr. Prospector
Forty Nine Deeds, b.c. 1999, Alydeed—Abrade, by Mr. Prospector
Patrol, b.c. 1999, Lear Fan—Maid for Walking (GB), by Prince Sabo
Remind, b.c. 2000, Deputy Minister—Watch the Time, by Timeless
  Moment
Stroll, dk.b/br.c. 2000, Pulpit—Maid for Walking (GB), by Prince Sabo
Urban Warrior, dk.b/br.c. 2001, Cape Town—Cluster, by Danzig

Claiborne Farm and The Gamely Corp. (William Haggin Perry)

Endear, b.f. 1982, Alydar—Chappaquiddick, by Relic
Herat, b.c. 1982, Northern Dancer—Kashan, by Damascus
Glow, dk.b/br.c. 1983, Northern Dancer—Glisk, by Buckpasser
Tile, ch.c. 1983, Honest Pleasure—Kashan, by Damascus
Wisla, b.f. 1983, Danzig—Gauri, by Sir Ivor
Bound, b.f. 1984, Nijinsky II—Special, by *Forli
Goa, ch.c. 1984, Coastal—Durga, by Tatan
Merce Cunningham, b.c. 1984, Nijinsky II—Foreseer, by Round Table
  (SW in England/France)
Asian, ch.f. 1985, Damascus—Bamesian, by Buckpasser
Digress, b.c. 1985, Topsider—Chappaquiddick, by Relic
Festive, b.c. 1985, Damascus—Swingtime, by Buckpasser
Lustra, b.c. 1985, Danzig—Glisk, by Buckpasser
Order, b.c. 1985, Damascus—Resume, by Reviewer
Bio, b.c. 1986, Linkage—Resume, by Reviewer
Gold Seam, b.c. 1986, Mr. Prospector—Ballare, by Nijinsky II
Hail Atlantis, b.f. 1987, Seattle Slew—Flippers, by Coastal
Jade Robbery, dk.b/br.c. 1987, Mr. Prospector—Number, by Nijinsky
  II (SW in France)
Slavic, b.c. 1987, Danzig—Bamesian, by Buckpasser

Yonder, dk.b/br.c. 1987, Seattle Slew—Far, by *Forli
Scan, b.c. 1988, Mr. Prospector—Video, by Nijinsky II
Lure, b.c. 1989, Danzig—Endear, by Alydar
Boundary, b.c. 1990, Danzig—Edge, by Damascus
Kashani, b.c. 1990, Danzig—Kashan, by Damascus (SW in France)
Grab, b.f. 1991, Danzig—Snitch, by Seattle Slew
Inflate, b.c. 1991, Forty Niner—Add, by Spectacular Bid
Numerous, b.c. 1991, Mr. Prospector—Number, by Nijinsky II
Walk Point, gr.g. 1991, Proper Reality—Steps, by Spectacular Bid
Ago, dk.b/br.c. 1992, Danzig—Far, by *Forli
Chequer, dk.b/br.g. 1992, Mr. Prospector—Number, by Nijinsky II
Limit, b.f. 1992, Cox's Ridge—Bound, by Nijinsky II
Spire, b.f. 1992, Topsider—Eaves, by Cox's Ridge
Applaud, ch.f. 1993, Rahy—Band, by Northern Dancer (SW in England)
Draw, b.c. 1993, Private Account—Wisla, by Danzig
Proper Blue, b.g. 1993, Proper Reality—Blinking, by Tom Rolfe (SW in England)

Claiborne Farm in various other partnerships
Assatis, b.c. 1985, Topsider—Secret Asset, by Graustark (SW in England/Italy)
Smackover Creek, b.c. 1985, Mr. Prospector—Grand Luxe, by Sir Ivor
White Mischief, ch.c. 1985, Roberto—Arachne, by Intentionally
Warrshan, b.c. 1986, Northern Dancer—Secret Asset, by Graustark (SW in England)
Razeen, b.c. 1987, Northern Dancer—Secret Asset, by Graustark (SW in England)
Bull Inthe Heather, ro.c. 1990, Ferdinand—Heather Road, by The Axe II
Nine Keys, b.f. 1990, Forty Niner—Clef d'Argent, by Key to the Mint
Tour, ch.f. 1990, Forty Niner—Fun Flight, by Full Pocket
Flight Forty Nine, ch.c. 1991, Forty Niner—Fun Flight, by Full Pocket
Roar, b.c. 1993, Forty Niner—Wild Applause, by Northern Dancer
Capilano, ch.f. 1994, Demons Begone—Bella Isabella, by Conquistador Cielo
District, b.c. 1997, Broad Brush—Eaves, by Cox's Ridge
Congrats, b.c. 2000, A.P. Indy—Praise, by Mr. Prospector
Yell, dk.b/br.f. 2000, A.P. Indy—Wild Applause, by Northern Dancer

Seth Hancock, solely and in partnership
Decidedly D, b.g. 1969, Decidedly—Little Sequoia, by Double Jay
Princess Doubleday, b.f. 1970, Hitting Away—Little Sequoia, by Double Jay
Silent Dignity, gr.c. 1976, Damascus—Croquet, by *Court Martial
Night Alert, b.c. 1977, Nijinsky II—Moment of Truth II, by Matador (SW in England/France/Ireland)
On Ack, b.c. 1977, Ack Ack—Spark, by Blade

Cherry Valley Farm (Seth Hancock)
I'm So Bad, b.c. 1984, Danzig—Betcha, by Riva Ridge
Milligan, b.g. 1993, Majestic Light—Millie Do, by Court Trial
Move, ch.f. 1994, Forty Niner—Tremor, by Tromos (GB)
Oath, dk.b/br.f. 1994, Known Fact—Vue, by Mr. Prospector
Buff, b.g. 1995, Pleasant Tap—Pedicure, by Mr. Prospector
Runawayfun, gr/ro.f. 1998, Runaway Groom—Rosedale, by Lost Code
Pomeroy, b.c. 2001, Boundary—Questress, by Seeking the Gold
Second Performance, dk.b/br. c. 2001, Theatrical (Ire)—Pedicure, by Mr. Prospector

Cherry Valley Farm and The Gamely Corp.
Trendy Gent, ch.c. 1981, Nijinsky II—Pucheca, by Tom Rolfe (SW in Norway)
Shadeed, b.c. 1982, Nijinsky II—Continual, by Damascus (SW in England)
Country Light, b.c. 1983, Majestic Light—Harbor Flag, by Hoist the Flag
Footy, b.f. 1984, Topsider—Obstetrician, by Dr. Fager
Buckbean, dk.b/br.c. 1985, Buckfinder—Pucheca, by Tom Rolfe
Deviled, b.f. 1987, Devil's Bag—Tableaux, by Round Table
Dotsero, b.c. 1987, Conquistador Cielo—Honest Joy, by Honest Pleasure
Rail, ch.c. 1989, Majestic Light—Berth, by Believe It
Scuffleburg, dk.b/br.c. 1989, Cox's Ridge—Tableaux, by Round Table
Footing, dk.b/br.f. 1991, Forty Niner—Footy, by Topsider
Packet, dk.b/br.f. 1991, Polish Navy—Harbor Flag, by Hoist the Flag

Cherry Valley Farm and Edward A. Cox Jr.
Bakharoff, b.c. 1983, The Minstrel—Qui Royalty, by Native Royalty (SW in England)
Sum, gr.f. 1984, Spectacular Bid—Qui Royalty, by Native Royalty
Broto, b.c. 1986, Danzig—Bosk, by Damascus
Demonry, dk.b/br.f. 1986, Devil's Bag—Qui Royalty, by Native Royalty
Majlood, dk.b/br.c. 1988, Danzig—Qui Royalty, by Native Royalty (SW in England)
Thyer, b.c. 1989, Nijinsky II—Qui Royalty, by Native Royalty (SW in Ireland)
Emperor Jones, dk.b/br.c. 1990, Danzig—Qui Royalty, by Native Royalty (SW in England)
Appointed One, b.f. 1992, Danzig—Qui Royalty, by Native Royalty

Cherry Valley Farm in various other partnerships
Honest Joy, ch.f. 1979, Honest Pleasure—Dimashq, by Damascus
Honest Glow, b.f. 1980, Honest Pleasure—Dimashq, by Damascus
Aneka, dk.b/br.f. 1981, Believe It—Kan Jive, by Mister Jive
Sintra, ro.f. 1981, Drone—Misty Plum, by Misty Day
Tide, b.f. 1982, Coastal—Go On With It, by Dewan
Aces, dk.b/br.f. 1994, Housebuster—Sweetbreads, by Secretariat

Arthur B. Hancock III
Dapper, b.c. 1970, *Forli—Punctilious, by Better Self (SW in Ireland)
Lullaby, b.f. 1974, Hawaii (SAf)—Sound of Success, by Successor
Hawaiian Sound, b.c. 1975, Hawaii (SAf)—Sound of Success, by Successor
Kitchen, b.f. 1979, Master Hand—Butter and Eggs, by Boston Doge
Tip Tap, ch.c. 1984, Tap Shoes—Border Try, by Saidam
Back to the Past, ro.f. 1986, Tilt Up—Back to Reality, by Native Charger
Festin, ch.c. 1986, Mat-Boy (Arg)—Felicidades, by Con Brio II (Bred in Argentina)
Galleria, b.f. 1988, Plugged Nickle—Plenteous, by Bold Forbes
Black Beauty D., b.f. 1989, Procida—Atacama, by Damascus (SW in Puerto Rico)
Logan's Mist, b.f. 1989, Brogan—Misty Gleam, by Gleaming
Point Spread, b.f. 1989, Capote—Mysterious Star, by Nijinsky II
Shezagorgousdancer, b.f. 1991, Brogan—Gorgeous Dee, by Dewan
Sonny's Bruno, b.g. 1991, Queen City Lad—Settlers Cabin, by Cabin
Strodes Creek, b.c. 1991, Halo—Bottle Top, by Topsider
Tuck's Honey Bear, b.c. 1991, Critique—My Honey Bun, by Queen City Lad

Boggle, b.g. 1992, Lively One—Dazzlin Florence, by Affirmed

She's a Lively One, dk.b/br.f. 1992, Lively One—Hope She's Bold, by Bold Forbes

Steprock, ro.g. 1992, Lively One—Bold Satin, by Bold Forbes

Bombay, b.g. 1993, Procida—Enfante, by Northern Baby

Dr. Ramsey (stp.), dk.b/br.g. 1994, Northern Baby—Insipid, by Sham

Swearingen, b.f. 1994, Deposit Ticket—Firey Affair, by Explodent

Wed by Proxy, dk.b/br.f. 1994, Procida—New Bride (Chi), by Nobloys (Fr)

Accomodator, b.g. 1995, His Majesty—Accommodate, by Honest Pleasure

Addinson (stp.), b.g. 1996, Northern Baby—Insipid, by Sham

Alannan, b.c. 1996, Conquistador Cielo—Dame Sybil, by Elocutionist

Karly's Harley, ch.g. 1996, Harlan—Impish, by Majestic Prince

Menifee, b.c. 1996, Harlan—Anne Campbell, by Never Bend

Big Hubie, ch.c. 1998, Boone's Mill—Run and Run (Ire), by Royal Academy

Blue Sky Baby, b.f. 1998, Skywalker—Protostar, by Procida

Lasting Light, dk.b/br.f. 1998, Skywalker—Wraith, by Herculean

Owsley, dk.b/br.f. 1998, Harlan—Insipid, by Sham

Chamrousse, dk.b/br.f. 1999, Peaks and Valleys—Loose Park, by Stop the Music

Dublino, dk.b/br.f. 1999, Lear Fan—Tuscoga, by Theatrical (Ire)

Electrode, ch.g. 1999, Devil's Bag—Maria Fumata (Chi), by Semenenko

Arthur Hancock III and Leone J. Peters

All Rain, b.c. 1978, One for All—Rain Dance, by Native Charger

December Sky, ro.f. 1978, Drone—Hollyhock, by Olden Times

Tap Shoes, ch.c. 1978, Riva Ridge—Bold Ballet, by Bold Bidder

Gato Del Sol, gr.c. 1979, Cougar II—Peacefully, by Jacinto

Punctilio, b.c. 1979, *Forli—Minstrelete, by Round Table

Valiant Cougar, b.g. 1983, Naskra—Valiente, by Cougar II

Risen Star, dk.b/br.c. 1985, Secretariat—Ribbon, by His Majesty

Arthur Hancock III and Roy Bowen

Bold Frond, b.g. 1979, Bold Forbes—Frond, by Hawaii (SAf)

Riverjoy, dk.b/br.c. 1981, Riverman—Joy Land, by Bold Ruler

Silvered, dk.b/br.f. 1987, Halo—Silvered Silk, by Fire Dancer

Arthur Hancock III, Roy Bowen, and Leone J. Peters

You're No Bargain, b.c. 1984, Riva Ridge—Suzest, by Fleet Nasrullah

American Patriot, b.c. 1985, Ack Ack—Suzest, by Fleet Nasrullah

Arthur Hancock III and Stonerside Ltd.

Truluck, b.c. 1995, Conquistador Cielo—Michelle Mon Amour, by Best Turn

Fusaichi Pegasus, b.c. 1997, Mr. Prospector—Angel Fever, by Danzig

No Matter What, ch.f. 1997, Nureyev—Words of War, by Lord At War (Arg) (SW in France)

E Dubai, b.c. 1998, Mr. Prospector—Words of War, by Lord At War (Arg)

Miss Jeanne Cat, b.f. 1999, Tabasco Cat—Few Choice Words, by Valid Appeal

Arthur Hancock III and William S. Farish

Lil Saucy, b.c. 1984, Sauce Boat—Table Vice, by Round Table

Slewfoot Seven, b.g. 1988, Lines of Power—Table Vice, by Round Table

Arthur Hancock III and Mrs. Mary Van Wyck

Welsh, b.f. 1976, Drone—Owe Everything, by Nashua

Cagey Cougar, b.c. 1979, Cougar II—Owe Everything, by Nashua

Arthur Hancock III in various other partnerships

Native Drone, dk.b/br.c. 1972, Drone—Miss Quickstep, by Native Dancer

Scrimshaw, dk.b/br.c. 1983, Apalachee—Crimson Lace, by First Landing

Lively Leah, b.f. 1988, Tilt Up—Roman Prospect, by Proudest Roman

Harlan, dk.b/br.c. 1989, Storm Cat—Country Romance, by Halo

Joyful Vigor, b.f. 1989, Vigors—Jillys Joy, by Bagdad

Ghost Power, dk.b/br.c. 1990, Wolf Power (SAf)—Wraith, by Herculean

Flora's Halo, dk.b/br.f. 1992, Banner Bob—Dustin's Halo, by Halo

Labamta Babe, dk.b/br.g. 1999, Skywalker—Bambina Linda (Arg), by Liloy (Fr)

Quest, ch.c. 1999, Seeking the Gold—Starlore, by Spectacular Bid

Knox, ch. c. 2001, Menifee—Count to Six, by Saratoga Six

Sources: *The Blood-Horse* Archives and The Jockey Club

Baby League

(b.f. 1935)

(Bubbling Over—*La Troienne, by *Teddy)

Property of Ogden Phipps after syndicate distribution of E.R. Bradley's Idle Hour Stock Farm estate in 1946: syndicate composed of Phipps, Robert Kleberg Jr. of King Ranch, and John Hay Whitney of Greentree Stud. At the time, the stakes-placed Baby League was in foal to War Admiral and produced her first foal for Phipps in 1947.

Bomb Dolly (b.f. 1941, Omaha) — bred by E.R. Bradley's Idle Hour Stock Farm
  CAROLOS (*Russia II) — bred by Mrs. D.P. Barrett
  Desert Wheat (Salmagundi) — bred by Mrs. D.P. Barrett
    High Wheat (High Finance) — bred by Bayne C. Welker
      BIG BAD TUNE (Riding Tune) — bred by Ralph W. Shebester
†**BUSHER** (ch.f. 1942, War Admiral) — bred by Idle Hour Stock Farm
  Miss Busher (*Alibhai) — bred by Maine Chance Farm
    Speed Bird II (Jet Pilot) — bred by Maine Chance Farm
      *AGOGO II (Never Say Die) — bred by Maine Chance Farm
      GEMINI SIX [GB] (Princely Gift) — bred by John H. Clark
      Ptarmigan (Nearctic) — bred by G.M. Bell
        SHERRIGAN (Sherry Prince) — bred by R.J. Bennett
      PILOT BIRD (Four-and-Twenty) — bred by Golden West Farms Ltd.
      PILOTSON (Nantwice) — bred by A & W Farms
    Bushfield (Jet Pilot) — bred by Maine Chance Farm
      Heamaw (Chieftain) — bred by Raymond R. Guest
        Runnun Tell (Tell) — bred by W.K. Taylor
          COUNTRY BORN (Native Born) — bred by L & M Farms
          Loshadka (Valdez) — bred by Jim H. Plemmons
            PRINCESS LAO (Laomedonte) — bred by Karolyn Goodman
        Tri Maw (Tri Jet) — bred by John S. Ferris & Dr. Jacques Ford
          SIGNIFICANT TRI (Irish Sur) — bred by Six Winters Farm
      Lorna Doone (Tom Rolfe) — bred by Raymond R. Guest
        HUARALINO (Habitat) — bred by unavailable
      Beaufield (Maribeau) — bred by Raymond R. Guest
        BEAU'S EAGLE (*Golden Eagle II) — bred by **Mr. & Mrs. John C. Mabee**
        Promising Girl (Youth) — bred by **Mr. & Mrs. John C. Mabee**
          MAN FROM ELDORADO (Mr. Prospector) — bred by **Mr. & Mrs. John C. Mabee**
          FAVORITE FUNTIME (Seeking the Gold) — bred by **Mr. & Mrs. John C. Mabee**
        Glamorous Beau (Assert [Ire]) — bred by **Mr. & Mrs. John C. Mabee**
          BEAU JINGLES (Riverman) — bred by **Mr. & Mrs. John C. Mabee**
      Little Tobago (Impressive) — bred by Raymond R. Guest
        PLAY ON (Stop the Music) — bred by Hillbrook Farm
        Why Did I (Foolish Pleasure) — bred by Hillbrook Farm
          JUST CUZ (Cormorant) — bred by Empire I Partnership
            COME ON GET HAPPY (Cure the Blues) — bred by Pamela S. Darmstadt-duPont
          I'll Redo It (Relaunch) — bred by Empire I
            BALLDO (Saint Ballado) — bred by Sabine Stable
  Popularity (*Alibhai) — bred by Maine Chance Farm
    BEVY OF ROSES (*Bernborough) — bred by L.B. Mayer
    TOP CHARGER (*Royal Charger) — bred by L.B. Mayer
    RED TULIP (Jet Pilot) — bred by Maine Chance Farm

BABY LEAGUE (Continued)

Native Lady (Raise a Native) — bred by **Leslie Combs II**
Our Great Love (Fleet Nasrullah) — bred by **Leslie Combs II**
†**COLOSO [MEX]** (Imperial Ballet) — bred by unavailable
GORILERO (Kandinsky) — bred by unavailable
Flower Lady (Fleet Nasrullah) — bred by **Leslie Combs II**
STRAIGHT FLOW (Going Straight) — bred by Donamire Farm
Cottoneaster (Summer Advocate) — bred by Donamire Farm
GLORY US (Northern Score) — bred by Triple C Farms
POSITION RYDER (Pole Position) — bred by Triple C Farms & Lang McLean
Shelephant (Bold Lad) — bred by **Leslie Combs II**
Miss Preppy (Rollicking) — bred by Helen M. Polinger & Milton Awalt
PREPORANT (Cormorant) — bred by Milton Ritzenberg
Betty's Bet (Bold Hour) — bred by **Leslie Combs II**
BRONZE COURT (*Tobin Bronze) — bred by Margaret O. duPont & Margaret V. Bloss
COME ON SASSA (Sassafras [Fr]) — bred by Pillar Stud
Popularity Plus (Ronsard) — bred by Barnett Serio
TWAS EVER THUS (Olden Times) — bred by Philip J. Torsney
JET ACTION (Jet Pilot) — bred by Maine Chance Farm
Bush Pilot (Jet Pilot) — bred by Maine Chance Farm
Whileaway (Summer Tan) — bred by Stevenwood Farm
Lady of Night (What a Pleasure) — bred by Stevenwood Farm
FROSTY AFFAIR (Icecapade) — bred by Tiger Lily Stables
ROSEMONT RISK (Exclusive Native) — bred by Mr. & Mrs. Albert Drohlich
ISLAND BANKING (Private Account) — bred by Paris Creek, Inc.
SUPER PLEASURE (What a Pleasure) — bred by Stevenwood Farm
FIVE STAR GENERAL (Lt. Stevens) — bred by Stevenwood Farm
Five Star's Sister (Lt. Stevens) — bred by Tom Chaffee
Five Gold Stars (Mining) — bred by Michael Anchel
KEEPING THE GOLD (Langfuhr) — bred by Nikolaus Bock
NEEDLE AND BALL (Tim Tam) — bred by Stevenwood Farm
STRIKING (b.f. 1947, War Admiral) — bred by **Ogden Phipps**
GLAMOUR (*Nasrullah) — bred by Wheatley Stable & **Ogden Phipps**
Brilliantly (Hill Prince) — bred by Wheatley Stable
Brilliant View (Summer Tan) — bred by Lewis C. Ledyard
September Dream (American Native) — bred by **Harbor View Farm**
SEPTEMBER TEN [VEN] (White Face) — bred by unavailable
Bright Machete (The Axe II) — bred by Lewis C. Ledyard
GET SWINGING (Get Around) — bred by T.M. Evans
CURRENT BLADE (Little Current) — bred by T.M. Evans
Heatherglow (The Axe II) — bred by Lewis C. Ledyard
Canonization (Native Heritage) — bred by **Harbor View Farm**
LADY SHIRL (That's a Nice) — bred by Irish Acres Farm
Glamorous (Exclusive Native) — bred by **Harbor View Farm**
LITTLE RAISIN (Little Current) — bred by Collins Partners #2
An Affinity (Raise a Native) — bred by **Harbor View Farm**
FORUM CLUB (Shelter Half) — bred by Burt Bacharach
ROYAL ASCOT (*Princequillo) — bred by **Ogden Phipps**
POKER (Round Table) — bred by **Ogden Phipps**
Intriguing (Swaps) — bred by **Ogden Phipps**
†**NUMBERED ACCOUNT** (Buckpasser) — bred by **Ogden Phipps**
PRIVATE ACCOUNT (Damascus) — bred by **Ogden Phipps**
Secret Asset (Graustark) — bred by **Ogden Phipps**
ASSATIS (Topsider) — bred by **Claiborne Farm** & Warner L. Jones Jr.
WARRSHAN (Northern Dancer) — bred by **Claiborne Farm** & Warner L. Jones Jr.'s Hermitage Farm
RAZEEN (Northern Dancer) — bred by **Claiborne Farm** & Hermitage Farm
Frankova (Nureyev) — bred by Darley Stud Management
FIRST FLEET (Woodman) — bred by Darley Stud Management
DANCE NUMBER (Northern Dancer) — bred by **Ogden Phipps**
Oscillate (Seattle Slew) — bred by **Ogden Mills Phipps**
MUTAKDDIM (Seeking the Gold) — bred by Deer Lawn Farm, Carloss, & Lamont
†**RHYTHM** (Mr. Prospector) — bred by **Ogden Mills Phipps**
GET LUCKY (Mr. Prospector) — bred by **Ogden Mills Phipps**
ACCELERATOR (A.P. Indy) — bred by **Ogden Mills Phipps**
Confidentiality (Lyphard) — bred by **Ogden Phipps**
CONFIDENTIAL TALK (Damascus) — bred by **Ogden Phipps**
Fascinating Trick (Buckpasser) — bred by **Ogden Phipps**
Northern Naiad [Fr] (Nureyev) — bred by Stavros Niarchos
†**GREY WAY** (Cozzene) — bred by Carlos Vittadini
DISTANT WAY (Distant View) — bred by Grundy Bloodstock
Sept a Neuf (Be My Guest) — bred by Stavros Niarchos

BABY LEAGUE (Continued)

      VICINALE [IRE] (Namaqualand) — bred by Allevamento Porto Medaglia

    Fast 'n Tricky (Lomond) — bred by International Thoroughbred Breeders Inc.

      SHISEIDO [IND] (Classic Tale [GB]) — bred by unavailable

    Political Intrigue (Deputy Minister) — bred by Kinghaven Farms Ltd.

      REDATTORE [BRZ] (Roi Normand) — bred by Haras Santa Ana do Rio Grande

  Special Account (Buckpasser) — bred by **Ogden Phipps**

    Bank On Love (Gallant Romeo) — bred by **Ogden Phipps**

      WELDNAAS (Diesis [GB]) — bred by W Lazy T Limited.

      Atyaaf (Irish River [Fr]) — bred by W Lazy T Limited

        RAISE A GRAND [IRE] (Grand Lodge) — bred by Churchtown House Stud

    GALLANT SPECIAL (Gallant Romeo) — bred by W.R. Hawn

    GALLANT SISTER (Vigors) — bred by W.R. Hawn

    Lyphard Gal (Lyphard) — bred by W.R. Hawn

      HERITAGE OF GOLD (Gold Legend) — bred by Georgia E. Hofmann

      GOLDEN OLDIE (Gold Legend) — bred by Double D Farm Corp.

    Awesome Account (Lyphard) — bred by Due Process Stable

      KASHGAR (Secretariat) — bred by JJG Partners & Due Process Stable

      ANGUILLA (Seattle Slew) — bred by Hill 'n' Dale Farms & Due Process Stable

    Kemp (Spectacular Bid) — bred by Due Process Stable

      HEAR THE BELLS (Deputy Minister) — bred by Due Process Stable

        WILD DEPUTY (Wild Again) — bred by Due Process Stable

        DEB'S HONOR (Affirmed) — bred by Due Process Stable

      JANE WHALEY (Deputy Minister) — bred by Jane M. Schosberg

      PERSONAL BID (Personal Flag) — bred by Jane M. Schosberg

    Tara's Number (Northern Dancer) — bred by Thomas P. Tatham

      MARY MCGLINCHY (Pleasant Colony) — bred by Dinwiddie Farms Limited Partnership

    Time Deposit (Halo) — bred by Oak Cliff Thoroughbred Bloodstock Ltd.

      TRESORIERE (Lyphard) — bred by Alec Head

  CUNNING TRICK (Buckpasser) — bred by **Ogden Phipps**

  HOW CURIOUS (Buckpasser) — bred by **Ogden Phipps**

  Playmate (Buckpasser) — bred by **Ogden Phipps**

    SINGLE THREAD (Damascus) — bred by Warner L. Jones Jr. & E.A. Cox Jr.

    WOODMAN (Mr. Prospector) — bred by Warner L. Jones Jr. & E.A. Cox Jr.

  The Cuddler (Buckpasser) — bred by **Ogden Phipps**

    VERIFICATION (Exceller) — bred by Belair Farm Ltd.

JAUNTY (*Ambiorix) — bred by **Ogden Phipps**

Artistically (*Ribot) — bred by **Ogden Phipps**

  ABSOLUTE [FR] (Luthier) — bred by Seven Hills Corporation

  ALPHABEL (Bellypha) — bred by Petra Bloodstock Agency Ltd.

  Abordable (Formidable) — bred by Petra Bloodstock Agency Ltd.

    ABOARD [FR] (Hero's Honor) — bred by Petra Bloodstock Agency Ltd.

  †**ABSURDE [FR]** (Green Desert) — bred by Petra Bloodstock Agency Ltd.

  Abstraite [GB] (Groom Dancer) — bred by Petra Bloodstock Agency Ltd.

    NAS NA RIOGH [IRE] (King's Theatre [Ire]) — bred by Eric Simian

BOUCHER (*Ribot) — bred by **Ogden Phipps**

Glorifying (*Ribot) — bred by **Ogden Phipps**

  Fire and Ice [Fr] (Reliance II) — bred by Alan Clore

    Nice and Icy [Fr] (Nice Havrais) — bred by Alan Clore

      ADAM [TUR] (Komando [Tur]) — bred by unavailable

    Be My Fire (Be My Guest) — bred by Alan Clore

      BE MY NAKAYAMA [JPN] (Ebros) — bred by Katsuyuki Nagata

Instant Beauty (*Pronto) — bred by **Ogden Phipps**

  Soy Bonita (Soy Numero Uno) — bred by Glencrest Farm

    Incomprensiva (Triomphe) — bred by Rog Acres Farm

      AZUCAR (Full Realization) — bred by unavailable

      CANA DULCE [DR] (Full Realization) — bred by unavailable

  Fun 'n Fame (Assert [Ire]) — bred by International Thoroughbred Breeders Inc.

    GOODWOOD [AUS] (Quest for Fame) — bred by T.J. Smith

Bonnie Blink (Buckpasser) — bred by **Ogden Phipps**

  Quick Glance (Time for a Change) — bred by **Ogden Phipps**

    SA TORRETA [ARG] (Southern Halo) — bred by Don Arcangel

So Chic (*Nasrullah) — bred by Wheatley Stable & **Ogden Phipps**

  Memsie (*Tulyar) — bred by **Ogden Phipps**

    Sierra Nevada (Warfare) — bred by James Cox Brady

      BARLOVENTO (Natidan) — bred by Potrero del Sur, Inc.

      Acrobata (Catullus) — bred by Potrero del Sur, Inc.

        †**EL PLATINO** (Gordie H.) — bred by Eduardo D. Maldonado

    GERABHER (*Herbager) — bred by James Cox Brady

    House Maid (Habitat) — bred by Airlie Stud

      MONA LISA (Henbit) — bred by Mrs. S.M. Rogers

BABY LEAGUE (Continued)

    FASHION VERDICT (*Court Martial) — bred by **Ogden Phipps**
    Court Circular (*Ambiorix) — bred by **Ogden Phipps**
      HIGH COURT [FR] (Hill Rise) — bred by Mrs. Pierre Wertheimer
      Spy Court [Fr] (Snob [Fr]) — bred by unavailable
      SOVEREIGN COURT (Sovereign Red [NZ]) — bred by unavailable
    ASH CREEK (Manado [Ire]) — bred by unavailable
    The Garden Club (*Herbager) — bred by **Ogden Phipps**; foundation mare for W.S. Farish
    Exotic Garden (Bold Ruler) — bred by Mrs. Peggy C. Neloy
      Mareve (*Hawaii) — bred by **W.S. Farish**
        Evolutionary (Silent Screen) — bred by Happy Valley Farm
        WESTERN LIL (Western Trick) — bred by Ridgemont Farms, Inc.
      LINES OF POWER (Raise a Native) — bred by **W.S. Farish**
      Visual Effects (Silent Screen) — bred by **W.S. Farish**
        Paid Assassin (To the Quick) — bred by **W.S. Farish**
        Quick Kill (Regal Search) — bred by Janet Erwin
        SCHMOOPY (Matchlite) — bred by Tracks-n-Time & Stonewood Farm
        SECRET SERVICE MAN (Shot Gun Scott) — bred by Janet Erwin
      Dunbarten Oaks (Raise a Native) — bred by **W.S. Farish**
      †**TEMPERENCE OAKS** (Temperence Hill) — bred by Tall Oaks Farm
    Dinner Meeting (Bold Ruler) — bred by **W.S. Farish**
      Dinner Music (Raise a Cup) — bred by **W.S. Farish**
      EAGLE CROWN (Beau's Eagle) — bred by Relatively Stable
      ANNIVERSARY WISH (Beau's Eagle) — bred by Relatively Stable
      WISHES AND ROSES (Greinton [GB]) — bred by Live Oak Stud
      APPROVANCE (With Approval) — bred by Live Oak Stud
    NOSTALGIA (Silent Screen) — bred by **W.S. Farish**
    Tacida (Vitriolic) — bred by Roy L. Bowen
      La Passionaria (Reform [GB]) — bred by unavailable
      KENARIA (Kenmare [Fr]) — bred by Baron Thierry Van Zuylen
      LA CARENE [FR] (Kenmare [Fr]) — bred by Mr. & Mrs. Philippe Chandioux
      LA CHAMANE [NZ] (Desert Sun [GB]) — bred by unavailable
    UP THE FLAGPOLE (Hoist the Flag) — bred by **W.S. Farish**
    FOLD THE FLAG (Raja Baba) — bred by **W.S. Farish**
    ALLIED FLAG (Danzig) — bred by **W.S. Farish**
    LONG VIEW (Damascus) — bred by **W.S. Farish**
      OVERVIEW (Kingmambo) — bred by **W.S. Farish**
      MUSIC CLUB (Dixieland Band) — bred by **W.S. Farish**
    PROSPECTORS DELITE (Mr. Prospector) — bred by **W.S. Farish**
    TOMISUE'S DELIGHT (A.P. Indy) — bred by **W.S. Farish**, James Elkins, & W.T. Webber Jr.
    DELTA MUSIC (Dixieland Band) — bred by **W.S. Farish**, James Elkins, & W.T. Webber Jr.
    MONASHEE MOUNTAIN (Danzig) — bred by **W.S. Farish**, James Elkins, & W.T. Webber Jr.
    ROCK SLIDE (A.P. Indy) — bred by **W.S. Farish**, James Elkins, & W.T. Webber Jr.
    †**MINESHAFT** (A.P. Indy) — bred by **W.S. Farish**, James Elkins, & W.T. Webber Jr.
    †**FLAGBIRD** (Nureyev) — bred by **W.S. Farish**
    TOP ACCOUNT (Private Account) — bred by **W.S. Farish**
    RUNUP THE COLORS (A.P. Indy) — bred by **W.S. Farish**
    BLUSHING CATHY (Blushing Groom [Fr]) — bred by **W.S. Farish**
    Hidden Garden (Mr. Prospector) — bred by **W.S. Farish**
    JAZZ CLUB (Dixieland Band) — bred by **W.S. Farish** & Joseph Jamail
    PLASTIC SURGEON (Dr. Fager) — bred by **Ogden Phipps**
    Mrs. Grundy (Stage Door Johnny) — bred by **Ogden Phipps**
    Sheckys' Choice (Shecky Greene) — bred by Hidaway Farm
    SHECKY CHOICE (Raised Socially) — bred by C.P. DuBose
    Finding Bucks (Buckfinder) — bred by Mr. & Mrs. F. Gill Aulick
    IMAGE OF APPROVAL (Grant Approval) — bred by Denise Herrin
    Best Dressed List (Buckpasser) — bred by **Ogden Phipps**
    NORTH VERDICT (Far North) — bred by Stephen D. Peskoff
    Fashionable Trick (Buckpasser) — bred by **Ogden Phipps**
    A Slick Chic (Chieftain) — bred by Dr. William O. Reed
    A SLIM CHIC (Angle Light) — bred by Muckler Stables
    HOUSE SPEAKER (King Pellinore) — bred by Keswick Stables
    Silk Brocade (The Minstrel) — bred by Keswick Stables
    SEMORAN (Phone Trick) — bred by Keswick Stables
    REASONABLE CHOICE (Hail to Reason) — bred by **Ogden Phipps**
    A Streaker (Dr. Fager) — bred by **Ogden Phipps**
    CAPTURE HIM (Mr. Prospector) — bred by Dr. William O. Reed
    Barely Dancin (Nureyev) — bred by Dr. William O. Reed
    GAVOTTE (Mining) — bred by Ridder Thoroughbred Stable
    Final Veil (Sadler's Wells) — bred by Dr. William O. Reed
    Zina [Ire] (Taufan) — bred by Mrs. G. Bronfman

BABY LEAGUE (Continued)

†**GOLDSIO [IRE]** (Goldmark) — bred by John Malone
SARTORIALY PERFECT (Gallant Romeo) — bred by **Ogden Phipps**
Dress Uniform (*Court Martial) — bred by **Ogden Phipps**
DIFFUSION [FR] (Habitat) — bred by Societe Aland
Flirting Lady (Swaps) — bred by **Ogden Phipps**
Laddisa (Sir Gaylord) — bred by Woodside, Neff, & Kennedy
Louisville [Fr] (Val de l'Orne) — bred by Societe Aland
LOUIS LE GRAND (Key to the Kingdom) — bred by Societe Aland
LE BELVEDERE (Miswaki) — bred by Societe Aland
YPHA (Lyphard) — bred by Juddmonte Farms
Louis d'Or (Mr. Prospector) — bred by Juddmonte Farms
DIPLOMATIC BAG (Devil's Bag) — bred by Juddmonte Farms
LAURIUS [FR] (Artaius) — bred by Fabien Saier
Love In Vain (Buckpasser) — bred by Woodside, Neff, & Kennedy
APALACHIAN AFFAIR (Apalachee) — bred by Kinghaven Farms Ltd.
STRONG AND STEADY (Steady Growth) — bred by Foxfield
STEADY RUCKUS (Bold Ruckus) — bred by Hill 'n' Dale Farm
MODRED (Round Table) — bred by Hexter Stables
BEAU BRUMMEL (Round Table) — bred by **Ogden Phipps**
PAS DE DEUX (Nijinsky II) — bred by **Ogden Phipps**
BASES FULL (*Ambiorix) — bred by Wheatley Stable
BOLD AND BRAVE (Bold Ruler) — bred by Wheatley Stable & **Ogden Phipps**
Pennant Star (Bold Ruler) — bred by Wheatley Stable & **Ogden Phipps**
Bases Loaded (Northern Dancer) — bred by Mrs. J. Walker Jr., G.Grayson, & R.S. Coleman
Dancing at Dawn (*Grey Dawn II) — bred by Mrs. Joseph Walker Jr. & David I. Durham III
THE NAME'S JIMMY (Encino) — bred by Triple D. Stables
RELIEF PITCHER [IRE] (Welsh Term [Ire]) — bred by Susan McKeon
Slide (First Landing) — bred by Mrs. J. Walker Jr., G.Grayson, & R.S. Coleman
DONA MARTHA [VEN] (Text) — bred by unavailable
Stolen Base (*Herbager) — bred by Wheatley Stable & **Ogden Phipps**; foundation mare for Frances Genter
ASK CLARENCE (Buckpasser) — bred by Frances A. Genter Stable
PASSING BASE (In Reality) — bred by Frances A. Genter Stable
BASIE (In Reality) — bred by Frances A. Genter Stable
JEANO (Fappiano) — bred by Frances A. Genter Stable
Seems To Me (Foolish Pleasure) — bred by Frances A. Genter Stable
Real Doll Dode (In Reality) — bred by Frances A. Genter Stable
FOR REAL ZEAL (Unreal Zeal) — bred by Cheryl A. Curtin
DON'T WORRY BOUT ME (Foolish Pleasure) — bred by Frances A. Genter Stable
I'LL GET ALONG (Smile) — bred by Frances A. Genter Stable
SMARTY JONES (Elusive Quality) — bred by Someday Farm
Cahill Miss (Cahill Road) — bred by Frank Stronach
MANZOTTINA (Manzotti) — bred by Becky Koester
COWBOY COP (Silver Deputy) — bred by Frank Stronach
Good Opportunity (Hail to Reason) — bred by Wheatley Stable & **Ogden Phipps**
Better Opportunity (Prince Regent) — bred by Collinstown Stud Farm Ltd.
One Better (Nebbiolo) — bred by Green Ireland Properties Ltd.
STACK ROCK [IRE] (Ballad Rock) — bred by Collinstown Stud Farm Ltd.
Northern Chance (Northfields) — bred by unavailable
NORTHERN PET (Petorius) — bred by Collinstown Stud Farm Ltd.
GRAND MORNING [IRE] (King of Clubs [GB]) — bred by Collinstown Stud Farm Ltd.
Opening (Dr. Fager) — bred by **Ogden Phipps**
Mimi's First (*Hawaii) — bred by Dr. & Mrs. R. Smiser West, Mackenzie Miller, & Shenstone Farm
SNEAKIN JAKE (Table Run) — bred by Murdock McPherson
Good Position (Bold Ruler) — bred by **Ogden Phipps**
Out Distance (*Forli) — bred by Keswick Stables
OUT OF THE BID (Spectacular Bid) — bred by Irving Pollack
LA POMPADOUR (*Vaguely Noble) — bred by Keswick Stables
OLD MAESTRO (Irish River [Fr]) — bred by Keswick Stables
No Opening (Buckpasser) — bred by **Ogden Phipps**
IRISH OPEN (Irish Tower) — bred by No Opening Partnership
November Breeze (Northern Baby) — bred by No Opening Partnership
LAKENHEATH (Colonial Affair) — bred by James Tafel
HITTING AWAY (*Ambiorix) — bred by **Ogden Phipps**
BATTER UP (Tom Fool) — bred by Wheatley Stable
Bravissimo (Bold Ruler) — bred by Wheatley Stable
Eclat (*Herbager) — bred by R.S. Evans
Noted Praise (Halo) — bred by R.S. Evans
LESTER [VEN] (York Minster) — bred by unavailable
Crafty Quarry (Crafty Prospector) — bred by Peter O. Lawson-Johnston
BARNABUS (Sea Salute) — bred by Bob Berger

BABY LEAGUE (Continued)

Front Stage (*Prince Taj) — bred by T.M. Evans
BRAVERY [AUS] (Zephyr Zip [NZ]) — bred by Doncaster No. 1 Breeding Syndicate
VALOURINA (Snippets [Aus]) — bred by Trans Media Group
BRAVE PRINCE (Kenmare [Fr]) — bred by S.F. Johnson
Bravo Native (Restless Native) — bred by T.M. Evans
CHEROKEE WONDER (Cherokee Colony) — bred by T.M. Evans
CHEROKEE'S BOY (Citidancer) — bred by Z.W.P. Stables
Mudville (Bold Lad) — bred by Wheatley Stable & **Ogden Phipps**
Facial (Crème dela Crème) — bred by Charles Infusion
FACE THE MOMENT (Timeless Moment) — bred by Harry T. Mangurian Jr.
REX'S PROFILE (Rexson) — bred by Harry T. Mangurian Jr.
IRON FACE (Iron Constitution) — bred by Harry T. Mangurian Jr.
Appealing Look (Valid Appeal) — bred by Harry T. Mangurian Jr.
SPECIAL MATTER (River Special) — bred by Penny Lewis
DARING YOUNG MAN (Bold Lad) — bred by Wheatley Stable & **Ogden Phipps**
MY BOSS LADY (Bold Ruler) — bred by Wheatley Stable & **Ogden Phipps**
LANDSCAPER (*Herbager) — bred by **Ogden Phipps**
How Pleasing (Tom Rolfe) — bred by **Ogden Phipps**
How Pleasant (Foolish Pleasure) — bred by Green Brothers Inc. & William R. Ryan
HOW RARE (Rare Brick) — bred by Robert C. Sims
His Squaw (Tom Rolfe) — bred by **Ogden Phipps**
RIVER SCAPE (Riverman) — bred by Virginia Kraft Payson
Sparkling (Bold Ruler) — bred by Wheatley Stable & **Ogden Phipps**
BUBBLING (Stage Door Johnny) — bred by **Ogden Phipps**
Lake Mist [Ire] (Kings Lake) — bred by Gilman Paper Company
SHANDON LAKE [IRE] (Darshaan [GB]) — bred by Don Brown
EFFERVESCING (*Le Fabuleux) — bred by **Ogden Phipps**
The Sweet Swinger (*Le Fabuleux) — bred by **Ogden Phipps**
SINOPTICO [ARG] (Southern Halo) — bred by La Quebrada
Sparkling Account (Private Account) — bred by **Ogden Phipps**
Sparkling Dixie (Dixieland Band) — bred by Ryedale Plantation
HOLY HOPE [BRZ] (Jarraar) — bred by Haras Pemale
CHELAN (Bertrando) — bred by 505 Farms
Bushleaguer (b.f. 1950, War Admiral) — bred by **Ogden Phipps**
SHAVETAIL (Jet Pilot) — bred by **Ogden Phipps**
Face Lift (*Herbager) — bred by Mossleigh Farms & W.S. Kilroy
TAKACHINO (Don) — bred by Mount Coote Waverton Stud
Plastic Surgery (Upper Case) — bred by unavailable
Binibini [Aus] (Solitary Hail) — bred by Gooree Park Stud
SARANGGANI [AUS] (Dazzling Account) — bred by Gooree Past Co.
CLONMEL [NZ] (Thunder Gulch) — bred by unavailable
So Fine [Ire] (Thatching [Ire]) — bred by Roncon Ltd.
NONE SO BRAVE [GB] (Dancing Brave) — bred by Sheikh Mohammed bin Rashid al Maktoum
Major Overhaul [Ire] (Known Fact) — bred by Roncon Ltd.
MAJOR PROCIDA (Procida) — bred by Michael Riordan
Halekulani (Thatching [Ire]) — bred by Barronstown Stud
Halekanoora [SAf] (Ahonoora [GB]) — bred by unavailable
HALE SAPIEHA [SAF] (Sapieha) — bred by Orangewood Stud
HARMONIZING (ch.g. 1954, Counterpoint) — bred by **Ogden Phipps**
La Dauphine (b.f. 1957, *Princequillo) — bred by **Ogden Phipps**; foundation mare for Leslie Combs II
Azeez (Nashua) — bred by **Leslie Combs II** & John W. Hanes
EMPEROR REX (Warfare) — bred by John M. Olin
AZIRAE (Raise a Native) — bred by John M. Olin
OBRATZSOY (His Majesty) — bred by John M. Olin
Cazeez (Cannonade) — bred by Warner L. Jones Jr.
Nameseeker (Run the Gantlet) — bred by Regal Oak Farm
LIL SNEEKER (Lil Tyler) — bred by Lawrence Littman
Zeeza (His Majesty) — bred by Warner L. Jones Jr.
Legal Miss (Law Society) — bred by Barbara T. Phillips & Swettenham Stud
BALI BEAUTY (Seven Zero) — bred by John D. Gunther & Ed Sweeney
AMERICAN JUSTICE (American Chance) — bred by John D. Guntheer & Ed Sweeney
Attract (Blushing Groom [Fr]) — bred by Warner L. Jones Jr.
NORTHERN CONDUCT (Northern Taste) — bred by Shadai Farm
JUNGLE ROAD (Warfare) — bred by **Leslie Combs II** & John W. Hanes
Dashua (Nashua) — bred by **Leslie Combs II** & Charles H. Wacker III
Wardasha (Warfare) — bred by Walter Haefner
Run Wardasha (Run the Gantlet) — bred by Walter Haefner's Moyglare Stud Farm Ltd.
EMPERORS PRIDE (Cut Above [GB]) — bred by Michael Corbett
SON OF WAR (Pragmatic) — bred by Michael Corbett
Dunia (Raise a Native) — bred by Douglas F. Lane

BABY LEAGUE (Continued)

| **FIERY FRED [SAF]** (Counter Action) — bred by The Alchemy
Courtesan (*Gallant Man) — bred by **Leslie Combs II** & Charles H. Wacker III
| **PUNCHLINE PATTY** (Two Punch) — bred by John Franks
GUILLAUME TELL (Nashua) — bred by **Leslie Combs II** & Charles H. Wacker III

◆

Big Hurry

(b.f. 1936)

(Black Toney—*La Troienne, by *Teddy)

Purchased privately from E.R. Bradley by Ogden Phipps for the 1944 foaling season. She produced two foals for Phipps that were obtained by Bieber-Jacobs and became foundation mares for that partnership: No Fiddling and Searching. From these mares descend the foundation mares Admiring (for Paul Mellon) and Regal Gleam (for Claiborne Farm).

Blue Line (b.f. 1941, Burgoo King) — bred by E.R. Bradley
| Inaname (Shut Out) — bred by Brookfield Farms
| | Ina Battle (Battlefield) — bred by Brookfield Farms
| | | Delta De (Delta Judge) — bred by Don R. Hardesty & W.P. Little
| | | | **TAYLORS PROMISE** (Promised City) — bred by Don R. Taylor
| | | | **MOONMON** (Maria's Mon) — bred by Dr. & Mrs. Stuart E. Brown II
| | | Inagarden (Farm to Market) — bred by Don R. Hardesty & W.P. Little
| | | **CHELO** (Daniel Boone) — bred by Mr. & Mrs. Harvey Vanier et al.
| | Boo's Babu (*Our Babu) — bred by Mr. & Mrs. John A. Bell
| | | Yessenia Eunice (Rio Dulce) — bred by unavailable
| | | **MR. MELQUIN** (Suavecito) — bred by unavailable
| | Inavale (*Royal Vale) — bred by Mr. & Mrs. John S. Grossman
| | | **BRASS** (*Fachendon) — bred by John Grossman
| | Manina (*Mangayte) — bred by Grossman & Gold
| | | **THE WHEEL TURNS** (Big Burn) — bred by Fast Breaking Stables
| | | | **VANA TURNS** (Wavering Monarch) — bred by Glencrest Farm
| | | | | **PETIONVILLE** (Seeking the Gold) — bred by Glencrest Farm
| | | | | **PIKE PLACE DANCER** (Seattle Dancer) — bred by Glencrest Farm
| | | | | **MIDNIGHT FOXTROT [GB]** (Kingmambo) — bred by Gainsborough Stud
| | | **QUEEN FOR THE DAY** (King Emperor) — bred by Michael Jablow
| | | | Princess Mum (Secreto) — bred by Calumet Farm
| | | | | **FAR EASTER** (Far Out East) — bred by Fiesta Stable
| | | | **DUKE OF MONMOUTH** (Secreto) — bred by Calumet Farm
| **ISASMOOTHIE** (Rosemont) — bred by Brookfield Farms
BE FEARLESS (b.c. 1942, Burgoo King) — bred by E.R. Bradley
†**BRIDAL FLOWER** (b.f. 1943, *Challenger II) — bred by E.R. Bradley
BEYERLEYBEY (War Admiral) — bred by **King Ranch**
Boda (War Admiral) — bred by **King Ranch**
| Bee Tree (Beau Max) — bred by **King Ranch**
| | Little Kate K. (Might) — bred by James C. Anthony
| | | **CORNELIA** (Bailjumper) — bred by John F. Kilfoil
| | | **BUDDY BRENTON** (Brent's Prince) — bred by John F. Kilfoil
| | Frolic and Fun (Jester) — bred by Mrs. Joseph Walker
| | | **SNAPPY CHATTER** (Rock Talk) — bred by Mrs. Joseph Walker
| | | **SMASHER** (Farewell Party) — bred by Mrs. Joseph Walker
| | | **Shurooq** (Affirmed) — bred by Gallagher's Stud
| | | | Maraatib [Ire] (Green Desert) — bred by Shadwell Estate
| | | | | **KHASAYL [IRE]** (Lycius) — bred by Shadwell Estate
| | | | | **MUKLAH [IRE]** (Singspiel) — bred by Shadwell Estate
| | Vichy (Restless Native) — bred by William C. MacMillen
| | | Prodigious [Fr] (Pharly) — bred by Mr. & Mrs. Paul L. Hexter
| | | | **MR. ADORABLE** (Blushing Groom [Fr]) — bred by Mr. & Mrs. Paul L. Hexter
| | | | **SUPER STAFF** (Secretariat) — bred by Rhydian Morgan Jones
| | | | **PUBLIC PURSE** (Private Account) — bred by Juddmonte Farms
| | | **SHEILA SHINE** (Full Pocket) — bred by Martin E. O'Boyle

BIG HURRY (Continued)

| | **CHANCE TO DANCE** (Stop the Music) — bred by David Mowat
Fauchon (Final Ruling) — bred by William C. MacMillen
    CHEROKEE FROLIC (Cherokee Fellow) — bred by George L. Onett
      Winelight (Green Dancer) — bred by Mark Friedfertig et al.
      | **JELLY ROLL JIVE** (Prosper Fager) — bred by Mr. & Mrs. Robert L. Billings
      Cherokee Darling (Alydar) — bred by Calumet Farm & Norman Thaw et al.
      | **KATIN** (Mountain Cat) — bred by Stoneworth Farm
      Cherokyfrolicflash (Green Dancer) — bred by M. Friedfertig et al.
      | **SMOK'N FROLIC** (Smoke Glacken) — bred by Cherokee Farms
      Proud Frolic (Proud Truth) — bred by Cherokee Farms
      | **CRUSADER'S SHIELD** (Crusader Sword) — bred by Cherokee Farms
      FABULOUS FROLIC (Green Dancer) — bred by Cherokee Farms
      LINDSAY FROLIC (Mt. Livermore) — bred by Cherokee Farms
        STORMY FROLIC (Summer Squall) — bred by Cherokee Farms
        SUPER FROLIC (Pine Bluff) — bred by Cherokee Farms
Royal Bride (*Princequillo) — bred by **King Ranch**
    FULL REGALIA (Middleground) — bred by **King Ranch**
    Stolen Scepter (Jacinto) — bred by **King Ranch**
      Shout the Crowd (Your Alibhai) — bred by Elcee-H Stable
      | **SCAMMS CAN** (Barachois) — bred by Segura Farm
      SHE CAN'T MISS (Duck Dance) — bred by Elcee-H Stable
      | **HORSAFIRE** (Hold Your Peace) — bred by Elcee-H Stable
      Steal the Line (Smiling Jack) — bred by Elcee-H Stable
      SMILING NEATLY (Neater) — bred by J.C. Pendray & Sons
Flower Faces (Nip and Tuck) — bred by **King Ranch**
    Wild Flower (Armageddon) — bred by **King Ranch**
      Violet Vale (Determine) — bred by **King Ranch**
      | **CANON LAW** (Canonero II) — bred by John & Barbara Dundee
      LAW OF THE LAND (Bombay Duck) — bred by Mrs. Julian Wilkirson
      Sinsemilla (Senate Whip) — bred by Stefanie Cagle
        FANCY DIANE (Royal Pavilion) — bred by C. Konecne & Tim Stalker
        AIR STALKER (Fancy Pavilion) — bred by C. Konecne & Tim Stalker
Capule (Middleground) — bred by **King Ranch**
    Carioca (Gala Performance) — bred by unavailable
    | **Flower Vase [Ire]** (Auction Ring) — bred by C.A. Moore
    | | **MAYFLOWER LASS** (Pilgrim) — bred by R.A. Kingwell
    | **CROFTITO [IRE]** (Crofter) — bred by C.A. Moore
    THE QUIET BIDDER (Auction Ring) — bred by unavailable
No Fiddling (b.f. 1945, King Cole) — foundation mare for Bieber-Jacobs Stable
    Miz Carol (Stymie) — bred by Bieber-Jacobs
      REGAL GLEAM (Hail to Reason) — bred by Bieber-Jacobs; foundation mare for Claiborne Farm
        Foreseer (Round Table) — bred by **Claiborne Farm** & William Haggin Perry
          PALMISTRY (*Forli) — bred by **Claiborne Farm**
            Conjuror (Nijinsky II) — bred by **Cherry Valley Farm** & The Gamely Corp.
            | **WICKED MAMA** (Devil's Bag) — bred by Rosemont Farm
            Albahaca (Green Dancer) — bred by William Roebling
              BONAIRE [GB] (Air Express [Ire]) — bred by Almagro de Actividades Commerciales
          Far (*Forli) — bred by **Claiborne Farm**
          | **YONDER** (Seattle Slew) — bred by **Claiborne Farm** & The Gamely Corp.
          | **AGO** (Danzig) — bred by **Claiborne Farm** & The Gamely Corp.
          †CAERLEON (Nijinsky II) — bred by **Claiborne Farm**
          VISION (Nijinsky II) — bred by **Claiborne Farm**
          VIDEO (Nijinsky II) — bred by **Claiborne Farm** & The Gamely Corp.
          | **SCAN** (Mr. Prospector) — bred by **Claiborne Farm** & The Gamely Corp.
          MERCE CUNNINGHAM (Nijinsky II) — bred by **Claiborne Farm** & The Gamely Corp.
        ROYAL GLINT (Round Table) — bred by **Claiborne Farm**
        Glisk (Buckpasser) — bred by A.B. Hancock
          Wink (*Forli) — bred by **Claiborne Farm**
            Dwell (Habitat) — bred by **Claiborne Farm** & The Gamely Corp.
            | **QUICK ACTION** (Alzao) — bred by Hullin & Co.
            | **KING LEON [IRE]** (Caerleon) — bred by Lowquest Ltd.
            | **DOOWALEY [IRE]** (Sadler's Wells) — bred by Mrs. A. Grim & David Jamison Bloodstock
            | **MISRAAH [IRE]** (Lure) — bred by Mrs. A. Grim & David Jamison Bloodstock
            Glister (Topsider) — bred by **Claiborne Farm** & The Gamely Corp.
            | **GLEAMING SKY [SAF]** (Badger Land) — bred by Maine Chance Farm Ltd.
            Tabyan (Topsider) — bred by S.C.E.A. du Haras de Manneville
              CAP JULUCA [IRE] (Mtoto) — bred by Mrs. N. Myers
          Hint (Nijinsky II) — bred by **Claiborne Farm**
            Heloise (Forty Niner) — bred by Indian Creek et al.
            | **PADLOCK** (Boundary) — bred by Adele Dilschneider
        GLOW (Northern Dancer) — bred by **Claiborne Farm** & The Gamely Corp.

BIG HURRY (Continued)

        LUSTRA (Danzig) — bred by **Claiborne Farm** & The Gamely Corp.
        Sole (Topsider) — bred by **Claiborne Farm** & The Gamely Corp.
           EXPENSIVE SLEW (Seattle Slew) — bred by Cypress Farms 1991
    Idle Gleam (*Pronto) — bred by **Claiborne Farm**
        Gleaming Water (*Pago Pago) — bred by Nancy S. Dillman
           MIDDLEFORK RAPIDS (Wild Again) — bred by J.T. & Robert Lundy
           MICHIGAN BLUFF (Skywalker) — bred by Bev Aiello
        L'EAU VIVRE (Fast Gold) — bred by Renato Gameiro
    Pucheca (Tom Rolfe) — bred by **Claiborne Farm**
        Tableaux (Round Table) — bred by **Cherry Valley Farm** & The Gamely Corp.
        DEVILED (Devil's Bag) — bred by **Cherry Valley Farm** & the Gamely Corp.
        SCUFFLEBURG (Cox's Ridge) — bred by **Cherry Valley Farm** & The Gamely Corp.
    Purace (*Forli) — bred by **Cherry Valley Farm** & The Marjory Corp.
        CHECKPASSER (Silver Buck) — bred by Robert L. Sarro et al.
    TRENDY GENT (Nijinsky II) — bred by **Cherry Valley** & The Gamely Corp.
    Gitana (Spectacular Bid) — bred by **Cherry Valley Farm** & The Gamely Corp.
        Kaydanna (L'Emigrant) — bred by Continental Thoroughbreds
           KAY DEE CLASSIC (Regal Classic) — bred by Owen Mullin
†**STRAIGHT DEAL** (Hail to Reason) — bred by Bieber-Jacobs
    REMINISCING (Never Bend) — bred by Ethel D. Jacobs and **Harbor View Farm (Louis Wolfson)**
        COMMEMORATE (Exclusive Native) — bred by Jacobs/**Wolfson**
        PERSEVERED (Affirmed) — bred by Jacobs/**Wolfson**
        PREMIERSHIP (Exclusive Native) — bred by Jacobs/**Wolfson**
    BELONGING (Exclusive Native) — bred by Ethel D. Jacobs and **Harbor View Farm**
    DESIREE (Raise a Native) — bred by Ethel D. Jacobs and **Harbor View Farm**
        ADORED (Seattle Slew) — bred by Jacobs/**Wolfson**
        COMPASSIONATE (Housebuster) — bred by Jacobs/**Wolfson**
    Affirmatively (Affirmed) — bred by Jacobs/**Wolfson**
    MAIS OUI (Lyphard) — bred by Jacobs/**Wolfson**/Seahorse Stables
THE ADMIRAL (br.c. 1946, War Admiral) — bred by **Ogden Phipps**
Dashing By (dk.b.f. 1948, Menow) — bred by **Ogden Phipps**; mare to **John W. Galbreath**
  Stealaway (Olympia) — bred by **John W. Galbreath**
    TRUE KNIGHT (Chateaugay) — bred by **John W. Galbreath**
    Kleptomaniac (Fleet Nasrullah) — bred by **John W. Galbreath**
        Nerves of Steal (Round Table) — bred by Peter Blum
           Excellent Spirit (Damascus) — bred by Peter Blum
           SPIRITUAL STAR (Soviet Star) — bred by **Mr. & Mrs. John C. Mabee**
        PUBLIC ACCOUNT (Private Account) — bred by Peter Blum
        Bossy Boots (Academy Award) — bred by Michael C. Byrne
           DRESSED FOR ACTION (Bold 'n Flashy) — bred by Michael C. Byrne
        Anything in Sight (Round Table) — bred by Peter Blum
        RIGAMAJIG (Majestic Light) — bred by Peter Blum
        Say Anything (Sovereign Dancer) — bred by Peter Blum
        SAID ENOUGH (Norquestor) — bred by Timber Creek Farm
        FLASHER (Majestic Light) — bred by Peter Blum
    SUSPICIOUS (Damascus) — bred by Peter Blum
    ASPIRING (Academy Award) — bred by Peter Blum
    Captured Moment (Graustark) — bred by **John W. Galbreath**
    MATTHEW'S MOMENT (Little Current) — bred by **John W. Galbreath**
    Darby Dame (His Majesty) — bred by **John W. Galbreath**
    PROUD N' APPEAL (Proud Appeal) — bred by David Popofsky
        SMART COUPONS (Gate Dancer) — bred by David Popofsky
        THERESA THE TEACHA (Cure the Blues) — bred by David Popofsky
Chicken Little (Olympia) — bred by **John W. Galbreath**
  SPANISH WAY (Roberto) — bred by Forty Oaks Ranch
  A LITTLE AFFECTION (King Emperor) — bred by **Harbor View Farm**
    LOVE AND AFFECTION (Exclusive Era) — bred by **Harbor View Farm**
        My Affection (Flying Paster) — bred by **Harbor View Farm**
        HAJJI'S HONOR (Honour and Glory) — bred by Fran & Stan Hodge
    †**ZOMAN** (Affirmed) — bred by **Harbor View Farm**
Affecting (Stevward) — bred by **Harbor View Farm**
  BLACK JACK MACK (Barrera) — bred by R. Richards Rolapp
  Call Me Al (Al Nasr [Fr]) — bred by R. Rolapp et al.
    WAFARE WARRIOR (Fort Chaffee) — bred by Wafare Farm & Touch and Go Stables
Affection Affirmed (Affirmed) — bred by **Harbor View Farm**
  RIVER DEEP (Riverman) — bred by Newgate Stud
  DREAMER (Zilzal) — bred by Newgate Stud
  Sweet Times [GB] (Riverman) — bred by Newgate Stud
    STONEMASON (Nureyev) — bred by White Oak Farm
  BELGRAVIA [GB] (Rainbow Quest) — bred by Newgate Stud

BIG HURRY (Continued)

Affirmed Ambience (Affirmed) — bred by **Harbor View Farm**
AFFIRMED AND READY (Great Above) — bred by C. Bowling & C. Thompson
GREAT CAPTAIN (dk.br.c. 1949, War Admiral) — bred by **Ogden Phipps**
SEARCHING (b.f. 1952, War Admiral) — bred by **Ogden Phipps**; foundation mare for Bieber-Jacobs Stable.
†AFFECTIONATELY (Swaps) — bred by Bieber-Jacobs
    †PERSONALITY (Hail to Reason) — bred by Bieber-Jacobs
ADMIRING (Hail to Reason) — Bred in Kentucky by Bieber-Jacobs. Admiring won the Arlington-Washington Lassie Stakes. She was purchased by the partnership of Charles Engelhard and Paul Mellon from the Bieber-Jacobs dispersal sale in 1966 for a then-record $310,000.
    Courting Days (Bold Lad) — bred by **Paul Mellon** & Cragwood Estates
        MAGESTERIAL (Northern Dancer) — bred by **E.P. Taylor**
    GLOWING TRIBUTE (Graustark) — bred by **Paul Mellon**
        HERO'S HONOR (Northern Dancer) — bred by **Paul Mellon**
        WILD APPLAUSE (Northern Dancer) — bred by **Paul Mellon**
            EASTERN ECHO (Damascus) — bred by **Paul Mellon**
            BLARE OF TRUMPETS (Fit to Fight) — bred by **Paul Mellon**
            ROAR (Forty Niner) — bred by **Claiborne Farm** & Adele Dilschneider
            Praise (Mr. Prospector) — bred by **Claiborne Farm** & Adele Dilschneider
                CONGRATS (A.P. Indy) — bred by **Claiborne Farm** & Adele Dilschneider
            YELL (A.P. Indy) — bred by **Claiborne Farm** & Adele Dilschneider
        Victoria Cross (Spectacular Bid) — bred by **Paul Mellon**
            ENGLAND EXPECTS (Topsider) — bred by **Paul Mellon**
            †MOZART [IRE] (Danehill) — bred by Newgate Stud
        GLOWING HONOR (Seattle Slew) — bred by **Paul Mellon**
        SEATTLE GLOW (Seattle Slew) — bred by **Paul Mellon**
        SEA HERO (Polish Navy) — bred by **Paul Mellon**
        CORONATION CUP (Chief's Crown) — bred by **Paul Mellon**
        MACKIE (Summer Squall) — bred by John R. Gaines
            MR. MELLON (Red Ransom) — bred by T.F. VanMeter & Vinery
    Fond Recollections (Arts and Letters) — bred by **Paul Mellon**
        State of Grace (Key to the Kingdom) — bred by **Paul Mellon**
            NORTH CARROLL (Baederwood) — bred by Mrs. Frank P. Wright
    Wealth of Nations (Key to the Mint) — bred by **Paul Mellon**
        Printing Press (In Reality) — bred by **Paul Mellon**
            Redeemer (Dixieland Band) — bred by Dr. & Mrs. R.S.West & Mr. & Mrs. Mackenzie Miller
                AMERICAN BOSS (Kingmambo) — bred by Candyland Farm
            LITE LIGHT (Majestic Light) — bred by Dr. & Mrs. R.S.West & Mr. & Mrs. Mackenzie Miller
            GAILY EGRET (Storm Cat) — bred by Green Gates Farm
            SADDAD (Gone West) — bred by Foxfield
†PRICELESS GEM (Hail to Reason) — bred by Bieber-Jacobs; acquired by A.B. Hancock Jr., agent, for $395,000 at Keeneland sale. Became foundation mare for Mrs. Allen Manning and family.
    †ALLEZ FRANCE (*Sea-Bird II) — bred by Bieber-Jacobs; sold to Daniel Wildenstein's Allez France Stable.
        Ave France (Seattle Slew) — bred by Allez France Stables
            AVEC LES BLEUS (Miswaki) — bred by Allez France Stables
        ACTION FRANCAISE (Nureyev) — bred by Allez France Stables
        Astina (Slew o' Gold) — bred by Allez France Stables
            PREMIUM POINT [IND] (Twist and Turn) — bred by unavailable
        ANDROID (Riverman) — bred by Allez France Stables
        ASTORG (Lear Fan) — bred by Allez France Stables
        AIRLINE (Woodman) — bred by Allez France Stables
    Lady Winborne (Secretariat) — bred by Mrs. Allen F. Manning
        AL MAMOON (Believe It) — bred by Haras de St. George Ltd.
        Lady Lady (Little Current) — bred by Wimborne Farm
            LOVAT'S LADY (Lord At War [Arg]) — bred by Wimborne Farm
        Benguela (Little Current) — bred by Wimborne Farm
            Catumbella (Diesis [GB]) — bred by Mill Ridge Farm and W Lazy T. Ltd.
                HONOR IN WAR (Lord At War [Arg]) — bred by Mill Ridge Farm and W Lazy T. Ltd.
        Light Ice (Arctic Tern) — bred by Wimborne Farm
            FROSTY WELCOME (With Approval) — bred by Wimborne Farm
        LA GUERIERE (Lord At War [Arg]) — bred by Wimborne Farm
            LASTING APPROVAL (With Approval) — bred by Wimborne Farm
        Caspian Tern (Arctic Tern) — bred by Wimborne Farm
            FLIGHT [GB] (Night Shift) — bred by Newgate Stud
        LOST SOLDIER (Danzig) — bred by Wimborne Farm
        BORN WILD (Wild Again) — bred by Wimborne Farm
        LORD OF WARRIORS (Lord At War [Arg]) — bred by Wimborne Farm & Diane L. Perkins
    Priceless Countess (*Vaguely Noble) — bred by Mrs. Allen Manning & Gaines-Johnson Prtnrshp
        ORDWAY (Salt Lake) — bred by Lantern Hill Farm
    Sans Prix (*Vaguely Noble) — bred by Mrs. Allen Manning & Gaines-Johnson Prtnrshp
        SPECIAL PRICE (Bering [GB]) — bred by Alec Head
Ambulance (b.f. 1954, *Ambiorix) — bred by **Ogden Phipps**

BIG HURRY (Continued)

St. Bernard (Hill Prince) — bred by **Ogden Phipps**
    PASS THE DRINK (Swaps) — bred by **Ogden Phipps**
Simplon Pass (Buckpasser) — bred by **Ogden Phipps**
    PASTOURELLES (Lypheor [GB]) — bred by New Hope Partnership
    ZUNO STAR (Afleet) — bred by New Hope Partnership
Nurse's Aid (Tom Fool) — bred by **Ogden Phipps**
    FOOLISH PRINCE (Blue Prince) — bred by Warner L. Jones Jr.
Head Nurse (Bold Ruler) — bred by **Ogden Phipps**
    On Duty (*Sea-Bird) — bred by **Tartan Farms**
        OPEN GATE (Dr. Fager) — bred by **Tartan Farms**
        PORTAGE (Riverman) — bred by **Tartan Farms**
    Bedside (*Le Fabuleux) — bred by **Tartan Farms**
        BROKEN PEACE (Devil's Bag) — bred by Allez France Stables
    Arrange the Silver (State Dinner) — bred by **Tartan Farms**
        JANE SCOTT (Copelan) — bred by **Tartan Farms**
    Amatilla (In Reality) — bred by **Tartan Farms**
        BRIAN'S BLUFF (Raja Baba) — bred by Chicago Exhibitors Corp.
        TIMELY BUSINESS (Diesis [GB]) — bred by Pennfield Farms
        ARCTICA (With Approval) — bred by John Toffan & Trudy McCaffery
    Take My Shift (Buckpasser) — bred by **Tartan Farms**
    Shifting Restless (Restless Wind) — bred by **Tartan Farms**
        ELECTRIC LADY (Inland Voyager) — bred by unavailable
KING OF THE CASTLE (Bold Ruler) — bred by **Ogden Phipps**
Allemande (b.f. 1955, Counterpoint) — bred by **Ogden Phipps**
MARKING TIME (To Market) — bred by Wheatley Stable & **Ogden Phipps**
    Timing (Bold Ruler) — bred by **Ogden Phipps**
    Moskee (Explodent) — bred by Dr. Wade Arledge
        DROUTH WILLOW (Premiership) — bred by Joe W. Arledge
        TYPE RYDER (Criminal Type) — bred by Joe W. Arledge III & Jurgen Arnemann
    Time Note (Buckpasser) — bred by **Ogden Phipps**
    Tidy [Fr] (Kashmir II) — bred by Jean-Luc Largardere
        TIDELIOSK [FR] (Hellios) — bred by S.N.C. Largardere Elevage
    †**RELAXING** (Buckpasser) — bred by **Ogden Phipps**
    CADILLACING (Alydar) — bred by **Ogden Phipps**
        STROLLING ALONG (Danzig) — bred by **Ogden Phipps**
        CAT CAY (Pleasant Colony) — bred by **Phipps Stable**
    †**EASY GOER** (Alydar) — bred by **Ogden Phipps**
    EASY NOW (Danzig) — bred by **Ogden Phipps**
    A Pretty Smile (Honest Pleasure) — bred by **Ogden Phipps**
    San Lo (Clever Trick) — bred by Glencrest Farm
        STAR CITY LIGHT (Risen Star) — bred by FF-HH Partnership
    Eye Catching (Alydar) — bred by Glencrest Farm & Calumet Farm
        LADY MELESI (Colonial Affair) — bred by Rutledge Farm
Processional (Reviewer) — bred by **Ogden Phipps**; foundation mare for Stonereath Farm
    Twitchet (Roberto) — bred by Mr. & Mrs. Darrell Brown
        EVANESCENT (Northern Jove) — bred by Kentucky Thoroughbred Associates
        TACTICAL ADVANTAGE (Forty Niner) — bred by Hidaway Farm
        TWEEDSIDE (Thunder Gulch) — bred by Dr. & Mrs. R.S. West & Mr. & Mrs. Mack Miller
    RAIN ON MY PARADE (Little Current) — bred by Stonereath Farm & Glenstone Farm
    Breaking Taboos (Devil's Bag) — bred by Stonereath Farm & Wayside Stables
        SHE'S TABOO (Excavate) — bred by J Adcock
    Sparkle in Her Eye (Miswaki) — bred by Stonereath Farm & Wayside Stables
        BLU TAXIDOO (Danzig Connection) — bred by Garrett Redmon

Businesslike

(br.f. 1939)

(Blue Larkspur—*La Troienne, by *Teddy)

Bred in Kentucky by E.R. Bradley's Idle Hour Farm. Property of Ogden Phipps after syndicate distribution of E.R. Bradley's Idle Hour Stock Farm estate in 1946: syndicate composed of Phipps, Robert Kleberg Jr. of King Ranch, and John Hay Whitney of Greentree Stud.

Rivers End (b.f. 1943, *Challenger II) — bred by E.R. Bradley
   Busy Flow (*Hairan) — bred by Danada Farm
     NATURAL FLOW (*Windy City II) — bred by Crown Crest Farm
   Spice Bandit (Bandit) — bred by Mrs. Mabel Frank
     Pappa's Spice (Pappa Steve) — bred by Charles Infusino
     STEP'N (Barbizon Streak) — bred by Kemling Bros.
Challenge Like (b.f. 1945, *Challenger II) — bred by E.R. Bradley
   Defiant (Sun Again) — bred by G.M. Humphrey
     JEFF D. (I Will) — bred by L.F. & W.W. Greathouse
     Marjon (Run For Nurse) — bred by L.F. & W.W. Greathouse
     KING L. B. (Harry's Secret) — bred by Mrs. John M. Smith
   I Like Blue (Blue Prince) — bred by G.M. Humphrey
   First I Like (First Landing) — bred by S.J. Langill
     SECRET HAVEN (Silent Screen) — bred by Verne H. Winchell
Busy Whirl (b.f. 1946, Whirlaway) — bred by E.R. Bradley
   Tushan (Bimelech) — bred by Clifford Mooers
     Dixie Do (*Deuce II) — bred by Claude Hudspeth
     DIXIE JAY (Praise Jay) — bred by Claude Hudspeth
   Lines Busy (*Alibhai) — bred by Clifford Mooers
     Busy Windsong (Bold Commander) — bred by A.B. Karsner
     DO THE BUMP (Marshua's Dancer) — bred by Archie Chase & Douglas M. Davis Jr.
     LI'L ARCH (Stage Director) — bred by Archie Chase & Douglas M. Davis Jr.
     Busy Wave (Bupers) — bred by Mrs. C.E. Buckley
     BORDER RULING (Court Ruling) — bred by Wesley R. Becker
     ABADASHA (Yorkville) — bred by Wesley R. Becker
     BUSY RULER (Orbit Ruler) — bred by Wesley R. Becker & Dean Short
BUSANDA (b.f. 1947, War Admiral) — bred by **Ogden Phipps**
   BUREAUCRACY (Polynesian) — bred by **Ogden Phipps**
   Finance (*Nasrullah) — bred by **Ogden Phipps**
     La Mesa (Round Table) — bred by Roger S. Braugh
       Corporate Queen (Truxton King) — bred by Braugh Ranches
         Cie Canadienne (Canadian Gil) — bred by Robert H. Walter
           ECUDIENNE (Cost Conscious) — bred by Mr. & Mrs. Robert H. Walter
         Royal Merger (Princely Native) — bred by Robert H. Walter
         BATROYALE (Batonnier) — bred by Robert H. Walter
     †**OUTSTANDINGLY** (Exclusive Native) — bred by **Harbor View Farm**
       Outlasting (Seattle Slew) — bred by **Harbor View Farm**
         FORTITUDE (Cure the Blues) — bred by **Harbor View Farm**
       SENSATION [GB] (Soviet Star) — bred by Sheikh Mohammed bin Rashin Al Maktoum's Darley Stud
         SYSTEMATIC [GB] (Rainbow Quest) — bred by Gainsborough Stud
       SUPERIORITY (Arazi) — bred by Darley Stud
     La Affirmed (Affirmed) — bred by Harbor View Farm
       CARESS (Storm Cat) — bred by **Harbor View Farm**
         SKY MESA (Pulpit) — bred by **Harbor View Farm**
       COUNTRY CAT (Storm Cat) — bred by **Overbrook Farm**
       Emmaus (Silver Deputy) — bred by **Harbor View Farm**
         WISEMAN'S FERRY (Hennessy) — bred by Nursery Place & Robert T. Manfuso
       BERNSTEIN (Storm Cat) — bred by Brushwood Stable
       DELLA FRANCESCA (Danzig) — bred by Brushwood Stale
     LOVELIER (Affirmed) — bred by **Harbor View Farm**
   Oak Cluster (*Nasrullah) — bred by **Ogden Phipps**
     OPEN HEARING (*Court Martial) — bred by **Ogden Phipps**
     Rules of the Game (Hitting Away) — bred by **Ogden Phipps**
       Fawncy Daughter (Forward Pass) — bred by John A. Morris
         ICE OVER (It's Freezing) — bred by Philip L. Mulholland & Woody Coyle
         BALDSKI'S HERO (Baldski) — bred by Farnsworth Farms & Sucher Stables
       ROMAN STARLET (Proudest Roman) — bred by Shoestring Stable
       Two Up (Upper Nile) — bred by Shoestring Stable
         ZIAD'S GAME (Ziad) — bred by Allan Bias & Melonie Bias

BUSINESSLIKE *(Continued)*

OUTDOORS (*Herbager) — bred by **Ogden Phipps**
LE NOTRE (*Herbager) — bred by **Ogden Phipps**
Twin Oaks (Double Jay) — bred by **Ogden Phipps**
  Mild Persuasion (Roman Line) — bred by Lin-Drake Farm
    Big Mad (Hail to Reason) — bred by Townsend B. Martin
      RESTANBEGONE (Restivo) — bred by Townsend B. Martin
      IDA LEWIS (Restivo) — bred by Townsend B. Martin
  Shade Princess (*Prince Taj) — bred by Lin-Drake Farm
    Magical Morn (Impressive) — bred by Mr. & Mrs. John Otto
      MAGIC NORTH (North Sea) — bred by John Otto
  FELLOW HEIR (Diplomat Way) — bred by Mr. & Mrs. Carey Rogers
  Twice Crowned (King's Bishop) — bred by Mr. & Mrs. Carey Rogers
    Sunny Princess (Sunny Clime) — bred by Mr. & Mrs. Michael T. Sutherland
      WINTER LEAF (Muhtafal) — bred by Michael T. Sutherland
    TREASURE COAST GEM (Mutakddim) — bred by North Highland Farm
TWICE CITED (Double Jay) — bred by Mrs. George G. Proskauer
  Second Ovation (*Le Fabuleux) — bred by Taylor's Purchase Farm
    Humdrum (Drum Fire) — bred by Kwik Lok Corporation
      ACCUMULATOR (Swing Music) — bred by Mr. & Mrs. Laurie S. Owen
    DANCING OVATION (Northern Jove) — bred by Northwest Farms
MANITOULIN (Tom Rolfe) — bred by Mrs. George G. Proskauer
Splendid Spree (Damascus) — bred by Westerly Stud Farm
  Splendid Ack Ack (Ack Ack) — bred by Lillie F. Webb
    SPLENDID ANN (Raise a Man) — bred by Northwest Farms
    LITTLE BAR FLY (Raise a Man) — bred by Northwest Farms
    BABY BARFLY (Son of Briartic) — bred by Northwest Farms
    ST. HELENS SHADOW (Septieme Ciel) — bred by Northwest Farms
    BARFIGHTER (Wild Again) — bred by Northwest Farms
  SPLENDID WAY (Apalachee) — bred by Lillie F. Webb
  SPLENDID SPRUCE (Big Spruce) — bred by Lillie F. Webb
Oak Leaf Cluster (Promised Land) — bred by Westerly Stud Farm
  Prominique (On the Sly) — bred by Kinderhill Farm
    Promise Star (Star de Naskra) — bred by Carelaine Farm & Jacqueline Phillips
      STARS AND SON (Son of Briartic) — bred by Larry O. Hillis
BUPERS (Double Jay) — bred by **Ogden Phipps**
†**BUCKPASSER** (Tom Fool) — bred by **Ogden Phipps**
Navsup (*Tatan) — bred by **Ogden Phipps**
  Naval (Broadway Forli) — bred by **Ogden Phipps**
    CHARTS (Mari's Book) — bred by William H. LaMaster
  Lightship (Majestic Light) — bred by **Ogden Phipps**
    FLOATING INTEREST (Lord Avie) — bred by George Strawbridge
  POLISH NAVY (Danzig) — bred by **Ogden Phipps**
AUDITING (b.c. 1948, Count Fleet) — bred by **Ogden Phipps**
His Duchess (b.f. 1950, *Blenheim II) — bred by **Ogden Phipps**
COMIC (Tom Fool) — bred by **Ogden Phipps**
Comic Relief (Tom Fool) — bred by **Ogden Phipps**
  Court Jestress (Crozier) — bred by Louis Lee Haggin
    ORDER IN COURT (Banderilla) — bred by Dr. J.R. Poirier
    CHIEF BANDITO (Banderilla) — bred by Fontainebleau Farm
    LUCKY TAURO (*El Centauro) — bred by unavailable
So Social (Tim Tam) — bred by **Ogden Phipps**
  SNOBISHNESS (*Forli) — bred by **Ogden Phipps**
  Quexine [Fr] (Sir Gaylord) — bred by **Ogden Phipps**
    QUEXION (Young Generation) — bred by unavailable
    QUEXIOSS [FR] (Ardross) — bred by Cotswold Bloodstock Ltd.
  Pretty Special (Riverman) — bred by Mrs. **Ogden Phipps**
    FOREVER COMMAND (Top Command) — bred by Mrs. **Ogden Phipps**
    MY BIG BOY (Our Hero) — bred by Mrs. **Ogden Phipps**
    Oh So Pretty (Spend a Buck) — bred by Allen Werneck & Hunter Farm
      REY CANELO [VEN] (Alhajras) — bred by unavailable
  OH SO CHOOSY (Top Command) — bred by Mrs. **Ogden Phipps**
  OH SO SNOBISH (Quadratic) — bred by Welcome Farm
WARD MCALLISTER (Bold Ruler) — bred by **Ogden Phipps**
Queen's Gambit (Bold Ruler) — bred by **Ogden Phipps**
  CHESS MOVE (Avatar) — bred by **Ogden Phipps**
  Whiffling (Wavering Monarch) — bred by Norman, Allen, & Phil Owens
    †**PRAIRIE BAYOU** (Little Missouri) — bred by Loblolly Stable
    FLITCH (Demons Begone) — bred by Loblolly Stable
HASTY REPLY (*Pronto) — bred by **Ogden Phipps**
BANNER GALA (Hoist the Flag) — bred by **Ogden Phipps**

BUSINESSLIKE (Continued)

Roberto's Social (Roberto) — bred by **Ogden Phipps**
| **EXCELLENCE ROBIN** (Polish Navy) — bred by Shadai Farm
SOCIAL BUSINESS (Private Account) — bred by **Ogden Phipps**
    SQUIRE JONES (Seeking the Gold) — bred by Hermitage Farm & C.R. McGaughey
    Miss Speed (Forty Niner) — bred by Warner L. Jones Farm Inc.
        WILD SPEED (Forest Wildcat) — bred by Victoria Farm
Duchess Rae (Hitting Away) — bred by Little M Farm
| **EL MORENO** (Noble Table) — bred by Roy Bowen
Cinto Tora (Jacinto) — bred by Roy Bowen & **A.B. Hancock III**
| **BEAU CENTAVO** (Beaudelaire) — bred by Saddler Trail Farm
Duchess dela Trois (Coastal) — bred by Roy Bowen & **A.B. Hancock III**
    Silent Duchess (Silent Screen) — bred by Box Arrow Farm
    | **DIRTY DUKE** (Nasty and Bold) — bred by J Adcock
    DUCHESS OF PEKISKO (Regal Classic) — bred by Box Arrow Farm
    DERBY DUKE (Salt Lake) — bred by Al Pruss
Flighty Duchess (Sunrise Flight) — bred by Little M Farm
| **BLUE RIBBON GIRL** (Star Envoy) — bred by Dr. & Mrs. David D. O'Neal
Topolina (Tropical Breeze) — bred by John Murrell
| **GOODBYEYOUALL** (Good Behaving) — bred by Don, Thomas, & Mrs. Charles C. Sturgill
Our First Pleasure (What a Pleasure) — bred by Kenneth Opstein
| **JOSH'S JOY** (Dr. Blum) — bred by Ashwood Thoroughbreds Inc.
Trevera (Peace Corps) — bred by Kenneth Opstein
    Tall Cotton (Al Hattab) — bred by Drew R. Maddux
    TABOO [ARG] (Ringaro) — bred by Santa Maria de Araras
Discriminate (b.f. 1954, Shut Out) — bred by **Ogden Phipps**
Weather Mate (Bagdad) — bred by J.S. Watkins
    Want More Wins (Three Bagger) — bred by W.M. Wickham
    | **VIBRANTE [VEN]** (Sun Cross) — bred by unavailable
Dot's Girl (*Tudor Grey) — bred by Barrett M. Morris
SOAKING SMOKING (Bucksplasher) — bred by Irish Acres Farm
Splashing Girl (Bucksplasher) — bred by Irish Acres Farm
    MINERS GAMBLE (Prospectors Gamble) — bred by J D Farms

Hildene
(b.f. 1938)

(Bubbling Over—Fancy Racket, by *Wrack)

Bred in Kentucky by Edward M. Simms' Xalapa Farm. C.T. Chenery purchased Hildene as a yearling from Simms' dispersal sale in 1939 for $750. She placed once from eight starts and earned $100.

MANGOHICK (b.g. 1944, Sun Beau) — bred by **C.T. Chenery**
†**HILL PRINCE** (b.c. 1947, *Princequillo) — bred by **C.T. Chenery**
First Flush (b.f. 1948, *Flushing II) — bred by **C.T. Chenery**
    Akobo (Bossuet) — bred by **C.T. Chenery**
      Minmognovich (Pan Dancer) — bred by Peter Fuller
      | †**KENTUCKY BLUE [DEN]** (Round Tower) — bred by Asger Jorgensen & Leif Hempel
      KENTUCKY QUEEN [DEN] (Jammed Red) — bred by Stud Hesthaven
    Windy Damsel (Warfare) — bred by W.T. Pascoe III & Marshall Wais
    PATRIOTAKI (Mangaki) — bred by Chrys & Sherwood
Acantha (Bossuet) — bred by **C.T. Chenery**
    Mercy Mine (*Court Martial) — bred by Mrs. George G. Proskauer
    NATIVE LOVIN (Exclusive Native) — bred by Marvin L. Warner
    | **EUTHALOS** (Youth) — bred by Mike G. Rutherford
    Explodin Love (Explodent) — bred by Mike G. Rutherford
    | **EXPOSE** (Flying Paster) — bred by John T.L. Jones Jr.
    | **TARA GLENN** (Ancestral [Ire]) — bred by Racetime Inc.
    Tempting Thought (Cox's Ridge) — bred by Mike G. Rutherford
    TEMPTER [ARG] (Southern Halo) — bred by La Quebrada
    TEMPURA [ARG] (Southern Halo) — bred by La Quebrada
    Exclusiva (Exclusive Native) — bred by Marvin L. Warner
    NORTH DIP (Explodent) — bred by Mike G. Rutherford
    Poppies and Wheat (Lost Code) — bred by NoName Ranch et al.
    APRIL'S LUCKY BOY (Ide) — bred by Kelly Thiesing
    Has No Mercy (Raise a Native) — bred by Marvin L. Warner
    TATE MAN (Boutinierre) — bred by Michael Stafford & Amy Stafford
    LAMMY (Abstract) — bred by Michael Stafford & Amy Stafford

HILDENE (Continued)

Thetis (Parnassus) — bred by **C.T. Chenery's Meadow Stud**
    Petite Greek (Beau Purple) — bred by Hobeau Farm
       QUINTAS GREEK LADY (Quintas) — bred by Richard D. Bokum
    Rhiannon (Beau Gar) — bred by Hobeau Farm
       RHIARADO (An Eldorado) — bred by David L. Poll
    Third Wife (Hydrologist) — bred by Jerome Castle Corporation
    EXPENSIVE DECISION (Explosive Bid) — bred by Edward L. Shapoff
BOLD EXPERIENCE (Bold Ruler) — bred by **Meadow Stud**
    UPPER CASE (Round Table) — bred by **Meadow Stud**
    All or None (Sir Gaylord) — bred by **Meadow Stud**
       Saturday Matinee (Silent Screen) — bred by **C.T. Chenery estate**
          HAMLET (Key to the Kingdom) — bred by Stephen DiMauro
       Give Her the Gun (*Le Fabuleux) — bred by **C.T. Chenery estate**
          Shawl Dance (Our Native) — bred by Oak Crest Farm
             NATIVE BRASS (Dixieland Brass) — bred by Denny Andrews
          ANOTHER FELIX (Miswaki) — bred by Cable Stables
    Aces Full (Round Table) — bred by **C.T. Chenery estate**
       PETITE ILE [IRE] (Ile de Bourbon) — bred by David Nagle
    MS. ROSS (Hoist the Flag) — bred by **C.T. Chenery estate**
    MAKE THE MAGIC (Raise a Native) — bred by Joseph Allen
Trollius (*Ambiorix) — bred by **Meadow Stud**
    Barclay Belle (Chieftain) — bred by Dr. W.O. Reed
       SOLO RIDE (Solo Landing) — bred by Broodmare Breeders Inc.
    Amtare (*Petare) — bred by Dr. W.O. Reed
    RACE THE WAVES (Sailing Along) — bred by Coffer Ranches
Bold Matron (Bold Ruler) — bred by **Meadow Stud**
    Beckley (*Tatan) — bred by **Meadow Stud**
       GILGIT [FR] (Kashmir II) — bred by Societe N.C. Maurin de Brignac & Luis Champion
    Due Dilly (Sir Gaylord) — bred by **Meadow Stud**
    SPIRIT LEVEL (Quadrangle) — bred by **C.T. Chenery estate**
    RING OF LIGHT (In Reality) — bred by **C.T. Chenery estate**
    DR. BLUM (Dr. Fager) — bred by **C.T. Chenery estate**
    Tempest Tost (First Landing) — bred by E.A. Seltzer & S.D. Petter Jr.
    Magic Toss (His Majesty) — bred by Edward Seltzer
          LADY COFFEE (Star Choice) — bred by Beaconsfield Farm
COPPER CANYON (Bryan G.) — bred by **Meadow Stud**
    Copernica (Nijinsky II) — bred by Charles W. Engelhard's Cragwood Estates
       Copperplate (Secretariat) — bred by **Paul Mellon**
          COPPER HORIZON (Pleasant Colony) — bred by T.M. Evans
       CRUSADER SWORD (Damascus) — bred by **Paul Mellon**
       Lucky Penny (Golden Act) — bred by Buck Chance Farm
          WAGE A PENNY (Valid Appeal) — bred by William C. Schettine
       Carmelita (Mogambo) — bred by Gaines Equine Investment
          FIVE TIMES A LADY (Quinton) — bred by B. Wayne Hughes
       COPPER BUTTERFLY (Blushing Groom [Fr]) — bred by Gaines Equine Investment
    Insilca (Buckpasser) — bred by Mrs. Charles W. Engelhard
       SILKEN DOLL (Chieftain) — bred by Corbin J. Robertson
          Chief Appeal (Valid Appeal) — bred by Corbin J. Robertson
             TURKAPPEAL (Turkoman) — bred by Four Horsemen's Ranch
          MEADOW SILK (Meadowlake) — bred by Corbin J. Robertson
             RUN PRODUCTION (Saint Ballado) — bred by Four Horsemen's Ranch
          Tropical Rain (Danzig Connection) — bred by Corbin J. Robertson
             TROPICAL WAY (Way West [Fr]) — bred by Dr. Luis E. Bonnet
          JUYUSH (Silver Hawk) — bred by Corbin J. Robertson
          †SILKEN CAT (Storm Cat) — bred by Ferme du Bois-Vert
             SPEIGHTSTOWN (Gone West) — bred by Aaron & Marie Jones
          Silken Light (Majestic Light) — bred by Corbin J. Robertson
             INCITATUS (Batonnier) — bred by Robert H. Walter Family Trust
       TURK PASSER (Turkoman) — bred by The Oaks Horse Farm Corporation
    Cherokee Phoenix (Nijinsky II) — bred by Mrs. Charles W. Engelhard
    CHEROKEE COLONY (Pleasant Colony) — bred by T.M. Evans
    RISEN COLONY (Pleasant Colony) — bred by T.M. Evans
Is You Class (Judger) — bred by **Cherry Valley Farm**
    POTRIMAGIC [ARG] (Potrillazo) — bred by La Madrugada
Quachita (Cox's Ridge) — bred by **Cherry Valley Farm**
    Pleasant Sue (Pleasant Colony) — bred by Fares Farm
       ANTEQUERA (Green Dancer) — bred by Justo Fernandez
    Copperband (Seeking the Gold) — bred by Fares Farm
       CARAMEL CUSTARD (Chief Honcho) — bred by Mr. & Mrs. Bertram R. Firestone
       BLUE HILLS (Cure the Blues) — bred by Mr. & Mrs. Bertram R. Firestone

HILDENE (Continued)

VIRGINIA DELEGATE (Bold Ruler) — bred by **Meadow Stud**
Hunting Pink (Jacinto) — bred by **Meadow Stud**
   Pink Screen (Silent Screen) — bred by Richard L. Duchossois
   †**CATIRE BELLO [VEN]** (Inland Voyager) — bred by unavailable
   LE BAG LADY (Sauce Boat) — bred by Richard L. Duchossois' Hill 'N Dale Farm
   Cat Hunt (*Cougar II) — bred by Four-D Partnership
   SR. SERENO (Lodz) — bred by Haras Santa Isabel
Long Stemmed Rose (Jacinto) — bred by **Meadow Stud**
   Such 'n Such (Ack Ack) — bred by Carelaine Stable
   KITTY TATCH (Vigors) — bred by Jonabell Farm
   SUCH CLASS (Vigors) — bred by Jonabell Farm
   BRILLIANT ROSE (Bold Reason) — bred by Funkhouser Industries
Satsuma (b.f. 1949, Bossuet) — bred by **C.T. Chenery**
   Sabana (Bryan G.) — bred by **C.T. Chenery**
   Will Hail (Hail to Reason) — bred by Bieber-Jacobs Stable
      Pilferer (No Robbery) — bred by Greentree Stud
         TELFERNER (Tell) — bred by J.C. Pollard
         LYIN TO THE MOON (Kris S.) — bred by Meadowbrook Farms
         State Tax (Caro [Ire]) — bred by North Ridge Farm
            TAXABLE DEDUCTION (Prized) — bred by Meadowbrook Farms
            TAX DANCER (Din's Dancer) — bred by Meadowbrook Farms
         SAVANNAH SLEW (Seattle Slew) — bred by North Ridge Farm
         ADMIRALTY (Strawberry Road [Aus]) — bred by **Allen Paulson**
         ASTRA (Theatrical [Ire]) — bred by **Allen Paulson**
      Accomodating (Raise a Native) — bred by Bieber-Jacobs Stable
         Aricia (Flag Raiser) — bred by Mrs. Guy Weisweiller
            Albinoua [Fr] (King of the Castle) — bred by Mme. Maurice Rohaut-Leger
            STAR MAITE [FR] (Kenmare) — bred by Haras du Mont Dit Mont
            Albala [Fr] (Arctic Tern) — bred by Mme. Maurice Rohaut-Leger
            UNEXPECTEDLY [AUS] (Geiger Counter) — bred by G T Moore Investments
         Arriance [Fr] (Gay Mecene) — bred by Skymarc Farm
            ARDANA [IRE] (Danehill) — bred by Ecurie Skymarc
      Cher Lafite (Bold Lad) — bred by J R K Ranch
      CLASSIC AMBITION (Cyane) — bred by Latour Farms
         Christofle (Cojak) — bred by Latour Farms
            NICHOLLE'S DEVIL (Devil His Due) — bred by Harvey Tenenbaum
         SONG OF AMBITION (Relaunch) — bred by Ronald K. Kirk
   †**CICADA** (Bryan G.) — bred by **Meadow Stud**
      CICADA'S PRIDE (Sir Gaylord) — bred by **Meadow Stud**
      Ancient Mystery (Restless Native) — bred by **C.T. Chenery estate**
         Bold and Dark (Bold Forbes) — bred by T.M. Evans
            CLASSIC PLAYBOY (Cure the Blues) — bred by Milfer Farm
PRINCE HILL (b.g. 1951, *Princequillo) — bred by **C.T. Chenery**
THIRD BROTHER (b.c. 1953, *Princequillo) — bred by **C.T. Chenery**
†**FIRST LANDING** (b.c. 1956, *Turn-to) — bred by **C.T. Chenery**

◆

Myrtlewood
(b.f. 1932)
(Blue Larkspur—*Frizeur, by Sweeper)

Bred by Brownell Combs; became a foundation mare for the Combs family. A champion on the race-track, Myrtlewood won 15 of 22 starts and earned honors as champion sprinter and champion handicap mare in 1936.

Crepe Myrtle (b.f. 1938, Equipoise) — bred by **Brownell Combs**
   †**MYRTLE CHARM** (Alsab) — bred by **Brownell Combs & Leslie Combs II**
      MYRTLE'S JET (Jet Pilot) — bred by Elizabeth Arden (Maine Chance Farm)
      Fair Charmer (Jet Action) — bred by Elizabeth Arden Inc.
         MY CHARMER (Poker) — bred by Fiege-Castleman
            SEATTLE SLEW (Bold Reasoning) — bred by Ben Castleman
            Clandestina (Secretariat) — bred by W.L. Jones, **W.S. Farish**, & W.S. Kilroy
               Night Secret (GB) (Nijinsky II) — bred by Swettenham Stud
                  DUSK DUEL (Kris) — bred by Darley Stud Management
               DESERT SECRET [IRE] (Sadler's Wells) — bred by Swettenham Stud
            LOMOND (Northern Dancer) — bred by Jones, **Farish**, & Kilroy

MYRTLEWOOD (Continued)

> **ARGOSY** (Affirmed) — bred by Jones, **Farish**, & Kilroy
> **SEATTLE DANCER** (Nijinsky II) — bred by Jones, **Farish**, & Kilroy
> Spinosa (Count Fleet) — bred by **Leslie Combs II**
> **MASKED LADY** (Spy Song) — bred by **Leslie Combs II & Brownell Combs**
> **WHO'S TO KNOW** (Fleet Nasrullah) — bred by **Leslie Combs II**
> **ANGEL ISLAND** (*Cougar II) — bred by **Leslie Combs II's Spendthrift Farm**
> **OUR REVERIE** (J.O. Tobin) — bred by **Spendthrift Farm**
> **SILKEN [GB]** (Danehill) — bred by Mr. & Mrs. K.J. Buchanan
> **SHARROOD** (Caro [Ire]) — bred by **Spendthrift Farm**
> **ISLAND ESCAPE** (Slew o' Gold) — bred by Woodstock Enterprises
> All Too True (Caro [Ire]) — bred by **Spendthrift Farm**
> **TRULY MET** (Mehmet) — bred by **Spendthrift Farm**
> Jolie Jolie (Sir Ivor) — bred by **Spendthrift Farm**
> **COUGARIZED** (*Cougar II) — bred by **Spendthrift Farm et al.**
> **JOLIE'S HALO** (Halo) — bred by Arthur Appleton
> **PLEASANT JOLIE** (Pleasant Colony) — bred by Arthur Appleton
> **MISTER JOLIE** (Valid Appeal) — bred by Arthur Appleton
> Confirmed Affair (Affirmed) — bred by **Spendthrift Farm**
> **THIS ONE'S FOR US** (Cox's Ridge) — bred by Kinderhill Corp.
> **RAISE A LADY** (Raise a Native) —bred by **Leslie Combs II**
MISS DOGWOOD (dk. b/br. f. 1939, *Bull Dog) — bred by **Brownell Combs**
> **SEQUENCE** (Count Fleet) — bred by **Brownell Combs**
> **NOORSAGA** (*Noor) — bred by **Brownell Combs & Leslie Combs II**
> **HERMOD** (*Royal Charger) — bred by **Brownell Combs & Leslie Combs II**
> Inviting (*My Babu) — bred by **Brownell Combs & Leslie Combs II**
> **WITHOUT PEER** (Crème dela Crème) — bred by **Leslie Combs II**
> **INTRIGUING HONOR** (Sham) — bred by **Spendthrift Farm**
> Bold Sequence (Bold Ruler) — bred by **Brownell Combs & Leslie Combs II**
> Surgery (Dr. Fager) — bred by R.S. Evans
> **LEFT COURT** (Valdez) — bred by R.S. Evans
> **SEWICKLEY** (Star de Naskra) — bred by R.S. Evans
> **SHARED INTEREST** (Pleasant Colony) — bred by R.S. Evans
> **FORESTRY** (StormCat) — bred by R.S. Evans
> **CASH RUN** (Seeking the Gold) — bred by R.S. Evans
> **BORN TO LEAD** (Mr. Leader) — bred by R.S. Evans
> **GOLD DIGGER** (Nashua) — bred by **Brownell Combs & Leslie Combs II**
> **MR. PROSPECTOR** (Raise a Native) — bred by **Leslie Combs II**
> **GOLD STANDARD** (*Sea-Bird) — bred by **Leslie Combs II**
> Myrtlewood Lass (*Ribot) — bred by **Leslie Combs II**
> Amelia Bearhart (Bold Hour) — bred by **Spendthrift Farm**
> **EXPLOSIVE RED** (Explodent) — bred by Richard D. Maynard
> **RUBY RANSOM** (Red Ransom) — bred by Richard D. Maynard
> **SACRED SONG** (Diesis [GB]) — bred by Jamm Ltd.
> **STRUT THE STAGE** (Theatrical [Ire]) — bred by Jamm Ltd.
> **CHIEF BEARHART** (Chief's Crown) — bred by Richard D. Maynard
> **LILLIAN RUSSELL** (Prince John) — bred by **Spendthrift Farm**
> Slew Princess (Seattle Slew) — bred by **Leslie Combs II**
> **SLEW GIN FIZZ** (Relaunch) — bred by Lee Pokoik et al.
> **IGOTRHYTHM** (Dixieland Band) — bred by Lee Pokoik et al.
> Kentucky Lill (Raise a Native) — bred by **Leslie Combs II**
> **LIL'S BOY** (Danzig) — bred by Gainsborough Farm
> Gold Mine (Raise a Native) — bred by **Spendthrift Farm**
> Heart of America (Northern Jove) — bred by **Leslie Combs II**
> **AMERICAN CHAMP** (Jolie's Halo) — bred by Arthur I. Appleton
> **ETIQUETTE** (Raja Baba) — bred by **Leslie Combs II**
> Tutta (In Reality) — bred by Frances A. Genter
> **QUEEN TUTTA** (Kris S.) — bred by E.P. Evans
> **Amiga** (*Mahmoud) — bred by **Brownell Combs**
> **CARRIER X** (Count Fleet) — bred by **Brownell Combs & Leslie Combs II**
> **DEDIMOUD** (Dedicate) — bred by **Brownell Combs & Leslie Combs II**
> **TUMIGA** (*Tudor Minstrel) — bred by **Brownell Combs & Leslie Combs II**
> **ALERT PRINCESS** (Raise a Native) — bred by **Leslie Combs II**
> **SHAM'S PRINCESS** (Sham) — bred by **Spendthrift Farm**
> **ETERNITY'S BREATH** (Nureyev) — bred by **Spendthrift Farm**
> Sweet Shampagne (Sham) — bred by **Spendthrift Farm**
> Change of Taste (Time for a Change) — bred by Mike Rost
> **EXTRA CHANGE** (Fiscal) — bred by Wanda M. Shepherd-Garst
> **A BID LACY** (Spectacular Bid) — bred by Mike Rost
> My Sweet Talker (Phone Trick) — bred by Dakotah Thoroughbred Farm
> **SWEETTRICKYDANCER** (Green Dancer) — bred by Eduardo Rojas

MYRTLEWOOD (Continued)

        NYMPH OF THE NIGHT (Magesterial) — bred by **Spendthrift Farm**

BERNWOOD (*Bernborough) — bred by **Brownell Combs**

BELLA FIGURA (Count Fleet) — bred by **Brownell Combs**

Ellenwood (*Shannon II) — bred by **Brownell Combs**

    Miss Boxwood (Mr. Busher) — bred by **Brownell Combs & Leslie Combs II**

        GREEK ROAD (Greek Money) — bred by C.A. & T.H. Asbury

    Arctica (*Arctic Prince) — bred by **Leslie Combs II**

        BEMO (Maribeau) — bred by Hickory Tree Farm

        PRINCE HAGLEY (Hagley) — bred by Stan Jones & Paul Perlin

    Diane W. (Jet Jewel) — bred by **Brownell Combs & Leslie Combs II**

        PARCHMENT (Royal Note) — bred by J.C. Metz

    PENINSULA PRINCESS (Crewman) — bred by **Leslie Combs II**

Miss Fleetwood (Count Fleet) — bred by **Brownell Combs & Leslie Combs II**

    Cold Morning (Mr. Busher) — bred by **Brownell Combs & Leslie Combs II**

        CUT CORNERS (Good Investment) — bred by **Leslie Combs II**

    Egret (*Tudor Minstrel) — bred by **Brownell Combs & Leslie Combs II**

        BOMBAY DUCK (Nashua) — bred by **Leslie Combs II**

    Sea Myrtle (Swoon's Son) — bred by **Leslie Combs II**

        SEMI PRINCESS (Semi-pro) — bred by **Leslie Combs II**

        SEA ROYALTY (Native Royalty) — bred by **Leslie Combs II**

†**DURAZNA** (b.f. 1941, Bull Lea) — bred by **Brownell Combs**

Manzana (Count Fleet) — bred by **Brownell Combs**

    Melon (*Heliopolis) — bred by **Leslie Combs II**

        HAND TO HAND (Warfare) — bred by **Leslie Combs II**

        STRAIGHT AS A DIE [ENG] (Never Bend) — bred by **Leslie Combs II**

    La Morlaye (Hafiz) — bred by **Brownell Combs & Leslie Combs II**

        LADY TRAMP (*Sensitivo) — bred by **Brownell Combs & Leslie Combs II**

        Indian Call (Warfare) — bred by **Leslie Combs II**

            ERWIN BOY (Exclusive Native) — bred by **Leslie Combs II**

            Cheyenne Birdsong (Restless Wind) — bred by **Leslie Combs II**

                SEQUOYAH (Gummo) — bred by Cardiff Stud Farms

                SHYWING (Wing Out) — bred by Cardiff Stud Farms

                COMPELLING SOUND (Seattle Slew) — bred by Cardiff Stud Farms

                CRESTON (Flying Paster) — bred by Cardiff Stud Farms

            DANCING PARTNER (Exclusive Native) — bred by **Spendthrift Farm**

            SIBERIAN EXPRESS (Caro [Ire]) — bred by **Spendthrift Farm**

        POPULAR DEMAND (*Sensitivo) — bred by **Leslie Combs II**

    JOURNALETTE (Summer Tan) — bred by **Brownell Combs & Leslie Combs II**

    TYPECAST (Prince John) — bred by Nuckols Brothers

        PRETTY CAST [JPN] (Cover Up Nisei) — bred by unavailable

    SOCIETY COLUMN (Sir Gaylord) — bred by Nuckols Brothers

        PRESENT THE COLORS (Hoist the Flag) — bred by Taylor's Purchase Farm

        Sarah Gamp (Hoist the Flag) — bred by Taylor's Purchase Farm

            Sly Sarah (On the Sly) — bred by Joanne Nor

                SALLY GIRL [ARG] (Fain [Arg]) — bred by Santa Maria de Araras

            ESTRELLA FUEGA (Lypheor [GB]) — bred by Centurion Farms Ltd.

                NEPENTHE (Broad Brush) — bred by Joanne Nor

            JEUNESSE DOREE (Kris S.) — bred by Haras Santa Maris de Araras

        Song Test (The Minstrel) — bred by Taylor's Purchase Farm

            DEL LA ROSA [GB] (Sabrehill) — bred by Stretchwood Park Stud

        SYLPH (Alleged) — bred by Taylor's Purchase Farm

            Neesah (Topsider) — bred by Juddmonte Farms

                ORFORD NESS [GB] (Selkirk) — bred by Juddmonte Farms

            ZINDARI (Private Account) — bred by Juddmonte Farms

            PRIVITY (Private Account) — bred by Juddmonte Farms

        LEADING COUNSEL (Alleged) — bred by Swettenham Stud et al.

        Lady Lavery (Northern Dancer) — bred by Swettenham Stud et al.

            FAVOURED NATIONS (Law Society) — bred by Lyonstown Stud

Querida (*Alibhai) — bred by **Brownell Combs & Leslie Combs II**

    Dear April (*My Babu) — bred by R.W. McIlvain

        HURRY UP DEAR (Dark Star) — bred by **Leslie Combs II**

        APRIL DAWN (*Gallant Man) — bred by **Leslie Combs II**

            SHARM A SHEIKH (New Policy) — bred by Allan Lazaroff

                VINNIE THE VIPER (Raise a Man) — bred by Dan J. Agnew

            PROFIT KEY (Desert Wine) — bred by Dan J. Agnew

        HIGHEST TRUMP (Bold Bidder) — bred by Morton W. Smith

        DANCE BID (Northern Dancer) — bred by John B. Crook

        NORTHERN PLAIN (Northern Dancer) — bred by John B. Crook

        Wasnah (Nijinsky II) — bred by **N.B. Hunt**

            BAHRI (Riverman) — bred by Shadwell Farm

            BAHHARE (Woodman) — bred by Shadwell Farm

MYRTLEWOOD (Continued)

           WINGLET (Alydar) — bred by **Allen Paulson**
             †**AJINA** (Strawberry Road [Aus]) — bred by **Allen Paulson**
           ROB'S SPIRIT (Theatrical [Ire]) — bred by **Allen Paulson**
    Royal Dowry (*Royal Charger) — bred by Mrs. R.W. McIlvain
       †**TUDOR QUEEN** (*King of the Tudors) — bred by Mrs. A.G. Daniels
       CHARGER'S STAR (Pia Star) — bred by Mrs. A.G. Daniels
    Quebabu (*My Babu) — bred by **Leslie Combs II**, John Hanes, & Walmac Farm
       BOLD MOVE (Bold Hour) — bred by Steve Arky
    MR. HINGLE (Bald Eagle) — bred by **Combs**, Hanes, & Walmac
    QUEEN JANINE (Tompion) — bred by Walmac Farm
Gallawood (b.f. 1943, *Sir Gallahad III) — bred by **Brownell Combs**
    Feronia (*Heliopolis) — bred by **Leslie Combs II**
    MR. BRICK (Johns Joy) — bred by Charles A. Dubois
    JERONIA (Johns Joy) — bred by Charles A. Dubois
    My Devotion (Crozier) — bred by Mrs. J.O. Burgwin
       TRUE DEVOTION [AUS] (Beau Sovereign) — bred by Coral Park Stud
    Gala Galla (*Royal Charger) — bred by **Leslie Combs II**
    Princess Tudor (*King of the Tudors) — bred by **Leslie Combs II**
    Dame d'Argent (Warfare) — bred by Virginia Condon & Alice Stoll
       SPOKESWOMAN (Elocutionist) — bred by Mr. & Mrs. Theodore Kuster
        TWOUNDER (Beau's Eagle) — bred by Busching, Reals, & Stevenson
       Fleur de Nord (Far North) — bred by Marablue Farm
       ALASKAN FROST (Copelan) — bred by Earl Hartley
       FROSTY TIP (Explosive Bid) — bred by Brylynn Farm
    WARDLAW (Decidedly) — bred by Virginia Condon & Alice Stoll
    PROFOUND CINDY (Apalachee) — bred by Virginia Condon & Alice Stoll
Spring Beauty (b.f. 1944, *Sir Gallahad III) — bred by **Brownell Combs**
    Spring Tune (Spy Song) — bred by **Brownell Combs & Leslie Combs II**
      Village Beauty (*My Babu) — bred by **Brownell Combs & Leslie Combs II**
      SQUABBLE (Never Bend) — bred by **Leslie Combs II**
      SILENT BEAUTY (Crème dela Crème) — bred by **Leslie Combs II**
       CATAHOULA (Never Bend) — bred by **Leslie Combs II**
       SUGAR CHARLOTTE (Wajima) — bred by **Spendthrift Farm**
      CROWN THE QUEEN (Swaps) — bred by **Leslie Combs II**
       HUGGLE DUGGLE (Never Bend) — bred by **Leslie Combs II**
       AL MUNDHIR (Seattle Slew) — bred by **Spendthrift Farm**
       Wife for Life (Dynaformer) — bred by Robert Neff
        HOLLYWOOD STORY (Wild Rush) — bred by Vinery
      VILLAGE SASS (Sassafras [Fr]) — bred by **Spendthrift Farm**
      SASSY CHIMES (Chimes Band) — bred by Fares Farm
      LADY WAYWARD (Dedicate) — bred by **Brownell Combs & Leslie Combs II**
      NEVER ASK (Never Bend) — bred by **Leslie Combs II**
      Village Gossip (Buckpasser) — bred by **Leslie Combs II**
      LIGHT OF NASHUA (Nashua) — bred by **Spendthrift Farm et al.**
    Aparoma (*My Babu) — bred by **Brownell Combs & Leslie Combs II**
      Myrtlewood Beauty (Never Bend) — bred by **Leslie Combs II**
      LOCUST BAYOU (Majestic Prince) — bred by **Spendthrift Farm**
  *YOUNG MAN'S FANCY II (Alycidon) — bred by **Leslie Combs II**
Moonflower (b.f. 1945, *Bull Dog) — bred by **Brownell Combs**
    Moon Relic (War Relic) — bred by **Brownell Combs & Leslie Combs II**
      MOON SHOT (Jet Pilot) — bred by Maine Chance Farm
    Timepiece (Eight Thirty) — bred by **Brownell Combs & Leslie Combs II**
      Frequently (Decidedly) — bred by George A. Pope
      JUMPING HILL (Hillary) — bred by George A. Pope
    MOON GLORY (*Norseman) — bred by **Brownell Combs & Leslie Combs II**
      AQUA VITE (Nashua) — bred by **Leslie Combs II** & John W. Hanes
    Moon Princess (*Princequillo) — bred by **Brownell Combs & Leslie Combs II**
    FINAL RETREAT (Yorktown) — bred by Nydrie Stud
Dragona (b.f. 1949, Bull Lea) — bred by **Brownell Combs**
    ROYAL ATTACK (*Royal Charger) — bred by **Leslie Combs II**

About the Author

Edward L. Bowen is considered one of Thoroughbred racing's most insightful and erudite writers. A native of West Virginia, Bowen grew up in South Florida, where he became enamored of racing while watching televised stakes from Hialeah.

Bowen entered journalism school at the University of Florida in 1960, then transferred to the University of Kentucky in 1963 so he could work as a writer for *The Blood-Horse*, the leading weekly Thoroughbred magazine. From 1968 to 1970, he served as editor of *The Canadian Horse*, then returned to *The Blood-Horse* as managing editor. He rose to the position of editor-in-chief before leaving the publication in 1993.

Bowen is president of the Grayson-Jockey Club Research Foundation, which raises funds for equine research. In addition to *Legacies of the Turf* (Vol. 2), Bowen has written fourteen books, including *Man o' War*, *At the Wire: Horse Racing's Greatest Moments*, and the first volume of *Legacies of the Turf*. Bowen has won the Eclipse Award for magazine writing and other writing awards. He lives in Versailles, Kentucky, with his wife, Ruthie, and son George. Bowen has two grown daughters, Tracy Bowen and Jennifer Schafhauser, and two grandchildren.

Other Titles from Eclipse Press

The Agua Caliente Story
American Classic Pedigrees
At the Wire *Horse Racing's Greatest Moments*
Baffert *Dirt Road to the Derby*
***The Blood-Horse* Authoritative Guide to Auctions**
The Calumet Collection *A History of the Calumet Trophies*
Care & Management of Horses
Country Life Diary *(revised edition)*
Crown Jewels of Thoroughbred Racing
Dynasties *Great Thoroughbred Stallions*
The Equid Ethogram *A Practical Field Guide to Horse Behavior*
Etched in Stone
Feeling Dressage
Four Seasons of Racing
Graveyard of Champions *Saratoga's Fallen Favorites*
Great Horse Racing Mysteries
Happy Trails
Handicapping for Bettor or Worse
Hoofprints in the Sand *Wild Horses of the Atlantic Coast*
Horse Racing's Holy Grail *The Epic Quest for the Kentucky Derby*
Investing in Thoroughbreds *Strategies for Success*
I Rode the Red Horse *Secretariat's Belmont Race*

The Journey of the Western Horse
Kentucky Derby Glasses Price Guide
Legacies of the Turf *A Century of Great Thoroughbred Breeders (Vol. 1)*
Lightning in a Jar *Catching Racing Fever*
Matriarchs *Great Mares of the 20th Century*
New Thoroughbred Owners Handbook
Old Friends *Visits with My Favorite Thoroughbreds*
Olympic Equestrian
Own a Racehorse Without Spending a Fortune *Partnering in the Sport of Kings*
Racing to the Table *A Culinary Tour of Sporting America*
Rascals and Racehorses *A Sporting Man's Life*
Ride of Their Lives *The Triumphs and Turmoils of Today's Top Jockeys*
Ringers & Rascals *The True Story of Racing's Greatest Con Artists*
Royal Blood
The Seabiscuit Story *From the Pages of he Blood-Horse Magazine*
Smart Horse *Understanding the Science of Natural Horsemanship*
Thoroughbred Champions *Top 100 Racehorses of the 20th Century*
Trick Training Your Horse to Success
Women in Racing *In Their Own Words*
Women of the Year

THOROUGHBRED
Legends®
S E R I E S

Affirmed and Alydar • **Citation** • **Damascus** • **Dr. Fager** • **Exterminator** • **Forego** • **Genuine Risk** • **Go for Wand John Henry** • **Kelso** • **Man o' War** • **Nashua** • **Native Dancer** • **Personal Ensign** • **Round Table** • **Ruffian** • **Seattle Slew** • **Secretariat** • **Spectacular Bid** • **Sunday Silence** • **Swaps** • **War Admiral**

To order these titles and to subscribe to *The Blood-Horse* magazine visit
ExclusivelyEquine.com or call **800-866-2361**.